Fungal Biology

Series Editors

Vijai Kumar Gupta
School of Natural Sciences
National University of Ireland Galway
Galway
Ireland

Maria G. Tuohy
School of Natural Sciences
National University of Ireland Galway
Galway
Ireland

About the Series

Fungal biology has an integral role to play in the development of the biotechnology and biomedical sectors. It has become a subject of increasing importance as new fungi and their associated biomolecules are identified. The interaction between fungi and their environment is central to many natural processes that occur in the biosphere. The hosts and habitats of these eukaryotic microorganisms are very diverse; fungi are present in every ecosystem on Earth. The fungal kingdom is equally diverse, consisting of seven different known phyla. Yet detailed knowledge is limited to relatively few species. The relationship between fungi and humans has been characterized by the juxtaposed viewpoints of fungi as infectious agents of much dread and their exploitation as highly versatile systems for a range of economically important biotechnological applications. Understanding the biology of different fungi in diverse ecosystems as well as their interactions with living and non-living is essential to underpin effective and innovative technological developments. This series will provide a detailed compendium of methods and information used to investigate different aspects of mycology, including fungal biology and biochemistry, genetics, phylogenetics, genomics, proteomics, molecular enzymology, and biotechnological applications in a manner that reflects the many recent developments of relevance to researchers and scientists investigating the Kingdom Fungi. Rapid screening techniques based on screening specific regions in the DNA of fungi have been used in species comparison and identification, and are now being extended across fungal phyla. The majorities of fungi are multicellular eukaryotic systems and therefore may be excellent model systems by which to answer fundamental biological questions. A greater understanding of the cell biology of these versatile eukaryotes will underpin efforts to engineer certain fungal species to provide novel cell factories for production of proteins for pharmaceutical applications. Renewed interest in all aspects of the biology and biotechnology of fungi may also enable the development of "one pot" microbial cell factories to meet consumer energy needs in the 21st century. To realize this potential and to truly understand the diversity and biology of these eukaryotes, continued development of scientific tools and techniques is essential. As a professional reference, this series will be very helpful to all people who work with fungi and should be useful both to academic institutions and research teams, as well as to teachers, and graduate and postgraduate students with its information on the continuous developments in fungal biology with the publication of each volume.

More information about this series at http://www.springer.com/series/11224

Bhim Pratap Singh • Vijai Kumar Gupta
Editors

Molecular Markers in Mycology

Diagnostics and Marker Developments

Springer

Editors
Bhim Pratap Singh
Department of Biotechnology
Mizoram University
Aizawl, Mizoram
India

Vijai Kumar Gupta
Department of Biochemistry
National University of Ireland Galway
Galway
Ireland

ISSN 2198-7777 ISSN 2198-7785 (electronic)
Fungal Biology
ISBN 978-3-319-34104-0 ISBN 978-3-319-34106-4 (eBook)
DOI 10.1007/978-3-319-34106-4

Library of Congress Control Number: 2016957633

Printed on acid-free paper

This Springer imprint is published by Springer Nature
The registered company is Springer International Publishing AG
The registered company address is: Gewerbestrasse 11, 6330 Cham, Switzerland

Preface

The fungal kingdom encompasses a massive diversity of taxa with wide-ranging ecologies, life cycles, and morphologies ranging from unicellular aquatic chytrids to large mushrooms. Before molecular methods came into existence, taxonomists considered this kingdom to be a member of the plant kingdom due to certain lifestyles like immobility and growth habitats. Molecular markers provide a better alternative to traditional morphological methods in identifying and characterizing fungi as well as understanding their evolution. Also, the accessibility of DNA-based methods has prompted to unwind the genetic capability of non-culturing fungi. The morphological methods used for identification mainly depend on spore color or microscopic features, whereas molecular markers are based on DNA polymorphism in the genomic organization. Phylogenetic studies reported in recent decades based on molecular markers have reshaped the classification system of Kingdom Fungi, which divided it into one subkingdom, seven phyla, and ten subphyla. Recent advances in molecular mycology have allowed for the identification and characterization of novel fungal species from unique environments.

Mycology is concerned with the systematic study of fungi, including their genetic and biochemical properties, their use to humans as a source of medicine and food, and their dangers such as poisoning and infections. In the 21st century, new insights into fungal taxonomy have been achieved with the development of DNA sequencing technologies and phylogenetic analysis based on molecular markers. Species-specific PCR strategies are developed for fast and precise detection and identification of respective fungal candidates.

This book provides a comprehensive overview of the applications and uses of different molecular markers in molecular mycology. It addresses the recent molecular markers employed to solve the present problems of identification and discusses the current approaches used in molecular characterization. In addition, this book depicts how cutting-edge sequencing procedures are at present exploited in biodiversity investigations of fungal groups in various zones around the world.

Aizawl, Mizoram, India Bhim Pratap Singh
Galway, Ireland Vijai Kumar Gupta

Contents

Contributors

Mashook Ali Dr.B. R. Ambedkar Centre for Biomedical Research, University of Delhi, Delhi, India

R. Anandham Department of Agricultural Microbiology, Agricultural College and Research Institute, Madurai, Tamil Nadu, India

K. Awasthi Department of Biotechnology, Delhi Technological University, Delhi, India

Preetida J. Bhetariya Department of Infectious Disease, University of Utah School of Medicine, Salt Lake City, UT, USA

Hillol Chakdar National Bureau of Agriculturally Important Microorganisms, Maunath Bhanjan, Uttar Pradesh, India

Rajeshkumar Kunhiraman Chandanadukkam National Fungal Culture Collection of India (NFCCI), Biodiversity and Palaeobiology Group, Agharkar Research Institute, Pune, Maharashtra, India

Anirudha Chattopadhyay Department of Plant Pathology, C.P. College of Agriculture, SD Agricultural University, S.K. Nagar, Gujarat, India

Kul Bhushan Division of Plant Pathology, Indian Agricultural Research Institute, New Delhi, India

Awadesh Dagar UP Council of Sugarcane Research, Shahjahanpur, Uttar Pradesh, India

Asmita Das Department of Biotechnology, Delhi Technological University, Delhi, India

Jawaharlal Nehru University, New Delhi, India

N.O. Gopal Department of Agricultural Microbiology, Agricultural College and Research Institute, Madurai, Tamil Nadu, India

Kamala Gupta Department of Botany, Government General Degree College, Singur, Hooghly, India

Department of Biological Sciences, Presidency University, Kolkata, India

Bhaskar Gupta Department of Biological Sciences, Presidency University, Kolkata, India

Department of Zoology, Government General Degree College, Singur, Hooghly, India

Zhenhu Hu School of Civil Engineering, Hefei University of Technology, Hefei, People's Republic of China

Aminul Islam Cymbio Pharma Pvt. Ltd., Yelahanka, Venkatala, Bangalore, India

M. Karthik Department of Agricultural Microbiology, Tamil Nadu Agricultural University, Coimbatore, Tamil Nadu, India

Department of Agricultural Chemistry, Chungbuk National University, Cheongju, Republic of Korea

Prem Lal Kashyap ICAR-National Bureau of Agriculturally Important Microorganisms, Mau, Uttar Pradesh, India

ICAR-Indian Institute of Wheat and Barley Research (IIWBR), Shimla, India

Selima Khatun Department of Botany, Government General Degree College, Singur, Hooghly, India

Sachin Khedekar Department of Plant Pathology, University of Agricultural Sciences, Dharwad, India

R. Krishnamoorthy Department of Agricultural Microbiology, Agricultural College and Research Institute, Madurai, Tamil Nadu, India

Sudheer Kumar ICAR-Indian Institute of Wheat and Barley Research, Karnal, Haryana, India

Pankaj Kumar Department of Molecular Biology and Genetic Engineering, Bihar Agricultural University, Sabour, Bhagalpur, Bihar, India

Ravi Ranjan Kumar Department of Molecular Biology and Genetic Engineering, Bihar Agricultural University, Sabour, Bhagalpur, Bihar, India

Mahesh Kumar Department of Molecular Biology and Genetic Engineering, Bihar Agricultural University, Sabour, Bhagalpur, Bihar, India

Ranjeet Ranjan Kumar Division of Biochemistry, Indian Agricultural Research Institute, Pusa, New Delhi, India

Dipendra Kumar Indian Agricultural Research Institute, Pusa, New Delhi, India

Pradeep Kumar Department of Biotechnology, Yeungnam University, Gyeongsan-si, Gyeongsangbuk-do, South Korea

Department of Biotechnology, Ben Gurion University of The Negev, Beer Sheva, Israel

Kavita Kumari UP Council of Sugarcane Research, Shahjahanpur, Uttar Pradesh, India

Vincent Vineeth Leo Molecular Microbiology and Systematic Laboratory, Department of Biotechnology, Mizoram University, Aizawl, Mizoram, India

Taruna Madan Innate Immunity Lab, National Institute for Research in Reproductive Health, Mumbai-12, India

Vineet Kumar Mishra Molecular Microbiology and Systematic Laboratory, Department of Biotechnology, Mizoram University, Aizawl, Mizoram, India

Iniya Kumar Muniraj Department of Agricultural Microbiology, Tamil Nadu Agricultural University, Coimbatore, India

Civil Engineering, College of Engineering and Informatics, National University of Ireland, Galway, Ireland

Naveen Pandey Arunanchal University of Studies, Namsai, Arunachal Pradesh, India

Ajit Kumar Passari Molecular Microbiology and Systematic Laboratory, Department of Biotechnology, Mizoram University, Aizawl, Mizoram, India

Peralam Yegneswaran Prakash Medical Mycology Division, Department of Microbiology, Kasturba Medical College – Manipal, Manipal University, Manipal, Karnataka, India

T. Prasad Jawaharlal Nehru University, New Delhi, India

Dharmendra Pratap Department of Genetics & Plant Breeding, Ch. Charan Singh University, Meerut, UP, India

N. Premalatha Department of Agricultural Microbiology, Agricultural College and Research Institute, Madurai, Tamil Nadu, India

Pallavi Rai ICAR-National Bureau of Agriculturally Important Microorganisms, Mau, Uttar Pradesh, India

Tongmin Sa Department of Agricultural Chemistry, Chungbuk National University, Cheongju, Republic of Korea

Ratul Saikia Biotechnology Group, Biological Science & Technology Division, CSIR-North East Institute of Science and Technology, Jorhat, Assam, India

Richa Salwal Department of Veterinary Microbiology, CSK-HPKV, Palampur, Himachal Pradesh, India

P. Usha Sarma Division of Plant Pathology, Indian Agricultural Research Institute, New Delhi, India

R.K. Sarma Biotechnology Group, Biological Science & Technology Division, CSIR-North East Institute of Science and Technology, Jorhat, Assam, India

Life Science Division, Institute of Advanced Study in Science and Technology (DST, Govt. India), Guwahati, Assam, India

Pratik Satya ICAR-Central Research Institute for Jute and Allied Fibres, Barrackpore, Kolkata, India

G. Selvakumar Department of Agricultural Chemistry, Chungbuk National University, Cheongju, Republic of Korea

M. Senthilkumar Department of Agricultural Microbiology, Tamil Nadu Agricultural University, Coimbatore, Tamil Nadu, India

Department of Agricultural Chemistry, Chungbuk National University, Cheongju, Republic of Korea

Md. Shamim Department of Plant Molecular Biology and Genetic Engineering, N.D. University of Agriculture and Technology, Kumarganj, Faizabad, U.P, India

Department of Molecular Biology and Genetic Engineering, Bihar Agricultural University, Sabour, Bhagalpur, Bihar, India

Vivek Sharma DST-Young Scientist, Department of Plant Pathology, CSK-HPKV, Palampur, Himachal Pradesh, India

K.N. Singh Department of Molecular Biology and Genetic Engineering, Bihar Agricultural University, Sabour, Bhagalpur, Bihar, India

Atul Singh UP Council of Sugarcane Research, Shahjahanpur, Uttar Pradesh, India

Surendra Pratap Singh UP Council of Sugarcane Research, Shahjahanpur, Uttar Pradesh, India

Bhim Pratap Singh Molecular Microbiology and Systematic Laboratory, Department of Biotechnology, Mizoram University, Aizawl, Mizoram, India

Atul Singha ICAR-Central Institute for Subtropical Horticulture, Lucknow, Uttar Pradesh, India

Alok Kumar Srivastava ICAR-National Bureau of Agriculturally Important Microorganisms, Mau, Uttar Pradesh, India

N.C. Talukdar Life Science Division, Institute of Advanced Study in Science and Technology (DST, Govt. India), Guwahati, Assam, India

Kapil Kumar Tiwari Department of Plant Molecular Biology & Biotechnology, C.P. College of Agriculture, S.D. Agricultural University, S.K. Nagar, Gujarat, India

Ajay Kumar Tiwari UP Council of Sugarcane Research, Shahjahanpur, Uttar Pradesh, India

Sivakumar Uthandi Department of Agricultural Microbiology, Tamil Nadu Agricultural University, Coimbatore, India

Liwen Xiao Department of Civil, Structural and Environmental Engineering, Trinity College Dublin, Dublin, Ireland

Mukesh Kumar Yadav Korea University College of Medicine, Korea University, Seoul, South Korea

Xinmin Zhan Civil Engineering, College of Engineering and Informatics, National University of Ireland Galway, Galway, Ireland

Chapter 1
Molecular Diversity and Detection of Endophytic Fungi Based on Their Antimicrobial Biosynthetic Genes

Vineet Kumar Mishra, Ajit Kumar Passari, Vincent Vineeth Leo, and Bhim Pratap Singh

1.1 Introduction

Endophytes comprise a diverse assemblage of ubiquitous microorganisms residing in the tissues of plants for at least a part of their lifecycle without causing any overt symptoms (Petrini 1991; Bacon and White 2000). Fossilized tissues of stems and leaves have suggested evolution of endophytes as early as higher plants from the time they first appeared on Earth (Redecker et al. 2000). The term endophyte was first coined in the nineteenth century by de Bary (1866) who described endophytes as any organism living in the plant tissues other than epiphytes found at the surface of plant organs (Petrini 1991). Carroll (1986) corrected this definition by describing endophytes as mutualists colonizing aerial part of the plant and excluded mycorrhizal fungi and pathogenic fungi. However, definition of Petrini (1991) have been most commonly used in the studies of endophytes who included all the organisms dwelling in plant organs as endophytes, spending at least some part in their life without apparently harming their host. Bills (1996) reported that endophytes and certain types of mycorrhizae, e.g. ectendomycorrhizae, ericoid mycorrhizae are indistinct. Henceforth, some root-inhabiting or mycorrhizal fungi associated with plant families Ericaceae and Orchidaceae have been referred to as endophytes (Bayman et al. 1997; Sun and Guo 2012). Endophytes were first described in *Lolium temulentum* (Freeman 1904) but, the milestone in the history of endophyte research was the discovery of the endophytic fungus *Neotyphodium coenophialum*, the causative agent of fescue toxicosis from grass *Festuca arundinacea* (Bacon et al. 1977). However, *N. coenophialum* was considered to be associated with increased tolerance of host to biotic and abiotic stresses (Schardl et al. 2004). The most commonly encounterd endophytes are fungi that form a multifarious group of microbes which

V.K. Mishra • A.K. Passari • V.V. Leo • B.P. Singh (✉)
Molecular Microbiology and Systematic Laboratory, Department of Biotechnology, Mizoram University, Aizawl, Mizoram 796004, India
e-mail: bhimpratap@gmail.com

B.P. Singh, V.K. Gupta (eds.), *Molecular Markers in Mycology*, Fungal Biology,
DOI 10.1007/978-3-319-34106-4_1

reside asymptomatically in the tissues of plants and has been found in every plant examined so far (Bills 1996; Arnold 2007; Kusari et al. 2012). Endophytic fungi have been widely studied from a broad range of host plant species spanning several geographical locations and different climatic zones (Petrini 1991; Stone et al. 2000; Arnold 2007; Guo 2010; Zhou et al. 2015). Endophytic fungi are ubiquitous and include a wide range of host orders, families, genera and species as diverse as mosses (Davey and Currah 2006), ferns (Swatzell et al. 1996), grasses (Müller and Krauss 2005), shrubs (Petrini et al. 1982), deciduous and coniferous trees (Guo et al. 2008; Sun et al. 2011), lichens (Suryanarayanan et al. 2005; Li et al. 2007) in diverse ecosystems including marine ecosystem (Flewelling et al. 2015). Endophytic fungi mostly belong to Ascomycota or their mitosporic fungi but members of Basidiomycota, Zygomycota, and Oomycota have also been reported (Guo 2010). Moreover, few entomopathogenic fungi *Beauveria bassiana*, *Metarhizium anisopliae* and *Lecanicillium lecanii* have been reported to colonize plant tissues as symptomless endophytes (Vidal and Jaber 2015) Because the plant tissues are multilayered and spatially arranged, they support a rich and diverse endophytic mycobiota that form specialized association with a variety of plant species and tissues (Guo 2010). Accordingly, existence of 1.5 million fungal species on earth has been estimated mainly based on the ratios of vascular plants to fungal species at 1:6, but endophytic mycobiota have not been seriously considered in the estimation (Hawksworth 1991). However, Petrini (1991) suggested that there should be more than 1 million species of endophytic fungi to be discovered and described based on the ratios of vascular plants to fungal species at 1:4. Endophytic fungi are producers of various bioactive secondary metabolites (Liu et al. 2008; Aly et al. 2010; Xu et al. 2010; Mousa and raizada 2013) including extracellular enzymes (Barrett et al. 2003; Correa et al. 2014) and volatile hydrocarbons (Strobel et al. 2008; Stadler and Schulz 2009). They also promote host growth and resistance to environmental stress (Ting et al. 2008; Saikkonen et al. 2010) and decompose litter (Purahong and Hyde 2011; Sun et al. 2011) which established them as important component of natural ecosystems.

1.2 Major Groups of Endophytic Fungi

Stone et al. (2000) classified endophytes into distict classes based on their isolation source into : (1) Endophytic Clavicipitaceae, (2) Fungal endophytes of dicots, (3) Endophytic fungi, (4) Other systemic fungal endophytes, (5) Fungal endophytes of lichens, (6) Endophytic fungi of bryophytes and ferns, (7) Endophytic fungi of tree bark, (8) Fungal endophytes of xylem, (9) Fungal endophytes of roots, (10) Fungal endophytes of galls and cysts, (11) Prokaryotic endophytes of plants (includes endophytic bacteria and actinomycetes) (Stone et al. 2000; Bills et al. 2004).

However, fungal endophytes were first ever categorized by Rodriguez et al. (2009) who broadly categorized them into two major groups clavicipitaceous endophytes

(C-endophytes) and nonclavicipitaceous endophytes (NC-endophytes) by differentiating them based on phylogeny, taxonomy, host plants and ecological functions. C-endophytes infect some grasses whereas the NC-endophytes can be recovered from asymptomatic tissues of nonvascular plants, ferns, conifers, and angiosperms. NC-endophytes were further divided into three classes founded on host range, type of tissue colonized, In planta colonization, In planta diversity, mode of transmission to the offspring and fitness benefits (Rodriguez et al. 2009).

C-endophytes (Class1 endophytes) signify a small number of evolutionary related clavicipitaceous species that are limited to some cool- and warm-season grasses (Bischoff and White 2005). The endophytic fungi typically occur within plant shoots and form intercellular infections (Rodriguez et al. 2009). Class 1 endophytes are vertically transmitted to offsprings via seed infections. Host plants generally harbor one dominant fungal isolate. Class 1 endophytes confer resistance to the host against drought, reduces herbivory by producing toxic compounds and increases host plant biomass (Rodriguez et al. 2009). *Epichloe festuce*, a C-endophytes is reported to produce bioactive compounds that inhibited the growth of other fungi in vitro (Yue et al. 2000).

NC-endophytes are extremely diverse group of endophytic fungi, representing primarily ascomycetous fungi with varied and often poorly defined ecological roles. NC endophytes have been recoverd from broad range of host plants from all terrestrial ecosystems varying from the tropics to the tundra. NC-endophytes are discriminated into three classes based on pattern of host colonization, means of transmission to offsprings, *in planta* biodiversity and ecological function (Rodriguez et al. 2009).

Class 2 endophytes encompass varity of species which mostly belong to Ascomycota and some members belong to basidiomycota. Members of class 2 endophytes are different from other classes of NC- endophytes because they colonize in both above- and below-ground tissues of plant, they transmit to the offspring through seed coat or rhizomes, have low abundance in rhizosphere and high infection frequencies. Several endophytic fungi of Class 2 confer resistance to host against fungal pathogens (Danielsen and Jensen 1999; Campanile et al. 2007) which reflect the production of secondary metabolites (Schulz et al. 1999). Endophytic *Fusarium oxysporum* and *Cryptosporiopsis* sp. were reported to protect barley (*Hordeum vulgare*) and larch (*Larix decidua*), respectively against virulent pathogens (Schulz et al. 1999).

Class 3 endophytes colonize only above ground tissues, form highly localized infections and transmit horizontally. Class 3 endophytic fungi are reported to be associated with tropical trees (Arnold et al. 2000; Gamboa and Bayman 2001) vascular and non vascular plants, conifers and woody angiosperms in varied geographical locations and climates (Stone 1988; Davis and Shaw 2008).

Class 4 endophyte comprises mycorrhizal fungi and 'dark septate endophytes' (DSE). DSE are a group of endophytes which are not described well taxonomically or ecologically. They are characterized as a functional group based on the presence of dark melanized septa, and their limitation to plant roots. DSE transmitted vertically and has little host specificity, found in high-stress environments and they seem to be ubiquitous in their occurrence and distribution across various ecosystems (Rodriguez et al. 2009).

1.3 Morphological Identification of Endophytic Fungi: Advantages and Disadvantages

Traditionally, classification systems of organisms are based on phenotypic approach i.e. by taking account of observable characteristics. Earlier, the identification of fungi and its taxonomy relied only on morphological characteristics under microscope. Traditional approaches of fungal endophyte identification involve two essential techniques i.e. direct microscopic observation and culture dependant techniques.

In the direct observation method, fungal endophytes colonizing within tissues are directly examined under a light or electron microscope. The advantage of this method is that it can detect fungi which can not be cultured on any growth media (Deckert et al. 2001). However, endophytic fungi residing inside the host tissues do not sporulate and reveal only hyphae and some non specific structure. Although, structure of hyphae and presence or absence of septa can provide some information about their identity, it is not sufficient to categorized fungal endophytes taxonomically (Deckert et al. 2001). Hence, in contrast to direct observation methods culturing endophytes in the appropriate growth media is still a well accepted technique which has been routinely employed in endophyte studies (Sun et al. 2011; Vieira et al. 2011; Sun and Guo 2012).

In cultivation dependent techniques, plant tissues are subjected to surface sterilization to remove all the epiphytic organisms from plant surface so that only endophytes can be isolated by incubating on specific growth media. Cultivation dependent techniques commonly involves (1) meticulous washing of the plant tissues with tap water to remove adhering soil particles and debris, (2) surface sterilization of plant tissues to kill any epiphytic microorganisms on the host surface, (3) isolation of endophytic fungi coming out from the plated tissues on specific growth media (4) purification of fungal isolates by repeated subculturing and sporulating the isolates under various incubation conditions, and (5) identification of the endophytic fungi based on morphological characteristics in cultures (Fig. 1.1) (Guo et al. 2008; Sun et al. 2011).

Cultivation dependent technique is very effective in rapid recovery of a large number of endophytic fungal isolates. The confirmation of isolated strains as endophytes can be established by imprinting the surface sterilized plant tissues on nutrient media or spreading an aliquot of water from the last wash of the sterilization protocol.

However, isolation of fungal endophytes is a method dependent process and it depends on type of surface sterilization technique used, incubation conditions,

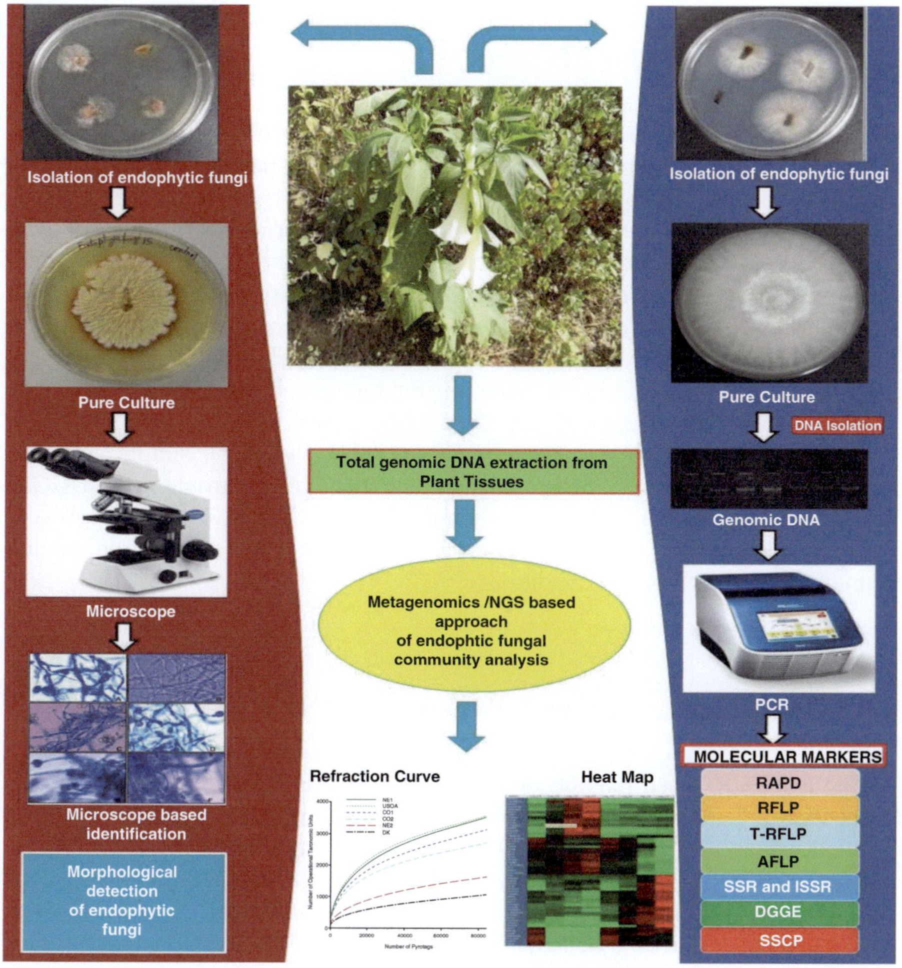

Fig. 1.1 Different approaches of detection and genetic diversity analysis of endophytic fungi

and whether the isolates sporulate. The major limitation of culture dependent technique are: (1) It is rather arduous and rigorous technique which is not suitable to compare large numbers of samples (Guo 2010), (2) The abundance of sterile endophyte isolates recovered posses a major problem, because they can not be identified to any taxonomic category since, they do not sporulate. However, a range of methods have been adopted to promote sporulation of isolates so as to overcome their shortcomings in culture (Taylor et al. 1999; Guo et al. 2008), (3) Some slow growth fungal isolates can be missed because they can be easily outcompeted by fast-growing species in artificial conditions. In order to overcome the potential technical bias, cultivation independent approaches, e.g., molecular techniques, to analyze endophytic fungal communities of plants are needed (Guo 2010).

1.4 Molecular Characterization of Endophytic Fungi Using Molecular Markers

Many fungal endophytes do not sporulate and therefore difficult to detect or identify by traditional morphholology based approaches. DNA based molecular markers have successfully employed in estimating genetic diversity of all fungal endophytes. This approach involves few common steps: (1) surface sterilization of plant tissue samples and incubation on specific media, (2) transfer of mycelia coming out from the surface sterilized tissues, (3) isolation of pure culture by frequent subculturing, (4) extraction of genomic DNA from pure culture and (5) characterization of endophytic fungi by molecular markers (Fig. 1.1). There are several DNA fingerprinting techniques, such as, random amplified polymorphic DNA (RAPD), restriction fragment length polymorphism (RFLP), terminal-RFLP (T-RFLP), amplified fragment length polymorphism (AFLP), simple sequence repeat (SSR), inter-SSR (ISSR), denaturing gradient gel electrophoresis (DGGE) and single-stranded conformation polymorphism (SSCP), which have been successfully applied to assess the genetic diversity of endophytic fungal population from diverse environmental conditions.

1.4.1 Random Amplified Polymorphic DNA (RAPD)

RAPD is a PCR based method in which a single primer of arbitrary nucleotide sequences targets unspecified genomic sequences in order to generate a genetic profile as an array of amplified DNA fragments, specific for the individual organism (Mbwana et al. 2006). DNA polymorphism among different individuals is thus detected as presence or, absence of amplicons among different individuals which can be used to construct genetic maps (Williams et al. 1990). RAPD markers have been used extensively in detection of different pathogens as well as in genetic diversity studies (Tejesvi et al. 2007). The advantage of using RAPD primers are that it is a rapid and cheap screening method, requires only little amount of genomic DNA and does not require specific restriction endonucleases, hybridization techniques or, radioactive assays. However, it has its own weak points since it has been criticized for several unwanted features such as non codominant inheritance, anonymous nature (Backeljau et al. 1995), lack of positional homology (Van de Zande and Bijlsma 1995) and especially inconsistent reproducibility (Staub et al. 1996; Singh 2015).

RAPD based genetic markers have been widely used for the analysis of genetic diversity of fungal endophytes in a wide range of host plants. Favaro et al. (2012) used RAPD marker to differentiate sugar endophytic fungi *Epicoccum nigrum* with morphologically similar colonies. Similar study was performed by Tiwari and Chittora (2013) in assessment of genetic diversity and distribution of endophytic fungal communities of 12 ecologically diverse group of *Alternaria solani* isolates associated with Karanja plants (*Pongamia pinnata* L.). 60.34% RAPD loci generated were polymorphic which proves efficacy of RAPD markers in polymorphism detection and identification of particular isolates.

1.4.2 Restriction Fragment Length Polymorphism (RFLP)

ITS RFLP analysis is a popular genetic analysis based on the use of restriction endonucleases, which recognizes specific restriction sites and cut DNA on these sites or makes cut elsewhere thereby generating polymorphism (Nicolotti et al. 2010). ITS rRNA gene is the most conserve gene frequently used gene in identification and barcoding of fungi. ITS-RFLP involves amplification and subsequent digestion of ITS gene which generates polymorphic bands using which inter and intra species variation can be determined (Singh 2015). Since, internal transcribed spacers include both conserve and variable domains, they are the most popular target site for fungal diagnostics using RFLP markers or sequencing approaches (Nicolotti et al. 2010). An RFLP analysis is inexpensive and easy to design but it has certain limitations since, SNPs may affect the same recognition sites. Moreover, the electrophoretic separation of RFLP fragments is a time consuming process. ITS-RFLP is a widely used and reproducible method used in identification of fungi. However, when applied to DNA extracted from wood it gives unreliable results like RAPD markers. This limitation can be avoided by including a cloning step prior to digestion (Nicolotti et al. 2010). Endophytic *Pestalotiopsis* and *Bartalinia robillardoides* strains isolated from the medicinal plants *Azadirachta indica*, *Holarrhena antidysenterica*, *Terminalia arjuna* and *Terminalia chebula* were analyzed using ITS –RFLP. The ITS-RFLP profiles revealed a high intraspecific polymorphism in *Pestalotiopsis* strains (Tejesvi et al. 2009) The population structure of endophytic fungi *Phialocephala fortinii* were studied with RAPD and single locus RFLP markers exhibited recombination, gene and genotype flow in *P. fortinii* (Guo 2010). In another study, microbial communities inhabiting *H. japonica* in different seasons were investigated using cultivation independent approaches, such as RFLP analysis (Gao et al. 2005).

1.4.3 Terminal RFLP (T-RFLP)

Terminal RFLP (t-RFLP) is an automated variant of RFLP which uses flurescent labeled PCR primers and high resolution capillary analyzer to detect digested fragments (Liu et al. 1997; Nicolotti et al. 2010). It is a high throughput technique for rapid profiling of diverse sequences of a single gene often used in studying microbial communities because it provides not only the taxonomic composition but the relative abundance of individual species in the community more rapidly at a lower cost (Camarinha-Silva et al. 2012; Shubin et al. 2014). Unlike RFLP, a primer used for PCR of marker gene is labeled with a fluorescent dye at 5′ termini so that only terminal digested fragments can be detected and quantified (Liesack and Dunfield 2004). This is a promising method for rapid investigation of many fungi. Uncultured basidiomycetes were reported to be detected directly from wood samples using T-RFLP (Raberg et al. 2005). Diversity of endophytic fungal communities associated with rhizomes of *Alpinia officinarum* were evaluated using T-RFLP technique which detect distinct 46 fragment peaks from fungal ITS rDNA sequences using HhaI or MspI digestion. In another study, microbial communities The T-RFLP profiles revealed influence of geographic location and growth year of *A. officinarum* on fungal endophytic communities in rhizomes (Shubin et al. 2014). In another study T-RFLP analysis revealed distinct microbial community structure from cryoconite holes on glaciers in Svalbard in comparison to soil from adjacent tundra sites (Edwards et al. 2013). Ericoid mycorrhizal fungal diversity from hair root samples of *Calluna vulgaris and Vaccinium macrocarpon* was investigated using T-RFLP of the ITS regions of rDNA in combination with sequence analysis (Hazard et al. 2014). Similar study was done on identification of mycorrhizal fungi using combination of T-RFLP with sequencing from *Pinus* (Dickie et al. 2010). Size and the relative intensity of terminal restriction fragments in a sample are highly reproducible, making this technique ideal for microbial community analysis (Liesack and Dunfield 2004).

1.4.4 Denaturing Gradient Gel Electrophoresis (DGGE)

Denaturing gradient gel electrophoresis (DGGE) is a molecular fingerprinting technique that separates closely related sequences by their differential mobilities in a gradient of denaturants (e.g. urea and formamide). Additionally, during electrophoresis the gel run at a high temperature (usually 60 ° C) to denature PCR products. Once denatured, the single stranded DNA fragments may keep migrating through the gel, but, the fragments should have to remain precisely at their position where they get denatured. To fix this problem a GC clamp (essentially, a string of 40–60 nucleotides composed of only guanine and cytosine) is used which gets attached to one of the PCR primers to prevent complete denaturation. PCR amplification using a GC clamp resulted in a product having high denaturing temperature at one end which denatures partially while running through DGGE gel. The fragment looked like Y shaped DNA, firmly placed at the position of denaturation (Roelfsema and Peters 2008). DGGE can also be used in combination with DNA oligonucleotide probes to increase the specificity of the analysis (Stephen et al. 1998).

DGGE is a robust method for the detection of point mutation and estimating microbial diversity (Jeewon and Hyde 2006). It has been effectively used to analyze diversity of fungal endophytic communities including obtaining taxonomical information by excising and sequencing bands and phylogenetic analysis (Götz et al. 2006; Sun and Guo 2012). Duong et al. (2006) used PCR-DGGE analysis coupled with sequencing of 18S rRNA gene to estimate fungal communities on leaves of *Magnolia liliifera* and recovered 14 operational taxonomical units (OTU) from different parts of studied leaf samples. Similar analysis of DGGE of endophytic fungi recovered from leaf and root tissues of *Bletilla ochracea* revealed eighteen OTUs of fungal endophytes from leaves and ten taxa from roots. A PCR- DGGE analysis of culturable fungal endophytes isolated from *Glycine max* reveals distribution of isolates more in the leaves as compared to roots (Fernandes et al. 2015). Yang et al. (2015) isolated endophytic fungi Phomopsis liquidambari from inner bark of *Bischofia polycarpa* and investigated nitrogen dynamics and abundance as well as composition of nitrogen cycling genes in rhizospheres treated with three levels of N (urea). PCR- DGGE was used to analyze the community of ammonia oxidizers and diazotrphs (Yang et al. 2015).

DGGE is widely used in estimating fungal diversity but the choice of nuclear ribosomal regions have always been controversial. Both the internal transcribed spacer (ITS) region and 18S ribosomal RNA genes have been coupled with DGGE fingerprinting to uncover fungal diversity (Anderson et al. 2003; Liu et al. 2015). Liu et al. (2015) compared ITS and 18S rRNA using DGGE under optimum experimental conditions, taking samples from both soil and aquatic environments. The results suggested ITS region to be more precise than 18S rRNA genes in fungal community analysis. Although DGGE has any advantages, it also has few shortcomings. The shorter fragments (<500 bp) of DNA have better resolution, consequently limiting the taxonomic information needed to identify fungi at the generic or species levels (Duong et al. 2006). Furthermore, even the most sensitive staining methods fail to estimate all the diversity present within a sample, mostly for rare fungi in the community (Anderson and Cairney 2004).

1.4.5 Simple Sequence Repeats (SSR) and Inter Simple Sequence Repeats (ISSR)

Simple sequence repeats (SSRs) also known as microsatellites are a powerful tool in taxonomic and population genetics studies. SSR or ISSR techniques originally used to assessed genetic diversity of plants or animals are now widely used in measuring fungal diversity (Liang et al. 2005; Guo 2010). Van der Nest et al. (2000) and Geistlinger et al. (2015) developed SSR markers for *F. oxysporum* and *Trichoderma* respectively to assses their genetic diversity. Barve et al. (2001) estimated the genetic variations in *F. oxysporum* f.sp. *ciceri* populations using oligonucleotide probes and restriction enzymes. Intraspecific genetic diversity among fungal endophytes of temperate pasture grasses were determined by developing SSR markers. Earlier, nuclear microsatellite markers for *Epichloë* sp. have been developed by Moon et al. (1999). De Jong et al. (2003) studied EST-derived simple sequence repeat (SSR) markers for pasture grass endophytes which revealed high level of polymorphism between *Neotyphodium* and *Epichloe* sp. and low level of polymorphism within *Neotyphodium coenophialum* and *N. lolii*. To compare population dynamics between *E. festucae* and its host plant *F. rubra* microsatellite markers have been developed (Cräutlein et al. 2014). ISSR markers are effective tools for studying population genetic characteristics and microbial diversity. Endophytic fungi *Phialocephala fortinii* and a nonsporulating mycelium having the same allozyme phenotypes were differentiated on the basis of ISSR analysis (Grünig et al. 2001; Guo 2010). ISSR markers are effective tools for studying population genetic characteristics and microbial diversity. SSR markers have also been used to estimate relationship among endophyte, its host and their geographical origin. Genetic diversity and population structure of *Alternaria alternata* recovered from *Pinus tabulaeformis* studied by using SSR marker revealed no relationship between genotypes of *A. alternata* and host tissue ages (Guo et al. 2004). Similarly, SSR markers were developed to study genetic diversity and gene flow between populations of endophytic fungi *Lasiodiplodia theobromae* exhibited no evidence of host specificity. Also, there was very high gene flow between populations of *L. theobromae* from different hosts (Mohali et al. 2005). In another study, 18 endophytic fungi *Guignardia mangiferae* strains from different host species were characterized using ISSR-PCR. The results showed that isolates did not correspond to the host species or the geographical origin (Rodrigues et al. 2004).

The SSR or ISSR techniques are fast, low in cost and easy to perform. It is similar to RAPD except the amplification conditions (especially anneling temperature) are more stringent and longer primers are required in this analysis. Since, microsatellites evolve and mutate more rapidly than other areas of genome, SSR markers are more reproducible and robust than RAPD markers (Liang et al. 2005; Guo 2010). SSR markers are more informative in revealing variation among closely related species (Aradhya et al. 2001). Hence, are considered as an ideal tool for studying genetic variation among endophytes.

1.4.6 Amplified Fragment Length Polymorphism (AFLP)

AFLP markers are a combination of RFLP and RAPD techniques in which the genomic DNA is digested by specific restriction endonuclease followed by selective amplification by PCR. AFLP fingerprints are produced without prior knowledge of sequence and are more reliable and robust than RAPD and RFLP techniques because the PCR amplification conditions are more stringent. Moreover the number of fragments detected can be controlled by selection of specific set of primers (Vos et al. 1995). AFLP is highly effective in detecting polymorphism even in very closely related species, where very less variation can be obtained by using RFLP techniques (Majer et al. 1996). It allows specific co amplification of many restriction fragments (60–100 bp) which can be visualized on polyacrylamide gel (Chen et al. 2012; Singh 2015). This marker has been widely used in estimating genetic diversity and identification of fungal phytopathogens (Majer et al. 1996; Aduramigba-Modupe et al. 2012). AFLP markers were used to estimate and compare genetic variation within and between eighteen isolates of *Alternaria brassicicola*, five isolates of *A. alternata*, and a single isolate of *Rhynchosporium secalis*. The AFLP analysis clearly distinguished all three isolates and revealed moderate level of genetic diversity within *A. brassicola* isolates even from different sites (Bock et al. 2002). *Gibberella gaditjirrii* sp. nov. isolated from *Heteropogon triticeus* and other grasses from the northern Australia were differentiated from morphologically similar species by using AFLP markers (Baayen et al. 2000). Host specificity of fungal endophytes *Neotyphodium* species symbiotic with three grass hosts *Festuca arundinacea*, *Festuca pratensis* and *Lolium perenne* was studied based on comparison drawn between hosts and their endophytes by AFLP. The results revealed high genetic variation between plant and fungal endophytes of amplified fragment length polymorphisms (AFLP) between hosts and their corresponding endophytes (Karimi et al. 2012). AFLP markers can produce meaningful phylogenies out of diverse group of fungi and can identify genetically similar closely related organisms groups of closely related individuals that are genetically and biologically similar. An analysis of AFLP in population of *Epichloe* and *Neotyphodium* was performed to assess the worth of this technique in evaluating phylogenetic relationships among these fungi. A total of 13 primer combinations were used to generate 963 unique restriction fragments. Phylogenetic relationship of endophytes estimated by AFLP was compared with sequencing of ribosomal ITS region. The results revealed significant difference between AFLP and rDNA trees due to discrepancies in the association of anamorphs (*Neotyphodium*) and teleomorphs (*Epichloe*) which may arise due to some mechanism of genetic recombination (Tredway et al. 1999).

1.4.7 Single Strand Conformation Polymorphism (SSCP)

Single-strand conformation polymorphism (SSCP) is defined as conformational dissimilarity of single-stranded nucleotide sequences of identical length (Singh 2015). It is a sensitive, rapid and inexpensive technique which estimates sequence variation in a sample set so that only an informative subset needs to be sequenced. Although SSCP has been widely applied in medical diagnosis, biomedical research and population genetics studies, it is not very popular technique when it comes to fungal genetic diversity studies (Sunnucks et al. 2000). Walsh et al. (1995) used SSCP for the identification of medically important opportunistic fungi. Identification of endomycorrhizal fungi was also studied by using PCR based SSCP (Kristiansen et al. 2001). The CE-SSCP (capillary electrophoresis single–stranded conformation polymorphism) and CEFLA (capillary electrophoresis fragment length polymorphism), have been considered very reliable in the community analysis of soil fungal communities (Zinger et al. 2008) and may also be used for endophytic community analysis from different environmental samples (Suryanarayanan 2011).

SSCP analysis was used in estimating fungal diversity in the endemic plant species of Tenerife (Canary Islands) and their relation to the rhizospheric soil of vegetation zones and other environmental factors. SSCP analysis using *Trichoderma* specific primers was also done which showed low diversity of *T.harzianum* (Zachow et al. 2009).

1.5 DNA Barcoding of Endophytic Fungi

Hawksworth (1991) estimated existence of about 1.5 million species of fungi on earth of which very less number of fungi are described. A universally acceptable, accurate, rapid and cost effective identification system is needed to identify such an estimate of unknown species. DNA barcoding employs a short standardized genomic sequence to identify species (Hebert et al. 2003a; Seifert et al. 2007). Barcoding can be established as optimal when the sequence is constant and unique to all species. Additionally, in the DNA barcode region intraspecific variation must exceed interspecific variation (Hebert et al. 2003a, b; Schoch et al. 2012). Cytochrome C oxidase subunit 1 (CO1) is well accepted barcode for animals and additionally was also used in studies of fishes and birds (Kerr et al. 2007; Hubert et al. 2008; Begerow et al. 2010). Consortium for the Barcode of Life (CBOL) has also adopted CO1as the default marker for barcoding of fungi. CO1 has been used successfully as a barcode in few genera such as *Penicillium* with sufficient species resolution (Seifert et al. 2007; Schoch et al. 2012) but, in some other cases results with CO1 were inconsistent. Also, an early divergent lineage of anaerobic, gut fungi Neocallimastigomycota lacked mitochondria (Bullerwell and Lang 2005; Schoch et al. 2012). Internal transcribed spacer (ITS) region, nuclear large ribosomal subunit, β-tubulin A (Ben A) gene and elongation factor 1-α have also been widely used in the identification of fungi and phylogenetic studies (Schoch et al. 2012) CBOL has come to a conclusion of naming ITS region as universal barcode marker for fungi as ITS has the highest probability of identifying broadest range of fungi successfully with defined barcode gap between intra and interspecific variation (Schoch et al. 2012). It is the most popular barcode in identification of endophytic fungi. Several studies have employed ITS region in identification of endophytic fungal community pyrosequencing coupled metagenomics study (U'ren et al. 2014; Zhang and Yao 2015). Endophytic fungi isolated from plants found in extreme region were also identified using ITS barcode region. For example, endophytic fungi were recovered from *Deschampsia Antarctica* Desv. which represents only one of the two vascular plants colonized the Antarctic continent. Twenty-six endophytic fungal isolates were recovered from 273 leaf fragments of 91 individual plants and identified by amplifying ITS region of rDNA which revealed *Alternaria* and *Phaeosphaeria* as most dominant genus. (Rosa et al. 2009). Endophytic fungi recovered from six plants (*Hedyotis diffusa*, *Trifoliumrepens*, *Digitaria ischaemum*, *Silene tenuis*, *Cynodon dactylon,* and *Alternanthera philoxeroides*) found in geothermal ecosystem in southwest China were also identified amplifying ITS region (Zhou et al. 2015).

1.6 Detection and Analysis of Endophytic Fungal Diversity Through Metagenomics and Next Generation Sequencing (NGS) Techniques

1.6.1 Metagenomics

The ever increasing demand to identify fungi and analysis of their community structure has also increased our understanding of fungal interaction with the environment. Moreover, molecular markers coupled with phylogenetic analysis have widened our knowledge about roles played by fungi in symbiotic (endophytes) and pathogenic interaction with plants, litter decomposition and nutrient cycling. However, a vast majority of fungi in nature can not be cultured and hence, cultivation independent approach of analyzing such fungi is needed. Cultivation independent high-throughput sequencing methods facilitate thorough analysis of fungal communities in large sample sets and offer information far more than previous techniques in terms of detailing and magnitude of data (Rastogi and Sani 2011).

Metagenomics is a culture independent approach that involves analysis of collective genomes of all microbes present in a environmental sample (Riesenfeld et al. 2004). Although, genetic fingerprinting techniques are widely used in assessing microbial diversity, they do not provide information beyond the genes that are being amplified (Rastogi and Sani 2011). However, in metagenomics, the entire genetic makeup of microbial communities in a sample can be sequenced and analyzed. Metagenomic investigations of microbial communities have been conducted on variety of habitats and niches including soil, ocean, plants, hot and also in human gasto-intestinal tract (Cuadros-Orellana et al. 2013). Metagenomic investigations involve constructing metagenomic libraries from environmental samples by (1) isolation of total DNA from the sample, (2) shotgun cloning of DNA into suitable vector (3) transformation of clones into suitable host and (4) screening out the positive clones. Small insert DNA fragment libraries (2–3 kb inserts) provide more coverage of metagenome of a given environmental samples than large inserts whereas large fragment libraries (100–200 kb) are mostly used while exploring multigene biochemical pathways (Riesenfeld et al. 2004; Rastogi and Sani 2011).

Metagenomic libraries can be screened by two approaches: (1) function driven analysis that involves expression of selected phenotypes and (2) sequence driven analysis involving high-throughput sequencing.

1.6.1.1 Function Based Screening Approach

Metagenomic library can be screened for isolation of genes encoding novel biomolecules based on metabolic activities of library containing clones. This strategy has the potential to discover and identify novel genes encoding novel biomolecules having known or novel functions (Ferrer et al. 2009; Gloux et al. 2011; Simon and Daniel 2011). Several important novel antibiotics (e.g., turbomycin, terragine), enzymes (e.g., cellulases, lipases, amylases), and proteins (e.g., antiporters) have been reported from soil metagenomic libraries (Rondon et al. 2000). However, rate of active metagenomic library clones expressing a phenotype is relatively low. Henne et al. (2000) reported that only one in 730,000 clones of soil metagenomic library showed lipolytic activity (Rastogi and Sani 2011).

Researchers have adopted three major function derived screening approaches to isolate novel biomolecules: (1) phenotypical detection of the desired activity, (2) heterologous complementation of host strains or mutants and (3) induced gene expression (Simon and Daniel 2011). In phenotypical detection the growth medium include dyes and chromophore-bearing derivatives of enzyme substrates, where specific metabolic capabilities of individual clones can be screened (Ferrer et al. 2009). Heterologous complementation of host strains or mutants is a simple and fast and highly selective screening approach of metagenomic libraries which requires target genes for growth under selective conditions (Simon et al. 2009). Substrate induced gene expression screening (SIGEX) was introduced by Uchiyama et al. (2005) which employed operon trap *gfp* expression vector in combination with fluorescence activated cell sorting. Uchiyama and Miyazaki (2010) introduced product induced gene expression (PIGEX), which can be triggerd by product formation and is also capable of expression detection of *gfp* (Simon and Daniel 2011).

1.6.1.2 Sequence Based Screening Approach

Sequence based screening approach throws light on many important genomic features on community structure such as traits acquired from distinctly related taxa through horizontal gene transfers, genomic organizations and redundancy of functions in a community (Handelsman 2004; Rastogi and Sani 2011). This approaches involves DNA probes or primers designed either from conserved regions of known genes or protein families. This leads to identification of novel variant of known proteins, genes functional classes of proteins can be identified. Nevertheless, this strategy has led to the successful identification of genes encoding novel enzymes such as nitrite reductases (Bartossek et al. 2010) [NiFe], hydrogenases (Maróti et al. 2009), chitinases (Hjort et al. 2010), glycerol dehydratases (Knietsch et al. 2003) and dimethylsulfoniopropionate degrading enzymes (Varaljay et al. 2010).

1.6.2 Next Generation Sequencing (NGS)

Metagenomics coupled with next generation sequencing techniques (NGS) resulted in generation of large sequence data sets originated from diverse environmental sources including soil, plants, ocean and human body. NGS analyses of environmental samples have opened a new window to get insight into massive diversity and functional dynamics of microbial community (Simon and Daniel 2011). NGS techniques with low cost, high speed and additional technical advantages have been extensively used in metagenomics (Edwards et al. 2006; Cuadros-Orellana et al. 2013). Metagenomic data analysis includes initially the assembly of DNA sequence reads into contigs followed by gene prediction. The next step involves used to prediction of the functional repertoire encoded in metagenome by protein coding gene and phylogenetic analysis (Raes and Bork 2008; Mende et al. 2012). Next-generation sequencing systems such as Roche 454 pyrosequencing, SOLiD/Ion Torrent PGM from Life Sciences, HeliScope/Helicos BioSciences and Genome Analyzer/HiSeq

2000/MiSeq from Illumina are much morefaster and less expensive than Sanger's dideoxy sequencing (Metzker 2010; Cuadros-Orellana et al. 2013).

In current scenario the most prominent high throughput sequencing methods used for metagenomics are pyrosequencing (Edwards et al. 2006; Turnbaugh et al. 2009) and recently introduced Illumina sequencing (Qin et al. 2010; Mende et al. 2012). The first commercial NGS system platform was launched in 2004 by Roche 454 based on pyrosequencing (Liu et al. 2012). Four hundred fifty four pyrosequencing involves detection of pyrophosphate released during nucleotide incorporation instead of incorporating dideoxynucleotides to terminate the chain amplification. Another NGS platform, Sequencing by Oligo Ligation Detection (SOLiD) was introduced by Applied Biosystems in 2006 which adopts two-base sequencing technology based on ligation sequencing. The Genome analyzer was first introduced by Solexa in 2006 but, soon the company was purchased by Illumina. This sequencer uses technology in which the library having fixed adaptors is denatured to single strands, grafted to the flowcell and followed by bridge amplification to form clusters containing clonal DNA fragments (Liu et al. 2012). Illumina later also introduced another sequencing platform HiSeq 2000. Liu et al. (2012) has compared all three NGS sequencing system and reported Illumina HiSeq 2000 has the biggest output, Roche 454 has the longest read length and SOLiD system has the highest accuracy. Ion Personal Genome Machine (PGM) was released by Ion Torrent in 2010 MiSeq was launched by Illumina. Both were small in size, with fast turnover rates but limited data throughput.

1.6.3 Analysis of Endophytic Fungal Communities by Metagenomic/Next Generation Sequencing

A common problem encountered in most metagenomic studies is the extraction of target molecule in high purity. The main challenge for metagenomic study of endophytic fungal community is separating microbial DNA from the host plant (Porras-Alfaro and Bayman 2011; Cuadros-Orellana et al. 2013). Owing to high abundance of plant DNA, fungal DNA is difficult to separate during extraction of total genomic DNA from plant tissues (Porras-Alfaro and Bayman 2011; Cuadros-Orellana et al. 2013). Unterseher et al. (2011) has strongly pointed out importance of sampling in fungal community analysis and asserted that ecology of microbial community strongly depends on depth of sampling. Analysis of 454 sequence data of plant associated fungal community addressed suitability of species abundance model into three goups namely, phyllosphere, ectomycorrhizal, and arbuscular mycorrhizal fungi. The first investigation of endophytic community using pyrosequencing was from micropropagated *Atriplex canescens* and *A. torreyi* var. *griffithsii* (Lucero et al. 2011), The analysis showed diverse strains interact in a manner which could alter host plant physiology. Gillevet et al. (2009) compared different fungal ITS contigs from *Spartina alterniflora* growing in a salt-marsh by either pyrosequencing or clone sequencing and demonstrated advantages of pyrosequencing over traditional sequencing approach. The analysis revealed that pyrrosequencing can scale greater depth both in terms of number of contigs and in revealing entire clades which could not be resolved by traditional sequencing. The common steps involved in pyrosequencing of endophytic fungi involved- (1) surface sterilization of plant samples (2) Extraction of total genomic DNA (3) PCR amplification of ITS r RNA (in most cases researchers have amplified ITS region but few studies also suggests use of D1/D2 region of 26S rRNA) and (4) Sequence analysis (Fig. 1.2).

Next generation sequencing, in particular pyrosequencing has been extensively used in community analysis of fungal endophytes from different habitats having different altitude and climatic conditions. U'ren et al. (2014), examined influence of tissue storage on commonly used extraction of DNA, primer design and 454 amplicon pyrosequencing in studies on endophytic and endolichenic fungi. The results revealed that tissue storage, selection of primer and other intrinsic biases can influence community richness and composition of fungal community structure. In another study diversity and distribution of endophytic fungal community associated with four vascular plants *Cassiope tetragona*, *Saxifraga cespitosa*, *Saxifraga oppositifolia* and *Silene acaulis* in high Arctic were assessed using 454pyrosequencing by amplifying ITS region with fungal specific primers. The results

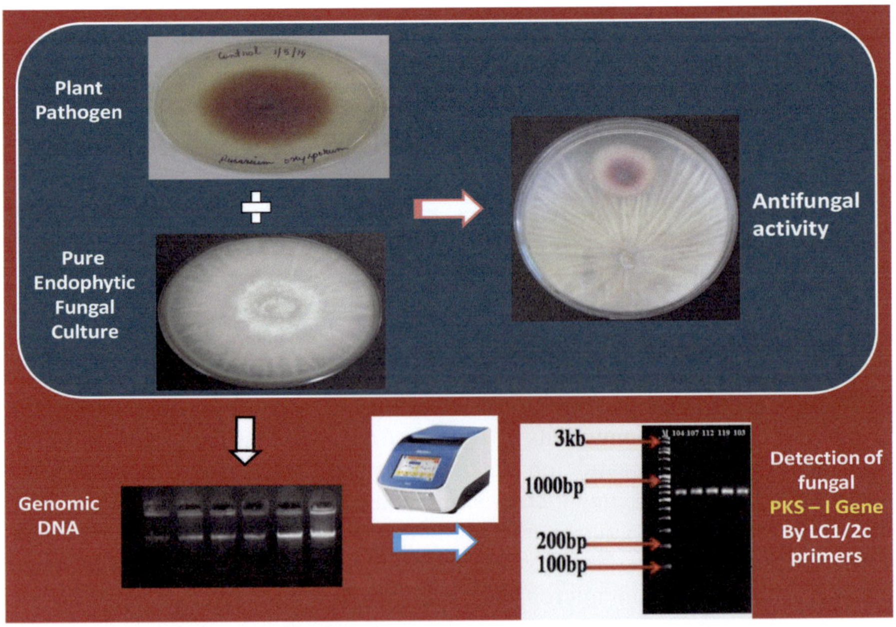

Fig. 1.2 Steps of detection of PKS I gene in endophytic fungi having antifungal activity by using LC1/2c primers

revealed high diversity of fungal communities. Ascomycota were most dominant class followed by Basidiomycota and Chytridiomycota. Helotiales, Pleosporales, Capnodiales, and Tremellales were the dominant orders (Zhang and Yao 2015). Endophytic fungal community analysis of *Metrosideros polymorpha*, from 13 sites in Hawaii along environmental gradient (rain and mean annual temperature) exhibited 4, 200 taxonomic units. The pyrosequencing analysis showed among site diversity contributed more than within site diversity to community richness. Also, the community composition of endophytic fungal communities associated with *M. polymorpha* correlated strongly with temperature and rainfall (Zimmerman and Vitousek 2012). A new progress in endophytic community analysis was made when Siddique and Unterseher (2016) demonstrated efficacy of cost effective and highly multiplexed paired end Illumina sequencing of full length ITS amplicons of endophytic fungal communities of European beech (*Fagus sylvatica*). Leaves of F. sylvatica were sampled along elongation gradient of about 1000 m. The results showed significant correlation of community composition with elevation surface. Kemler et al. (2013) used Ion Torrent PGM with semiconductor sequencing technology in community analysis of endophytes in *Eucalyptus grandis* (Kemler et al. 2013). ITS1 ribosomal RNA of endophytic community was sequenced and the sequence analysis showed the dominance of Ascomycetes followed by Dothideomycetes (Kemler et al. 2013). Gottel et al. (2011) demonstrated that root endophytic communities are distinct from that of rhizospheric communities. Microbial communities comprising root endophytes and rhizospheric microorganism of *Populus deltoids* from both upland and bottomland sites in Tennessee were analyzed through 454 pyrosequencing using V4 region of 16S rRNA for bacteria and D1/D2 region of fungal 28S rRNA genes. Hierarchical clustering of OUTs together with hest map analysis and relative abundance patterns exhibited that most of the microbial OUTs are likely to be dominant either in root or rhizospheric samples (Gottel et al. 2011).

1.7 Antimicrobial Biosynthetic Genes of Endophytic Fungi

Endophytic fungi synthesize diverse classes of antimicrobial secondary metabolites including alkaloids, terpenoids, polyketides, phenylpropanoids, aliphatic compounds, and peptides (Mousa and Raizada 2013). Endophytes are known to produce compounds that protect the plant from various biotic stresses. Further, endophytic fungi have been reported to produce many compounds also produced by their host. Polyketides and terpenoids are the most purified antimicrobial secondary metabolites synthesized by endophytic fungi (Mousa and raizada 2013). Bioinformtic analysis has revealed two highly conserved genes polyketide synthases (PKS) and non ribosomal peptide synthases (NRPS) that are putatively involved in biosynthesis of many secondary metabolites (Brakhage and Schroeckh 2011).

1.7.1 Fungal Polyketide Synthases (PKS) Genes

Polyketides always had the reputation of being one of the most abundant classes of secondary metabolites widely distributed in plants, fungi, bacteria and marine organisms (Cox 2007). The polyketides are an enormous group of structurally diverse natural products that have found commercial application in pharmaceutical, agriculture and chemical industries. They play important role as antibiotics (erythromycin, tetracycline, grahamimycin and patulin), antifungal (griseofulvin and monocerin) immunosupressors (rapamycin), antitumorals (daunorubicin, doxorubicin), anticholesteremic (statins e.g. Lovastatin and Compactin) and antiparasitics (avermictin) (Cox 2007; Hertweck 2009; Rojas et al. 2012; Zhou et al. 2011). Fungal polyketides comprise a range of compounds such as the mycotoxins aurofusarin (Shibata et al. 1966), aflatoxin (Bhatnagar et al. 2003), ochratoxins, fuminosins, zearalenone, tetrahydroanthraquinones (Mousa and Raizada 2013) and spore pigments (Mayorga and Timberlake 1992; Watanabe et al. 1998; Gaffoor et al. 2005). Fungal polyketides present as monocyclic aromatic compounds (orsellinic acid and 6-methylsalicylic acid), polyclic aromatic compounds (citrinin, deoxyherqueinone, and norsolorinic acid) as well as non aromatic compounds (macrolides e.g. decarestrictine D, lovastatin). Several other polyketide metabolites comprised of an aromatic ring attached to a highly reduced moiety (zearalenone), whereas some metabolites arises from oxidative metabolism of preformed aromatic polyketides (aflatoxin B1 from norsolorinic acid). However, there are other types of metabolites which contain a part of polyketide derived structure and remainder comes from other biosynthetic pathways (e.g. terpenoid humulene in the case of xenovulene A) (Cox and Simpson 2009).

Endophytic fungi also produce variety of natural products having wide range of bioactivities including antibacterials (6-O-methylalaternin, altersolanol A, xanalteric acids I and II), antifungal (Nodulosporins, Pestalachloride B, CR377, Chaetoglobosins A and C), antiparasitic and anti leishmania (palmarumycin CP18), anti brine shrimp larvae (chaetomugilin A and D, Chaetoglobosins A and C) and antiviral (pestalotheol C) (Mousa and Raizada 2013). Polyketide synthases (PKSs) are multimodular enzymes that biosynthesize polyketides by catalyzing the polymerisation of acyl-CoA thioesters (Amnuaykanjanasin et al. 2009). Bacterial polyketide synthases are modular in which each condensation cycle is catalyzed by distinct module comprising all the required catalytic domains (Staunton and Weissman 2001) whereas fungal typeI PKSs are iterative. Fungal type I PKSs can be represented by mammalian fatty acid synthases (FASs) whereas type III PKS which do not correspond with any known FAS (Cox 2007). Type I PKSs are multimodular enzymes comprising distinct catalytic domains that are required to catalyze each step of polyketide biosynthesis. Type I PKS contains three major domains: β-ketosynthase (KS), acyltransferase (AT) and an acyl-carrier protein (ACP) along with some additional domains such as β-ketoacyl reductase (KR), dehydratase (DH), enoyl reductase (ER), methyl transferase (MT), methyl esterase (ME), and thioesterase (TE) (Amnuaykanjanasin et al. 2005). Type II PKS complex involves individual proteins with distinct functions which synthesizes typically aromatic polyketides iteratively. Type III PKSs are characterized by chalcone synthases and stillbene synthases. Though like type I PKSs, type III PKS also follow the modular mechanism of polyketide biosynthesis but, without the acyl carrier protein (ACP) domain (Rojas et al. 2012). Type I PKS are found in bacteria, actinomycetes as well as fungi, type II PKS are found in bacteria, whereas, type III PKS are found in plants, bacteria and fungi (Cox 2007).

1.7.1.1 Detection of Fungal Type I PKS

Fungal type I PKSs are described as iterative which catalyze the condensation of subunits into polyketide backbone. They can be classified into two broad categories non reducing PKSs and reducing PKSs based on the compounds they synthesize. Non reducing (NR) polyketide synthases synthesize polyketides without any chemical reduction in their structure. Reducing PKSs synthesize compounds with several chemical reductions in polyketide structure. They have KR domain (catalyzes β-keto reduction; keto to hydroxyl group), DH domain (dehydration; hydroxyl to enoyl group) and ER domain (enoyl reduction; enoyl to hydroxyl group). The condensation reaction is usually followed by reduction or dehydration steps follow to create a polyketide compound having alcohol, ketone or alkene at specific portions of the chain. Reducing type polyketide synthases are divided into highly reduced (HR) PKS and partially reduced (PR) PKS. PR PKSs include KR and DH domain and are also known as MSA Synthase (Bingle et al. 1999), given that most of them involved in synthesis of Methylsalicylic acid (MSA). HR PKSs mostly contain all three reducing domains (KR, DH and ER). They are further subdivided into four subclades (I, II, III and IV) based on their characteristic domain structures (Kroken et al. 2003; Amnuaykanjanasin et al. 2005; Cox 2007).

Molecular genetics and genomics have facilitated the understanding of catalytic function of polyketide synthases and fatty acid synthases up to a great extent. Structural complexity of polyketides is due to use and control of reductive, dehydrative and methylating steps of PKS biosynthesis. This must be due to differences in PKS protein sequence and structure. Bingle et al. (1999) realised that difference in the protein sequence should also reflect in DNA sequence and PCR primers can be designed to detect specific fungal PKS genes responsible for synthesis of particular type of polyketide (Cox 2007). Several approaches have been made to detect PKS genes by developing PCR primers based on conserved regions of KS, AT and MT domains. Bingle et al. (1999) designed LC series degenerate primers based upon most conserved KS region of fungal PKSs to selectively amplify KS of NR-type PKSs (LC1/2c) as well as PR-type PKSs (LC3/5c). Later, Nicholson et al. (2001) designed degenerate KS series primers to amplify KS domain responsible for biosynthesis of highly reduced metabolites. Availability of lovastatin nonaketide synthase (LNKS) and lovastatin diketide synthase (LDKS) DNA sequences has also made it possible to design the primers for *C*MeT domain (Table 1.1) (Nicholson

Table 1.1 Primers used in detection of PKS I gene in fungi

Target	Primer	Primer sequence (5′ → 3′)	Expected band size	Reference
KS domain	LC1 (F)	GAYCCIMGITTYTTYAAYATG	750 b.p.	Bingle et al. (1999); Amnuaykanjanasin et al. (2005)
	LC2c (R)	GTICCIGTICCRTGCATYTC		
	LC3 (F)	GCIGARCARATGGAYCCICA	680 b.p.	
	LC5c (R)	GTIGAIGTIGCRTGIGCYTC		
	KS 3 (F)	TTYGAYGCIGCITTYTTYAA	~750 b.p.	Nicholson et al. (2001); Amnuaykanjanasin et al. (2005)
	KS 4 (R)	RTGRTTIGGCATIGTIATICC		
	XKS1 (F)	TTYGAYGCIBCITTYTTYRA		Amnuaykanjanasin et al. (2005)
	XKS2 (F)	CRTTIGYICCICYDAAICCAAA		
KS-AT	KAF1 (F)	GARKSICAYGGIACIGGIAC		Amnuaykanjanasin et al. (2005)
	KAF2 (F)	GARGCICAYGCIACITCIAC		
	KAR1 (R)	CCAYTGIGCICCRTGICCIGARAA		
	KAR2 (R)	CCAYTGIGCICCYTGICCIGTRAA		
MT	CMeT1	GARATIGGIGSIGGIACIGG	300–400 b.p.	Nicholson et al. (2001); Amnuaykanjanasin et al. (2005)
	CMeT2	ATIARYTTICCICCIGGYTT		

Degenerate bases are described as: *B* C/G/T, *D* A/G/T, *I* inosine, *K* G/T, *M* A/C, *R* A/G, *S* C/G, *Y* C/T

et al. 2001; Cox 2007). Sequence analysis of the PCR product obtained by these selective primers from various filamentous fungi revealed new PKS genes which clustered with other PKS gene sequences of expected classes (Kroken et al. 2003). In the recent past, this approach has been widely accepted and applied to detect a particular PKS type from fungi distributed in wide ecological niches (Cox 2007). PKS gene based clades of NR, PR and HR PKSs are being used in phylogenetic analysis to categorize endophytic fungi (Rojas et al. 2012), entomopathogenic fungi (Amnuaykanjanasin et al. 2005, 2009) and lichenenized fungi (Schmitt et al. 2008) in diversity studies (Rojas et al. 2012). PKS based classification has also been used in genomic and metagenomic (Ginolhac et al. 2004) analysis. Kroken et al. (2003) analyzed fungal genomes and demonstrated that PKS amino acid sequences can be used in phylogenomic analysis.

1.7.1.2 Fungal Type III PKS

Type III polyketide synthases are homodimeric enzymes that catalyzes an array of reactions including priming of starter substrate, condensation of extender substrates, ring closure and aromatization of the polyketide chain (Li et al. 2011). Type III PKSs were believed to exist exclusively in plant and bacteria but recent genome projects have revealed their presence also in filamentous fungi. Chalcone synthase

(CHS) and stilbene synthases are most studied plant-specific type III PKS (Lucia and Martin 2010). CHS catalyzes biosynthesis of naringenin chalcone which is a common precursor of all flavonoids produced by plants (Austin and Noel 2003; Li et al. 2011). Type III PKS genes csy A, B, C and D were discovered in *Aspergillus oryzae* (Seshime et al. 2005; Hashimoto et al. 2014b). Recently fungal type III PKS *Neurospora crassa* 2′oxoalkylresorcylic acid synthase (ORAS), *Aspergillus niger* AnPKS and *Botrytis cinerea* BPKS have been described (Funa et al. 2007; Lucia and Martin 2010; Hashimoto et al. 2014a, b). ORAS catalyzes biosynthesis of aromatic pentaketide alkyl resorcylic acid (Funa et al. 2007) whereas, AnPKS and BPKS catalyzes synthesis of tri and tetra ketide pyrones, penta and hexa ketide resorcylic acid and resorcinol (Hashimoto et al. 2014a, b).

1.7.1.3 Detection of Polyketide Synthase Gene in Endophytic Fungi

Polyketide synthase (PKS) genes have been detected from endophytic fungi in a number studies (Table 1.1) which primarily involve initial screening of antimicrobial activity of fungal isolates prior to detection of PKS gene. Diversity of PKS genes from 63 endophytic fungi recovered from *Annona squamosa* was studied using three pair of primers specific for NR, PR and HR β-ketoacyl synthase (KS) domain (Lin et al. 2010). Similarly, Rojas et al. (2012), have studied diversity of PKS genes from sugarcane fungal endophytes by detecting both β-ketoacyl synthase (KS) and C-methyltransferase (CMT) domains. They sequenced the PKS genesa and performed phylogenetic analysis based on deduced amino acid sequences to determine the genetic potential of endophytic fungi associated with sugarcane to biosynthesis different polyketide chemistry. An endophytic fungi *Phomopsis liquidambaris* isolated from *Cryptolepis buchanani* Roem. was investigated for type I PKS gene using specific degenerate primers (Rao et al. 2015). Miller et al. (2012) isolated total genomic DNA from Chinese anticancer herbs and detected 10 PKS genes from 30 medicinal herbs using specific primers which suggest possibility of production of bioactive natural compounds by fungal endophytes in medicinal herbs. In another study, endophytic fungi recovered from wild rice (*Oryza rufipogon* Griff.) from China were investigated for their antagonistic activity against rice pathogens and presence of PKS genes (Wang et al. 2015). Non reducing (NR) PKS genes were also detected in lichenized fungi of order Pertussariales (Schmitt et al. 2005).

1.7.2 Fungal Non Ribosomal Peptide Synthases (NRPS)

Nonribosomal peptide synthetases (NRPSs) are enzymes which catalyzes synthesis of small bioactive pepetide in bacteria and fungi independent of ribosomes (Mootz et al. 2002; Reiber et al. 2005; Stach et al. 2007). Nonribosomal peptide synthetases are multimodular enzymes (up to 2.3 MDa molecular mass) encoded by NRP synthatase genes usually present as part of multiple gene clusters (Keller et al. 2005; Stach et al. 2007). NRPS encoding genes are known to occur in plenty in fungi and bacteria but to the best of our knowledge there are no reports of their presence in plants or animals (Bushley and Turgeon 2010). The substrates for multimodular enzyme involves the D and L forms of the 20 amino acids involved in ribosomal protein synthesis along with non-proteinogenic amino acids such as imino acids, ornithine, as well as hydroxy acids such as α-aminoadipic and α-butyric acids (Finking and Marahiel 2004). Bacterial NRPS complexes are characteristically composed of more than one polypeptide (Mootz and Marahiel 1997; Keating et al. 2000) whereas, fungal NRPS complexes comprise of a single, large polypeptide. However, ergopeptine synthetase in *Claviceps purpurea* is an exception are which consists of two separate proteins (Correia et al. 2003). Therefore, most fungal NRPS complexes are encoded by a single, intronless ORF in the producer organism genome in spite of their often considerable size (Weber et al. 1994; Yuan et al. 2001; Slightom et al. 2009).

NRP synthatases synthesize astounding diversity of complex oligopeptides through linear condensation of proteinogenic and non-proteinogenic amino acids (Marahiel et al. 1997; Miller et al. 2012; Johnson et al. 2007a). NRPS are consists of series of modules and each single module consists of a minimum of three core domains: adenylation domain (A), thiolation (T) or peptidyl carrier protein (PCP) domain and condensation (C) domain arranged in the order C-A-T. At the time of peptide assembly, adenylation domain (A) recognizes and activates the substrate amino acids by adenylation. Adjacent thiolation domain (T) anchors the activated substrate to a 4′ phosphopantetheine cofactor with a thioester linkage and transported the substrate to a condensation (C) domain which catalyzes peptide bond formation by the condensation of the amino group of one amino acid to the carboxyl group of the adjacent amino acid (Stachelhaus et al. 1998; Bushley and Turgeon 2010; Slightom et al. 2009). Fungal NRPSs may also have two additional modifying domains such as N-methyl transferase (MT) and epimerisation (E) domains. Methylation (M) domains catalyze N-methylation of amino acids by transfer of a methyl group from S-adenosylmethionine to an α-amino of the amino acid (Weber et al. 1994; Haese et al. 1993) whereas, epimerization (E) domains which converts L-amino acids to D amino acids (Stachelhaus and Marahiel 1995). Some fungi also use other mechanism that involves a dedicated D-amino acid biosynthesis pathway for introduction of D-amino acids (Hoffmann et al. 1994; Slightom et al. 2009). In bacterial NRPS complexes peptide chain release is mostly effected by either hydrolysis or internal cyclisation catalyzed by thioesterase domain (Bushley and Turgeon 2010). A very few fungal NRPSs such as ACV synthatases have a thioesterase (T) domain at their C-terminus which catalyzes the release of the final peptide product. The chain release is carried out by three mechanisms, two of which are found also in bacterial systems: (1) a terminal C domain, which catalyzes the release of peptide chain by inter- or intra-molecular amide bond formation (Keating et al. 2001) and (2) a thioesterase NADP (H) dependent reductase (R) domain which catalyzes reduction with NADPH to form an aldehyde (Bushley and Turgeon 2010). The third mechanism was reported in the case of fungal ergot alkaloid synthesis is that which involve in nonenzymatic cyclization by formation of a diketopiperazine ring (Keating et al. 2001; Walzel et al. 1997; Johnson et al. 2007a). Some additional enzymes other than NRPSs may modify the final product by glycosylation, hydroxylation, acylation, or halogenation (Samel et al. 2008). The number and arrangement of modules of fungal NRPSs typically, resolves the length and structure of the final peptide product (Johnson et al. 2007a). The study of fungal genome sequences of *Cochliobolus heterostrophus*, *Claviceps purpurea* and *Fusarium graminearum* has led to the identification of multiple ORFs predicted to encode NRP synthetase genes (Haarmann et al. 2005; Lee et al. 2005; Varga et al. 2005; Stach et al. 2007).

Oligopeptides biosynthesized by NRPSs are involves in production of antibiotics (penicillin, fusaricidin and cephalosporin) and immunosuppressants (cyclosporine A) (Miller et al. 2012). NRPSs plays important role in fungal morphology, reproductive and pathogenic development, cell surface properties, stress management, nutrient procurement and niche adaptation (Marahiel et al. 1997; Challis et al. 2000; Lee et al. 2005; Oide et al. 2006, 2007; Johnson et al. 2007a; Bushley and Turgeon 2010).

1.7.2.1 Detection of NRPS Gene in Endophytic Fungi

Other than filamentous fungi, NRPS complexes encoded by different NRPS genes in fungal endophytes associated with grass as well as medicinal plants have also been reported to synthesise bioactive metabolites (Table 1.2). Whole genome analysis of closely related asexual and sexual species of grass fungal endophytes *Epichloe/Neotyphodium* sp. has revealed presence of 12 NRPS genes (Johnson et al. 2007a). These endophytes have been reported to uniquely synthesize peramine, a pyrrolopyrazine metabolite which protect host plants from insect herbivory (Rowan 1993; Clay and Schardl 2002; Tanaka et al. 2005). NRPS genes are involved in both Ergovaline and Peramine biosynthetic pathways suggests that it provides competitive abilitiy to *Neotyphodium* and *Epichloe* in symbiosis with grass host (Johnson et al. 2007a). Degenerate primers designed to identify members of the NRPS gene family from *Neotyphodium* and *Epichloë* species have revealed a minimum of 12/13 NRPS genes (Johnson et al. 2007b) This distribution pattern of NRPS genes among different lineages of *Neotyphodium/Epichloë* species proposes of a common ancestor which may have contributed most of the complement of genes and might have been lost during the evolution of these fungi (Johnson et al. 2007b). Another gene (aba1) encoding the NRPS complex responsible for the synthesis of the cyclic peptide antibiotic Aureobasidin A (AbA) was reported from phylloplane yeast *Aureobasidium pullulans* (Slightom et al. 2009). Genome mining of a mycotoxin, acetylaszonalenin isolated from *Neosartorya fischeri* was studied by Yin et al. (2009) on the information available from its structure. A putative biosynthetic cluster was identified containing three genes coding for a putative non-ribosomal peptide synthetase (AnaPS), a prenyltransferase (AnaPT), and an acetyltransferase (AnaAT) through genome mining (Nikolouli and Mossialos 2012).

Table 1.2 Primers used in detection of NRPS gene in fungi

Target	Primer	Primer sequence (5′ → 3′)	Expected band size	Reference
Adenylation domain of fungal NRPSs	RJ016-F	Tayggnccnacnga	~300 bp	Johnson et al. (2007a, b); Miller et al. (2012)
	RJ016-R	Arrtcnccngtyttrta		
A3 (YTSGTTGxPKG) (**core motif**)	AUG003-A3FWD- forward	5′-CCGGCACCACCggnaarcchaa-3′	AUG003–AUG007- 1100 bp AUG005–AUG007- 580 bp AUG003–AUG006- 480	Slightom et al. (2009); Zhou et al. (2011)
A3 (YTSGTTGxPKG) (**core motif**)	AUG004-A3FWD2-forward	5′-TCACCTCCGGCACCachggnaarcc-3′		
A7 (YR/KTGDL) (**core motif**)	AUG005-A7FWD- forward	5′-GTCCACGGACGGATGTACarrachggvga-3′		
A7 (YR/KTGDL) (**core motif**)	AUG006-A7RVS - reverse	5′-CCGGACCATGTCGccngtbykrta-3′		
T (LGGH/DSL/I) (**core motif**)	AUG007- ThioRvs	5′-GCTGCATGGCGGTGATGswrtsnccbcc-3′		

Degenerate bases are described as: *B* C/G/T, *D* A/G/T, *I* inosine, *K* G/T, *M* A/C, *R* A/G, *S* C/G, *Y* C/T

Acknowledgements BPS is thankful to University Grant Commission (UGC) for financial support as major research project. VKM is thankful to Department of Science and Technology (DST), New Delhi for providing fellowship under DST INSPIRE fellowship (IF 130374). Authors are also thankful to the Department of Biotechnology, New Delhi for establishment of DBT-BIF centre and DBT-State Biotech hub in the Department of Biotechnology, Mizoram University which has been used for the present study.

References

Aduramigba-Modupe AO, Asiedu R, Odebode AC, Owolade OF. Genetic diversity of *Colletotrichum gloeosporioides* in Nigeria using amplified fragment length polymorphism (AFLP) markers. Afr J Biotechnol. 2012;11(33):8189–95.

Aly AH, Debbab A, Kjer J, Proksch P. Fungal endophytes from higher plants: a prolific source of phytochemicals and other bioactive natural products. Fungal Divers. 2010;41:1–16.

Amnuaykanjanasin A, Punya J, Paungmoung P, Rungrod A, Tachaleat A, Pongpattanakitshote S, Cheevadhanarak S, Tanticharoen M. Diversity of type I polyketide synthase genes in the wood-decay fungus Xylaria sp. BCC 1067. FEMS Microbiol Lett. 2005;251(1):125–36.

Amnuaykanjanasin A, Ponghanphot S, Sengpanich N, Cheevadhanarak S, Tanticharoen M. Insect-specific polyketide synthases, potential PKS-NRPS hybrids, and novel PKS clades in tropical fungi. Appl Environ Microbiol. 2009;75(11):3721–32.

Anderson IC, Cairney JWG. Diversity and ecology of soil fungal communities: increased understanding through the application of molecular techniques. Environ Microbiol. 2004;6:769–79.

Anderson IC, Campbell CD, Prosser JI. Potential bias of fungal 18S rDNA and internal transcribed spacer polymerase chain reaction primers for estimating fungal biodiversity in soil. Environ Microbiol. 2003;5(1):36–47.

Aradhya MK, Chan HM, Parfitt DE. Genetic variability in the pistachio late blight fungus, *Alternaria alternata*. Mycol Res. 2001;105:300–6.

Arnold AE. Understanding the diversity of foliar fungal endophytes: progress, challenges, and frontiers. Fungal Biol Rev. 2007;21:51–66.

Arnold AE, Maynard Z, Gilbert GS, Coley PD, Kursar TA. Are tropical fungal endophytes hyperdiverse? Ecol Lett. 2000;3:267–74.

Austin MB, Noel JP. The chalcone synthase superfamily of type III polyketide synthases. Nat Prod Rep. 2003;20(1):79–110.

Baayen RP, O'Donnell K, Bonants PJM, Cigelnik E, Kroon LPNM, Roebroeck EJA, Waalwijk C. Gene genealogies and AFLP analysis in the *Fusarium oxysporum* complex identify monophyletic and non-monophyletic formae speciales causing wilt and rot diseases. Phytopathology. 2000;90:891–900.

Backeljau T, De Bruyn L, De Wolf H, Jordaens K, Van Dongen S, Verhagen R, Winnepenninckx B. Random amplified polymorphic DNA (RAPD) and parsimony methods. Cladistics. 1995;11:119–30.

Bacon CW, White JF. Microbial endophytes. New York: Marcel Deker Inc; 2000.

Bacon CW, Porter JK, Robbins JD, Luttrell ES. *Epichloe typhina* from toxic tall fescue grasses. Appl Environ Microbiol. 1977;34:576–81.

Barrett AJ, Rawlings ND, Woessner JF. The handbook of proteolytic enzymes. 2nd ed. London: Academic; 2003.

Bartossek R, Nicol GW, Lanzen A, Klenk HP, Schleper C. Homologues of nitrite reductases in ammonia-oxidizing archaea: diversity and genomic context. Environ Microbiol. 2010;12:1075–88.

Barve MP, Haware MP, Sainani MN, Ranjekar PK, Gupta VS. Potential of microsatellites to distinguish four races of *Fusarium oxysporum* f.sp. *ciceri* prevalent in India. Theor Appl Genet. 2001;102:138–47.

Bayman P, Lebrón LL, Tremblay RL, Lodge DJ. Variation in endophytic fungi from roots and leaves of Lepanthes (Orchidaceae). New Phytol. 1997;135:143–9.
Begerow D, Nilsson H, Unterseher M, Maier W. Current state and perspectives of fungal DNA barcoding and rapid identification procedures. Appl Microbiol Biotechnol. 2010;87:99–108. doi:10.1007/s00253-010-2585-4.
Bhatnagar DK, Ehrlich C, Cleveland TE. Molecular genetic analysis and regulation of aflatoxin biosynthesis. Appl Microbiol Biotechnol. 2003;61:83–93.
Bills GF. Isolation and analysis of endophytic fungal communities from wood plants. In: Redlin SC, Carris LM, editors. Endophytic fungi in grasses and woody plants: systematics, ecology, and evolution. St. Paul: APS Press; 1996. p. 31–65.
Bills JK, Christensen M, Powell M, Thorn G. In: Mueller GM, Bills GF, Foster M, editors. Biodiversity of fungi: endophytic fungi. California: Elsevier Academic Press; 2004. p. 241.
Bingle LE, Simpson TJ, Lazarus CM. Ketosynthase domain probes identify two subclasses of fungal polyketide synthase genes. Fungal Genet Biol. 1999;26:209–23.
Bischoff JF, White Jr JF. Evolutionary development of the Clavicipitaceae. In: Dighton J, White JF, Oudemans P, editors. The fungal community: its organization and role in the ecosystem. Boca Raton: Taylor & Francis; 2005. p. 505–18.
Bock CH, Thrall PH, Brubaker CL, Burdon JJ. Detection of genetic variation in *Alternaria brassicicola* using AFLP fingerprinting. Mycol Res. 2002;106:428–34.
Brakhage AA, Schroeckh V. Fungal secondary metabolites – Strategies to activate silent gene clusters. Fungal Genet Biol. 2011;48:15–22.
Bullerwell CE, Lang BF. Fungal evolution: the case of the vanishing mitochondrion. Curr Opin Microbiol. 2005;8:362–9.
Bushley KE, Turgeon BG. Phylogenomics reveals subfamilies of fungal nonribosomal peptide synthetases and their evolutionary relationships. BMC Evol Biol. 2010;10:26.
Camarinha-Silva A, Wos-Oxley ML, Jáuregui R, Becker K, Pieper DH. Validating T-RFLP as a sensitive and high-throughput approach to assess bacterial diversity patterns in human anterior nares. FEMS Microbiol Ecol. 2012;79(1):98–108.
Campanile G, Ruscelli A, Luisi N. Antagonistic activity of endophytic fungi towards *Diplodia corticola* assessed by in vitro and in planta tests. Eur J Plant Pathol. 2007;117:237–46.
Carroll GC. The biology of endophytism in plants with particular reference to woody plants. In: Fokkema NJ, van den Heuvel J, editors. Microbiology of the phyllosphere. Cambridge (UK): Cambridge University Press; 1986. p. 205–22.
Challis GL, Ravel J, Townsend CA. Predictive, structure-based model of amino acid recognition by nonribosomal peptide synthetase adenylation domains. Chem Biol. 2000;7(3):211–24.
Chen M, Chang Y, Zhang J, Ding J, Sun Q. A genetic linkage map of Japanese scallop *Mizuhopecten yessoensis* based on amplified fragment length polymorphism (AFLP) and microsatellite (SSR) markers. Afr J Biotechnol. 2012;11(46):10517–26.
Clay K, Schardl C. Evolutionary origins and ecological consequences of endophyte symbiosis with grasses. Am Nat. 2002;160:S99–127.
Correa RCG, Rhoden SA, Mota TR, Azevedo JL, Pamphile JA, de Souza CGM, Polizeli Mde L, Bracht A, Peralta RM. Endophytic fungi: expanding the arsenal of industrial enzyme producers. J Ind Microbiol Biotechnol. 2014;41:1467–78.
Correia T, Grammel N, Ortel I, Keller U, Tudzynski P. Molecular cloning and analysis of the ergopeptine assembly system in the ergot fungus *Claviceps purpurea*. Chem Biol. 2003;10:1281–92.
Cox RJ. Polyketides, proteins and genes in fungi: programmed nano-machines begin to reveal their secrets. Org Biomol Chem. 2007;5:2010–26.
Cox RJ, Simpson TJ. Fungal type I polyketide synthase. Methods Enzymol. 2009;459:49–78.
Cräutlein MV, Korpelainen H, Helander M, Ohberg A, Saikkonen K. Development and characterization of nuclear microsatellite markers in the endophytic fungus *Epichloë festucae* (clavicipitaceae). Appl Plant Sci. 2014;2(12):1400093.
Cuadros-Orellana S, Leite LR, Smith A, Medeiros JD, Badotti F, Fonseca PLC, Vaz ABM, Oliveira G, Goes-Neto A. Assessment of fungal diversity in the environment using metagenomics: a decade in review. Fungal Genom Biol. 2013;3:110. doi:10.4172/2165-8056.1000110.

Danielsen S, Jensen DF. Fungal endophytes from stalks of tropical maize and grasses: isolation, identification, and screening for antagonism against *Fusarium verticillioides* in maize stalks. Biocontrol Sci Techn. 1999;9:545–53.

Davey ML, Currah RS. Interactions between mosses (*Bryophyta*) and fungi. Can J Bot. 2006;84:1509–19.

Davis EC, Shaw AJ. Biogeographic and phylogenetic patterns in diversity of liverwort-associated endophytes. Am J Bot. 2008;95:914–24.

De Bary A. Morphologie und physiologie der pilze, flechten, und myxomyceten. Hofmeister's handbook of physiological botany, vol. II. Leipzig: Engelmann; 1866.

De Jong EZ, Guthridge KM, Spangenberg GC, Forster JW. Development and characterization of EST-derived simple sequence repeat (SSR) markers for pasture grass endophytes. Genome. 2003;46:277–90.

Deckert RJ, Melville LH, Peterson RL. Structural features of a *Lophodermium* endophyte during the cryptic life-cycle phase in the foliage of *Pinus strobus*. Mycol Res. 2001;105:991–7.

Dickie IA, Bolstridge N, Cooper JA, Peltzer DA. Co-invasion by Pinus and its mycorrhizal fungi. New Phytol. 2010;187(2):475–84.

Duong LM, Jeewon R, Lumyong S, Hyde KD. DGGE coupled with ribosomal DNA gene phylogenies reveal uncharacterized fungal phylotypes. Fungal Diver. 2006;23:121–38.

Edwards RA, Rodriguez-Brito B, Wegley L, Haynes M, Breitbart M, et al. Using pyrosequencing to shed light on deep mine microbial ecology. BMC Genomics. 2006;7:57.

Edwards BD, Alexandre MA, Sara MR, Tristram DL, Irvine-Fynn BS, Gareth W. Griffith A distinctive fungal community inhabiting cryoconite holes on glaciers in Svalbard Arwyn. Fungal Eco. 2013;6:168–76.

Favaro LCDL, Sebastianes FLDS, Araujo WL. *Epicoccum nigrum* P16, a sugarcane endophyte, produces antifungal compounds and induces root growth. PLoS One. 2012;7(6):e36826.

Fernandes EG, Pereira OL, Silva CC, Bento CBP, Queiroz MV. Diversity of endophytic fungi in *Glycine max*. Microbiol Res. 2015;181:84–92.

Ferrer M, Beloqui A, Timmis KN, Golyshin PN. Metagenomics for mining new genetic resources of microbial communities. J Mol Microbiol Biotechnol. 2009;16:109–23.

Finking R, Marahiel MA. Biosynthesis of nonribosomal peptides. Annu Rev Microbiol. 2004;58:453–88.

Flewelling AJ, Currie J, Gray CA, Johnson JA. Endophytes from marine macroalgae: promising sources of novel natural products. Curr Sci. 2015;109(88):88–111.

Freeman EM. The seed-fungus of *Lolium temulentum* L., the Darnel. Phil Trans R Soc B. 1904;196:1–27.

Funa N, Awakawa T, Horinouchi S. Pentaketide resorcinol acid synthesis by type III polyketide synthase from *Neurospora crassa*. J Biol Chem. 2007;282:14476–81.

Gaffoor I, Brown DW, Plattner R, Proctor RH, Qi W, Trail F. Functional analysis of the polyketide synthase genes in the filamentous fungus *Gibberella zeae* (anamorph *Fusarium graminearum*). Eukaryot Cell. 2005;4(11):1926–33.

Gamboa MA, Bayman P. Communities of endophytic fungi in leaves of a tropical timber tree (*Guarea guidonia*: Meliaceae). Biotropica. 2001;33:352–60.

Gao XX, Zhou H, Xu DY, Yu CH, Chen YQ, Qu LH. High diversity of endophytic fungi from the pharmaceutical plant, Heterosmilax japonica Kunth revealed by cultivation-independent approach. FEMS Microbiol Lett. 2005;249:255–66. doi:10.1016/j.femsle.2005.06.017.

Geistlinger J, Zwanzig J, Heckendorff S, Schellenberg I. SSR Markers for Trichoderma virens: their evaluation and application to identify and quantify root-endophytic strains. Diversity. 2015;7:360–84. doi:10.3390/d7040360.

Gillevet PM, Sikaroodi M, Torzilli AP. Analyzing salt marsh fungal diversity: comparing ARISA fingerprinting with clone sequencing and pyrosequencing. Fungal Ecol. 2009;2:160–7.

Ginolhac A, Jarrin C, Gillet B, Robe P, Pujic P, Tuphile K, Bertrand H, Vogel TM, Perriere G, Simonet P, Nalin R. Phylogenetic analysis of polyketide synthase I domains from soil metagenomic libraries allows selection of promising clones. Appl Environ Microbiol. 2004;70: 5522–7.

Gloux K, Berteau O, El Oumami H, Beguet F, Leclerc M, Dore J. A metagenomic β-glucuronidase uncovers a core adaptive function of the human intestinal microbiome. Proc Natl Acad Sci U S A. 2011;108:4539–46.

Gottel NR, Castro HF, Kerley M, Yang Z, Pelletier DA, Podar M, Karpinets T, Uberbacher E, Tuskan GA, Vilgalys R, Doktycz MJ, Schadt CW. Distinct microbial communities within the endosphere and rhizosphere of Populus deltoides roots across contrasting soil types. Appl Environ Microbiol. 2011;77(17):5934–44. doi:10.1128/AEM.05255-11.

Götz M, Nirenberg H, Krause S, Wolters H, Draeger S, Buchner A, Lottmann J, Berg G, Smalla K. Fungal endophytes in potato roots studied by traditional isolation and cultivation-independent DNA-based methods. FEMS Microbiol Ecol. 2006;58:404–13.

Grünig CR, Sieber TN, Holdenrieder O. Characterisation of dark septate endophytic fungi (DSE) using inter-simple-sequence-repeat-anchored polymerase chain reaction (ISSR-PCR) amplification. Mycol Res. 2001;105:24–32.

Guo LD. Molecular diversity and identification of endophytic fungi. In: Gherbawy Y, Voigt K, editors. Molecular identification of fungi. Berlin: Springer; 2010. p. 277–96.

Guo LD, Xu L, Zheng WH, Hyde KD. Genetic variation of *Alternaria alternata*, an endophytic fungus isolated from *Pinus tabulaeformis* as determined by random amplified microsatellites (RAMS). Fungal Divers. 2004;16:53–65.

Guo LD, Huang GR, Wang Y. Seasonal and tissue age influences on endophytic fungi of *Pinus tabulaeformis* (Pinaceae) in Dongling mountain, Beijing. J Integr Plant Biol. 2008;50:997–1003.

Haarmann T, Machado C, Lubbe Y, Correia T, Schardl CL, Panaccione DG, Tudzynski P. The ergot alkaloid gene cluster in *Claviceps purpurea*: extension of the cluster sequence and intra species evolution. Phytochemistry. 2005;66:1312–20.

Haese A, Schubert M, Herrman M, Zocher R. Molecular characterization of the enniatin synthetase gene encoding a multifunctional enzyme catalyzing N-methyldepsipeptide formation in *Fusarium scirpi*. Mol Microbiol. 1993;7:905–14.

Handelsman J. Metagenomics: application of genomics to uncultured microorganisms. Microbiol Mol Biol Rev. 2004;68:669–85.

Hashimoto M, Nonaka T, Fujii I. Fungal type III polyketide synthases. Nat Prod Rep. 2014a;31:1306–17.

Hashimoto M, Koen T, Takahashi H, Suda C, Kitamoto K, Fujii I. *Aspergillus oryzae* CsyB catalyzes the condensation of two β-ketoacyl-CoAs to form 3-acetyl-4-hydroxy-6-alkyl-α-pyrone. J Biol Chem. 2014b;289(29):19976–84.

Hawksworth DL. The fungal dimension of biodiversity: magnitude, significance, and conservation. Mycol Res. 1991;95:641–55.

Hazard C, Gosling P, Mitchell DT, Doohan FM, Gary D. Bending Diversity of fungi associated with hair roots of ericaceous plants is affected by land use. FEMS Microbiol Ecol. 2014;87: 586–600.

Hebert PDN, Cywinsk A, Ball SL, deWaard JR. Biological identifications through DNA barcodes. Proc Biol Sci. 2003a;270:313–21.

Hebert PDN, Ratnasingham S, deWaard JR. Barcoding animal life: cytochrome c oxidase subunit 1 divergences among closely related species. Proc Biol Sci. 2003b;270 Suppl 1:S96–9.

Henne A, Schmitz RA, Bomeke M, Gottschalk G, Daniel R. Screening of environmental DNA libraries for the presence of genes conferring lipolytic activity on Escherichia coli. Appl. Environ. Microbiol. 2000;66:3113–16.

Hertweck C. The biosynthetic logic of polyketide diversity. Angew Chem Int Ed Engl. 2009;48: 4688–716.

Hjort K, Bergstrom M, Adesina MF, Jansson JK, Smalla K, Sjoling S. Chitinase genes revealed and compared in bacterial isolates, DNA extracts and a metagenomic library from a phytopathogensuppressive soil. FEMS Microbiol Ecol. 2010;71:197–207.

Hoffmann K, Schneider-Scherzer E, Kleinkauf H, Zocher R. Purification and characterization of eukaryotic alanine racemase acting as key enzyme in cyclosporine biosynthesis. J Biol Chem. 1994;269:12710–4.

Hubert N, Hanner R, Holm E, Mandrak NE, Taylor E, Burridge M, Watkinson D, Dumont P, Curry A, Bentzen P, Zhang J, April J, Bernatchez L. Identifying Canadian freshwater fishes through DNA barcodes. PLoS One. 2008;3:e2490.

Jeewon R, Hyde KD. Diversity and detection of fungi from environmental samples: traditional versus molecular approaches. In: Varma A, Oelmuller R, editors. Advanced techniques in soil microbiology, Soil Biology Series. Berlin: Springer; 2006. p. 1–15.

Johnson R, Voisey C, Johnson L, Pratt J, Fleetwood D, Khan A, Bryan G. Distribution of NRPS gene families within the *Neotyphodium/Epichloe* complex. Fungal Genet Biol. 2007a;44: 1180–90.

Johnson RD, Voisey CR, Pratt JM, Johnson LJ, Bryan G. Identification of NRPS gene families from *Neotyphodium/Epichloë* endophytic fungi that form mutualistic associations with cool season grasses. In: Popay AJ, Thom ER, editors. Proceedings of the 6th international symposium on fungal endophytes of grasses, Grassland research and practice series no. 13. New Zealand: New Zealand Grassland Association; 2007b.

Karimi S, Mirlohi A, Sabzalian MR, Sayed Tabatabaei BE, Sharifnabi B. Molecular evidence for Neotyphodium fungal endophyte variation and specificity within host grass species. Mycologia. 2012;104(6):1281–90.

Keating TA, Marshall CG, Walsh CT. Vibrobactin biosyntheis in *Vibrio cholerae*: VibH in an amide synthase homologous to nonribosomal peptide synthase condensation domains. Biochemistry. 2000;39:15513–21.

Keating TA, Ehmann DE, Kohli RM, Marshall CG, Trauger JW, Walsh CT. Chain termination steps in nonribosomal peptide synthetase assembly lines: directed acyl-S-enzyme breakdown in antibiotic and siderophore biosynthesis. Chembiochem. 2001;2(2):99–107.

Keller NPG, Turner G, Bennet JW. Fungal secondary metabolism – From biochemistry to genomics. Nat Rev Microbiol. 2005;3:937–47.

Kemler M, Garnas J, Wingfield MJ, Gryzenhout M, Pillay K-A, Slippers B. Ion Torrent PGM as tool for fungal community analysis: a case study of endophytes in *Eucalyptus grandis* reveals high taxonomic diversity. PLoS One. 2013;8(12):e81718. doi:10.1371/journal.pone.0081718.

Kerr KCR, Stoeckle MY, Dove CJ, Weigt LA, Francis CM, Hebert PDN. Comprehensive DNA barcode coverage of North American birds. Mol Ecol Notes. 2007;7:535–43.

Knietsch A, Bowien S, Whited G, Gottschalk G, Daniel R. Identification and characterization of coenzyme B_{12}-dependent glycerol dehydratase- and diol dehydratase-encoding genes from metagenomic DNA libraries derived from enrichment cultures. Appl Environ Microbiol. 2003;69:3048–60.

Kristiansen KA, Taylor DL, Kjoller R, Rasmussen HN, Rosendahl S. Identification of mycorrhizal fungi from single pelotons of *Dactylorhiza majalis* (Orchidaceae) using single-strand conformation polymorphism and mitochondrial ribosomal large subunit DNA sequences. Mol Ecol. 2001;10(8):2089–93.

Kroken S, Glass NL, Taylor JW, Yoder OC, Turgeon BG. Phylogenomic analysis of type I polyketide synthase genes in pathogenic and saprobic ascomycetes. Proc Natl Acad Sci. 2003;100:15670–5.

Kusari S, Hertweck C, Spiteller M. Chemical ecology of endophytic fungi: origins of secondary metabolites. Chem Biol. 2012;19:792–8.

Lee BN, Kroken S, Chou DY, Robbertse B, Yoder OC, Turgeon BG. Functional analysis of all nonribosomal peptide synthetases in *Cochliobolus heterostrophus* reveals a factor, NPS6, involved in virulence and resistance to oxidative stress. Eukaryot Cell. 2005;4:545–55.

Li WC, Guo SY, Guo LD. Endophytic fungi associated with lichen *Physcia stellaris* using different surface sterilization methods. J Fungal Res. 2007;5:202–6.

Li J, Luo Y, Lee JK, Zhao H. Cloning and characterization of a type III polyketide synthase from *Aspergillus niger*. Bioorg Med Chem Lett. 2011;21:6085–9.

Liang Y, Guo LD, Ma KP. Population genetic structure of an ectomycorrhizal fungus *Amanita manginiana* in a subtropical forest over 2 years. Mycorrhiza. 2005;15:137–42.

Liesack W, Dunfield PF, Spencer JFT, Ragout de Spencer AL. T-RFLP analysis: a rapid fingerprinting method for studying diversity, structure, and dynamics of microbial communities. In: Environmental microbiology: methods and protocols. Methods in biotechnol. Totowa: Humana Press; 2004. p. 23–37.

Lin X, Huang YJ, Zheng ZH, Su WJ, Qian XM, Shen YM. Endophytes from the pharmaceutical plant Annona squamosa: isolation, bioactivity, identification and diversity of its polyketide synthase gene. Fungal Divers. 2010;41:41–51.

Liu WT, Marsh TL, Cheng H, Forney LJ. Characterization of microbial diversity by determining terminal restriction fragment length polymorphisms of genes encoding 16S rRNA. Appl Environ Microbiol. 1997;63:4516–22.

Liu L, Liu SC, Jiang LH, Chen XL, Guo LD, Che YS. Chloropupukeananin, the first chlorinated pupukeanane derivative and its precursors from *Pestalotiopsis fici*. Org Lett. 2008;10:1397–400.

Liu L, Li Y, Li S, Hu N, He Y, Pong R, Lin D, Lu L, Law M. Comparison of next-generation sequencing systems. J Biomed Biotechnol. 2012;2012:1–11. doi:10.1155/2012/251364.

Liu J, Yu Y, Cai Z, Bartlam M, Wang Y. Comparison of ITS and 18S rDNA for estimating fungal diversity using PCR-DGGE. World J Microbiol Biotechnol. 2015;31(9):1387–95. doi:10.1007/s11274-015-1890-6.

Lucero ME, Unc A, Cooke P, Dowd S, Sun SL. Endophyte microbiome diversity in micropropagated *Atriplex canescens* and *Atriplex torreyi* var. *griffithsii*. PLoS One. 2011;6:e17693.

Lucia M, Martin G. Type III polyketide synthases in lichen mycobionts. Fungal Biol. 2010;114:379–85.

Majer D, Mithen R, Lewis BG, Vos P, Oliver RP. The use of AFLP fingerprinting for the detection of genetic variation in fungi. Mycol Res. 1996;100:1107–11.

Marahiel MA, Stachelhaus T, Mootz HD. Modular peptide synthetases involved in nonribosomal peptide synthesis. Chem Rev. 1997;97(7):2651–73.

Maróti G, Tong Y, Yooseph S, Baden-Tillson H, Smith HO, Kovács KL, Frazier M, Venter JC, Xu Q. Discovery of [NiFe] hydrogenase genes in metagenomic DNA: cloning and heterologous expression in *Thiocapsa roseopersicina*. Appl Environ Microbiol. 2009;75(18):5821–30. doi:10.1128/AEM.00580-09.

Mayorga ME, Timberlake WE. The developmentally regulated *Aspergillus nidulans* WA gene encodes a polypeptide homologous to polyketide and fatty acid synthases. Mol Gen Genet. 1992;235:205–12.

Mbwana J, Bölin I, Lyamuya E, Mhalu F, Lagergard T. Molecular characterization of *Haemophilus ducreyi* isolates from different geographical locations. J Clin Microbiol. 2006;44(1):132–7.

Mende DR, Waller AS, Sunagawa S, Järvelin AI, Chan MM, et al. Assessment of metagenomic assembly using simulated next generation sequencing data. PLoS One. 2012;7(2):e31386. doi:10.1371/journal.pone.0031386.

Metzker ML. Sequencing technologies – The next generation. Nat Rev Genet. 2010;11:31–46.

Miller KI, Qing C, Sze DMY, Neilan BA. Investigation of the biosynthetic potential of endophytes in traditional Chinese anticancer herbs. PLoS One. 2012;7(5):e35953. doi:10.1371/journal.pone.0035953.

Mohali S, Burgess TI, Wingfield MJ. Diversity and host association of the tropical tree endophyte *Lasiodiplodia theobromae* revealed using simple sequence repeat Markers. Forest Pathol. 2005;35:385–96.

Moon CD, Tapper BA, Scott B. Identification of *Epichloë* endophytes in plants by microsatellite-based PCR fingerprinting assay with automated analysis. Appl Environ Microbiol. 1999;65:1268–79.

Mootz HD, Marahiel MA. The tyrocidine biosynthesis operon of *Bacillus brevis*: complete nucleotide sequence and biochemical characterization of functional internal adenylation domains. J Bacteriol. 1997;179:6843–50.

Mootz HD, Schwarzer D, Marahiel MA. Ways of assembling complex natural products on modular nonribosomal peptide synthetases. Chembiochem. 2002;3:490–504.

Mousa WK, Raizada MN. The diversity of anti-microbial secondary metabolites produced by fungal endophytes: an interdisciplinary perspective. Front Microbiol. 2013;4:1–18.

Müller CB, Krauss J. Symbiosis between grasses and asexual fungal endophytes. Curr Opin Plant Biol. 2005;8:450–6.

Nicholson TP, Dawson M, Lazarus CM, Simpson TJ, Cox RJ. Design and utility of oligonucleotide gene probes for fungal polyketide synthases. Chem Biol. 2001;8:157–78.

Nicolotti G, Gonthier P, Guglielmo F. Advances in detection and identification of wood rotting fungi in timber and standing trees. In: Gherbawy Y, Voigt K, editors. Molecular identification of fungi. Berlin/Heidelberg: Springer; 2010. p. 251–76.

Nikolouli K, Mossialos D. Bioactive compounds synthesized by non-ribosomal peptide synthetases and type-I polyketide synthases discovered through genome-mining and metagenomics. Biotechnol Lett. 2012;34:1393–403. doi:10.1007/s10529-012-0919-2.

Oide S, Moeder W, Krasnoff S, Gibson D, Haas H, Yoshioka K, Turgeon BG. NPS6, encoding a nonribosomal peptide synthetase involved in siderophore-mediated iron metabolism, is a conserved virulence determinant of plant pathogenic ascomycetes. Plant Cell. 2006;18(10):2836–53.

Oide S, Krasnoff SB, Gibson DM, Turgeon BG. Intracellular siderophores are essential for ascomycete sexual development in heterothallic *Cochliobolus heterostrophus* and homothallic *Gibberella zeae*. Eukaryot Cell. 2007;6(8):1339–53.

Petrini O. Fungal endophyte of tree leave. In: Andrew JA, Hirano SS, editors. Microbial ecology of leaves. New York: Springer; 1991. p. 179–97.

Petrini O, Stone J, Carroll FE. Endophytic fungi in evergreen shrubs in western Oregon: a preliminary study. Can J Bot. 1982;60:789–96.

Porras-Alfaro A, Bayman P. Hidden fungi, emergent properties: endophytes and microbiomes. Annu Rev Phytopathol. 2011;49:291–315.

Purahong W, Hyde KD. Effects of fungal endophytes on grass and non-grass litter decomposition rates. Fungal Divers. 2011;47:1–7.

Qin J, Li R, Raes J, Arumugam M, Burgdorf KS, et al. A human gut microbial gene catalogue established by metagenomic sequencing. Nature. 2010;464:59–65.

Raberg U, Högberg NOS, Land CJ. Detection and species discrimination using rDNA T-RFLP for identification of wood decay fungi. Holzforsch. 2005;59:696–702.

Raes J, Bork P. Molecular eco-systems biology: towards an understanding of community function. Nat Rev Microbiol. 2008;6:693–9.

Rao HCY, Santosh P, Rakshith D, Satish S. Molecular characterization of an endophytic *Phomopsis liquidambaris* CBR-15 from *Cryptolepis buchanani* Roem. and impact of culture media on biosynthesis of antimicrobial metabolites. 3. Biotech. 2015;5:165–73. doi:10.1007/s13205-014-0204-2.

Rastogi G, Sani RK. Molecular techniques to assess microbial community structure. In: Ahmad I, editor. Microbes and microbial technology. New York: Springer; 2011. p. 29–57.

Redecker D, Kodner R, Graham LE. Glomalean fungi from the Ordovician. Science. 2000;289:1920–1.

Reiber K, Reeves EP, Neville C, Winkler R, Gebhardt P, Kavanagh K, Doyle S. The expression of three nonribosomal peptide synthetases in *Aspergillus fumigatus* is mediated by the availability of free iron. FEMS Microbiol Lett. 2005;248:83–91.

Riesenfeld CS, Goodman RM, Handelsman J. Uncultured soil bacteria are a reservoir of new antibiotic resistance genes. Environ Microbiol. 2004;6(9):981–9.

Rodrigues KF, Sieber TN, Grunig CR, Holdenrieder O. Characterization of Guignardia mangiferae isolated from tropical plants based on morphology, ISSR-PCR amplifications and ITS1–5.8 S-ITS2 sequences. Mycol Res. 2004;108:45–52.

Rodriguez RJ, White JF, Arnold AE, Redman RS. Fungal endophytes: diversity and functional roles. New Phytol. 2009;182(2):314–30.

Roelfsema J, Peters DM. Denaturing Gradient Gel Electrophoresis (DGGE). In: Walker J, Rapley R, editors. Molecular biomethods handbook. Totowa, NJ: Humana Press; 2008. p. 107–15. http://link.springer.com/protocol/10.1385/1-59259-870-6:079.

Rojas JD, Sette LD, de Araujo WL, Lopes MSG, da Silva LF, Furlan RLA, Padilla G. The diversity of polyketide synthase genes from sugercane-derived fungi. Microb Ecol. 2012;63:565–77.

Rondon MR, August PR, Bettermann AD, Brady SF, Grossman TH, Liles MR, Loiacono KA, Lynch BA, MacNeil IA, Minor C, Tiong CL, Gilman M, Osburne MS, Clardy J, Handelsman J, Goodman RM. Cloning the soil metagenome: a strategy for accessing the genetic and functional diversity of uncultured microorganisms. Appl Environ Microbiol. 2000;66:2541–7.

Rosa LH, Vaz ABM, Caligiorne RB, Campolina S, Rosa CA. Endophytic fungi associated with the Antarctic grass *Deschampsia antarctica* Desv. (*Poaceae*). Polar Biol. 2009;32:161–7. doi:10.1007/s00300-008-0515-z.

Rowan DD. Lolitrems, peramine and paxilline: mycotoxins of the ryegrass/endophyte interaction. Agr Ecosyst Environ. 1993;44:103–22.

Saikkonen K, Saari S, Helander M. Defensive mutualism between plants and endophytic fungi? Fungal Divers. 2010;41:101–13.

Samel SA, Marahiel MA, Essen LO. How to tailor non-ribosomal peptide products – New clues about the structures and mechanisms of modifying enzymes. Mol Biosyst. 2008;4(5):387–93.

Schardl CL, Leuchtmann A, Spiering MJ. Symbioses of grasses with seedborne fungal endophytes. Annu Rev Plant Biol. 2004;55:315–40.

Schmitt I, Kautz S, Martín MP, Lumbsch HT. Diversity of non-reducing polyketide synthase genes in the Pertusariales (lichenized Ascomycota): a phylogenetic perspective. Phytochemistry. 2005;66(11):1241–53.

Schmitt I, Kautz S, Lumbsch HT. 6-MSAS-like polyketide synthase genes occur in lichenized ascomycetes. Mycol Res. 2008;112:289–96.

Schoch CL, Seifert KA, Huhndorf S, Robert V, Spouge JL, Levesque CA, Chen W, Fungal Barcoding Consortium. Nuclear ribosomal internal transcribed spacer (ITS) region as a universal DNA barcode marker for Fungi. Proc Natl Acad Sci U S A. 2012;109(16):6241–6.

Schulz B, Rommert AK, Dammann U, Aust HJ, Strack D. The endophyte-host interaction: a balanced antagonism? Mycol Res. 1999;10:1275–83.

Seifert KA, Samson RA, Dewaard JR, Houbraken J, Lévesque CA, et al. Prospects for fungus identification using CO1 DNA barcodes, with *Penicillium* as a test case. Proc Natl Acad Sci U S A. 2007;104:3901–6.

Seshime Y, Juvvadi PR, Fujii I, Kitamoto K. Discovery of novel superfamily of type III polyketide synthases in *Aspergillus oryzae*. Biochem Biophys Res Commun. 2005;331:253–60.

Shibata S, Morishita E, Takeda T, Sakata K. The structure of aurofusarin. Tetrahedron Lett. 1966;7:4855–60.

Shubin L, Juan H, RenChao Z, ShiRu X, YuanXiao J. Fungal endophytes of *alpinia officinarum* rhizomes: insights on diversity and variation across growth years, growth sites, and the inner active chemical concentration. PLoS One. 2014;9(12):e115289. doi:10.1371/journal.pone.0115289.

Siddique AB, Unterseher M. A cost-effective and efficient strategy for Illumina sequencing of fungal communities: A case study of beech endophytes identified elevation as main explanatory factor for diversity and community composition. Fungal Ecol. 2016;20:175–85.

Simon C, Daniel R. Metagenomic analyses: past and future trends. Appl Environ Microbiol. 2011;77(4):1153–61. doi:10.1128/AEM.02345-10.

Simon C, Herath J, Rockstroh S, Daniel R. Rapid identification of genes encoding DNA polymerases by function-based screening of metagenomic libraries derived from glacial ice. Appl Environ Microbiol. 2009;75:2964–8.

Singh BP. Molecular markers used in fungal diversity analysis: an overview. In: Choudhury H, editor. Biology, biotechnology and sustainable development. Delhi: Research India Publication; 2015. p. 167–82.

Slightom JL, Metzger BP, Luu HT, Elhammer AP. Cloning and molecular characterization of the gene encoding the Aureobasidin A biosynthesis complex in *Aureobasidium pullulans* BP-1938. Gene. 2009;431:67–79.

Stach D, Neville C, Doyle S. Nonribosomal peptide synthesis in *Aspergillus fumigatus* and other fungi. Microbiology. 2007;153:1297–306.

Stachelhaus T, Marahiel MA. Modular structure of peptide synthetases revealed by dissection of the multifunctional enzyme GrsA. J Biol Chem. 1995;270:6163–9.

Stachelhaus T, Mootz HD, Bergendahl V, Marahiel MA. Peptide bond formation in nonribosomal peptide biosynthesis catalytic role of the condensation domain. J Biol Chem. 1998;273:22773–81.
Stadler M, Schulz B. High energy biofuel from endophytic fungi? Trends Plant Sci. 2009;14(7):353–5.
Staub J, Bacher J, Poetter K. Sources of potential errors in the application of random amplified polymorphic DNAs in cucumber. Horti Sci. 1996;31:252–66.
Staunton J, Weissman KJ. Polyketide biosynthesis: a millennium review. Nat Prod Rep. 2001;18(4):380–416.
Stephen JR, Kowalchuk GA, Bruns MAV, McCaig AE, Phillips CJ, Embley TM, Prosser JI. Analysis of β-subgroup proteobacterial ammonia oxidizer populations in soil by denaturing gradient gel electrophoresis and hierarchical phylogenetic probing. Appl Environ Microbiol. 1998;64:2958–65.
Stone JK. Fine structure of latent infection by *Rhabdocline parkeri* on Douglas fir, with observation on uninfected epidermal cells. Can J Bot. 1988;66:45–54.
Stone JK, Bacon CW, White JF. An overview of endophytic microbes: endophytism defined. In: Bacon CW, White Jr JF, editors. Microbial endophytes. New York: Marcel Dekker Inc; 2000. p. 3–5.
Strobel GA, Knighton B, Kluck K, Ren Y, Livinghouse T, Griffin M, Spakowicz D, Sears J. The production of myco-diesel hydrocarbons and their derivatives by the endophytic fungus *Gliocladium roseum* (NRRL 50072). Microbiology. 2008;154:3319–28. doi:10.1099/mic.0.2008/022186-0.
Sun X, Guo LD. Endophytic fungal diversity: review of traditional and molecular techniques. Mycology. 2012;3(1):65–76.
Sun X, Guo LD, Hyde KD. Community composition of endophytic fungi in Acer truncatum and their role in decomposition. Fungal Divers. 2011;47:85–95.
Sunnucks P, Wilson AC, Beheregaray LB, Zenger K, French J, Taylor AC. SSCP is not so difficult: the application and utility of single-stranded conformation polymorphism in evolutionary biology and molecular ecology. Mol Ecol. 2000;9:1699–710.
Suryanarayanan TS. Diversity of fungal endophytes in tropical trees. In: Pirttilä AM, Frank AC, editors. Endophytes of forest trees: biology and applications. Forestry sciences. Vol. 80. Springer, Netherlands; 2011. p. 67–80.
Suryanarayanan TS, Thirunavukkarasu N, Hariharan GN, Balaji P. Occurrence of non-obligate microfungi inside lichen thalli. Sydowia. 2005;57:119–29.
Swatzell LJ, Powell MJ, Kiss JZ. The relationship of endophytic fungi to the gametophyte of the fern *Schizaea pusilla*. Int J Plant Sci. 1996;157:53–62.
Tanaka A, Tapper BA, Popay A, Parker EJ, Scott B. A symbiosis expressed non-ribosomal peptide synthetase from a mutualistic fungal endophyte of perennial ryegrass confers protection to the symbiotum from insect herbivory. Mol Microbiol. 2005;57(4):1036–50. doi:10.1111/j.1365-2958.2005.04747.x.
Taylor JE, Hyde KD, Jones EBG. Endophytic fungi associated with the temperate palm, trachycarpus fortunei, within and outside its natural geographic range. New Phytol. 1999;142:335–46.
Tejesvi MV, Kini KR, Prakash HS, Subbiah V, Shetty HS. Genetic diversity and antifungal activity of species of *Pestalotiopsis* isolated as endophytes from medicinal plants. Fungal Divers. 2007;24:37–54.
Tejesvi MV, Tamhankar SA, Kini KR, Rao VS, Prakash HS. Phylogenetic analysis of endophytic *Pestalotiopsis* species from ethnopharmaceutically important medicinal trees. Fungal Divers. 2009;38:167–83.
Ting ASY, Meon S, Kadir J, Radu S, Singh G. Endophytic microorganisms as potential growth promoters of banana. BioControl. 2008;53:541–53.
Tiwari K, Chittora M. Assessment of genetic diversity and distribution of endophytic fungal communities of *Alternaria solani* isolates associated with the dominant Karanja plants in Sanganer Region of Rajasthan. Springerplus. 2013;2:313.

Tredway LP, White Jr JF, Gaut BS, Reddy PV, Richardson MD, Clarke BB. Phylogenetic relationships within and between epichloe and neotyphodium endophytes as estimated by AFLP markers and rDNA sequences. Mycol Res. 1999;103(12):1593–603.

Turnbaugh PJ, Hamady M, Yatsunenko T, Cantarel BL, Duncan A, Ley RE, Sogin ML, Jones WJ, Roe BA, Affourtit JP, Egholm M, Henrissat B, Heath AC, Knight R, Gordon I. A core gut microbiome in obese and lean twins. Nature. 2009;457:480–4.

U'ren JM, Riddle JM, Monacell JT, Carbone I, Miadlikowska J, Arnold AE. Tissue storage and primer selection influence pyrosequencing-based inferences of diversity and community composition of endolichenic and endophytic fungi. Mol Ecol Resour. 2014;14:1032–48. doi:10.1111/1755-0998.12252.

Uchiyama T, Miyazaki K. Product-induced gene expression (PIGEX): a product-responsive reporter assay for enzyme screening of metagenomic libraries. Appl Environ Microbiol. 2010;76(21):7029–35. doi:10.1128/AEM.00464-10.

Uchiyama T, Abe T, Ikemura T, Watanabe K. Substrate induced gene expression screening of environmental metagenome libraries for isolation of catabolic genes. Nat Biotechnol. 2005;23(1):88–93.

Unterseher M, Jumpponen A, Opik M, Tedersoo L, Moora M, Dorman CF, Schnittler M. Species abundance distributions and richness estimations in fungal metagenomics- lessons learned from community ecology. Mol Ecol. 2011;20:275–85.

Van de Zande L, Bijlsma R. Limitations of the RAPD technique in phylogeny reconstruction in *Drosophila*. J Evol Biol. 1995;8:645–56.

van der Nest MA, Steenkamp ET, Wingfield BD, Wingfield MJ. Development of simple sequence repeat (SSR) markers in Eucalyptus from amplified inter-simple sequence repeats (ISSR). Plant Breed. 2000;119:433–6.

Varaljay VA, Howard EC, Sun S, Moran MA. Deep sequencing of a dimethylsulfoniopropionate degrading gene (*dmdA*) by using PCR primer pairs designed on the basis of marine metagenomic data. Appl Environ Microbiol. 2010;76:609–17.

Varga J, Kocsube S, Toth B, Mesterhazy A. Nonribosomal peptide synthetase genes in the genome of *Fusarium graminearum* causative agent of wheat head blight. Acta Biol Hung. 2005;56: 375–88.

Vidal S, Jaber LR. Entomopathogenic fungi as endophytes: plant–endophyte–herbivore interactions and prospects for use in biological control. Curr Sci. 2015;109(46):46–54.

Vieira PDDS, Souza-Motta CMD, Lima D, Torres JB, Quecine MC, Azevedo JL, de Oliveira NT. Endophytic fungi associated with transgenic and non-transgenic cotton. Mycology. 2011;2: 91–7.

Vos P, Hogers R, Bleeker M, Reijans M, van de Lee T, Homes M, Frijters A, Pot J, Peleman J, Kuiper M, Zabeau M. AFLP: a new technique for DNA fingerprinting. Nucleic Acids Res. 1995;23(21):4407–14.

Walsh TJ, Francesconi A, Kasai M, Chanock SJ. PCR and single-strand conformational polymorphism for recognition of medically important opportunistic fungi. J Clin Microbiol. 1995;33(12):3216–20.

Walzel B, Riederer B, Keller U. Mechanism of alkaloid cyclopeptide synthesis in the ergot fungus *Claviceps purpurea*. Chem Biol. 1997;4(3):223–30.

Wang Y, Gao BL, Li XX, Zhang ZB, Yan RM, Yang HL, Zhu D. Phylogenetic diversity of culturable endophytic fungi in Dongxiang wild rice (*Oryza rufipogon* Griff), detection of polyketide synthase gene and their antagonistic activity analysis. Fungal Biol. 2015;119(11):1032–45.

Watanabe A, Ono Y, Fujii I, Sankawa U, Mayorga ME, Timberlake WE, Ebizuka Y. Product identification of polyketide synthase coded by *Aspergillus nidulans wA* gene. Tetrahedron Lett. 1998;39:7733–6.

Weber G, Schoergendorfer K, Schneider-Scherzer E, Leitner E. The peptide synthetase catalyzing cyclosporine production in *Tolypocladium niveum* is encoded by a giant 45.8-kilobase open reading frame. Curr Genet. 1994;26:120–5.

Williams JGK, Kubelik AR, Livak KJ, Rafalski JA, Tingey SV. DNA polymorphisms amplified by arbitrary primers are useful as genetic markers. Nucleic Acids Res. 1990;18(22):6531.

Xu J, Ebada SS, Proksch P. *Pestalotiopsis* a highly creative genus: chemistry and bioactivity of secondary metabolites. Fungal Divers. 2010;44:15–31.

Yang B, Wang XM, Ma HY, Yang T, Jia Y, Zhou J, Dai CC. Fungal endophyte *Phomopsis liquidambari* affects nitrogen transformation processes and related microorganisms in the rice rhizosphere. Front Microbiol. 2015;6:982. doi:10.3389/fmicb.2015.00982.

Yin WB, Grundmann A, Cheng J, Li SM. Acetylaszonalenin biosynthesis in *Neosartorya fischeri* Identification of the biosynthetic gene cluster by genomic mining and functional proof of the genes by biochemical investigation. J Biol Chem. 2009;284(1):100–9.

Yuan WM, Gentil GD, Budde AD, Leong SA. Characterization of the *Ustilago maydis* sid2gene, encoding a multidomain peptide synthetase in the ferrichrome biosynthetic gene cluster. J Bacteriol. 2001;183:4040–51.

Yue C, Miller CJ, White JFJ, Richardson M. Isolation and characterization of fungal inhibitors from *Epichloë festucae*. J Agric Food Chem. 2000;48:4687–92.

Zachow C, Berg C, Muller H, Meincke R, Komon-Zelazowska M, Druzhinina IS, Kubicek CP, Berg G. Fungal diversity in the rhizosphere of endemic plant species of Tenerife (Canary Islands): relationship to vegetation zones and environmental factors. ISME J. 2009;3:79–92.

Zhang T, Yao YF. Endophytic fungal communities associated with vascular plants in the high arctic zone are highly diverse and host plant specific. PLoS One. 2015;10(6):e0130051. doi:10.1371/journal.pone.0130051.

Zhou K, Zhang X, Zhang F, Li Z. Phylogenetically diverse cultivable fungal community and polyketide synthase (PKS), non-ribosomal peptide synthase (NRPS) genes associated with the south China sea sponges. Microb Ecol. 2011;62:644–54. doi:10.1007/s00248-011-9859-y.

Zhou WN, White Jr JF, Soares M, Torres MS, Zhou ZP, Li HY. Diversity of fungi associated with plants growing in geothermal ecosystems and evaluation of their capacities to enhance thermotolerance of host plants. J Plant Growth Regul. 2015;10:305–14. doi:10.1080/17429145.2015.1101495.

Zimmerman NB, Vitousek PM. Fungal endophyte communities reflect environmental structuring across a Hawaiian landscape. Proc Natl Acad Sic. 2012;109(32):13022–7.

Zinger L, Gury J, Alibeu O, et al. CE-SSCP and CE-FLA, simple and high-throughput alternatives for fungal diversity studies. J Microbiol Methods. 2008;72:142–53.

Chapter 2
Molecular Markers and Their Use in Taxonomic Characterization of Trichoderma spp

Vivek Sharma and Richa Salwal

2.1 Introduction

Trichoderma was introduced more than 200 years ago in 1794 by Persoon and its teleomorphs (Persoon 1794). *Hypocrea* was described in 1825 (Fries 1825). The members of *Trichoderma* are cosmopolitan, distributed in diverse habitats from dead/decaying organic matter (Samuels 1996; Klein and Eveleigh 1998) to endophytes (Mulaw et al. 2013). The species of *Trichoderma* also are important to human, both as a producer of commercial enzymes (Seidl 2006), biocontrol agents (Sharma et al. 2012; Vinale et al. 2008; Sharma and Shanmugam 2012; Shanmugam and Sharma 2008; Sharma et al. 2013; Sharma and Salwan 2015) and as plant growth promoters (Kaewchai 2009; Samuels and Hebbar 2015). In general, species of *Trichoderma* are characterized by rapid growth, conidia that are generally smooth, rarely ornamented, typically ellipsoidal to oblong, rarely globose, with green or hyaline or rarely yellow in color on repetitively branched ans complex conidiophore's structure. The chlamydospores of *Trichoderma* are typically globose to subglobose and are formed within or at the tips of hyphae (Samuels 1996). Ascospore pigmentation (Tode 1791), fruiting anatomy and other morphological features have been used for taxonomic identification of anamorphs by Dingley (1955).

Molecular markers have great potential in classification of microbes at species level, particularly when morphological distinctions are rare and difficult to observe. Also in genera where morphological features are or may be influenced by the environment, species classification is difficult to assess. Molecular markers are a powerful tool and will play a critical role in fungal taxonomy.

V. Sharma (✉)
DST-Young Scientist, Department of Plant Pathology, CSK-HPKV, Palampur, Himachal Pradesh 176062, India
e-mail: ankvivek@gmail.com

R. Salwal
Department of Veterinary Microbiology, CSK-HPKV, Palampur, Himachal Pradesh 176062, India

B.P. Singh, V.K. Gupta (eds.), *Molecular Markers in Mycology*, Fungal Biology,
DOI 10.1007/978-3-319-34106-4_2

2.2 Taxonomy of Trichoderma

In present, the genus *Trichoderma/Hypocrea* approximately consists of 171 species growing in diverse habitat where its asexual stage includes *Acremonium, Gliocladium, Trichoderma* and *Verticillium* (http://www.gbif.org/species). In the beginning, only four species were included in the genus and consequently *Trichoderma viride* was only proved as *Trichoderma*. In later studies, *T. viride* was revealed as the asexual stage of its teleomorph *Hypocrea rufa* (Tulasne and Tulasne 1860). In 1969, *Trichoderma* strains were categories into nine 'species aggregates' (Rifai 1969). Bissett in 1991, categorized, *Trichoderma* into five sections: Longibrahiatum, Saturnisporum, Pachybasium, Hypocreanum, Trichoderma and Lone lineage (ISTH.info).

2.2.1 Section Longibrahiatum

The members of the section Longibrahiatum are characterized by warted conidia. This section was introduced by Doi in 1987 (Doi et al. 1987) and consequently revaluated by several workers initially included *Trichoderma saturnisporum* and *Trichoderma ghanense*. Based on ITS1 and ITS2 sequence analysis and phenetic characters, teleomorphs for this section were also included (Kuhls et al. 1996, 1997; Samuels 1996; Samuels et al. 1998; Turner et al. 1997). According to international subcommission of *Trichoderma/Hypocrea* taxonomy presently this section includes clade-14 and contained members; *Hypocrea cerebriformis, Hypocrea jecorina/Trichoderma reesei, Hypocrea novaezelandiae, Hypocrea orientalis, Hypocrea patella, Hypocrea poronoidea, Hypocrea pseudokoningii/Trichoderma pseudokoningii, Hypocrea schweinitzii/Trichoderma citrinoviride, Trichoderma effusum, Trichoderma konilangbra, Trichoderma longibrachiatum, Trichoderma sinensis*. The section, Longibrahiatum section was recognized as monophyletic and teleomorphs for four pairs; *H. schweinitzii/T. citrinoviride; H. pseudokoningii/T. pseudokoningii; H. jecorina/T. reesei and H. orientalis/T. longibrachiatum* were also identified. In later studies, the members of the section Saturnisporum, containing, *T. saturnisporum* and *T. ghanense* (Doi et al. 1987), were merged with Longibrahiatum section, and synonymy of *T. ghanense* with *T. parceramosu* was also revealed. Based on molecular data, Saturnisporium section of *Trichoderma* containing tuberculate and colorless ascospores was categorized within the *H. schweinitzii* complex (Kuhls et al. 1997; Samuels et al. 1998).

2.2.2 Section Pachybasium

The polyphyletic section Pachybasium identified by Bisset in 1991 is composed of variable shape conidia (Kindermann et al. 1998; Lieckfeldt and Seifert 2000; Kullnig-Gradinger et al. 2002) and contains anamorphs for majority of the described

Hypocrea/Trichoderma species. Species in this section are known to produce compact pustules with generally pyramidal branching, with or without fertile or sterile conidiophores. The phialides are typically short, wide and generally produced in crowded clusters. The section Pachybasium is further divided into two distinct phylogenetic sub groups (Kindermann et al. 1998; Kullnig-Gradinger et al. 2002; Chaverri et al. 2003). The first subgroup includes type species of sect. Pachybasium i.e. *T. hamatum, T. atroviride, T. koningii, T. minutisporum, T. piluliferum, T. polysporum, T. pubescens, T. strigosum* and *T. viride* while second subgroup consists of *T. crassum, T. fertile, T. flavofuscum, T. harzianum, T. longipile*, etc. Bissett (1991) expanded the concept of "Pachybasium" to include species with white conidiophores similar to *T. hamatum*. Seifert (1985) and other authors (Rehner and Samuels 1994) found that *Hypocrea* species with Gliocladium-like anamorphs are phylogenetically distinct from the Gliocladium type. The morphological features of gliocladium-like anamorphs with Gliocladium revealed that formers are distinct and hence could be excluded from *Gliocladium* with few exceptions. From genomic analysis it was also observed that *T. virens*, also known as *Gliocladium virens* is an anamorph of *H. virens* and is closely related to *T. harzianum* (Chaverri et al. 2003).

This section is differentiated into six clades where teleomorphs relation for several species has been revealed. The harizanum clade contains *Hypocrea atrogelatinosa, Hypocrea catoptron/Trichoderma catoptron, Hypocrea cinnamomea/Trichoderma cinnamomeum, Hypocrea lixii/Trichoderma harzianum, Hypocrea straminea/Trichoderma stramineum, Hypocrea tawa, Hypocrea thailandica/Trichoderma thailandicum, Trichoderma aggressivum, Trichoderma cerinum, Trichoderma pleuroticola, Trichoderma pleurotum, Trichoderma tomentosum, Trichoderma velutinum* and clade virens includes *Hypocrea crassa/Trichoderma crassum, Hypocrea virens/Trichoderma virens.* The third clade constituting strictipiis contains *Hypocrea aureoviridis var. macrospora, Hypocrea cuneispora, Hypocrea strictipilosa/Trichoderma strictipilis, Trichoderma longipile*. The clade stromaticum includes *Trichoderma rossicum, and Trichoderma stromaticum* where as clade -pachybasioides include *Hypocrea lacuwombatensis, Hypocrea minutispora/Trichoderma minutisporum, Hypocrea pachybasioides/Trichoderma polysporum, Hypocrea parapilulifera, Hypocrea pilulifera/Trichoderma piluliferum, Hypocrea stellata* respectively.

2.2.3 Section Hypocreanum

This section with polyphyletic distribution and verticillium or acremonium-like anamorphs was considered to be the anamorph of *H. lacteal* by Bissett (1991). The species closely related to *H. lactea*, such as *H. citrina H. pulvinata* and *H. sulphurea* were assigned anamorphs with T. sect. Hypocreanum like morphology. Further this section is differentiated into five clades- Ceramica, Chlorospora, Psychrophyla, Megalocitrina and Citrina which includes *Hypocrea ceramica/Trichoderma ceramicum, Hypocrea estonica/Trichoderma estonicum, Hypocrea aureoviridis/Trichoderma aureoviride, Hypocrea candida/Trichoderma candidum, Hypocrea chlorospora/Trichoderma chlorosporum, Hypocrea costaricensis, Hypocrea cremea/Trichoderma cremeum, Hypocrea sinuosa/Trichoderma sinuosum, Hypocrea surrotunda/Trichoderma surrotundum, Hypocrea thelephoricola/Trichoderma thelephoricolum, Hypocrea crystalligena, Hypocrea psychrophila, Hypocrea pulvinata, Hypocrea sulfurea Hypocrea megalosulphurea, Hypocrea citrinaHypocrea virescentiflava.*

2.2.4 Section Trichoderma

The section of *Trichoderma* is monophyletic and differentiated into two clades- viridae and Pachybasium or Hamatum but has not been monographed so far (Chaverri and Samuels 2003). First section includes *Hypocrea atroviridis/Trichoderma atroviride, Hypocrea koningii/Trichoderma koningii, Hypocrea muroiana, Hypocrea rufa/ Trichoderma viride, Hypocrea stilbohypoxyli, Hypocrea viridescens/Trichoderma viridescens, Trichoderma austrokoningii, Trichoderma erinaceum, Trichoderma intricatum, Trichoderma koningiopsis, Trichoderma ovalisporum, Trichoderma petersenii, Trichoderma rogersonii, Trichoderma strigosum, Trichoderma taiwanense* while section hamatum contained, *Hypocrea neorufa, Hypocrea pezizoides, Trichoderma asperellum, Trichoderma hamatum, Trichoderma pubescens* and *Trichoderma theobromicola.*

2.2.5 Lone Lineages in Trichoderma

The section further differentiated into two sections. The section Lutea includes, *Hypocrea lutea, Hypocrea melanomagnum* and *Trichoderma brevicompactum* whereas Lone lineages includes *Hypocrea chromosperma/Trichoderma chromospermum, Hypocrea gelatinosa/Trichoderma gelatinosus, Hypocrea nigrovirens, Hypocrea phyllostachydis/Trichoderma phyllostahydis, Hypocrea sulawesensis, Hypocrea voglmayrii, Trichoderma helicum* and *Trichoderma spirale.*

Keeping in view the diversity of *Trichoderma/Hypocrea* strains/species, there is a need of concordant and monophyletic groups based classification system for *Trichoderma* taxonomy. The extensive sampling, rigorous analytical approaches and use of multilocus based markers can play a major role in resolving nodes and distinct identification of *Trichoderma* species/strains.

2.3 Molecular Markers

The introduction of DNA sequencing opened a new era for fungal systematics by providing a set of independently derived data that could be analyzed in tandem with classically derived data. The role of molecular biology in resolving fungal taxonomy started in 1990s, with ribosomal RNA genes amplification through PCR

(White et al. 1990). The use of multi-gene in taxonomic classification of fungi has increased over the years (Lutzoni et al. 2001). So far most commonly used molecular markers in fungal phylogenetics are either non coding nuclear and mitochondrial ribosomal rDNA (18S, 28S, ITS, IGS, mtSSU, mtLSU), or introns of several protein-encoding genes nuclear protein-coding genes, such as RNA polymerases (*RPB1*and *RPB2)*, β-tubulin, γ-actin, ATP synthase (*ATP6*), and elongation factor EF-1α (*TEF1*α).

2.3.1 Non Coding Markers

The non-coding internal transcribed spacer (ITS), 18S and 28S region of genomic DNA are well explored for taxonomic identification of fungi. The rDNAs genes of fungi, similar to eukaryotes is composed of 18S, 5.8S, 28S rRNA transcribed by RNA polymerase I as a 35S to 40S precursor, with both internal and external transcribed spacers (ITS and ETS). The large part (LSU) contains 28S, 5.8S & 5S rRNAs whereas smaller subunit (SSU) contains the 18S rRNA (SSU rRNA). The genomic organization of rRNA gene consists of main rRNA transcription unit (rDNA) and the 5S rRNA genes (5S rDNA). These rRNA genes may be located alone in the genome, in tandem repeats, either with each other or to other genes and are transcribed as a single entity by RNA polymerase I. The processing of cistron during posttranscription removes two internal transcribed spacers, including the 5.8S genes, are usually known as ITS region. The 18S small subunit rRNA gene (SSU) is also commonly used in phylogenetics whereas 28S large subunit rRNA gene (LSU) either alone or in combination with ITS is capable of discriminating species. The D1/D2 region of LSU is adopted for characterizing species long before the concept of DNA barcoding was promoted (Schoch et al. 2012).

The spacers are spliced out of the transcript and between each cluster is a non-transcribed or intergenic spacer (NTS or IGS) which separate the repeats from each other. The total length of one DNA repeat is between 7.7 and 24 kb. The splitting of cistron through posttranscriptional processes and removal of two internal transcribed spacers ie 5.8S gene, also known as ITS region and other fragment of 18S nuclear ribosomal corresponding to small subunit rRNA gene (SSU) are routinely used in phylogenetics where as larger subunit corresponding to 28S rRNA gene (LSU) is used for the differentiation of species either alone or in combination with ITS. The popularity of rDNA as a molecular marker lies in is its multiple copy, non-protein nature, and treatment as a single-locus gene. The 18S and 28S markers are recommended for taxonomic studies at family and generic level where as internal transcribed spacers (ITS) is useful at the species level resolution (Bridges et al. 2005).

The first single gene based phylogenetic analysis for the genera of the Hypocreales was done using ribosomal DNA sequence data (Spatafora and Blackwell 1993; Rehner and Samuels 1994, 1995). So far the use of ITS in phylogenetic studies of species in *Trichoderma* has been used by several workers eg the molecular systematics of the Hypocreales was reassessed (Lieckfeldt et al. 1998;

Lieckfeldt and Seifert 2000) and similarly, ITS markers have been used for *Trichoderma* sect. Longibrahiatum (Kuhls et al. 1997; Samuels et al. 1998). Kindermann et al. (1998) analyzed the phylogeny of species in T. sect Pachybasium while Dodd et al. (2000) also used ITS based markers for studying the relationship of species of *Hypocrea/Trichoderma*. Lieckfeldt et al. (2001) also studied the taxonomoic position of *H. aureoviridis*/T. *aureoviride* using DNA and morphological features.

2.3.2 Protein Coding Genes

The molecular markers encoding single or low-copy nuclear protein are used in taxonomic characterization of wide range of fungi including the phyla Chytridiomycota, Zygomycota, Ascomycota and Basidiomycota (Helgason et al. 2003; Froslev et al. 2005; James et al. 2006; Liu et al. 1999; O'Donnell et al. 2001; Thon and Royse 1999; Voigt and Wostemeyer 2001; Matheny 2005). A number of protein coding genes like endochitinase 42 (*ech42*) and RNA-polymerase subunit 2 (*rpb2*) with significant intra- and inter- specific variability from *Trichoderma* have been used for taxonomic studies with limited success (Kullnig-Gradinger et al. 2002). Fragments of the calmodulin (*cal1*) and the actin-encoding genes (*act1*) had also been used in *T. harzianum/H. lixii*. Attempts on the use of histon 3A and β-tubulin gene sequences have proven worthwhile for phylogenetic analysis in *Fusarium* and other genera but comparatively less variation in the histon 3A gene and presence of multiple heterologous copies of the tub1 gene, has led to conclusion that both genes are not applicable for taxonomy of *Trichoderma*. The largest intron of tef1 has been proven good for taxonomic resolution of closely related taxa such as the *H. lixii, T. harzianum, T. aggressivum, T. tomentosum, T. cerinum, T. velutinum, H. tawa* of *T. harzianum/H. lixii* species clade or *T. viride, T. atroviride, T. koningii* of *H. rufa* clade. However, the large introns in tef1 are less suited for resolving the phylogenetic elations of more distantly related species due to ambiguous alignments.

2.4 The Genealogical Concordance Phylogenetic Species Recognition (GCPSR)

Taxonomic studies based on GCPSR, requires the analysis of phylogenetic trees using several unlinked genes, and implies that the phylogenetic position of a true species will be concordant. Kullnig-Gradinger et al. (2002) used four different loci (ITS1 and ITS2, mitssuDNA, short fifth tef1α intron, a fragment of ech42 large exon) to assess a global phylogeny of the genus. However, a stringent clade to clade concordance was not possible for most of the species because of insufficient phylogenetic resolution by the genes used or absence of reference database for each gene. Role of ITS1 and ITS2, and other DNA fragments in *H. lixii/T. harzianum*, has been used for seven phylogenetic lineages resolution which was concordant but is not easily fulfilled across all species of *Trichoderma*.

2.5 Discussion

The molecular markers have played vital roles in resolving the taxonomic positions and phylogenetic status of *Trichoderma* and its teleomorphs and. The members of the *Trichoderma/Hypocrea* are primarily classified using morphological species recognition (MSR) concept and the species delimitations vary from one fungal group to another are often influenced by scarcity of sampling and lack of detailed biological knowledge. The comparative studies based on phenotypic and DNA data may help in delineated the status of asexual/sexual life cycle of *Trichoderma/Hypocrea*. The detailed analysis of various loci suggests (Spatafora et al. 2006) that the simultaneous usage of (i) tef1 large intron and last large exon (Figs. 2.1 and 2.2), rpb2 II subunt (Figs. 2.3 and 2.4), ech42 last large exon and ITS1 region (Figs. 2.5 and 2.6) as diagnostic regions may lead to the most reliable phylogeny. In studies, based on multilocus independent loci including ITS1 (Figs. 2.5 and 2.6) and, the fifth intron of translation elongation factor 1-alpha (*tef1*) (Figs. 2.1 and 2.2), a partial exon of endochitinase 42 (*ech42*), and the small subunit of the mitochondrial rRNA-encoding gene sequence, the phylogeny for a number of *Trichoderma* species has been described (Taylor et al. 2000) and strikingly, the protein-coding genes commonly used in fungal systematics, such as *rpb* subunit II (Figs. 2.3 and 2.4) *tef1*α, β-tubulin, and γ-actin are not found among the best performing genes.

Further single-copy protein-coding genes such as *rpb* subunit II are promising for yielding well resolved and highly supported phylogenies (Liu and Hall 2004; Reeb et al. 2004; Crespo et al. 2007; Lumbsch et al. 2004; Hibbett et al. 2007) compared to *tef*. Other protein-coding genes, such as the tubulins which are present in the genome in multiple copies have the potential of being phylogenetically misleading (Landvik et al. 2001). In general, slow evolving loci are more suitable for reconstruction of deep phylogenetic relationships, while loci with high rates of evolution are better suited for the reconstruction of more recent evolutionary events. Ribosomal loci with high and heterogeneous rates of change, such as ITS, IGS and mtSSU rDNA can be used to distinguish taxa at the genus and species level. However, the non-coding regions of these loci are prone to significant length variation across species, making alignment of distantly related taxa problematic. Fast evolving ribosomal genes are therefore less useful in large scale concatenated analyses involving higher-level phylogenetic relationships. Molecular systematics are constantly searching for loci that are conserved enough to produce reliable alignments, and at the same time have sufficient variability to yield well resolved and well supported phylogenies.

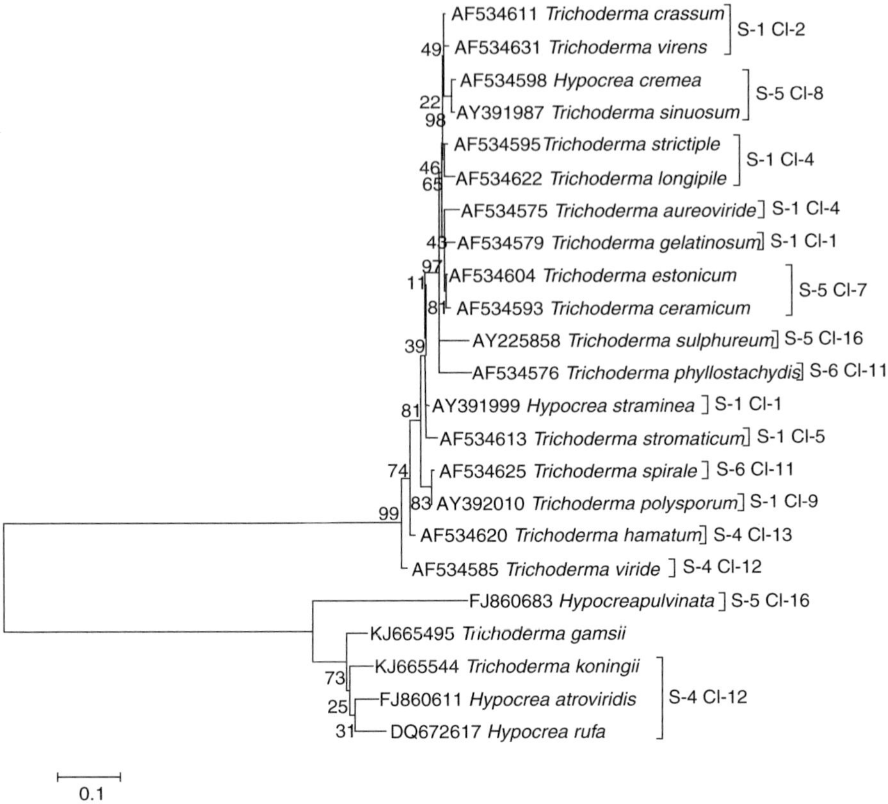

Fig. 2.1 Evolutionary relationships based on translation elongation factor (*tef*) of Trichoderma/*Hyprocrea* isolates inferred using the Neighbor-Joining method. The tree is drawn to scale, with branch lengths in the same units as those of the evolutionary distances used to infer the phylogenetic tree

While DNA-barcode will enable many researchers to reliably identify their *Trichoderma* strains, the limited accessibility or financial capability to use DNA-based methods, definitely demands alternatives. The ITS regions have been used in past by several workers (Kuhls et al. 1997; Kindermann et al. 1998; Lieckfeldt et al. 1998, 2001; Dodd et al. 2000), because of its presence in >90 copies per genome and easy amplification and better resolution (Figs. 2.5 and 2.6). Overall, ribosomal markers had fewer problems with PCR amplification than protein-coding markers. Based on overall performance in species discrimination, ITS are generally considered superior to LSU in species discrimination and had a more clearly defined barcode gap compared to SSU and LSU. Further overall, bootstrap based (Felsenstein 1985) phylogenetic studies (Rzhetsky and Nei 1992; Tamura et al. 2004, 2013) conducted using maximum parsimony and methods using Subtree-Pruning-Regrafting (SPR) algorithm are observed to better for taxonomic differentiations (Nei and Kumar 2000) than neighbor joining methods

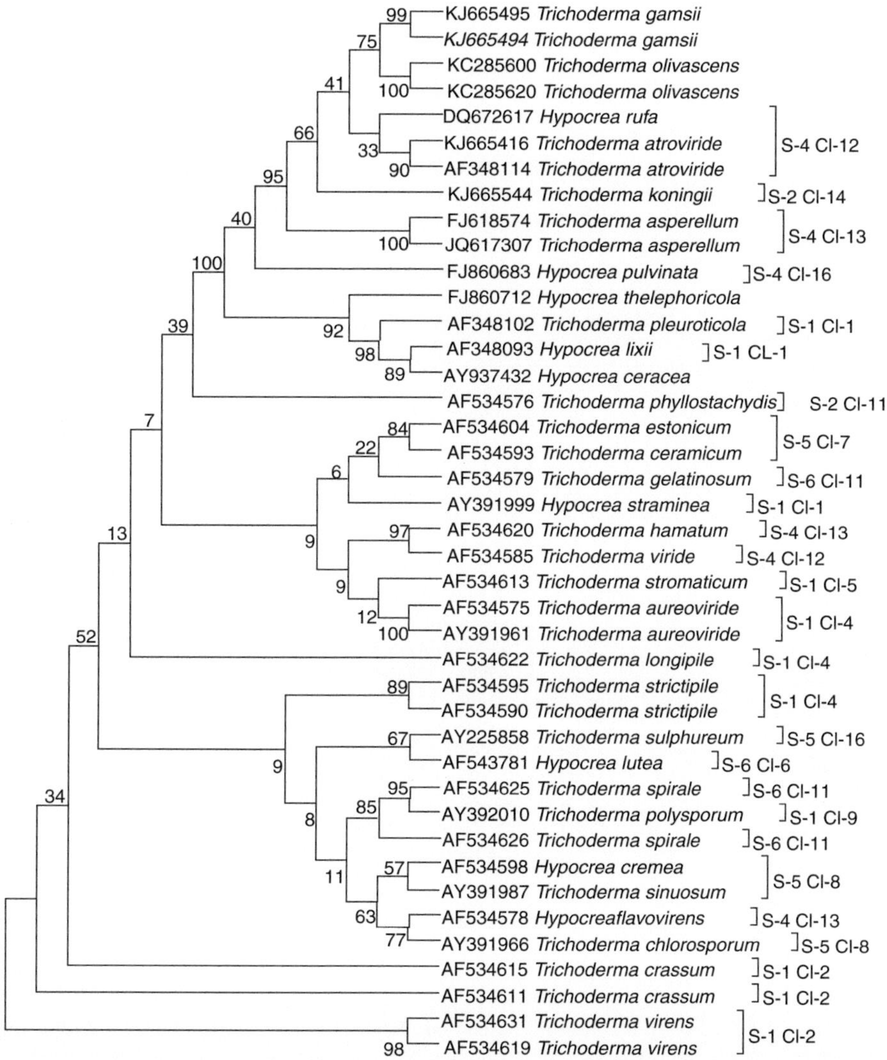

Fig. 2.2 The evolutionary relationships based on translation elongation factor (*tef*) of Trichoderma/*Hyprocrea* inferred using Maximum Parsimony method. The analysis involved 41 nucleotide sequences. All positions containing gaps and missing data were eliminated. Evolutionary analyses were conducted in MEGA 6.0

(Saitou and Nei 1987) (Figs. 2.1, 2.2, 2.3, 2.4, 2.5, and 2.6). The overall success reported for correct species identification using ITS is comparable with the success reported for the two-marker plant barcode system (0.73 vs. 0.70) (Schoch et al. 2012). ITS combines the highest resolving power for discriminating closely related species with a high PCR and sequencing success rate across a broad range of Fungi (Schoch et al. 2012).

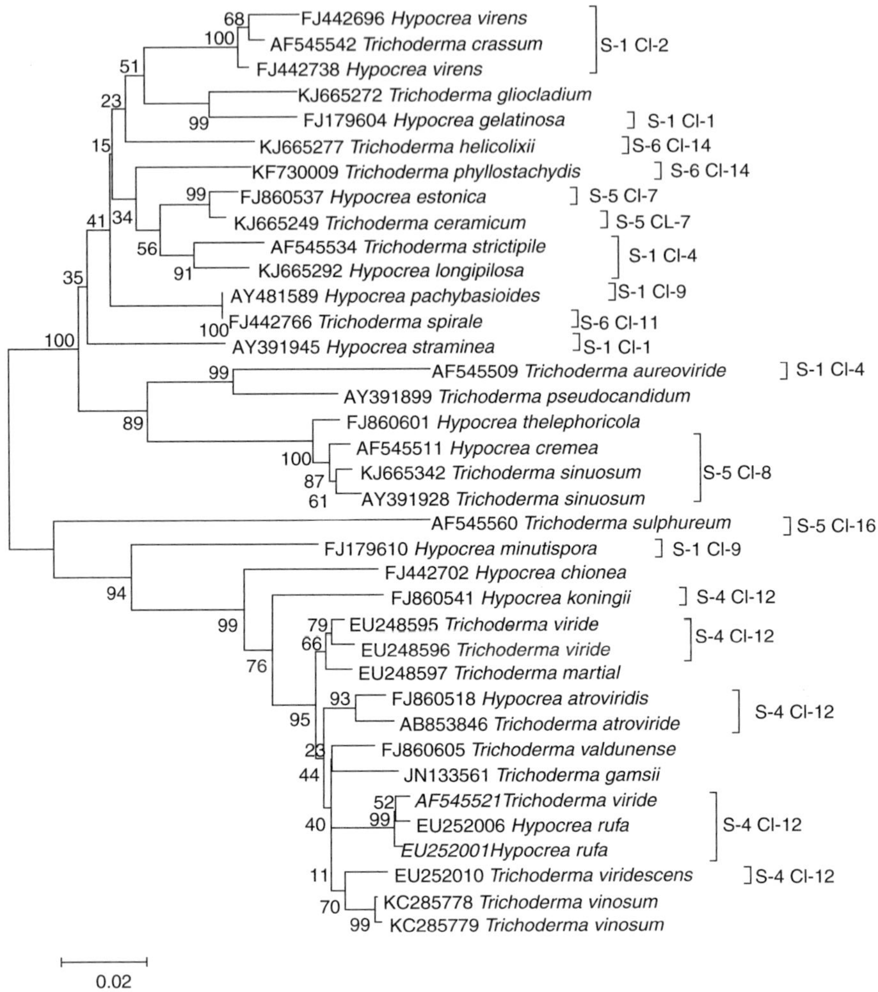

Fig. 2.3 Evolutionary relationships based on RNA polymerase subunit II (*rpb2)* of *Trichoderma/Hyprocrea* isolates inferred using the Neighbor-Joining method The tree is drawn to scale, with branch lengths (Dopazo 1994) in the same units as those of the evolutionary distances used to infer the phylogenetic tree

A serious drawback of the use of ITS1 and ITS2 is that it provides only poor phylogenetic resolution in some clades, particularly Pachybasium B (Kullnig-Gradinger et al. 2002; Chaverri et al. 2003). Even though slightly lower success compared to Basidiomycota has been reported for filamentous Ascomycota (0.75) (0.79) still the performance of ITS in taxonomic studies is close to protein-coding markers RPB1. Also the molecular markers have been used for understanding the relationship of *Gliocladum* and *Trichoderma* (Rehner and Samuels 1994).

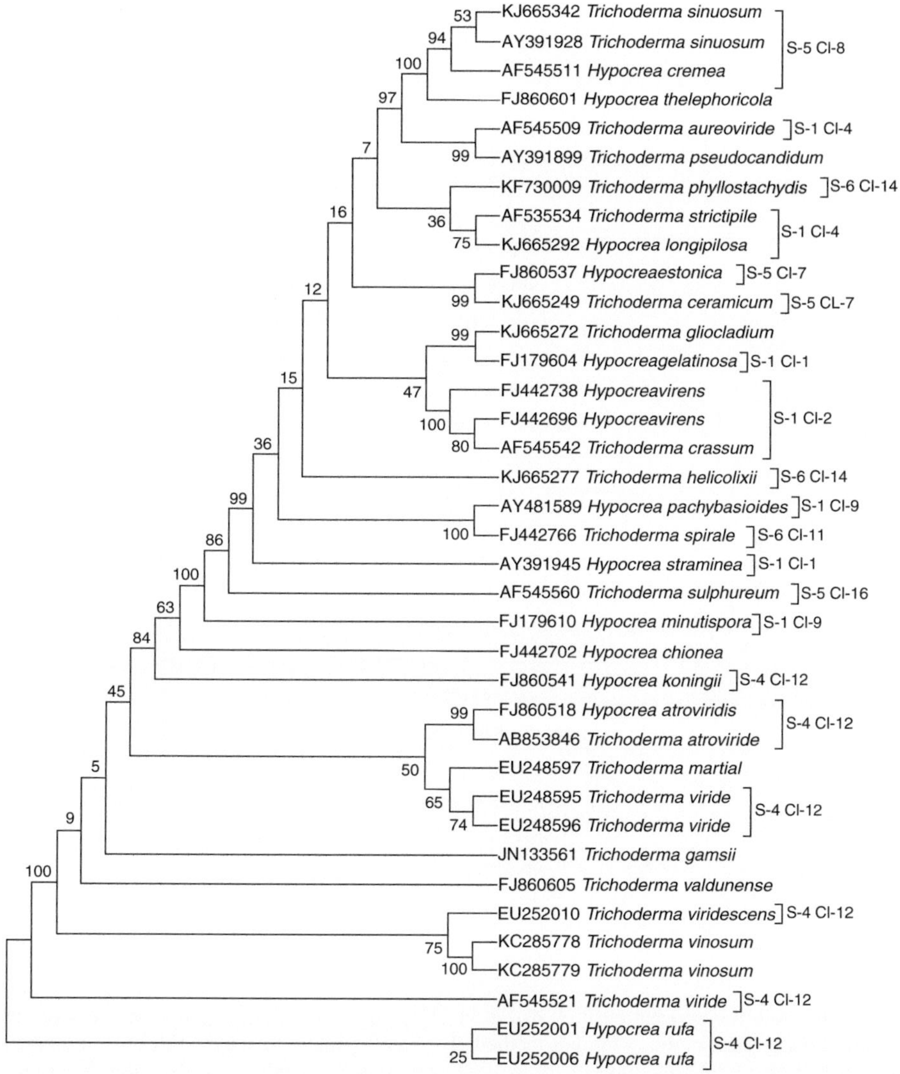

Fig. 2.4 The evolutionary relationships based on RNA polymerase subunit II (*rpb2)* of *Trichoderma/Hyprocrea* isolates inferred using Maximum Parsimony method. There were a total of 342 positions in the final dataset. Evolutionary analyses were conducted in MEGA 6.0

Still a lot of research needs to be done for resolving the relation of clades and sections. The development and use of reference DNA barcodes based on standardized 500- to 800-bp sequences to identify species must be derived from expertly identified vouchers deposited in biological collections with online metadata and validated by available online sequence chromatograms. Interspecific

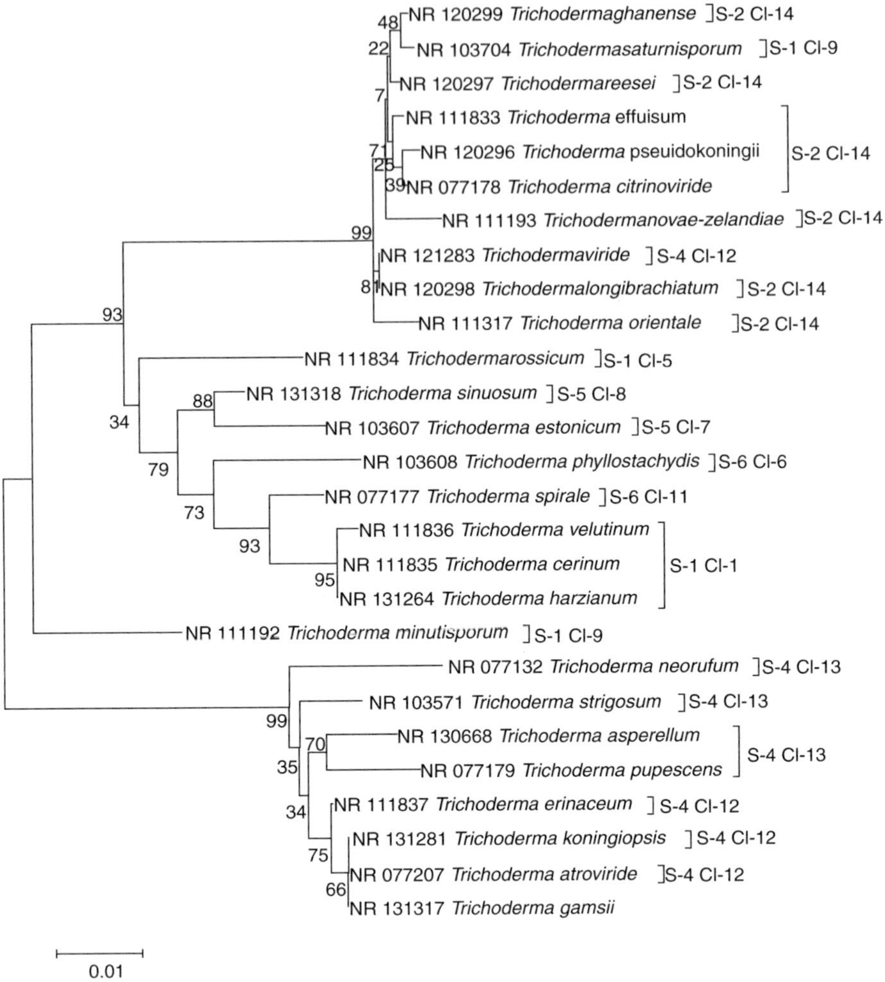

Fig. 2.5 Evolutionary relationships based on ITS region of *Trichoderma/Hypocrea* isolates inferred using the Neighbor-Joining method. The tree is drawn to scale, with branch lengths (Dopazo 1994) in the same units as those of the evolutionary distances used to infer the phylogenetic tree using MEGA 6.0

variation should exceed intraspecific variation and barcoding is optimal when a sequence is constant and unique to one species. The development of modern tools likes DNA barcode program or the genealogical concordance phylogenetic species recognition (GCPSR) (Taylor et al. 2000) involving phylogenetic studies based on several genes is an attractive alternative or complement to the morphological species concept, so that the phylogenetic position of a species will be consistent and monophyletic. However a search for new molecular markers is strongly recommended to achieve distinct taxonomic position of the genus *Trichoderma.*

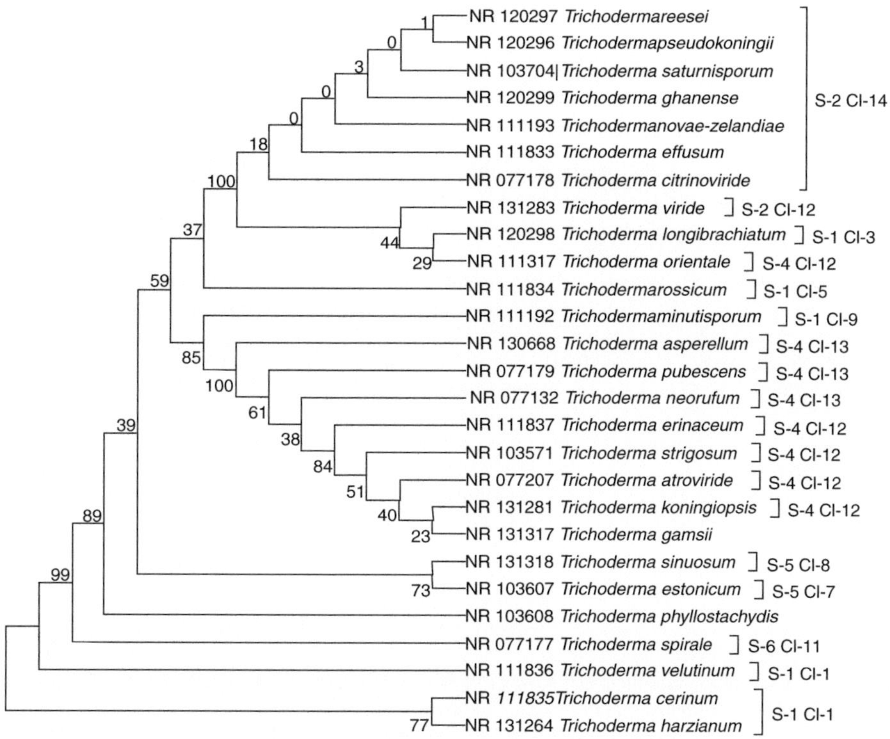

Fig. 2.6 The evolutionary history based on ITS region of *Trichoderma*/*Hyprocrea* isolates inferred using the Maximum Parsimony method. The MP tree was obtained using the Subtree-Pruning-Regrafting (SPR) algorithm with search level 1 in which the initial trees were obtained by the random addition of sequences (10 replicates). All positions containing gaps and missing data were eliminated. Evolutionary analyses were conducted in MEGA 6.0

References

Bissett J. A revision of the genus *Trichoderma*. II. Infrageneric classification. Can J Bot. 1991;69:2357–72.

Bridges PD, Spooner BM, Roberts PJ. The impact of molecular data in fungal systematic. Adv Bot Res. 2005;42:34–67.

Chaverri P, Castlebury LA, Samuels GJ, Geiser D. Multilocus phylogenetic structure within the *Trichoderma harzianum*/*Hypocrea lixii* complex. Mol Phylogenet Evol. 2003;27:302–13.

Chaverri P, Samuels GJ. *Hypocrea Trichoderma* (Ascomycota, Hypocreales, Hypocreaceae): species with green ascospores. Stud Mycol. 2003;48:1–119.

Crespo A, Lumbsch HT, Mattsson JE, Blanco O, Divakar PK, Articus K,Wiklund E, Bawingan PA, Wedin M. Testing morphology-based hypotheses of phylogenetic relationships in Parmeliaceae (Ascomycota) using three ribosomal markers and the nuclear RPB1 gene. Mol Phylogenet Evol. 2007;44:812–24.

Dingley JM. Notes on *Hypocrea citrina* (Pers.) Fr., *H. lacteal* Fr., *H. pulvinata* Fuckel and *H. sulphurea* (Schw.) Sacc. Kew Bull. 1955;9:575–8.

Dodd SL, Crowhurst RN, Rodrigo AC, Samuels GJ, Hill RA, Stewart A. Examination of *Trichoderma* phylogenies derived from ribosomal DNA sequence data. Mycol Res. 2000; 104:23–34.

Doi Y, Abe Y, Sugyama J. *Trichoderma* sect. *Saturnisporum*, sect. nov. and *Trichoderma ghanense*, sp. nov. Bull Natl Sci Mus Ser B. 1987;12:1–9.
Dopazo J. Estimating errors and confidence intervals for branch lengths in phylogenetic trees by a bootstrap approach. J Mol Evol. 1994;38:300–4.
Felsenstein J. Confidence limits on phylogenies: An approach using the bootstrap. Evol. 1985;39:783–91.
Fries EM. Systema Orbis Vegetabilis (in Latin). Lundin: Typographia Academica; 1825. p. 104.
Froslev TG, Matheny PB, Hibbett DS. Lower level relationships in the mushroom genus Cortinarius (Basidiomycota, Agaricales): a comparison of RPB1, RPB2, and ITS phylogenies. Mol Phylogenet Evol. 2005;37:602–18.
Helgason T, Watson IJ, Young JPW. Phylogeny of the Glomerales and Diversisporales (Fungi: Glomeromycota) from actin and elongation factor 1-alpha sequences. FEMS Microbiol Lett. 2003;229:127–32.
Hibbett DS, Binder M, Bischoff JF, Blackwell M, Cannon PF, Eriksson OE, et al. A higher-level phylogenetic classification of the fungi. Mycol Res. 2007;111(Pt 5):509–47.
James TY, Kauff F, Schoch CL, Matheny PB, Hofstetter V, Cox CJ, Celio G, Gueidan C, Fraker E, Miadlikowska J, Lumbsch HT, et al. Reconstructing the early evolution of fungi using a six-gene phylogeny. Nature. 2006;443:818–22.
Kaewchai S. Mycofungicides and fungal biofertilizers. Fungal Divers. 2009;164(2):81–9.
Kindermann J, El-Ayouti Y, Samuels GJ, Kubicek CP. Phylogeny of the genus *Trichoderma* based on sequence analysis of the internal transcribed spacer region 1 of the rDNA clade. Fungal Genet Biol. 1998;24:298–309.
Klein D, Eveleigh DE. Ecology of *trichoderma*. In: Kubicek CP, Harman GE, editors. Trichoderma and gliocladium, Basic biology, taxonomy and genetics, vol. 1. London: Taylor and Francis Ltd.; 1998. p. 57–74.
Kuhls K, Lieckfeldt E, Samuels GJ, Kovacs W, Meyer W, Petrini O, Gams W, Borner T, Kubicek CP. Molecular evidence that the asexual industrial fungus *Trichoderma reesei* is a clonal derivative of the ascomycete *Hypocrea jecorina*. Proc Natl Acad Sci U S A. 1996;93:7755–60.
Kuhls K, Lieckfeldt E, Samuels GJ, Borner T, Meyer W, Kubicek CP. Revision of *Trichoderma* sect. Longibrachiatum including related teleomorphs based on analysis of ribosomal DNA internal transcribed spacer sequences. Mycologia. 1997;89:442–60.
Kullnig-Gradinger CM, Szakacs G, Kubicek CP. Phylogeny and evolution of the fungal genus *Trichoderma*, a multigene approach. Mycol Res. 2002;106:757–67.
Landvik S, Eriksson OE, Berbee ML. Neolecta – a fungal dinosaur? Evidence from beta-tubulin amino acid sequences. Mycologia 2001;93:1151–63.
Lieckfeldt E, Seifert KA. An evaluation of the use of ITS sequences in the taxonomy of the Hypocreales. Stud Mycol. 2000;45:35–44.
Lieckfeldt E, Samuels GJ, Borner T, Gams W. *Trichoderma koningii*: neotypification and Hypocrea teleomorph. Can J Bot. 1998;76:1507–22.
Lieckfeldt E, Kullnig CM, Kubicek CP, Samuels GJ, Borner T. *Trichoderma aureoviride*: phylogenetic position and characterization. Mycol Res. 2001;105:313–22.
Liu YJ, Hall BD. Body plan evolution of ascomycetes, as inferred from an RNA polymerase II phylogeny. Proc Natl Acad Sci U S A. 2004;101:4507–12.
Liu YJ, Whelen S, Hall BD. Phylogenetic relationships among ascomycetes: evidence from an RNA polymerase II subunit. Mol Biol Evol. 1999;16:1799–808.
Lumbsch HT, Schmitt I, Palice Z, Wiklund E, Ekman S, Wedin M. Supraordinal phylogenetic relationships of Lecanoromycetes based on a Bayesian analysis of combined nuclear and mitochondrial sequences. Mol Phylogenet Evol. 2004;31:822–32.
Lutzoni F, Pagel M, Reeb V. Major fungal lineages are derived from lichen symbiotic ancestors. Nature. 2001;411:937–40.
Matheny PB. Improving phylogenetic inference of mushrooms with RPB1 and RPB2 nucleotide sequences (inocybe; agaricales). Mol Phylogenet Evol. 2005;35:1–20.
Mulaw TB, Druzhinina IS, Kubicek CP, Atanasova L. Novel endophytic Trichoderma spp. isolated from Healthy *Coffea Arabica* roots are capable of controlling coffee tracheomycosis. Diversity. 2013;5:750–66.

Nei M, Kumar S. Molecular evolution and phylogenetics. New York: Oxford University Press; 2000.

O'Donnell KL, Ward TJ, Lutzoni F, Benny GL. Evolutionary relationships among mucoralean fungi (Zygomycota): evidence for family polyphyly on a large scale. Mycologia. 2001;93:286–96.

Persoon CH. Disposita methodica fungorum. Romer's Neues Mag Bot. 1794;1:81–128.

Reeb V, Lutzoni F, Roux C. Multilocus phylogenetic circumscription of the lichen-forming fungi family acarosporaceae and its position within the ascomycota. Mol Phylogenet Evol. 2004;32: 1036–60.

Rehner SA, Samuels GJ. Taxonomy and phylogeny of Gliocladium analysed from nuclear large subunit ribosomal DNA sequences. Mycol Res. 1994;98(6):625–34.

Rehner SA, Samuels GJ. Molecular systematics of the hypocreales: a teleomorph gene phylogeny and the status of their anamorphs. Can J Bot. 1995;73 Suppl 1:816–23.

Rifai MA. A revision of the genus *Trichoderma*. Mycol Pap. 1969;116:1–116.

Rzhetsky A, Nei M. A simple method for estimating and testing minimum evolution trees. Mol Biol Evol. 1992;9:945–67.

Saitou N, Nei M. The neighbor-joining method: a new method for reconstructing phylogenetic trees. Mol Biol Evol. 1987;4:406–25.

Samuels GJ. *Trichoderma*: a review of biology and systematics of the genus. Mycol Res. 1996;100:923–35.

Samuels GJ, Hebbar PK. The *Trichoderma* manual. Their identification and application in agriculture. St Paul, MN: American Phytopathological Society Press. 2015. 196, 55 illustr. (some col.). ISBN 978-0-89054-484-6.

Samuels G, Petrini O, Kuhls K, Lieckfeldt E, Kubicek CP. The *Hypocrea schweinitzii* complex and Trichoderma sect. Longibrachiatum. Stud Mycol. 1998;41:1–54.

Schoch CL, Seifert KA, Huhndorf S, Robert V, Spouge JL, Levesque CA. Nuclear ribosomal internal transcribed spacer (ITS) region as a universal DNA barcode marker for Fungi. Proc Natl Acad Sci U S A. 2012;109(16):6241–6.

Seidl DIV. Molecular and physiological investigations of biocontrol by the genus *Hypocrea/Trichoderma*. Dissertation, 2006. Austria: Vienna University of Technology. http://www.isth.info/files/DissertationVS.pdf

Seifert KA. A monograph of Stilbella and some allied hyphomycetes. Studies in Mycology. 1985; 27:1–235.

Spatafora JW, Johnson D, Rourke BO, Spotts R, Rauhut A, Hewitt D, et al. A five-gene phylogeny of Pezizomycotina. Mycologia. 2006;98(6):1018–28.

Spatafora JW, and Blackwell M. Molecular systematics of unitunicate perithecial ascomycetes: The Clavicipitales-Hypocreales connection. Mycologia. 1993;85:912–22.

Sharma V, and Shanmugam V. Purification and characterization of an extracellular 24 kDa chitobiosidase from the mycoparasitic fungus *Trichoderma saturnisporum*. Journal Basic Microbiol. 2012;324–31.

Sharma V, Bhandari P, Singh B, Bhatacharya A, and Shanmugam V. Chitinase expression due to reduction in fusaric acid level in an antagonistic *Trichoderma harzianum* S17TH. Indian Journal of Microbiology 2013;53(2):214–20.

Sharma V, and Salwan R. Plate assay for the detection of total and specific chitinase activity of fungi. Indian Journal of Applied Microbiology, 2015;18(2):1–6.

Shanmugam V, Sharma V. Genetic Relatedness of Trichoderma Isolates Antagonistic against Fusarium oxysporum f.sp. dianthi Inflicting Carnation Wilt. Folia Microbiol. 2008;53(2), 130–38.

Tamura K, Nei M, Kumar S. Prospects for inferring very large phylogenies by using the neighbor-joining method. Proc Natl Acad Sci U S A. 2004;101:11030–5.

Tamura K, Stecher G, Peterson D, Filipski A, Kumar S. MEGA6: molecular evolutionary genetics analysis version 6.0. Mol Biol Evol. 2013;30:2725–9.

Taylor JW, Jacobson DJ, Kroken S, Kasuga T, Geiser DM, Hibbett DS, Fisher MC. Phylogenetic species recognition and species concepts in fungi. Fungal Genet Biol. 2000;31:21–32.

Thon MR, Royse DJ. Partial beta-tubulin gene sequences for evolutionary studies in the Basidiomycotina. Mycologia. 1999;91:468–74.

Tode HJ. Fungi Mecklenburgenses selecti. Fasc. 1791;2:64, 17 pl.

Tulasne L, Tulasne R. De quelques Spheries fungicoles, a propos d'un memoire de M. Antoine de Bary sur les Nyctalis. Ann Sci Natl Bot. 1860;13:5–19.

Turner D, Kovacs W, Kuhls K, Lieckfeldt E, Peter B, Arisan-Atac I, Strauss J, Samuels GJ, Borner T, Kubicek CP. Biogeography and phenotypic variation in *Trichoderma* sect. Longibrachiatum and associated *Hypocrea* species. Mycol Res. 1997;101:449–59.

Vinale F, Sivasithamparam K, Ghisalberti EL, Marra R, Woo SL, Lorito M. *Trichoderma*–plant–pathogen interactions. Soil Biol Biochem. 2008;40:1–10.

Voigt K, Wostemeyer J. Phylogeny and origin of 82 zygomycetes from all 54 genera of the Mucorales and Mortierellales based on combined analysis of actin and translation elongation factor EF-1a genes. Gene. 2001;270:113–20.

White TJ, Bruns TD, Lee S, Taylor JW. Amplification and direct sequencing of fungal ribosomal RNA genes for phylogenetics. In: Innis MA, Gelfand DH, Sninsky J, White TJ, editors. PCR protocols: a guide to methods and applications. San Diego: Academic; 1990. p. 315–22.

Chapter 3
Molecular Diversity of Oleaginous Fungi in Irish Soil and Their Potential for Biodiesel Production

Iniya Kumar Muniraj, Sivakumar Uthandi, Liwen Xiao, Zhenhu Hu, and Xinmin Zhan

3.1 Introduction

Among the three macromolecules (Lipids, proteins and carbohydrates) in microbial systems, lipids are an important structural component of the microbial cell membrane. Some microorganisms (algae, yeast, fungi and bacteria) can accumulate more than 20 % of lipids inside their cells as triacylglycerol (TAG) under certain conditions. Those microorganisms are called as oleaginous microorganisms. The oils thus produced are popularly called single cell oils (SCOs) (Ratledge 2004). Microbial lipids are one of the major feedstocks for production of the second generation biodiesel (Shi et al. 2011; Li et al. 2008, 2011; Meng et al. 2009; Kumar et al. 2011; Coradini et al. 2015). The number of oleaginous microorganims currently in use for

---

I.K. Muniraj
Department of Agricultural Microbiology, Tamil Nadu Agricultural University, Coimbatore 641003, India

Civil Engineering, College of Engineering and Informatics, National University of Ireland Galway, Galway, Ireland

S. Uthandi (✉)
Department of Agricultural Microbiology, Tamil Nadu Agricultural University, Coimbatore 641003, India
e-mail: usivakumartnau@gmail.com

L. Xiao
Department of Civil, Structural and Environmental Engineering, Trinity College Dublin, Dublin, Ireland

Z. Hu
School of Civil Engineering, Hefei University of Technology, Hefei, People's Republic of China

X. Zhan
Civil Engineering, College of Engineering and Informatics, National University of Ireland Galway, Galway, Ireland

B.P. Singh, V.K. Gupta (eds.), *Molecular Markers in Mycology*, Fungal Biology,
DOI 10.1007/978-3-319-34106-4_3

energy production is limited; for example, of the 600 yeasts only 25 % are known to produce lipids and among 60,000 fungal species less than 50 % are known to accumulate lipids (Ratledge and Wynn 2002). Although the numbers seem large, the number of oleaginous microorganims in use is limited, since the microbial lipid production research and application until now is focused on production of special lipids rich in poly unsaturated fatty acids (such as gamma linolenic acid and arachidonic acid) (Ratledge 1993, 2004; Shene et al. 2010; Sakuradani and Shimizu 2009; Patnayak and Sree 2005; Kendrick and Ratledge 1992). Oleaginous fungi used for these special lipids production are mainly cultures belonging to the order, Mucorales (Ratledge 1992). Such mucaraceous oleaginous fungi for biodiesel production has a limited scope as their microbial oils would contain a high linolenic acid content, which does not satisfy the European standards (EN14214 and 14213) for biodiesel (Vicente et al. 2004). Hence search for newer organisms with potential oil yielding capacity becomes essential. While searching for hyper oleaginous microbes, their environment in which these organisms grow and produce oils, their adaptability are to be considered.

Feedstocks for the second generation biodiesel include biomass, agricultural and forestry residues, and biowastes (Li et al., 2009; Sims et al. 2010; Wei et al. 2009). In order to utilize these residues for lipid production, oleaginous microorganims should have respective enzymes to breakdown complex organic molecules, like cellulose, starch, etc. Otherwise enzymes to break these macromolecules to simpler monmers need to be added exogenously, thus increasing the lipid production cost. It is reported that when oleaginous fungi *Mortierella isabelliana ATHUM* 2935 and *Cunninghamella echinulata* ATHUM 4411 were grown on renewable carbon sources (starch, pectin and lactose) their lipid yields were greatly influenced by secreted enzymes (amylase and polygaluctronase) (Papanikolaou et al. 2007). Hence, it is of particular interest to isolate more and more oleaginous microorganims with the capability to secrete enzymes (Meng et al. 2009). Oleaginous yeasts with xylose assimilating capability for utilizing ligno-cellulosic wastes for the biodiesel production was attempted (Zikou et al. 2013). Furthermore, the oleaginous microorganisms must be safe to human beings, local ecology and local environment. Hence, exploring the diversity of oleaginous microorganims form Irish soils and their lipid production capability was studied. In addition, lipid production on complex carbon sources and their respective fatty acid content is also described. Finally biodiesel properties of lipids produced by oleaginous microorganisms were evaluated and their feasibility was reported in this chapter.

3.2 Diversity of Oleaginous Fungi

A total of 247 fungal cultures were isolated from soil samples collected from various parts Ireland and the places of soil sample collection is depicted in Fig. 3.1. Of which, fifty isolates were screened for lipid production on the medium containing

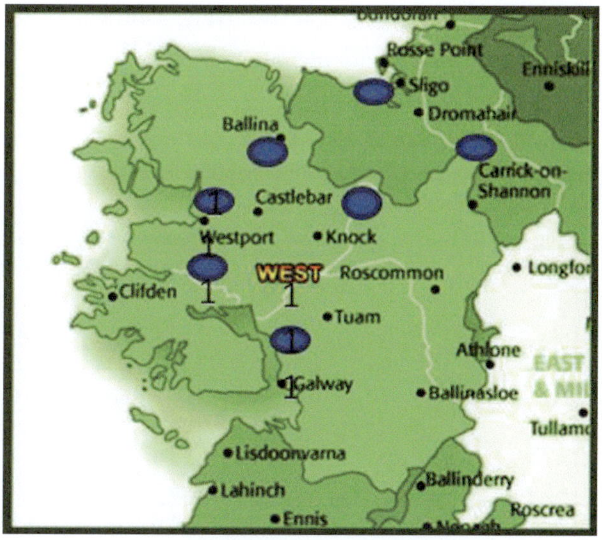

Fig. 3.1 Soil sample collection from various parts of Western Ireland

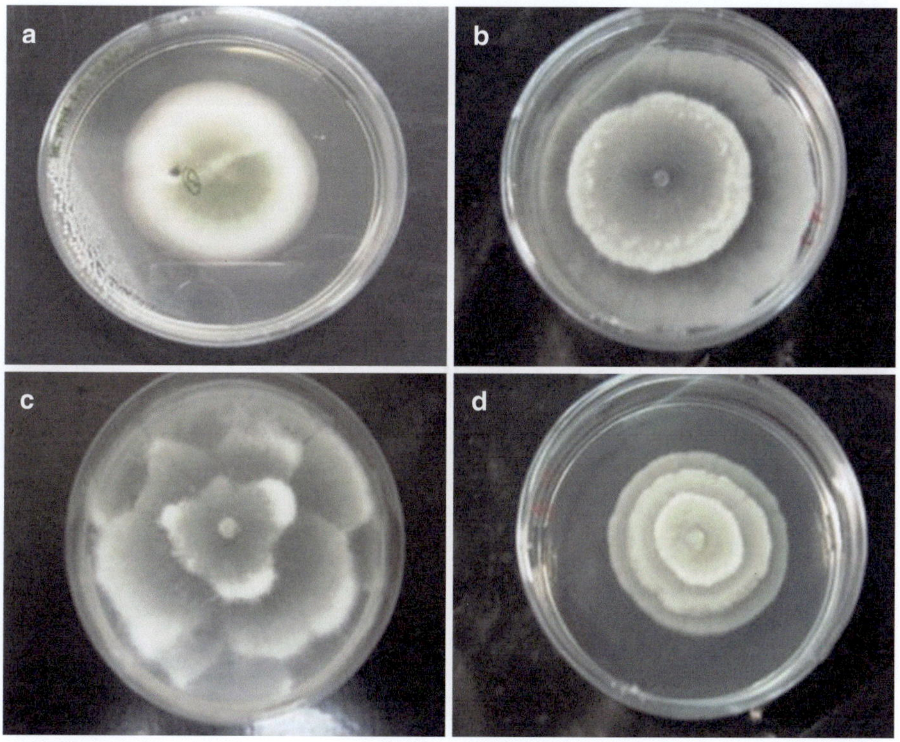

Fig. 3.2 Examples of morphological variations of fungi isolated from Irish soils (**a**) Isolate I17-1; (**b**) Isolate 2A31; (**c**) Isolate I19-2; (**d**) Isolate I5-10

3 % glucose or 3 % starch, respectively. Morphological variations among these isolates were observed (Fig. 3.2).

Table 3.1 Characterization of oleaginous fungi grown on the glucose containing medium (cultivation for 7 days)

Isolate name	X (g/L)	L_{max} (g/L)	Lipid content (%)	Consumed Glucose (g/L)	$Y_{L/C}$ (g lipid/100 g glucose)
I1-1	9.8±0.2	2.0±0.3	20.2	25.5±3.1	7.8
I1-2	7.6±0.4	1.5±0.4	20.1	26.0±0.4	5.9
I1-3	5.8±0.3	1.7±0.2	29.2	25.8±2.2	6.6
I1A1	5.2±0.5	1.2±1.0	23.2	11.0±1.0	11.0
I1A3	9.5±0.5	2.4±0.5	25.5	21.9±1.7	11.0
I1A10	9.7±0.1	2.5±1.0	25.3	25.8±2.5	9.5
I1A5	7.5±0.4	2.0±0.6	26.2	15.7±3.1	12.5
I2A2	4.4±0.2	1.5±0.8	34.0	21.1±0.8	7.1
I2A3	4.6±0.1	1.5±0.2	33.3	28.9±3.4	5.3
I2A31	5.7±0.1	1.6±1.0	28.9	27.3±4.1	6.0
I5-3	4.1±0.7	1.2±0.2	30.2	26.4±3.4	4.7
I5-5	4.8±1.0	1.4±0.8	29.7	23.7±1.9	6.0
I5-6	4.9±0.3	1.4±0.8	29.4	29.0±0.7	5.0
I5-7	7.2±0.4	1.9±0.8	27.1	24.1±0.4	8.1
I5-8	9.4±0.8	3.5±0.2	36.9	27.5±0.7	12.6
I5-10	5.9±0.3	1.8±0.4	30.3	25.4±1.9	7.1
I8-1	11.8±0.4	2.6±0.6	22.3	26.0±4.2	10.2
I10-1	10.2±0.2	2.4±0.3	23.5	26.7±3.2	9.0
I10-2	6.2±0.4	1.5±0.2	24.7	28.6±0.7	5.4
I13-1	5.8±0.3	1.3±0.1	22.0	26.9±1.2	4.7
I14-2	6.5±0.2	1.5±0.1	22.9	27.5±3.6	5.4
I14-4	8.7±0.7	1.8±0.4	20.7	21.1±2.7	8.5
I14-5	9.3±0.5	2.5±0.7	27.2	24.3±2.8	10.4
I15-1	6.3±0.9	1.4±0.3	22.3	14.6±0.6	9.6
I16-3	9.3±0.3	4.1±0.5	44.7	29.3±2.4	14.0
I16-4	7.3±0.1	1.9±0.2	25.4	24.8±0.2	7.5
I17-1	5.7±0.4	2.4±0.3	42.6	28.7±0.3	8.5
I17-2	7.3±0.5	1.8±0.2	24.5	27.9±1.2	6.4
I18-1	7.2±0.9	2.9±0.2	39.6	22.3±3.4	12.8
I19-2	6.5±0.5	1.3±0.5	20.2	16.4±2.1	8.0

3.3 Growth and Lipid Production by Selected Isolates in the Glucose Containing Medium

Among the fifty fungal isolates screened for lipid production, 30 isolates were able to convert glucose into lipids with the lipid contents in biomass ranging from 20 to 44 % within 7 days of cultivation. Therefore, these 30 isolates were designated as oleaginous. Glucose supported biomass growth (4.4 to 11.8 g biomass yield/L medium) and was consumed by all the 30 cultures (11.0–29.3 g glucose consumed/L medium) (Table 3.1). The maximum lipid content in biomass was observed in the isolate I16-3, up to 44.3 %. The lipid yield with respect to glucose consumed, $Y_{L/C}$ (g lipids/g glucose consumed), $Y_{L/C}$ values were in the range of 4.7–14.0 g lipids/100 g of glucose consumed. The higher $Y_{L/C}$ values demonstrate that, higher substrate conversion efficiency of the isolated cultures. The isolate I16-3 had the maximum $Y_{L/C}$ of 14.0 g of lipids yield per 100 g of glucose consumed. Eight isolates (I1A1, I1A3, I1A5, I5-8, I8-1, I14-5, I16-3 and I18-1) had $Y_{L/C}$ higher than 10 g of lipids yield per 100 g of glucose consumed. In the study, glucose was not completely utilized by any of the selected fungi; hence occurrence of lipid turnover was not observed. With the glucose concentration of 30 g/L the maximum $Y_{L/C}$ values obtained for *Mortierella isabelliana* and *Cunninghiamella echinulata* were 10.6 and 11.3 g lipids/100 g of glucose consumed, respectively. Our present study identified, oleaginous fungi with higher $Y_{L/C}$ values than the already reported values. For example, *Z.moelleri* MUCLL1430 had $Y_{L/C}$ of 4.23 g lipids/100 g of glucose consumed, *M.ramannia* ATHUM2922 had 5.0 g lipids/100 g of glucose consumed, *R.stolonifer* LGAM (9) 1 had 4.5 g lipids/100 g of glucose consumed, *R.stolonifer* BPIL 1676 had 2.8 g lipids/100 g of glucose consumed, and *Mucor rouxianus* CBS120-08 had 1.2 g lipids/100 g of glucose consumed when cultivated on glucose (Kavadia et al. 2001).

3.4 Growth and Lipid Production by Selected Isolates in the Starch Containing Medium

All the selected 50 isolates were grown on the medium containing 3 % starch to test their capability in lipid accumulation with starch as the carbon substrate. After 7days of incubation 19 cultures were able to accumulate lipids; interestingly all the 19 isolates were among the 50 which were able to accumulate lipids in the medium containing 3 % glucose (Ahmed et al. 2006). Other 11 isolates did not accumulate lipids in the starch containing medium; this might be because glucose is a simple carbon source. A noticeable amount of biomass was produced (4.8–15.0 g/L) and the maximum biomass value obtained (among the 19 cultures) in the starch containing medium was higher than in the glucose containing medium (15.0 against 11.8 g/L), confirming that starch is a good substrate carbon for oleaginous fungi (Papanikolaou et al. 2007; Ahmed et al. 2006). All the 19 isolates utilized starch above 15 g/L (Table 3.1). The lipid content in biomass ranged from 19 % to 39 %; the maximum lipid content (39 %) was observed in the same isolate I16-3. Interestingly, the isolate had a slightly higher lipid yield (4.3 g/L) when grown on 3 % starch medium than 3 % glucose medium (4.1 g/L). Moreover, the maximum lipid yield obtained by Isolate I16-3 (4.3 g/L) is comparable with other researchers' data. For instance, 3.7 and 3.8 g/L of lipids were produced by *Mortierella isabelliana* and *Cunninghiamella echinulata* when grown on 3 % starch (Papanikolaou et al. 2007). 3.4 g/L microbial lipids were produced when *Aspergillus oryzae* cultured on 4 % starch. Our isolates possessed a higher lipid yield than reported yield.

In this study, seven isolates (I1-1, I1-3, I13-1, I15-1, I16-3, I17-1 and I19-2) had $Y_{L/C}$ values above 10 g lipids/100 g of glucose consumed and a maximum value of 19.3 g lipids/100 g of glucose consumed was observed in the isolate I16-3 (Table 3.1). $Y_{L/C}$ values of 14.9 and 15.2 g lipids/100 g of glucose consumed were obtained when *Mortierella isabelliana* and *Cunninghiamella echinulata* cultured on the medium containing 3 % starch. Much lower $Y_{L/C}$ of 3.5 g lipids/100 g of starch consumed was obtained when *Aspergillus oryzae* grown on 4 % starch. These results suggest that Isolate I16-3 is a potential starch utilizing culture; this is probably because of its high amylase activity.

Amylase is the primary enzyme responsible for starch hydrolysis. It is reported that, when complex carbon sources are used for microbial lipid production, the

Table 3.2 Characterization of isolated oleaginous fungi cultured in the starch containing medium (cultivation for 7 days)

Isolate name	X (g/L)	L_{max} (g/L)	Lipid content (%)	Consumed Starch (g/L)	$Y_{L/C}$ (g lipid/100 g starch)	Amylase secretion (IU/mL)
I1-1	10.5±0.4	3.1±0.8	29.5	26.9±1.4	11.5	24.6±0.3
I1-3	15.0±1.0	3.5±0.6	23	27.5±1.7	12.6	22.1±0.4
I2A3	5.8±0.9	1.7±0.4	29.1	23.6±0.4	7.2	20.2±1.6
I5-3	7.8±0.9	2.0±0.3	25.1	20.9±1.0	9.4	22.2±0.1
I5-5	5.4±0.8	1.3±0.2	23.7	22.9±0.7	5.6	24.3±1.9
I5-8	5.2±0.4	1.8±0.4	34.4	25.5±0.5	7.0	25.7±1.2
I5-10	8.9±0.4	2.3±0.6	25.3	25.8±1.4	8.7	27.9±1.4
I13-1	11.7±0.6	2.6±0.2	22.6	15.7±0.6	16.8	28.4±1.6
I14-2	7.0±0.2	2.3±0.8	33.5	27.6±0.7	8.5	24.5±1.4
I14-4	5.6±0.8	2.0±0.4	35.3	26.6±0.2	7.4	27.5±1.2
I14-5	6.9±0.5	2.2±0.1	32.5	28.6±0.6	7.8	28.4±1.6
I15-1	10.9±0.6	2.5±0.8	22.5	21.1±0.6	11.7	30.2±2.0
I10-1	10.5±0.9	2.2±0.5	20.5	24.3±2.4	8.9	32.3±1.3
I16-3	11.0±1.1	4.3±0.7	39.4	22.5±0.7	19.3	34.6±2.3
I16-4	9.5±0.8	2.0±0.8	20.8	27.9±0.2	7.1	27.5±0.4
I17-2	8.4±0.3	2.3±0.6	27.6	25.5±0.5	9.1	26.9±0.3
I17-1	12.4±0.6	3.7±0.5	29.6	24.8±2.9	14.8	30.8±0.6
I18-1	7.8±0.5	2.1±0.3	27.2	24.8±1.8	8.6	24.3±0.8
I19-2	8.9±0.3	2.2±0.4	24.6	21.1±2.2	10.4	22.9±0.3

Table 3.3 Screening of oleaginous fungi with amylase secretion

Isolate name	Width of the clearing zone (mm)	Width of the biomass (mm)	Hydrolytic capacity
I1-1	10±2	10±3	1.00
I1-3	5±2	8±1	0.63
I2A3	3±1	10±3	0.30
I5-3	3±1	5±4	0.60
I5-5	2±2	5±5	0.40
I5-8	9±1	5±2	1.80
I5-10	5±2	5±1	1.00
I13-1	7±1	4±3	1.75
I14-2	4±3	7±2	0.57
I14-4	5±2	10±1	0.50
I14-5	2±3	7±3	0.29
I15-1	12±1	10±2	1.20
I15-4	3±2	2±1	1.50
I16-3	20±2	10±3	2.00
I16-4	3±3	10±3	0.30
I16-5	6±2	4±4	1.50
I17-1	9±3	1±5	9.00
I18-1	7±2	8±2	0.88
I19-2	3±1	3±1	1.00

capability of oleaginous fungi in secretion of amylase is critical to obtain high lipid yields (Papanikolaou et al. 2007). Hence, in this chapter, all 50 isolates were screened for amylase secretion on plates and the positive ones were tested with the lipid producing medium containing 3 % starch. The results showed that, among the 30 isolates screened, 19 isolates were positive (+ve) for amylase secretion with the width of the clearing zones ranging from 2 to 20 mm and the width of the biomass ranging from 3 to 40 mm (Table 3.3). The result shows that the isolates which were positive were all oleaginous. The isolate I16-3 achieving the highest lipid yield possessed the highest amylase activity with a 20 mm clearing zone in plates and 34 U/mL of amylase was secreted in the liquid medium (Tables 3.2 and 3.3).

In order to use low-cost starchy substrates as raw materials for microbial lipid production, it is important to use oleaginous fungi with amylase secretion capability so as to reduce the lipid production cost. However, in the literature very few oleaginous microbes with amylase secretion capability for utilization of starchy wastes have been reported; they were *Aspergillus oryzae*, *M. isabelliana ATHUM 2935* (0.12 IU/mL) and *C. echinulata* ATHUM 4411 (0.19 IU/mL) (Papanikolaou et al. 2007). Other studies on use of starch based wastes for microbial lipid production either utilize commercial amylase or followed an initial step of cultivating amylase positive cultures on starchy waste to hydrolyse starch and then employed oleaginous microorganims on hydrolysate for lipid production. For instance, alpha amylase and gluco amylase were added to hydrolyze cassava starch for lipid production by micro algae *Chlorella protothecoide*. Li et al. (2010) used amylase secreting

Streptomyces fibuligera to produce crude amylase, which was then added to cassava starch hydrolysate for lipid production by *Rhodotorula mucilaginosa* TJY15a. In the present study, 19 oleaginous fungi were observed to have amylase secretion, which can be useful for low-cost lipid production by using starchy wastes.

3.5 Molecular Identification of Oleaginous Fungi

Sequencing of 18S rDNA of the total 30 isolates from 50 oleaginous isolates shows that they belonged to seven different genus: *Trichoderma* sp., *Penicillium* sp., *Aspergillus* sp., *Mortierella* sp., *Zygomycetes* sp., *Acremonium* sp. and *Umbiliopsis* sp. (Fig. 3.3). Among the seven genus, 13 isolates showed 99 % similarity with already reported sequences of reference strains identified by BLAST analysis and the sequences were submitted to the NCBI gene bank (Acc. No: **JF895924**, **JF895925**, **JF895926**, **JF895927**, **JF895928**, **JF895929**, **JF895930**, **JF912414**, **JF912415**, **JF912416**, **JF912417**, **JF912418** and **JF912419**). Out of the 30 isolates, seven, ten, seven and three belongs to *Trichoderma* sp., *Penicillium* sp., *Aspergillus* sp. and *Zygomycetes* sp., respectively; one each of *Mortierella* sp., *Acremonium* sp. and *Umbliopsis* sp. was identified (Table 3.4). Most of the studies in the literature related to oleaginous fungi used the Mucorales order; very few, to our best knowledge, only one genus *Aspergillus* which is non mucaraceous fungi have been used for lipid production. In addition, oleaginous fungi belonging to the Mucorales order are always used for production of lipids with special long- chain poly unsaturated fatty Acids (PUFA), such as gamma linolenic acid (GLA) and arachidonic acid (AA). These fatty acids have nutritional importance. The present study explored fungi which were capable of producing lipids and did not belong to the Mucorales order. Non-mucarales fungi are attractive alternative fungi for microbial lipid production; *Aspergillus* a known oleaginous fungus belongs to the Deutromycetes order, *Trichoderma* belongs to the Hypocreales order, and *Penicillum* belongs to the Eurotiales order. Furthermore, their amylase secretion could be an additional advantage when using starch wastes for low-cost lipid production. *Acremonium* sp. is well known for hydrolysis of lignocellulosic materials (Beopoulos et al. 2009), while in this study, it is found that *Aceremonium* sp. (I18-1) possessed starch utilizing capability and oleaginous nature. This is reported in the first time to the best of our knowledge. Therefore, this culture might be used for simultaneous saccharification and fermentation (SSF) of lignocellulosic materials for lipid production (Liu et al., 2010, 2011). This could be an attractive option for producing the second generation biofuels. When the microbial oil is used for biodiesel applications, European standards (EN14214 and 14213) regulate that the PUFA content in the microbial oil should not be more than 12 %. Our study has explored non mucaraceous oleaginous fungi whose PUFA contents presumably would not exceed the limit.

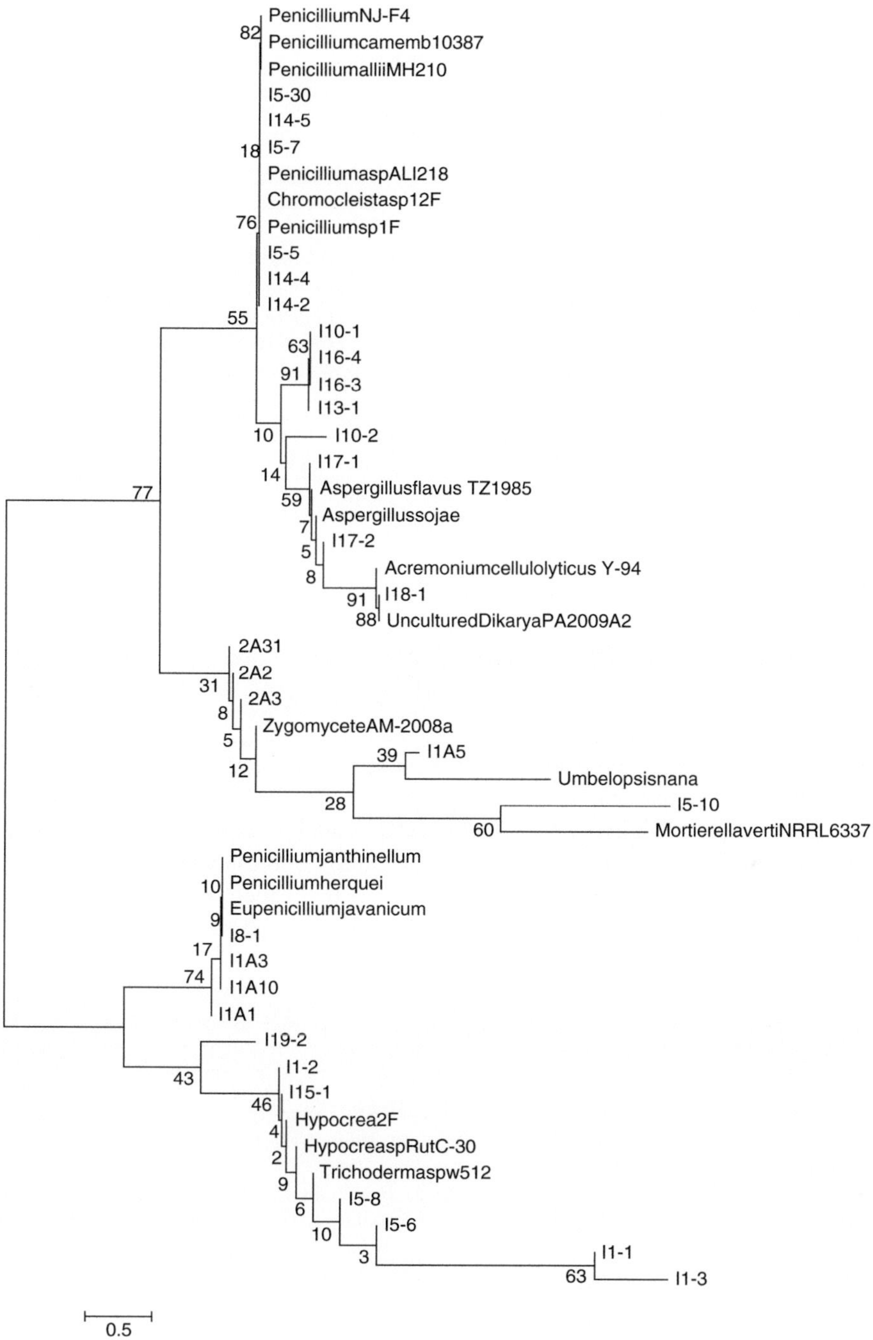

Fig. 3.3 Phylogentic analysis of 18S rDNA of isolated oleaginous fungi

Table 3.4 Blast sequence analysis of isolated oleaginous fungi

		NCBI submission		
Isolate name	Closest match in NCBI database	Isolate name	Gene bank accession No.	Sequence similarity (%)
I17-1, I17-2, I13-1, I10-1, I16-3, I16-4, I10-2	*Aspergillus* sp.	I16-3,I17-1	JF 895924, JF912418	99
I14-2, I14-4, I5-7, I5-5, I14-5, I8-1, IA10, IA3, IA1	*Penicillum* sp.	I1A1,I5-5,I8-1,I14-4,I19-2	JF 895926, JF895928, JF895930, JF912416, JF912419	99
I19-2, I15-1, I1-2, I1-1, I5-8, I1-3, I5-6	*Trichoderma* sp.	I1-1,I1-3,	JF895925, JF912414	99
I2A31, I2A2, I2A3	*Zygomycetes* sp.	I2A3	JF912415	99
I1A5	*Umbleopsis* sp.	I1A5	JF895927	99
I5-10	*Mortierella* sp.	I5-10	JF895929	99
I18-1	*Acremoium* sp.	I18-1	JF912418	99

3.6 Conclusion

In this chapter molecular diversity of indigenous oleaginous fungi from Irish soils was carried out and the capability of those fungi in lipid production with the media containing glucose and starch was studied. Fifty out of 247 isolates were screened with the medium containing 3 % glucose or 3 % starch. Best performing isolates were identified based 18S rDNA sequencing. Many non mucaraceous oleaginous fungi with amylase secretion capability were reported, which could be used for low cost lipid production using starchy waste materials as carbon substrates.

References

Ahmed SU, Singh SK, Pandey A, Kanjilal S, Prasad RBN. Effects of various process parameters on the production of gamma-linolenic acid in submerged fermentation. Food Technol Biotechnol. 2006;44(2):283–7.

Beopoulos A, Cescut J, Haddouche R, Uribelarrea JL, Molina-Jouve C, Nicaud JM. Yarrowia lipolytica as a model for bio-oil production. Prog Lipid Res. 2009;48(6):375–87. doi:10.1016/j.plipres.2009.08.005.

Coradini A, Anschau A, Vidotti A, Reis É, da Cunha Abreu Xavier M, Coelho R, et al. Microorganism for bioconversion of sugar hydrolysates into lipids. In: Kamm B, editor. Microorganisms in biorefineries. Berlin/Heidelberg: Springer; 2015. p. 51–78.

Kavadia A, Komaitis M, Chevalot I, Blanchard F, Marc I, Aggelis G. Lipid and gamma-linolenic acid accumulation in strains of zygomycetes growing on glucose. J Am Oil Chem Soc. 2001;78(4):341–6.

Kendrick A, Ratledge C. Lipids of selected molds grown for production of n-3 and n-6 polyunsaturated fatty acids. Lipids. 1992;27(1):15–20.
Kumar I, Ramalakshmi MA, Sivakumar U, Santhanakrishnan P, Zhan XM. Single cell oil production from Mortierella sp for generation of biodiesel feedstock- a feasibility study. Afr J Microbiol Res. 2011;5(24):4105–11.
Li P, Miao X, Li R, Zhong J. In situ biodiesel production from fast-growing and high oil content Chlorella pyrenoidosa in rice straw hydrolysate. J Biomed Biotechnol. 2011;2011:141207.
Li Q, Du W, Liu D. Perspectives of microbial oils for biodiesel production. Appl Microbiol Biotechnol. 2008;80(5):749–56.
Lu Y, Ding Y, Wu Q. Simultaneous saccharification of cassava starch and fermentation of algae for biodiesel production. J Appl Phycol. 2011;23(1):115–21.
Lu Y, Zhai Y, Liu M, Wu Q. Biodiesel production from algal oil using cassava (Manihot esculenta Crantz) as feedstock. J Appl Phycol. 2010;22(5):573–8.
Meng X, Yang JM, Xu X, Zhang L, Nie QJ, Xian M. Biodiesel production from oleaginous microorganisms. Renew Energy. 2009;34(1):1–5.
Papanikolaou S, Galiotou-Panayotou M, Fakas S, Komaitis M, Aggelis G. Lipid production by oleaginous Mucorales cultivated on renewable carbon sources. Eur J Lipid Sci Technol. 2007;109(11):1060–70.
Patnayak S, Sree A. Screening of bacterial associates of marine sponges for single cell oil and PUFA. Lett Appl Microbiol. 2005;40(5):358–63.
Ratledge C, Wynn JP. The biochemistry and molecular biology of lipid accumulation in oleaginous microorganisms. Adv Appl Microbiol. 2002;51:1–51.
Ratledge C. Fatty acid biosynthesis in microorganisms being used for Single Cell Oil production. Biochimie. 2004;86(11):807–15.
Ratledge C. Single cell oils – have they a biotechnological future? Trends Biotechnol. 1993;11(7):278–84.
Ratledge C. Tailor-made oils and fats possibilities for microbial oils. Revue Francaise Des Corps Gras. 1992;39(11–12):325–32.
Sakuradani E, Shimizu S. Single cell oil production by Mortierella alpina. J Biotechnol. 2009; 144(1):31–6.
Shene C, Leyton A, Esparza Y, Flores L, Quilodran B, Hinzpeter I, et al. Microbial oils and fatty acids: effect of carbon source on docosahexaenoic acid (C22:6 N-3, Dha) production by thraustochytrid strains. J Soil Sci Plant Nutr. 2010;10(3):207–16.
Shi S, Valle-Rodriguez JO, Siewers V, Nielsen J. Prospects for microbial biodiesel production. Biotechnol J. 2011;6(3):277–85.
Sims REH, Mabee W, Saddler JN, Taylor M. An overview of second generation biofuel technologies. Bioresour Technol. 2010;101(6):1570–80.
Vicente G, Martinez M, Aracil J. Integrated biodiesel production: a comparison of different homogeneous catalysts systems. Bioresour Technol. 2004;92(3):297–305.
Wei Z, Xu C, Li B. Application of waste eggshell as low-cost solid catalyst for biodiesel production. Bioresour Technol. 2009;100(11):2883–5.
Zikou E, Chatzifragkou A, Koutinas AA, Papanikolaou S. Evaluating glucose and xylose as cosubstrates for lipid accumulation and γ-linolenic acid biosynthesis of *Thamnidium elegans*. J Appl Microbiol. 2013;2013:1020–32.

Chapter 4
Real-Time Polymerase Chain Reaction (PCR) Based Identification and Detection of Fungi Belongs to Genus *Fusarium*

Mukesh Kumar Yadav and Bhim Pratap Singh

4.1 Introduction

The widespread genus *Fusarium* is of great importance, since members of this genus are implicated in animal, human and is important plant pathogens, produces a wide range of secondary metabolites that are hazardous to agricultural products, wildlife, livestock and humans. *Fusarium* is a diverse genus consisting of an array of species including saprobic, pathogenic and toxin producing species, is of vital importance since it is responsible for damping-off, root rot, and vascular wilt in a multitude of economically important plant species (Nelson et al. 1994). In human, *Fusarium* species causes a broad spectrum of infections including superficial infections, such as keratitis and onychomycosis, as well as locally invasive and disseminated infections in immune-compromised patients (Nucci and Anaissie 2007; Al-Hatmi et al. 2014; van Diepeningen et al. 2015a, and b). Furthermore, in immune-competent individuals the *Fusarium* species causes allergic diseases, such as sinusitis (Wickern 1993), and the fusarium toxin ingestion through food causes mycotoxicosis (Nelson et al. 1994).

Many *Fusarium* species identified based on morphological characters proved to be species complexes, with little to no morphological differences, rather than single species. Based on morphological characteristics and multi-locus sequence typing (MLST) the genus *Fusarium* has been classified into various species complex such as the *Fusarium solani* species complex (FSSC), the *Fusarium oxysporum* species complex (FOSC) and *Fusarium fujikuroi* species complex (FFSC). Less frequently

M.K. Yadav (✉)
Department of Otorhinolaryngology-Head and Neck Surgery & Institute for Medical Device Clinical Trials, Korea University college of Medicine, Korea University, Seoul, South Korea
e-mail: mukiyadav@gmail.com

B.P. Singh
Molecular Microbiology and Systematic Laboratory, Department of Biotechnology, Aizawl, Mizoram University, Aizawl, Mizoram 796004, India

B.P. Singh, V.K. Gupta (eds.), *Molecular Markers in Mycology*, Fungal Biology,
DOI 10.1007/978-3-319-34106-4_4

encountered are members of the *Fusarium incarnatum-equiseti* (FIESC), *Fusarium dimerum* (FDSC) and *Fusarium chlamydosporum* species complexes (FCSC) or species such as *Fusarium sporotrichioides* (Nelson et al. 1994; O'Donnell et al. 2008, 2009, 10a; Migheli et al. 2010).

Fusarium isolates has long been identified based on morphological characteristics such as colony color, colony morphology, growth characteristics, pigmentation, number of conidia, type of hypae etc., and for many laboratories, this is still the standard practice for identification. These methods are time-consuming and require extensive taxonomical knowledge, as well as accurate identification of fungi by visual examination of such morphological criteria is very difficult and erroneous, which complicate timely disease management decisions. Many *Fusarium* species look similar in culture and the growth media and growth conditions used influence growth rate and colony pigmentation, formation of differentiating structures and the dimensions of these structures. Moreover, the formation of specific structures may take long time under controlled conditions. Often the morphological methods may not be able to identify the colonies obtained from patients or tissues prove degenerate in their colony morphology or pionnotal (Leslie and Summerell 2006; De Hoog et al. 2011).

In recent years, molecular methods are increasingly used in the fungal identification and diagnosis that has emerged as a possible answer to the problems associated with existing morphological identification systems. One of the molecular techniques frequently used in detection and identification of pathogenic fungi is polymerase chain reaction (PCR). PCR allows rapid detection and identification of pathogens and overcomes most of the limitations of classical approaches. An advanced technique of PCR is the real-time PCR, which can measure the target pathogen in real time and save the time and more accurate than PCR. Molecular methods based on PCR and real-time PCR can avoid many of the drawbacks associated with classical methods of *Fusarium* diagnostics and can also improve our understanding of *Fusarium* species detection in different conditions. In general, the PCR and real-time PCR based molecular techniques are more specific, sensitive and accurate than traditional methods. Early and reliable detection is crucial for implementation of disease control strategies, and for the disease management. These assays can function as a basis for clinical labs, regulatory personnel, and other diagnosticians to adapt or implement for detection of human, plant and animal pathogenic *Fusarium* species. In the present chapter, we present overview on the basic technique of real-time PCR for the accurate quantification, detection and diagnosis of plant and human pathogenic *Fusarium* species.

4.2 PCR Based Identification and Detection of *Fusarium*

PCR has revolutionized the detection of pathogens and PCR based methods are now widely used for identification of a variety of pathogens because of its rapid, sensitive and specific nature. Many PCR-based approaches have been reported for identification of plant-pathogenic and human pathogenic *Fusarium* species. *Fusarium*species has been differentiated by designing PCR primers using either mycotoxin related genes, ribosomal DNA, other genes such as topoisomerase, beta tubulin, alfa elongation factor or unique DNA bands from RAPD analysis (Edel et al. 1997; Yadav et al. 2011; O'Donnell et al. 2010b). PCR based assays are the most widely used method for the detection of pathogenic fungi through DNA amplification in laboratories. The target DNA exponential amplification makes it possible to detect even small amounts of fungal pathogen. In past, many PCR-based assays have been used for the identification of fungal infections, but only few assays are commercially available, and most of them are used as in-house validation assay. Recent guidelines (Tortorano et al. 2014) recommend the use of PCR-based diagnostic tools when available for confirmation of infection. The agriculturally important *Fusarium* species such as *Fusarium equiseti* and *Fusarium sporotrichioides* has been identified and detected using a PCR assay targeting the intergenic spacer sequences (IGS) (Jurado et al. 2005). And the same assay can be used for detection and identification of clinical *Fusarium* species such as *Fusarium equiseti* and *Fusarium sporotrichioides*. The three human pathogenic fungi, *Aspergillus*, *Candida* and *Fusarium oxysporum* has been identified using PCR method targeting the ITS2 region (Landlinger et al. 2009). However, this method cannot distinguish between the *Fusarium* spp. that having similar size ITS amplicon. The author's further increases the sensitivity of this assay utilizing a semi-nested PCR (nPCR) and makes it usable in blood samples with a low pathogen load. Similarly, Ahmad et al. (2010) and Sugawara et al. (2013) developed an PCR assay based on the ribosomal ITS1-5.8S-ITS2 region to detect *Fusarium* in bronchoalveolar lavage (BAL), serum and/or blood samples, however, the DNA detection limit was up to genus level. If the aim of the assay is to detect several pathogens simultaneously, then the multiplex PCR can be used. Various multiplex PCR assays have been designed for the detection and identification of plant-pathogen and human-pathogen fusarium species. Lau et al. (2008) developed a multiplex tandem PCR on cultures based on ITS, beta-tubulin and elongation factor 1-α regions to distinguish between different genera of human pathogens on one side and fungal pathogens like *Candida*, *Cryptococcus* and *Fusarium* on other side (Lau et al. 2008). *Fusarium verticillioides* and *Fusarium subglutinans* are pathogen of cereals such as maize, and causes infection in human, has been differentiated using a multiplex PCR assay based on galactose oxidase B (*gaoB*) (Faria et al. 2012).

4.2.1 Conventional PCR

Although PCR-based techniques are rapid, highly sensitive and specific, they might suffer from robustness (van der Wolf et al. 2001). An important limitation of PCR-based identification assays is that the technique is not immediately quantitative. Although it is comparatively easy to quantify the amount of a PCR product produced as a result of a successful PCR amplification, it is difficult to tell the amount of target DNA initially present at the start of the reaction. This is because the reaction rate is exponential; as a result, slight variations in the amplification procedure can generate different amounts of final product from the same amount of starting material. Although, target DNA can be quantified using competitive PCR (Nicholson et al. 1998), this method is labor intensive. However, many of these limitations can be overcome by using advance techniques like real-time PCR which are increasingly being used for routine pathogen identification.

4.3 Real-Time Polymerase Chain Reaction (RT-PCR)

To address the need for robust quantification, the technique of real-time quantitative PCR was developed. Real-time PCR could measure the amount of DNA after each cycle via fluorescent dyes that yield increasing fluorescent signal in direct proportion to the number of PCR product molecules (amplicons) generated. Data collected in the exponential phase of the reaction yield quantitative information on the starting quantity of the amplification target. Fluorescent reporters used in real-time PCR include double-stranded DNA (dsDNA) – binding dyes, or dye molecules attached to PCR primers or probes that hybridize with PCR product during amplification. The fluorescence change over the course of the reaction is measured by an instrument that combines thermal cycling with fluorescent dye scanning capability. And by plotting fluorescence against the cycle number, the real-time PCR instrument generates an amplification plot that represents the accumulation of product over the duration of the entire PCR reaction. The real-time PCR technology provides escalating opportunities to identify phytopathogenic fungi and has been used in several studies for detection and identification of various *Fusarium* species. It can more accurately quantify the extent of pathogen biomass in the host tissue and, with multiplex formats, enables simultaneous detection of different pathogens (Lievens et al. 2003). The main advantage of real-time PCR assay over end-point quantitative PCR is that the amplification products can be monitored as they are accumulated in the exponential phase (Schena et al. 2004), thus allowing precise measurement of fungal DNA content in the reaction. And the dynamic range of real-time PCR is usually six to eight orders of the magnitude rather than two for competitive assays. Furthermore, postreaction processing like gel-electrophoresis is not required and thereby the real time detection of fluorescence saves time and higher throughput is possible. By using probes with different fluorescent reporter dyes, amplification of more PCR products can be used to detect different strains, polymorphisms, or even single point mutations in a single tube.

4.4 Strategy for Designing Real-Time PCR Assay for Detection of *Fusarium*

The first step for successful development of any real-time PCR assay is the search of unique DNA sequences for designing PCR primers for specific amplification of target pathogen and internal control. Usually housekeeping genes are used as internal control, which shows no or little variation among isolates. Second requirement is the selection of chemistry, which depends on the utility of the designed assay. Many real-time PCR fluorescent chemistries exist, but the most widely used are the TaqMan probe, SYBR green dye-based and molecular beacon. The probe based assay are costly than SYBR green assay, however, the sensitivity and specificity of probe based assay are higher in compare to SYBR green dye based assays. The validation of the assay can be performed using reference strains, genetically closely related isolates and host DNA. Furthermore, for quantification of pathogen in soil, plant or in clinical samples standard DNA of known concentration is required for construction of standard curve.

4.4.1 Taqman Probe Chemistry

The Taqman probe consists of two types of fluorophores, the reporter dye is found on the 5′ end of the probe and the quencher at the 3′ end. In unbounded state the probe is stabilized by a phenomenon called FRET: fluorescent resonance energy transfer. In FRET, the emissions of a fluorescent dye can be strongly reduced by the presence of another dye, often called the quencher, in close proximity. Once the TaqMan probe has bound to the template DNA after denaturation and the reaction cools, the primers anneal to the DNA. *Taq* polymerase then adds nucleotides and removes the Taqman probe from the template DNA by 5′-3′exonucelase activity. This separates the quencher from the reporter, and allows the reporter to give off its emit energy (Fig. 4.1a). This is then quantified using a computer. The more times the denaturing and annealing takes place, the more opportunities there are for the Taqman probe to bind and, in turn, the more emitted light is detected.

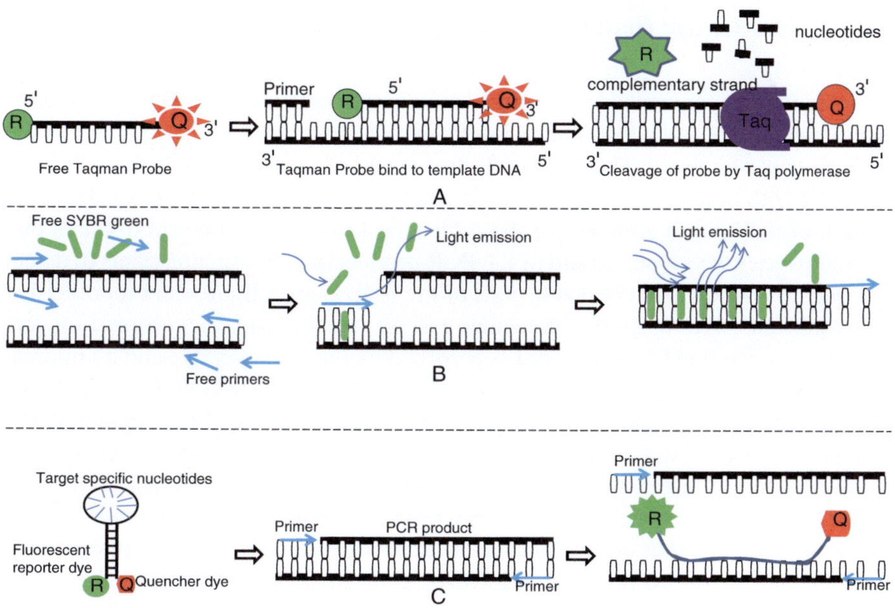

Fig. 4.1 Schematic representation of real time PCR chemistry. (**a**) Taqman probe chemistry. (**b**) SYBR Green real time PCR chemistry PCR chemistry. (**c**) Molecular beacon real-time

4.4.2 SYBR Green

SYBR green dye based real time PCR assay designing is an attractive choice for specificity assessment because it does not add cost to the experiment and can be done right in the PCR reaction vessel. However, Real time PCR assay using SYBR green are comparative less sensitive and specific in compare to probe based assay. SYBR Green I dye is a fluorescent DNA binding dye, binding to the minor groove of any double-stranded DNA. Excitation of DNA-bound SYBR Green dye produces a much stronger fluorescent signal compared to unbound dye. At the beginning of amplification when no reaction taking place, the reaction mixture contains the denatured DNA, the primers, and the dye. After annealing of the primers, a few dye molecules can bind to the double strand. DNA binding results in a dramatic increase of the SYBR Green molecules to emit light upon excitation. During elongation, more and more dye molecules bind to the newly synthesized DNA (Fig. 4.1b). If the reaction is monitored continuously, an increase in fluorescence is viewed in real-time. Upon denaturation of the DNA for the next heating cycle, the dye molecules are released and the fluorescence signal falls. Fluorescence measurement at the end of the elongation step of every PCR cycle is performed to monitor the increasing amount of amplified DNA. Together with a melting curve analysis performed subsequently to the PCR, the SYBR Green format provides an excellent tool for specific product identification and quantification.

4.4.3 Molecular Beacon

A molecular beacon consists of short segments of single-stranded DNA, a reporter and quencher dye and a loop containing bases that are complementary to detect the PCR product of interest. The nine bases on one end of the molecular beacon that can base pair with nine bases forms the stalk of a hairpin structure (Fig. 4.1c). The loop portion of the hairpin structure of the molecular beacon is composed of bases that are complementary to one strand of the PCR product the investigator wants to detect and quantify. Attached to opposite ends of the beacon are a fluorescent reporter dye and a quencher dye. When the molecular beacon is in the hairpin conformation, any fluorescence emitted by the reporter is absorbed by the quencher dye and no fluorescence is detected. As the PCR continues, the newly synthesized PCR products are denatured by high temperatures. As each strand of the product are separated, the molecular beacon also is denatured so the hairpin structure is disrupted. As the temperatures cool for the next round of primer annealing, the molecular beacon is capable of forming base pairs with the appropriate strand of the PCR product. Any molecular beacons that do not bind to PCR product reform the hairpin structures and thus are unable to fluoresce. However, molecular beacons that bind to PCR product remove the ability for the quencher to block fluorescence from the reporter dye. Therefore, as PCR product accumulates, there is a linear increase in fluorescence. Molecular beacon is good choice when designing multiplex real time PCR. The variation in reporter dye and quenching dye allows detecting multiple targets simultaneously.

4.4.4 Primer Target Genes

The most important step for designing the real-time PCR is the selection of the target gene and designing specific primers. Many PCR and real-time PCR assays have been design based on PCR primers using sequences of the internal transcribed spacer (ITS) regions of ribosomal DNA (rDNA) for diagnosis of *Fusarium* species. The rDNA genes subunits of (18S, 5.8S, and 28S) of fungi have evolved slowly and are useful for identification of differentiation up to genera or above levels. Whereas, the two ITS regions and intergenic spacer (IGS) of the nuclear r-DNA evolved fast and could be used for differentiation up to species (White et al. 1990). Since rDNA is highly repeated in fungi, primers developed based on rDNA could be highly sensitive. There are several reports where 18S rDNA sequences have been used for the detection and identification of fungi (Kappe et al. 1996; Smit et al. 1999; Borneman and Hartin 2000). And the fusarium species infecting the black pepper, red pepper, corn, and cereal samples has been detecting using a Taqman probe targeting 18S rDNA sequences. Since ITS regions are highly conserved among some genetically related fungi there may be only few base (even one) pair differences among ITS sequences of these species. In such circumstances, allele-specific PCR primers designed based on single base-pair change could be used to detect these related fungal species. *Fusarium udum*, a pathogen of pigeon pea has been detected in plant tissue and soil using real-time PCR assay based on single nucleotide variation in primers (Yadav et al. 2011). Because the terminals of primers are of vital importance for a successful amplification and the 3′-end position in the primer affects PCR amplification more dramatically than mismatches at other positions. In an allele-specific PCR assay, one or both PCR primers are designed to match the desired bases of target sequence and mismatch others in non-target species at the 3′ end of primers.

When the rDNA sequence variation or conservation is not suitable for developing primers for a target pathogen the species-specific PCR primers could also be designed based on other conserved genes such as the β-tubulin, elongation factor 1α (EF-1α) (Scauflaire et al. 2012), Histone3, and ascomycete mating-type genes. Such genes may have little variation in their translated amino acid sequences, but their third codon position and intron regions appear to have relatively high rates of nucleotide substitutions. O'Donnell et al. (2007) used nucleotide sequence data from the RNA polymerase II second largest subunit (RPB2) for the designing of probe for the identification of medically important *Fusarium* species. Yadav et al. (2011) and Hatsch et al. (2004) identified 11 *Fusarium* species through primer designed using genes encoding Topoisomerase-II and Cellobiohydrolase-C genes and found that despite their short size, the two genes showed higher variability between the species than the r-DNA cluster. RAPD markers also have been used for designing real-time PCR primers (Lievens et al. 2007).

Food and feed contamination by mycotoxins is of great concern because many mycotoxins are carcinogenic and they are not easily removed during food processing. Although toxin abundance does not correlate with fungal contamination, it is linked to the toxigenic properties of each microbial strain. PCR primers designed targeting genes involved in mycotoxigenesis can be used to detect toxigenic strains, subsequently, real-time PCR assays can be developed for quantitative detection of toxigenic fungi in food. Negative results indicate that a sample should be virtually free of mycotoxins, only the positive samples need to be analyzed for the presence of mycotoxins using physico-chemical standard methods. Enniatin a fusarium toxin is synthesized by multifunctional enzyme enniatinsynthetase encoded by*esyn1* gene. The enniatin-producing *Fusarium* species has been identified and detected by using real-time PCR assay by designing primers targeting *esyn1* gene (Kulik et al. 2007).

4.5 Real-Time PCR Based Detection of Fusarium Species

4.5.1 Detection of Fusarium oxysporum

Fusarium oxysporum includes morphologically indistinguishable pathogenic, non-pathogenic and biocontrol strains. The plant-pathogenic strains cause diseases such as vascular wilt, yellows, root rot and damping-off in a wide variety of economically important crops. *Fusarium oxysporum* species alone probably causes more economic damage to agricultural crops than any other pathogen. In spite of the broad host range of the species as a whole, individual strains usually infect only a single or a few plant species. These individual fungal strains usually show a high level of host specificity and based on the plant species they can infect, they have been classified into more than 120 *formae speciales* (Armstrong and Armstrong et al. 1981). Therefore, the identification and detection of *fusarium oxysporum* is very difficult using morphological methods. With the advancement of technology various real-time PCR assays has been developed targeting various genes and fluorescence chemistry. The *Fusarium oxysporum chrysanthemi* causes disease in paris daisy has been identified using real time PCR (Pasquali et al. 2004). The real time PCR assay was based on Taqman chemistry and could successfully identified infected plants as early as the fifth day after artificial inoculation, although the plants remained symptomless until the 13th day after inoculation. This early detection of pathogen helps growers to make the strategy for prevention and control spread of infection. Real time PCR assay has been used to detect *Fusarium oxysporum niveum* and *Mycosphaerell amelonis* pathogens directly from soil samples (Zhang et al. 2005). *Fusarium oxysporum vasinfectum* is another important pathogen of cotton. Abd-Elsalam et al. (2006) designed a real-time PCR based on the 16S and 23S rRNA genes to detect *Fusarium oxysporum vasinfectum* in infected cotton plant. The authors calmed that the assay could detect as low as 200 fg of *Fusarium oxysporum vasinfectum* genomic DNA in infected cotton roots, without amplification from other related fungi and host plant structures. Another real time PCR assay for detection and quantification of *Fusarium oxysporum vasinfectum* from cotton plants and soil has been designed by Zambounis et al. (2007). The *Fusarium oxysporum vasinfectum* detection method was combination of two specific real-time PCR-based assays based on the single nucleotide polymorphism at the 5′ portion of the rDNA-IGS regions. The cucumber pathogen *Fusarium oxysporum cucumerinum* and *Fusarium oxysporum radicis-cucumerinum* has been detected using real time PCR assay based on RAPD marker (Lievens et al. 2007).

4.5.2 Specific Detection of Biocontrol **Fusarium oxysporum**

Real-time PCR has also been designed for the specific detection of biocontrol *fusarium oxysporum*. Fusarium wilts induced by *formae speciales* of *Fusarium oxysporum* are still one of the most difficult soil-borne diseases to control. And the *Fusarium oxysporum* strain 47 (Fo47) is a protective strain that effectively controls Fusarium wilts of several plant species, especially tomato. However, there are no morphological difference in biocontrol Fo47 and other pathogenic fusarium oxysporum that hinders the accurate identification and detection. The classical isolation techniques cannot distinguish Fo47 from the pathogenic strain as they belong to the same species. The detection and quantification of the biological control strain at the surface and in the root requires a highly specific method. Edel-Hermann et al. (2011) design a real time PCR based on SYBR green chemistry for detection of biocontrol Fo47 from root samples of tomato. The specific PCR primers were designed from a sequence-characterized amplified region (SCAR) marker. The strain-specific marker development relies on finding unique DNA sequences that differentiate the target organisms from others fungi and host DNA. The authors identified a unique fragment using PCR fingerprinting with consensus primer that was not specific for Fo47. In a second step, comparison of the sequences generated from consensus primers were aligned and specific primer for Fo47 was designed based on single nucleotide variation at 3 end of primers. The assay was specific and could accurately quantify Fo47 in roots in the absence or presence of the pathogen and in the absence or in presence of the native microbial communities.

4.5.3 Detecting **Fusarium** *Causing Common Bean Seeds Infection*

Fusarium oxysporum f. sp. *phaseoli* (Fop) is a devastating pathogen that can cause significant economic losses and can be introduced into fields through infested *Phaseolus vulgaris* (common bean) seeds. The spread of the disease could be control by testing the contaminated seeds. A real time PCR method has been developed for the rapid, accurately and sensitively detection of Fop in common bean seeds (de Sousa et al. 2015). The real-time PCR method developed by de Sousa et al., was based on SYBR Green and TaqMan qPCR assay and the primers were designed targeting the Fop virulence factor *ftf1*. Under optimized conditions, both qPCR assays detected Fop infection at low levels (0.25 %); however, the results suggest the TaqMan assay was more reliable at quantification than the SYBR Green assay.

4.5.4 *Detection of* Fusarium *Causing Panama Disease of Banana*

Fusarium wilt (Panama disease), is one of the most devastating diseases of banana (*Musa* spp.) caused by *Fusarium oxysporum* f. sp. *cubense* race 4. Early and accurate detection of Foc TR4 is essential to protect banana industry and guide banana planting. For rapid identification and quantification of *Fusarium oxysporum* f. sp. *cubense* race 4 from soil, Peng et al. (2014) developed a real-time PCR (Peng et al. 2014). The real-time fluorescence loop-mediated isothermal amplification (RealAmp) was based on RAPD marker primers which could specifically detect and distinguish R4 isolates from other related species. The detection sensitivity of the Real-Amp assay was approx. $3 \cdot 82 \times 10^3$copies of plasmid DNA or 10^3 of spores per gram in artificially infested soil, indicating that the method was highly tolerant to inhibitor substances in soil. Another real-time PCR assay (Real-Amp) based on SYBR Green fluorescent dye for detection and quantification of Foc TR4 from soil was developed by Zhang et al. (2013). The detection limit of the Real-Amp assay was approximately 0.4 pg/μl plasmid DNA when mixed with extracted soil DNA or 10^3 spores/g of artificial infested soil, and no cross-reaction with other relative pathogens were observed. The sensitivity of Real-Amp assay was tested with genomic DNA of Foc TR4, and specificity was confirmed by testing both artificially and naturally infested samples and no cross reaction was detected. The assay was suitable for the quantification of the soil-borne pathogen DNA of Foc TR4 in naturally infested samples.

4.5.5 *Detection of* Fusarium *Producing Mycotoxin on Maize*

Fusarium proliferatum, *Fusarium subglutinans*, *Fusarium temperatum* and *Fusarium verticillioides* are pathogens of maize, causing ear rot and stalk rot. Along with the crop losses and reduction of seed quality, these *Fusarium* species produce mycotoxins that accumulate in plant tissues and when consumed pose serious problems for both human and animal health. The *Fusarium proliferatum*, *Fusarium subglutinans*, *Fusarium temperatum* and *Fusarium verticillioides* belong to the *Gibberella fujikuroi* species complex (GFSC) with little variability in morphological character. The rapid identification of these *Fusarium* species is critical to evaluate quality of feed samples or to assess their occurrence in fields during the growing season. Scauflaire et al. (2012), developed a multiplex real-time PCR assay using hybridization probes for the specific identification, detection, and quantification of *Fusarium proliferatum*, *Fusarium subglutinans*, *Fusarium temperatum*, and *Fusarium verticillioides* that produces mycotoxin on maize. The primers and hybridization probe was designed using two different genes; the one set of primer were designed to target the translation elongation factor 1α (EF-1α) gene of *F. subglutinans* and *Fusariumtemperatum* and, another set of primers and probes were designed to target the calmodulin (*cal*) gene of *Fusarium proliferatum* and *Fusarium verticillioides*. The real-time PCR assays specificity was confirmed for the specific amplification of four *Fusarium* species, giving no amplification with DNA from other closely related fungal species recovered from maize. And the real time assays was able to detect 5 pg and 50 pg of *Fusarium* DNA in simplex and multiplex conditions respectively (Scauflaire et al. 2012).

4.5.6 *Detection of* Fusarium *Infecting Wheat*

Fusarium head blight caused by *Fusarium graminearum* is one of the economically important diseases of maize, wheat, barley, rice and other small grain cereals (Sarver et al. 2011). Often the infecting pathogen contaminate the food grains with Trichothecene mycotoxins and estrogenic compounds (Kim et al. 2005), which inhibit protein synthesis and can interfere with immune function in humans (Gutleb et al. 2002). *Fusarium graminearum* could be detected from infected wheat using real-time PCR assay developed by Demeke et al. (2010). A SYBR green based real-time PCR assay is also available for detection of a range of *Fusarium* species (*F. graminearum*, *F. culmorum*, *F. poae*, *F. langsethiae*, *F. sporotrichioides*, *F. equiseti*, *F. tricinctum*, *F. avenaceum*, *F. verticillioides*, *F. subglutinans* and *F. proliferatum*) from wheat samples (Nicolaisen et al. 2009). The primers were designed for the amplification EF1α gene and the real-time assay was able to detect as low as 0.1 pg DNA. The assays were intended for quantification of individual *Fusarium* species in plant samples. Using these assays it is possible to monitor the presence of *Fusarium* species in a large number of samples when performing the analysis on high through-put real-time PCR machines. Furthermore, the authors also tested the assay onfield with wheat samples, the results showed good correlation between the occurrence of *Fusarium* species determined by real-time PCR and the toxin content of the individual samples.

4.5.7 *Detection of* **Fusarium** *Causing* **Fusarium** *Wilt of Pigeonpea*

Pigeonpea (*Cajanuscajan* L. *Mill* sp.) is an important pulse crop cultivated in the tropics and sub-tropics. Crop yield is significantly reduced due to wilt disease caused by *Fusarium udum* Butler, with estimated yield losses of US$36 million in India and $5 million in eastern Africa. The detection and identification of the pathogen is prerequisite for disease management. For the detection and identification of *Fusarium udum* from infected pigeonpea plants, Yadav et al. (2011) have designed a real-time PCR assay based on Topoisomerase-II gene sequence data from *Fusarium udum* and other related *Fusarium* species. To reduce the cost of the assay, SYBR green chemistry was used and to impart the specificity, the primers were designed from signature sequence of Topoisomerase-II gene (Fig. 4.2). Moreover, the primers were designed based on single nucleotide polymorphism at 3′ end of primes. The single nucleotide variability in the aligned sequence was kept at 3′ end of the primer, because the terminals of primers are of vital importance for a successful amplification, and the 3′ end position in the primer affects PCR amplification more dramatically than mismatches at other positions. *Fusarium udum* pure culture DNA and infected plant sample DNA amplification using the designed real-time assay produced a reproducible fluorescence with a melting peak of *Fusarium udum* amplicon at 81.25 °C, which shows the specificity of the primers. The real-time PCR assay showed a lowest detection of 0.1 pg genomic DNA (Fig. 4.3). The assay was more sensitive, accurate and less time consuming for detection of *Fusarium udum* in infected plants root. The real time PCR assay could detect fungal DNA in the roots of symptomless infected pigeon pea plants at an early stage, however the assay was unable to detect pathogen in stem and leave's sample DNA.

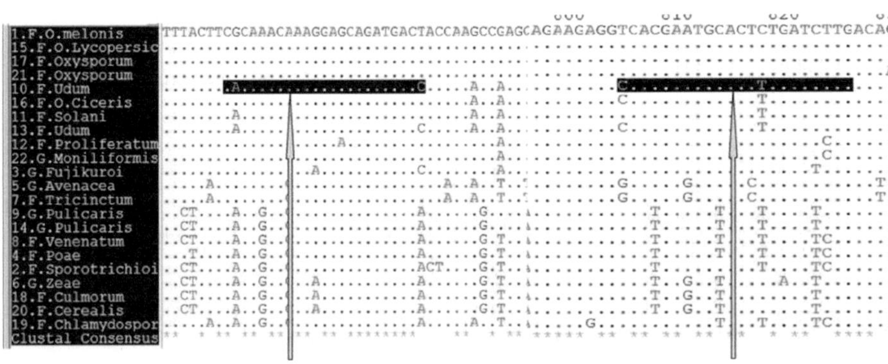

Fig. 4.2 Multiple sequence alignment of *Fusarium udum* with closely related *Fusarium* spp. L1 and L2 sequence-specific locations. Forward primer (FUTF) was designed from L1 region and Reverse primer (FUTR) was designed from L2 region (Yadav et al. 2011)

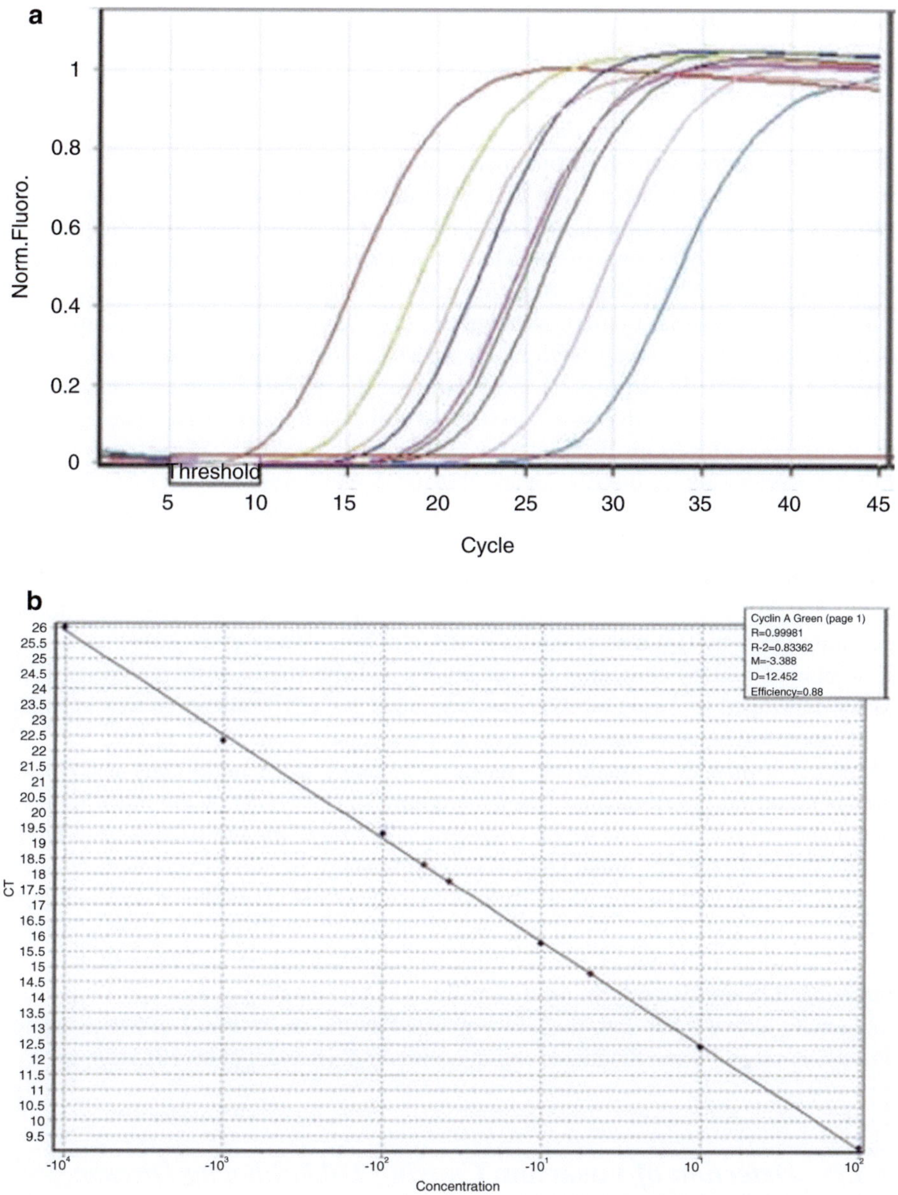

Fig. 4.3 (**a**) Real-time PCR amplification of *Fusarium udum* DNA, (1:10serial diluted from 10 to 0.1 ng/μl) and infected plant's root sample DNA with SYBR Green assay targeted on Topoisomerase-II gene using FUTF and FUTR primers (Yadav et al. 2011). (**b**) Standard curve obtained with diluted *Fusarium udum* DNA with SYBR Green assay targeted on Topoisomerase-II gene using FUTF and FUTR primers. Plot represents Ct versus the log of DNA (Yadav et al. 2011)

4.5.8 *Detection of* Fusarium *Causing Sudden Death Syndrome (SDS) in Soybean*

Fusarium virguliforme (syn. *Fusarium solani* f. sp. *glycines*) is an important fungi that causes sudden death syndrome (SDS) in soybean, and for the disease management the accurately detectection and quantification of *Fusarium virguliforme* in samples such as plant root tissue and soil is extremely valuable. SDS diagnosis is difficult because symptoms of disease can be inconsistent or some time similar to several other soybean diseases. Additionally, the identification and quantification of *Fusarium virguliforme* by traditional dilution plating of soil or ground plant tissue is difficult due to the slow growth rate and plastic morphology of *Fusarium virguliforme*. A real-time PCR assay has been used for the quantification and detection of *Fusarium virguliforme* from plant tissue and soil (Wang et al. 2014). The real-time PCR assay developed by Wang et al. (2014) was based on TaqMan chemistry and primers were target on the ribosomal DNA (rDNA) intergenic spacer (IGS) region for the accurately identification and quantification *Fusarium virguliforme* from soil and plant tissue. The specificity of the assay was checked with co-amplification of closely related *Fusarium* spp. and commonly encountered soil-borne fungal pathogens. The qPCR assay detection limit was 100 g of pure *Fusarium virguliforme* genomic DNA or 100 macroconidia in 0.5 g of soil (Wang et al. 2014). Recently, Kandel et al. (2015) compared the sensitivity and specificity of the TaqMan qPCR assay developed by Wang et al. (2014) and other six qPCR assays in five independent laboratories using the same set of DNA samples from fungi, plants, and soil. The six real-time PCR assay were designed to amplify the multi-copy gene-based assays targeting the ribosomal DNA intergenic spacer (IGS) or the mitochondrial small subunit (mt SSU), or single-copy gene (*FvTox1*). The multi-copy gene-based assays targeting the ribosomal DNA IGS or tmtSSU showed relatively high sensitivity (0.05–5 pg) compared with a single-copy gene (*FvTox1*)-based assay (5–50 pg). However, the specificity of real time assay targeting the *FvTox1* gene was highest (100 %) followed by two IGS. They further concluded that real time PCR assay targeting IGS consistently highest sensitivity (LOD = 0.05 pg) and specificity and inclusivity above 94 % and, thus, is suggested as the most useful qPCR assay for *Fusarium virguliforme* diagnosis and quantification (Kandel et al. 2015).

4.5.9 *Detection of* Fusarium *Causing Pokkahboeng Disease of Sugarcane*

Fusarium verticillioides and *Fusarium proliferatum* causes pokkahboeng disease of sugarcane, which causes significant yield losses. The loss of crop can be decrease by timely disease management and prevent the spreading of sugarcane pokkahboeng. Therefore, the rapid and accurate detection and identification of the pathogen is required. Lin et al. (2014) developed a species-specific TaqMan real-time PCR for rapid and accurate detection of *Fusarium* species causing pokkahboeng of sugarcane. The primers and probes were designed for amplification of rDNA-ITS regions. The specificity and sensitivity of real-time PCR assay was evaluated using 84 isolates of *Fusarium* species and several other fungal pathogens as well as with sugarcane endophyte. And the sensitivity of the TaqMan real-time PCR assay was approximately 10 pg of fungal DNA, and assay successfully detected the pokkahboeng in the field-grown sugarcane.

4.5.10 Detection of Mycotoxin Producing Fusarium

Fusarium species are known to produce mycotoxins such as deoxynivalenol (DON), zearalenone (ZEA), nivalenol (NIV), T-2 toxin, HT-2 toxins, moniliformin (MON), beauvericin (BEA), and enniatins (ENNs) which contaminates food grain including wheat, and when consumed may pose health risks to humans and animals. According to Commission Regulation (EC) No 856/2005 (2005), the maximum level for DON should be 1250 ppb in unprocessed cereals other than durum wheat, oats and maize. The mycotoxins levels in cereals are generally detected using chemical methods, which are often expensive and time-consuming. Another problem in mycotoxin quantification is that the mycotoxin content of raw cereals may change during processing. Another method of detecting *Fusarium* toxins in malting barley is by analyzing the contaminated kernels by using plating method, which often is laborious and time-consuming, and no information about toxigenic species could be obtained. The rapid and simple quantification method for toxigenic *Fusarium* species could evaluate the mycotoxin risk in cereals used in the cereal-based industry. Real-time PCR assay could be used to detection mycotoxin and the mycotoxin producing fungi and to establish the correlation between the mycotoxin and fungi (Bluhm et al. 2004; Reischer et al. 2004; Waalwijk et al. 2004; Yli-Mattila et al. 2008). The mycotoxin producing *Fusarium* species and the level of mycotoxin has been quantified using a combined application of the real-time PCR assay and liquid chromatography/electrospray ionisation-tandem mass spectrometry (HPLC/ESI-MS/MS) (Lindblad et al. 2013) The real time PCR was used for the quantification of the DNA from seven species and the toxin levels evaluated using liquid chromatography/electrospray ionisation-tandem mass spectrometry (HPLC/ESI-MS/MS). The method enabled detection of many fungal metabolites, including DON, ZEA, NIV, T-2 toxin, HT-2 toxins, MON, BEA, and ENNs. The authors found that *Fusarium graminearum, Fusarium culmorum, Fusarium poae* and *Fusarium avenaceum* were present in almost all samples. The strongest correlations between mycotoxin and *Fusarium* DNA levels were found between *Fusarium avenaceum* and ENNs. Similarly, the presence of toxigenic fusarium species (*Fusarium poae, Fusarium graminearum, Fusarium langsethiae, Fusarium culmorum, Fusarium tricinctum, Fusarium sporotrichioides and Fusarium avenaceum*) in oat and the level of mycotoxin (DON, ZEA, NIV, T-2 toxin, HT-2 toxins, MON, BEA and ENNs) has been quantified using combination of real-time PCR assay and liquid chromatography/electrospray ionisation-tandem mass spectrometry (Fredlund et al. 2013).

Trichothecene is another important mycotoxin produced by *Fusarium graminearum* in number of cereals. The trichothecene could be detected by using real time PCR by designing primers for PCR amplification of trichothecene encoding *Tri5*gene. Vegi and Wolf-Hall (2013) developed a multiplex real-time PCR assay for quantification of trichothecene mycotoxins producing *fusarium graminearum.* The multiplex real-time PCR was based on TaqMan probes chemistry and could simultaneously detect and quantify mycotoxigenic *Fusarium, Penicillium* and *Aspergillus* species in cereal grains. Primers and probes were designed targeting three gene, the trichothecene synthase (*Tri5*) gene in trichothecene-producing *Fusarium,* rRNA gene in *Penicillium verrucosum,* and polyketide synthase gene (*PKS*) in *Aspergillus ochraceus.* The minimum detection limit of real time PCR assay was 3 pg of genomic DNA and was highly specific when tested with related fungal isolated. Using this multiplex real-time PCR assay the DON- and OTA-producing fungi can be detected and quantified in a single reaction tube.

4.5.11 Detection of Fusarium *Causing Fusariosis*

In humans, *Fusarium solani* species complex (FSSC) are known to cause fusariosis, followed by *Fusarium oxysporum* species complex (FOSC) and *Gibberella fujikuroi* species complex (GFSC) in immunocompromised patients (Lortholary et al. 2010). And the early diagnosis is extremely important in starting an appropriate antifungal treatment in immunocompromised patients with disseminated fusariosis. The classical method for diagnosis of fusariosis has been based on fungal isolation and identification by observation of morphological characteristics of fungal cultures. However, the accurate identification of *Fusarium* species is very difficult due to the variability of cultural characteristic such as shape and size of conidia, and colony color and texture among isolates and the absence of characteristic macroconidia production. Often, the morphological identification at the species level is very difficult, time consuming and not accurate and requiring extensive microbiological experience and expertise. Muraosa et al. (2014) developed a real-time PCR assay based on the cycling probe technology (CPT) for the molecular detection of *Fusarium* species and distinguishing FSSC from clinical sample. The authors designed primers and probes targeting 28S ribosomal RNA gene. The assay was based on two real-time PCR reactions: the *Fusarium* genus-specific assay, which is designed for the sensitive detection of *Fusarium* species, and the FSSC-specific assay, which is designed for the specific detection of FSSC. The real time PCR assay showed a high sensitivity and reproducibility and was able to detect single copy of standard DNA with no cross reaction with closely related fungal species. Furthermore, the authors evaluated the efficacy of the real-time PCR assay in detecting *Fusarium* pathogen in clinical sample using mouse model and in human blood. The real-time PCR assay was able detect 3 germinated conidia per 3 ml of whole blood and was could detect pathogen in directly in clinical samples of fusariosis.

4.5.12 Detection of Fusarium Solani *from Clinical Sample*

Fusarium solani is the most common clinically important fusarium species that causes human infections, followed by *Fusarium oxysporum, Fusarium verticilliodes* and *Fusarium proliferatum*. Moreover, the *F. solani* has undergone extensive taxonomic re-evaluation and it has been found that this species is a complex referred to as the *F. solani* species complex (FSSC) with no difference in susceptibility profile among these species. In tissue sections, the fusarium species appear as septated, branched hyphae, and can be easily misidentified as *Aspergillus* spp., *Scedosporium* spp. or other pathogenic fungi. And the diagnosis based on the isolation of the fungus in culture and the detection of fungal elements by microscopy is not accurate. PCR-based methods are suitable alternatives to classical methods of identifying the etiologic agent to species in culture and clinical samples. Bernal-Martínez et al. (2012), develop a rapid and sensitive real time PCR assay for the detection of *Fusarium* spp. and validated using mouse models of invasive infection. The duplex real-time polymerase chain reaction assay was designed for specific amplification of ITS1 region of rDNA which have advantage as this target is present in multiple copy numbers against other single copy genes such as β-tubulin or Elongation Factor 1, which improves the sensitivity of the PCR assay. The assay was based on two different fluorescent labelled molecular beacon probes, which enabled the specific and quantitative detection of *Fusarium solani* (Fsol) DNA and non-*Fusarium solani* (Fuspp) DNA without the need for post-amplification process. The non *Fusarium solani* species that the PCR-based assay could detect are: *F. oxysporum, Fusarium verticilloides, Fusarium proliferatum, F. polyphialidicum, Fusarium sacchari* and *Fusarium reticulatum*. The authors claimed that the RT-PCR assay was specific (100 %) for both *fusarium solani* and *non-fusarium solani* detection, with good sensitivity and reproducibility. The duplex RT-PCR technique has a detection limit 10 fg of DNA. The validation of the assay was carried out using animal models of infection. And the sensitivity in clinical samples from the murine model of *Fusarium solani* was 93.9 % for lung tissues and 86.7 % for serum and the sensitivity of the *Fusarium oxysporum* murine model was 87.0 % for lung tissues and 42.8 % for serum.

4.6 Limitations of Real-Time PCR

However, like all other molecular methods based on DNA amplification, a major drawback of the system is that it is unable to distinguish between viable and dead propagules. Similarly, multiplexing in real-time PCR is limited by the number of different fluorescent dyes available. In addition, the initial and running costs of a real-time PCR system are several times more than a normal PCR system. However, considering the many benefits of the real-time PCR technology compared to normal PCR, the use of real-time PCR is still advantageous. Higgins et al. (2003) developed a portable real-time PCR instrument for performing diagnostic assays directly in the field. Such rapid real-time PCR diagnosis could result in taking appropriate and timely control measures than possible with traditional methods of pathogen identification. Therefore, the resulting losses due to diseases as well as the cost of disease management could be greatly reduced.

References

Abd-Elsalam KA, Asran-Amal A, Schnieder F, Migheli Q Verreet JA. Molecular detection of Fusarium oxysporum f.sp. vasinfectum in cotton roots by PCR and real-time PCR assay. J Plant Dis Protect. 2006;113:14–19.

Ahmad S, Khan ZU, Theyyathel AM. Development of a nested PCR assay for the detection of *Fusarium solani* DNA and its evaluation in the diagnosis of invasive fusariosis using an experimental mouse model. Mycoses. 2010;53:40–7.

Al-Hatmi AM, Bonifaz A, de Hoog G, Vazquez-Maya L, Garcia-Carmona K, Meis JF, van Diepeningen AD. Keratitis by *Fusarium temperatum*, a novel opportunist. BMC Infect Dis. 2014;14:588.

Armstrong GM, Armstrong JK (1981) Formae specialis and races of Fusarium oxysporum causing wilt diseases. In: Fusarium: Diseuses, Biology and Taxonomy, Pennsylvania State University Press, University Park (US) pp 391–99.

Bernal-Martínez L, Buitrago MJ, Castelli MV, Rodríguez-Tudela JL, Cuenca-Estrella M. Detection of invasive infection caused by *Fusarium solani* and non-*Fusarium solani* species using a duplex quantitative PCR-based assay in a murine model of fusariosis. Med Mycol. 2012;50:270–5.

Bluhm BH, Cousin MA, Woloshuk CP. Multiplex real-time PCR detection of fumonisin-producing and trichothecene-producing groups of *Fusarium* species. J Food Prot. 2004;67:536–43.

Borneman J, and Hartin RJ. PCR primers that amplify fungal rRNA genes from environmental samples. Appl. Environ. Microbiol. 2000;66:4356–60.

De Hoog GS, Guarro J, Gené J, Figueras MJ. Atlas of clinical fungi, vol. 3. Utrecht: CD-ROM, CBS-KNAW Fungal Biodiversity Centre; 2011.

de Sousa MV, Machado JC, Simmons HE, Munkvold GP. Real-time quantitative PCR assays for the rapid detection and quantification *of Fusarium oxysporum* f. sp. *phaseoli* in *Phaseolus vulgaris* (common bean) seeds. Plant Pathol. 2015;64:478–88.

Demeke T, Gräfenhan T, Clear RM, Phan A, Ratnayaka I, Chapadps J, Patrick SK, Gaba D, Levesque CA, Seifert KA. Development of a specific TaqMan® real-time PCR assay for quantification of *Fusarium graminearum* clade 7 and comparison of fungal biomass determined by PCR with deoxynivalenol content in wheat and barley. Int J Food Microbiol. 2010;141:45–50.

Edel V, Steinberg C, Gautheron N, Alabouvette C. Evaluation of restriction analysis of polymerase chain reaction (PCR)- amplified ribosomal DNA for the identification of *Fusarium* species. Mycol Res. 1997;101:179–87.

Edel-Hermann V, Aimé S, Cordier C, Olivain C, Steinberg C. Development of a strain-specific real-time PCR assay for the detection and quantification of the biological control agent Fo47 in root tissues. FEMS Microbiol Lett. 2011;322:34–40.

Faria CB, Abe CA, da Silva CN, Tessmann DJ, Barbosa-Tessmann IP. New PCR assays for the identification of *Fusarium verticillioides*, *Fusarium subglutinans*, and other species of the *Gibberellafujikuroi* complex. Int J Mol Sci. 2012;13:115–32.

Fredlund E, Gidlund A, Sulyok M, Börjesson T, Krska R, Olsen M, Lindblad M. Deoxynivalenol and other selected Fusarium toxins in Swedish oats – occurrence and correlation to specific Fusarium species. Int J Food Microbiol. 2013;167:276–83.

Gutleb AC, Morrison E, Murk AJ. Cytotoxicity assays for mycotoxins produced by Fusarium strains: a review. Environ Toxicol Pharmacol. 2002;11:309–20.

Hatsch D, Phalip V, Jeltsch JM. Use of genes encoding cellobiohydrolase-C and topoisomerase II as targets for phylogenetic analysis and identification of Fusarium, Research in Microbiology 2004;155:290–96.

Higgins JA, Nasarabadi S, Karns JS, Shelton DR, Cooper M, Gbakima A, Koopman RP. A handheld real time thermal cycler for bacterial pathogen detection. Biosens Bioelectr 2003;18:1115–23.

Jurado M, Vázquez C, Patiño B, González-Jaén MT. PCR detection assays for the trichothecene-producing species *Fusarium graminearum*, *Fusarium culmorum*, *Fusarium poae*, *Fusarium equiseti* and *Fusarium sporotrichioides*. Syst Appl Microbiol. 2005;28:562–8.

Kandel YR, Haudenshield JS, Srour AY, Islam KT, Fakhoury AM, Santos P, Wang J, Chilvers MI, Hartman GL, Malvick DK, Floyd CM, Mueller DS, Leandro LF. Multilaboratory comparison of quantitative PCR assays for detection and quantification of Fusarium virguliforme from soybean roots and soil. Phytopathology. 2015;105:1601–11.

Kappe R, Fauser C, Okeke CN, Maiwald M. Universal fungus-specific primer systems and group-specific hybridization oligonucleotides for 18S rDNA. Mycoses. 1996;39:25–30.

Kim YT, Lee YR, Jin J, Han KH, Kim H, Kim JC, Lee T, Yun SH, Lee YW. Two different polyketide synthase genes are required for synthesis of zearalenone in *Gibberellazeae*. Mol Microbiol. 2005;58:1102–13.

Kulik T, Pszczółkowska A, Fordoński G, Olszewski J. PCR approach based on the esyn1 gene for the detection of potential enniatin-producing Fusarium species. Int J Food Microbiol. 2007;116:319–24.

Landlinger C, Baskova L, Preuner S, Willinger B, Buchta V, Lion T. Identification of fungal species by fragment length analysis of the internally transcribed spacer 2 region. Eur J Clin Microbiol Infect Dis. 2009;28:613–22.

Lau A, Sorrell TC, Lee O, Stanley K, Halliday C. Colony multiplex-tandem PCR for rapid, accurate identification of fungal cultures. J Clin Microbiol. 2008;46:4058–60.

Leslie JF, Summerell BA. The *Fusarium* laboratory manual. Oxford: Blackwell Publishing Ltd.; 2006.

Lievens B, Claes L, Vakalounakis DJ, Vanachter ACRC, Thomma BPHJ. A robust identification and detection assay to discriminate the cucumber pathogens Fusarium oxysporum f.sp. cucumerinum and f.sp. radicis-cucumerinum. Environ Microbiol. 2007;9:2145–61.

Lievens B, Brouwer M, Vanachter ACRC, Lévesque CA, Cammue BPA, Thomma BPHJ. Design and development of a DNA array for rapid detection and identification of multiple tomato vascular wilt pathogens. FEMS Microbiol Lett. 2003;223:113–22.

Lin Z, Xu S, Que Y, Wang J, Comstock JC, Wei J, McCord PH, Chen B, Chen R, Zhang M. Species-specific detection and identification of Fusarium species complex, the causal agent of sugarcane Pokkah Boeng in China. PLoS One. 2014;9:e104195.

Lindblad M, Gidlund A, Sulyok M, Börjesson T, Krska R, Olsen M, Fredlund E. Deoxynivalenol and other selected *Fusarium* toxins in Swedish wheat – occurrence and correlation to specific Fusarium species. Int J Food Microbiol. 2013;167:284–91.

Lortholary O, Obenga G, Biswas P, Caillot D, Chachaty E, Bienvenu AL, Cornet M, Greene J, Herbrecht R, Lacroix C, Grenouillet F, Raad I, Sitbon K, Troke P, French Mycoses Study Group. International retrospective analysis of 73 cases of invasive fusariosis treated with voriconazole. Antimicrob Agents Chemother. 2010;54:4446–50.

Migheli Q, Balmas V, Harak H, Sanna S, Scherm B, Aoki T, O'Donnell K. Molecular phylogenetic diversity of dermatologic and other human pathogenic fusarial isolates from hospitals in Northern and Central Italy. J Clin Microbiol. 2010;48:1076–84.

Muraosa Y, Schreiber AZ, Trabasso P, Matsuzawa T, Taguchi H, Moretti LM, Mikami Y, Kamei K. Development of cycling probe-based real-time PCR system to detect *Fusarium* species and *Fusarium solani* species complex (FSSC). Int J Med Microbiol. 2014;304:505–11.

Nelson PE, Dignani MC, Anaissie EJ. Taxonomy, biology, and clinical aspects of Fusarium species. Clin Microbiol Rev. 1994;7:479.

Nicholson P, Simpson DR, Weston G, Rezanoor HN, Lees AK, Parry DW, Joyce D. Detection and quantification of Fusarium culmorum and Fusarium graminearum in cereals using PCR assays. Physiol Mol Plant Pathol. 1998;53:17–37.

Nicolaisen M, Suproniené S, Nielsen LK, Lazzaro I, Spliid NH, Justesen AF. Real-time PCR for quantification of eleven individual *Fusarium* species in cereals. J Microbiol Meth. 2009;76:234–40.

Nucci M, Anaissie E. Fusarium infections in immunocompromised patients. Clin Microbiol Rev. 2007;20:695.

O'Donnell K, Sutton DA, Fothergill A, McCarthy D, Rinaldi MG, Brandt ME, Zhang N, Geiser DM. Molecular phylogenetic diversity, multilocus haplotype nomenclature, and in vitro antifungal resistance within the *Fusarium solani* species complex. J Clin Microbiol. 2008;46:2477–90.

O'Donnell K1, Sarver BA, Brandt M, Chang DC, Noble-Wang J, Park BJ, Sutton DA, Benjamin L, Lindsley M, Padhye A, Geiser DM, Ward TJ. Phylogenetic diversity and microsphere array-based genotyping of human pathogenic Fusaria, including isolates from the multistate contact lens-associated U.S. keratitis outbreaks of 2005 and 2006. J Clin Microbiol. 2007 Jul;45:2235–48.

O'Donnell K, Gueidan C, Sink S, Johnston PR, Crous PW, Glenn A, Riley R, Zitomer NC, Colyer P, Waalwijk C, Lee TV, Moretti A, Kang S, Kim HS, Geiser DM, Juba JH, Baayen RP, Cromey MG, Bithell S, Sutton DA, Skovgaard K, Ploetz R, Corby Kistler H, Elliott M, Davis M, Sarver BA. A two-locus DNA sequence database for typing plant and human pathogens within the *Fusarium oxysporum* species complex. Fungal Genet Biol. 2009;46:936–48.

O'Donnell K, Sutton DA, Rinaldi MG, Gueidan C, Crous PW, Geiser DM. Novel multilocus sequence typing scheme reveals high genetic diversity of human pathogenic members of the *Fusarium incarnatum, F. equiseti* and *F. chlamydosporum* species complexes within the United States. J Clin Microbiol. 2010a;47:3851–61.

O'Donnell K, Sutton DA, Rinaldi MG, et al. Internet-accessible DNA sequence database for identifying Fusaria from human and animal infections. J Clin Microbiol. 2010b;48:3708–18.

Pasquali M, Marena L, Fiora E, Piatti P, Gullino ML, Garibaldi A. Real-time PCR for the identification of a highly pathogenic group of Fusarium oxysporum f.sp. chrysanthemi on Argyranthemum frutescens L. J Plant Pathol. 2004;86:51–57.

Peng J, Zhang H, Chen F, Zhang X, Xie Y, Hou X, Li G, Pu J. Rapid and quantitative detection of *Fusarium oxysporum* f. sp. *cubense* race 4 in soil by real-time fluorescence loop-mediated isothermal amplification. J Appl Microbiol. 2014;117:1740–9.

Reischer GH, Lemmens M, Farnleitner A, Adler A, Mach RL. Quantification of Fusarium graminearum in infected wheat by species specific real-time PCR applying a TaqMan probe. J Microbiol Methods. 2004;59:141–6.

Sarver BAJ, Ward TJ, Gale LR, Broz K, Corby Kistler H, Aoki T, Nicholson P, Carter J, O'Donnell K. Novel *Fusarium* head blight pathogens from Nepal and Louisiana revealed by multilocus genealogical concordance. Fungal Genet Biol. 2011;48:1096–107.

Scauflaire J, Godet M, Gourgue M, Liénard C, Munaut F. A multiplex real-time PCR method using hybridization probes for the detection and the quantification of *Fusarium proliferatum, F. subglutinans*, *F. temperatum*, and *F. verticillioides*. Fungal Biol. 2012;116:1073–80.

Schena L, Franco N, Ippolito A, Gallitelli D. Real-time quantitative PCR: a new technology to detect and study phytopathogenic and antagonistic fungi. Eur J Plant Pathol. 2004;110: 893–908.

Smit E, Leeflang P, Glandorf B, van Elass JD, Wernars K. Analysis of fungal diversity in the wheat rhizosphere by sequencing of cloned PCR-amplified genes encoding 18S rRNA and temperature gradient gel electrophoresis. Appl. Environ. Microbiol. 1999;65:2614–21.

Sugawara Y, Nakase K, Nakamura A. Clinical utility of a panfungal polymerase chain reaction assay for invasive fungal diseases in patients with haematologic disorders. Eur J Haematol. 2013;90(4):331–9.

Tortorano AM, Richardson M, Roilides E. ESCMID & ECMM joint guidelines on diagnosis and management of hyalohyphomycosis: *Fusarium* spp, *Scedosporium* spp, and others. Clin Microbiol Infect. 2014;20:27–46.

van der Wolf JM, van Beckhoven JRCM, Bonanats PJM, Schoen CD (2001) New technologies for sensitive and specific routine detection of plant pathogenic bacteria. In: de Boer, SH Ed Plant pathogenic bacteria Kluwer, Dordrecht, pp 75–77.

van Diepeningen AD, Brankovics B, Iltes J, van der Lee TA, Waalwijk C. Diagnosis of *Fusarium* infections: approaches to identification by the clinical mycology laboratory. Curr Fungal Infect Rep. 2015a;9(3):135–43.

van Diepeningen AD, Feng P, Ahmed S, Sudhadham M, Bunyaratavej S, de Hoog GS. Spectrum of *Fusarium* infections in tropical dermatology evidenced by multilocus sequencing typing diagnostics. Mycoses. 2015b;58(1):48–57.

Vegi A, Wolf-Hall CE. Multiplex real-time PCR method for detection and quantification of mycotoxigenic fungi belonging to three different genera. J Food Sci. 2013;78:70–6.

Waalwijk C, van der Heide R, de Vries I, van der Lee T, Schoen C, Costrel-de Corainville G, Häuser-Hahn I, Kastelein P, Kohl J, Lonnet P, Demarquet T, Kema GHJ. Quantitative detection of Fusarium species in wheat using TaqMan. Eur J Plant Pathol. 2004;110:481–94.

White TJ, Bruns T, Lee S, Taylor J. Amplification and direct sequencing of fungal ribosomal RNA genes for phylogenetics. Mycologia. 1990;18:315–22.

Wang J, Jacobs JL, Byrne JM, Chilvers MI. Improved diagnoses and quantification of *Fusarium virguliforme*, causal agent of soybean sudden death syndrome. Phytopathology. 2014;105:378–87.

Wickern GM. Fusarium allergic fungal sinusitis. J Allergy Clin Immunol. 1993;92:624.

Yadav MK, Babu BK, Saxena AK, Singh BP, Singh K, Arora DK. Real-time PCR assay based on topoisomerase-II gene for detection of *Fusarium udum*. Mycopathologia. 2011;171:373–81.

Yli-Mattila T, Paavanen-Huhtala S, Jestoi M, Parikka P, Hietaniemi V, Gagkaeva T, Sarlin T, Haikara A, Laaksonen S, Rizzo A. Real-time PCR detection and quantification of *Fusarium poae, F. graminearum, F. sporotrichioides* and *F. langsethiae* in cereal grains in Finland and Russia. Arch Phytopathol Plant Prot. 2008;41:243–60.

Zambounis AG, Paplomatas E, Tsaftaris AS. Intergenic spacer–RFLP analysis and direct quantification of Australian Fusarium oxysporum f sp vasinfectum isolates from soil and infected cotton tissues. Plant Dis. 2007; 91:1564–73.

Zhang X, Zhang H, Pu J, Qi Y, Yu Q, Xie Y, Penj J. Development of a real-time fluorescence loop-mediated isothermal amplification assay for rapid and quantitative detection of *Fusarium oxysporum f. sp. cubense* tropical race 4 in soil. PLoS One. 2013;8:e82841.

Zhang Z, Zhang J, Wang Y, Zheng X. Molecular detection of Fusarium oxysporum f.sp. niveum and Mycosphaerella melonis in infected plant tissues and soil. FEMS Microbiol Lett 2005;249:39–47.

Chapter 5
DNA Barcoding for Diagnosis and Monitoring of Fungal Plant Pathogens

Prem Lal Kashyap, Pallavi Rai, Sudheer Kumar, Hillol Chakdar, and Alok K. Srivastava

5.1 Introduction

Fungal diseases are taking heavy toll on crop production every year, resulting in huge economic losses throughout the world (Kashyap et al. 2017; Ceasar and Ignacimuthu 2012; Kashyap et al. 2011; Mann et al. 2008). Moreover, this adversely affects the international trade and transport across countries due to high risk of spread of plant diseases. The effective and timely management of such disease, to study epidemiology and implementation of quarantine regulations requires the ability to diagnose and monitor fungal plant pathogens. Thus, an accurate method for the detection and identification of such agents is fundamental to rapid diagnostics of crop infection. Further, reliable fungal species identification is plays an important role in studying biology and ecology, because all biological aspects of any given fungal isolate in an ecosystem can only be attributed meaningfully via an unambiguous identifier like a species or race name (Kress et al. 2015). The conventional methods in fungal taxonomy are mainly based on morphological and physiological traits (Sharma et al. 2015;

The original version of this chapter was revised. An erratum to this chapter can be found at DOI 10.1007/978-3-319-34106-4_17

P.L. Kashyap (✉)
ICAR-National Bureau of Agriculturally Important Microorganisms,
Mau, Uttar Pradesh 275103, India

ICAR-Indian Institute of Wheat and Barley Research (IIWBR), Regional Station,
Flowerdale, Shimla 171002, India
e-mail: plkashyap@gmail.com

P. Rai (✉) • H. Chakdar (✉) • A.K. Srivastava (✉)
ICAR-National Bureau of Agriculturally Important Microorganisms,
Mau, Uttar Pradesh 275103, India

S. Kumar
ICAR-Indian Institute of Wheat and Barley Research, Karnal, Haryana 132001, India

B.P. Singh, V.K. Gupta (eds.), *Molecular Markers in Mycology*, Fungal Biology,
DOI 10.1007/978-3-319-34106-4_5

Miller et al. 2009; Atkins and Clark 2004). The morphology-based procedures are usually time consuming and may not always provide resolution to the species level (Zhao et al. 2011; Packer et al. 2009). Additionally, these characteristics are strongly influenced by environmental conditions. Therefore, it is needed to develop molecular taxonomy methods, which are independent to morphological and physiological characters of fungi. Current techniques for the molecular taxonomy of fungal plant pathogens include DNA fingerprinting (RAPD, AFLP, DGGE, SSCP, microsatellites, RFLP, microsatellites etc.), species-specific PCR (SS-PCR) using conserved housekeeping genes, unigene sequencing, multilocus sequence analysis (MLSA) and DNA hybridization (Sharma et al. 2017; Rai et al. 2016a; Kashyap et al. 2016a, b; Fang and Ramasamy 2015; Singh et al. 2014; Kumar et al. 2013; Kashyap et al. 2013; Gao and Zhang 2013; De Boer and López 2012; Kang et al. 2010; Sundelin et al. 2009; McCartney et al. 2003). However, some of these techniques are complicated, time-intensive, costly and tedious to use. Under such circumstances, DNA barcoding emerged as a species identification technology and holds promise for the reliable, quick, and accurate for the detection of phytopathogenic fungi.

DNA barcoding is a universal typing system to ensure rapid and accurate identification of any living entity including fungus. This technique allows the species characterization of organisms using a short DNA sequence from a standard and agreed-upon position in the genome (Xu 2016). In simple words, a DNA barcode represents a unique DNA sequence pattern 400–800 nucleotides in length that can be quickly processed from thousands of specimens or cultures and unambiguously analyzed by computer programs to identify species (Crous et al. 2015; Gao and Zhang 2013). At present, there is no universally accepted standard portion in genome for DNA barcoding of Fungi. In April 2011, the Fungal Working Group met in Amsterdam to discuss and evaluate the data generated with six markers from all major lineages of fungi. From the various markers evaluated, the internal transcribed spacer (ITS) appeared to be the main candidate because of its broad utility as a species marker in taxonomic and ecological studies and the ease of amplification across the kingdom. The ITS was subsequently proposed as a standard barcode for fungal plant pathogens. In spite of the fact that the ITS region cannot accurately identify species in many genera of plant-pathogenic fungi (e.g. *Alternaria, Botryosphaeria, Cercospora, Diaporthe, Fusarium* etc.), it always gets the user to at least the generic level (Sharma et al. 2015). Alternatively, other housekeeping genes with higher variability are being more extensively used to develop diagnostics for fungi, including nuclear genes such as β-tubulin (Aroca et al. 2008; Mostert et al. 2006; Fraaije et al. 2001), translation elongation factor 1 alpha (TEF-1α) (Rai et al. 2016b; Kristensen et al. 2005; Knutsen et al. 2004; Geiser et al. 2004), calmodulin (Mulè et al. 2004), RNA polymerase genes (RPB1 and RPB2) (O'Donnell et al. 2013), mini-chromosome maintenance complex component 7 (MCM7) (de Beer et al. 2014), 60S ribosomal protein RPL10 (de Beer et al. 2014), avirulence genes (Lievens et al. 2009), and mitochondrial genes such as the multicopy cox I and cox II and their intergenic region (Bilodeau et al. 2014; Nguyen and Seifert 2008; Seifert et al. 2007; Martin and Tooley 2003). The details of the genes as DNA barcode for diagnosis and monitoring of fungal plants pathogen are given in Table 5.1. Mating

Table 5.1 Important developments in DNA barcodes for the diagnosis and monitoring of fungal plant pathogens

Year	DNA barcode region/loci	Title/description	References
1999	Histone H 3	Differentiation of *Fusarium subglutinans* f. sp. *pini* by histone gene sequence data	Steenkamp et al. (1999)
2000	MAT-1 and MAT-2	PCR-Based identification of MAT-1 andMAT-2 in the *Gibberella fujikuroi* species complex	Steenkamp et al. (2000)
2002	ITS	Polymerase chain reaction assays for the detection and discrimination of the soybean rust pathogens *Phakopsora pachyrhizi* and *P. Meibomiae*	Frederick et al. (2002)
2002	β-tubulin and histone H3	β-Tubulin and histone H3 gene sequences distinguish *Cryphonectria cubensis* from South Africa, Asia, and South America	Myburg et al. (2002)
2004	Cellobiohydrolase-C and topoisomerase II	Use of genes encoding cellobiohydrolase-C and topoisomerase II as targets for phylogenetic analysis and identification of *Fusarium*	Hatsch et al. (2004)
2004	Calmodulin gene sequences	Specific detection of the toxigenic species *Fusarium proliferatum* and *F. oxysporum* from asparagus plants using primers based on calmodulin gene sequences	Mulè et al. (2004)
2006	ITS	Identification and quantification of pathogenic Pythium spp. from soils in eastern Washington using real-time polymerase chain reaction	Schroeder et al. (2006)
2006	ITS	Oligonucleotide array for identification and detection of *Pythium* species	Tambong et al. (2006)
2007	TEF1-α gene	Identification of *Fusarium solani* f. sp. *cucurbitae* race 1 and race 2 with PCR and production of disease-free pumpkin seeds	Mehl and Epstein (2007)
2007	ITS and partial β-tubulin	PCR-based strategy to detect and identify species of *Phaeoacremonium* causing grapevine diseases	Aroca and Raposo (2007)
2008	β-tubulin	A biomarker for the identification of four *Phaeoacremonium* species using the β-tubulin gene as the target sequence	Aroca et al. (2008)
2008	ITS	Identification and quantification of *Rhizoctonia solani* and *R. oryzae* using real-time polymerase chain reaction.	Okubara et al. (2008)

(continued)

Table 5.1 (continued)

Year	DNA barcode region/loci	Title/description	References
2009	ITS	PCR-Based assays for the detection of *Puccinia horiana* on chrysanthemums	Pedley (2009); Alaei et al. (2009)
2009	ITS	Detection of rDNA ITS polymorphism in *Rhizoctonia solani* AG 2-1 isolates	Pannecoucque and Hofte (2009)
2009	ITS and β-tubulin	Molecular tools to investigate *Rhizoctonia solani* distribution in soil	Budge et al. (2009)
2010	TEF1 and RPB2	Specific PCR detection of four quarantine *Fusarium* species in Korea	Hong et al. (2010)
2010	TEF-1α and IGSof the nuclear ribosomal operon	A molecular diagnostic for tropical race 4 of the banana fusarium wilt pathogen	Dita et al. (2010)
2010	ITS	PCR assays for the sugarcane rust pathogens *Puccinia kuehnii* and *P. melanocephala* and detection of a SNP associated with geographical distribution in *P. Kuehnii*	Glynn et al. (2010)
2010	ITS1 and ITS2	PCR-based assays to detect and quantify *Phomopsis sclerotioides* in plants and soil	Shishido et al. (2010)
2010	TEF1-α gene	A molecular based strategy for rapid diagnosis of toxigenic *Fusarium* species associated to cereal grains from Argentina	Sampietro et al. (2010)
2011	Histone-3 gene	Rapid detection and quantification of *Fusarium udum* in soil and plant samples using real-time PCR	Mesapogu et al. (2011)
2011	ITS	Real-time PCR detection and discrimination of the southern and common corn rust pathogens (*Puccinia polysora* and *Puccinia sorghi*)	Crouch and Szabo (2011)
2011	ITS, large subunit (28S) ribosomal RNA gene, CO1 and NAD 6	DNA barcoding in the rust genus *Chrysomyxa* and its implications for the phylogeny of the genus	Feau et al. (2011)
2011	ITS and CO1	DNA barcoding of oomycetes with cytochrome c oxidase subunit I and internal transcribed spacer	Robideau et al. (2011)
2011	Topoisomerase-II	Real-time PCR assay based on topoisomerase-II gene for detection of *Fusarium udum*.	Yadav et al. (2011)
2012	β-tubulin	A real-time PCR assay for detection and quantification of *Verticillium dahliae* in spinach seed	Duressa et al. (2012)

(continued)

Table 5.1 (continued)

Year	DNA barcode region/loci	Title/description	References
2012	Ribosomal DNA (rDNA) intergenic spacer (IGS)	Development of an assay for rapid detection and quantification of *Verticillium dahliae* in soil	Bilodeau et al. (2012)
2012	β-tubulin	Detection and quantification of *Rhizoctonia cerealis* in soil using real-time PCR	Guo et al. (2012)
2013	IGSregion of rDNA	Development and evaluation of a TaqMan real-time PCR assay for *Fusarium oxysporum* f. sp. *spinaciae*	Okubara et al. (2013)
2013	ITS and COX	Membrane-based oligonucleotide array developed from multiple markers for the detection of many *Phytophthora* species.	Chen et al. (2013)
2013	β-tubulin	A method to detect and quantify *Phaeomoniella chlamydospora* and *Phaeoacremonium aleophilum* DNA in grapevine-wood samples.	Pouzoulet et al. (2013)
2013	β-tubulin	Rapid detection and quantification of *Alternaria solani* in tomato	Kumar et al. (2013)
2014	Mitochondrial genome (ATP synthase subunits 9 and reduced nicotinamide adenine dinucleotide ubiquinone oxireductase subunits 9)	Development of a multiplex assay for genus- and species-specific detection of *Phytophthora* based on differences in mitochondrial gene order.	Bilodeau et al. (2014)
2014	Mini-chromosome maintenance complex component 7 (MCM7), 60S ribosomal protein RPL10 (60S), nuclear ribosomal DNA large subunit (LSU)	Redefining *Ceratocystis* and allied genera	de Beer et al. (2014)
2015	ITS	Development of a DNA macroarray for the detection and identification of fungal pathogens causing decline of young grapevines	Úrbez-Torres et al. (2015)
2015	ITS	Development of molecular assays for detection of *Stenocarpella maydis* and *Stenocarpella macrospora* in corn	Romero and Wise (2015)
2016	ITS	Internal transcribed spacer sequence analysis of *Puccinia helianthi* Schw. and its application in detection of sunflower rust	Guo et al. (2016)
2016	ITS	Oomycete-specific ITS primers for identification and metabarcoding	Riit et al. (2016)

type genes also show high diversity and fast evolutionary rate and could be used for inter- and intra-species differentiation (Kashyap et al. 2015). The DNA barcoding could be used for large-scale fungal identification and to assign unknown fungal individuals to species thus can enhance discovery of new species. The identification of organisms at different life cycle stages will become possible, where morphological identification is not realizable (Gilmore et al. 2009). As a result, DNA barcoding will provide not only a strong link between conventional and molecular taxonomy, but also a diagnostic tool for plant pathogens. Furthermore, this is also very helpful to assess the inoculums load, exotic fungal specimen or propagule, natural host range, and spread, which is very crucial for development of successful disease management programs. DNA barcode markers could provide valuable information to understand species boundaries, community ecology, functional trait evolution, trophic interactions, and biodiversity. The application of next-generation sequencing (NGS) technology will greatly expand the versatility of DNA barcodes across the Tree of Life, habitats, and geographies as new methodologies (Kress et al. 2015; Joly et al. 2014). Briefly, the chapter attempts to provide current updates on DNA barcoding and its applications in diagnosis and monitoring of fungal plant pathogens.

5.2 DNA Barcoding: A Concept

The concept of DNA barcoding proposes that effective and broad identification systems based on sequence based diversity in short and well defined gene regions. This concept was initially proposed by Hebert (2003). In his pioneer work on biological identifications through DNA barcodes, species identification using a short segment of DNA from a standardized region of the genome was described and highlighted the application of this tool to identify different species. DNA barcodes allow non – expert to identify species, even from small, damaged or industrially processed material. Principally, the aim of a barcode system is to be applicable for all kingdoms of eukaryotic life and to simplify recognition of cryptic species. In laboratory, species-level identification by DNA barcoding is usually adapted by the recovery of a short DNA sequence from a standard part of the genome. The sequence of barcode from each unknown specimen was then compared with a library of reference barcode sequences obtained from individuals of recognized identity. In DNA barcoding, species are separated by standardized barcode gap analyses or phylogenetic tree-building methods. A barcode gap exists if the minimum interspecific variation is bigger than the maximum intraspecific variation. Alternatively, phylogenetic neighbour joining analysis based on Kimura two-parameter distances is a standard method for phylogenetic analysis of fungal plant pathogens. At present, information on DNA barcoding for fungi is still limited and much more work is required.

5.3 Strategies for DNA Barcoding of Fungal Plant Pathogens

Two strategies could be adopted for DNA barcoding of fungal plant pathogens: (i) verifying the reliability of existing structural and functional gene sequences available in GenBank for use as barcodes, and (ii) determining other available barcodes of species that cannot be identified by universal fungal DNA barcode like ITS. For fungal plant pathogen for which the ITS can be used as the barcode, research interest lies in extensively verifying the reliability and authenticity of the ITS, which should be easy to amplify, and show low intra-specific and high inter-specific divergence. The methodological studies on the species are common and must thus be suitable and valid for selecting primers, determining threshold values, sequencing, phylogenetic distance matrix statistical analysis, and species identification methods (Gao and Zhang 2013). At first, the acquired reliable DNA barcode sequence data should be linked as closely as possible with the homologous voucher specimens or credible fungal species. Second, for the identification methods for phytopathogenic fungi using DNA barcoding *sensu stricto*, divergent thresholds or barcoding gaps should correspond to each individual species rather than being uniform values. The ideal situation would be to have ten or more specimens per species to cover intra-specific variability (Begerow et al. 2010). Second, the fungi remained unidentified using ITS as a barcode, other potential molecular markers should be determined. However, the lack of information can be troublesome. Mining the extensive sequencing data, such as that for key genes or the whole genome, should be a priority. Next-generation sequencing (NGS) was recently applied to discover, validate, and assess genetic markers and to describe species boundaries (Davey et al. 2011). Though, for the development and identification of short unique signature sequences, NGS approach is a costly affair. However, as the cost of NGS declines and user-friendly bioinformatics pipelines are established, thus it may become economical.

5.4 Workflow for Generating DNA Barcodes

The workflow for generating DNA barcodes for individual species entails two basic steps: (i) building the DNA barcode library of known species; and (ii) matching the DNA barcode sequence of an unknown sample against the barcode library for identification (Fig. 5.1). The DNA barcode library is a collection of DNA sequences associated with verified taxonomic identification and ideally with voucher specimens. A comprehensive DNA barcode library is a recognized limiting factor because of the overwhelming number of species of fungi already described by taxonomists. In this respect, herbarium specimens are critical source of tissue for generating DNA barcodes with known vouchers. Genomic DNA is extracted from the fungi, and the barcode portion of the ITS gene or other agreed upon segment of genome is amplified by PCR. The primer pairs or locus used for generating fungal DNA

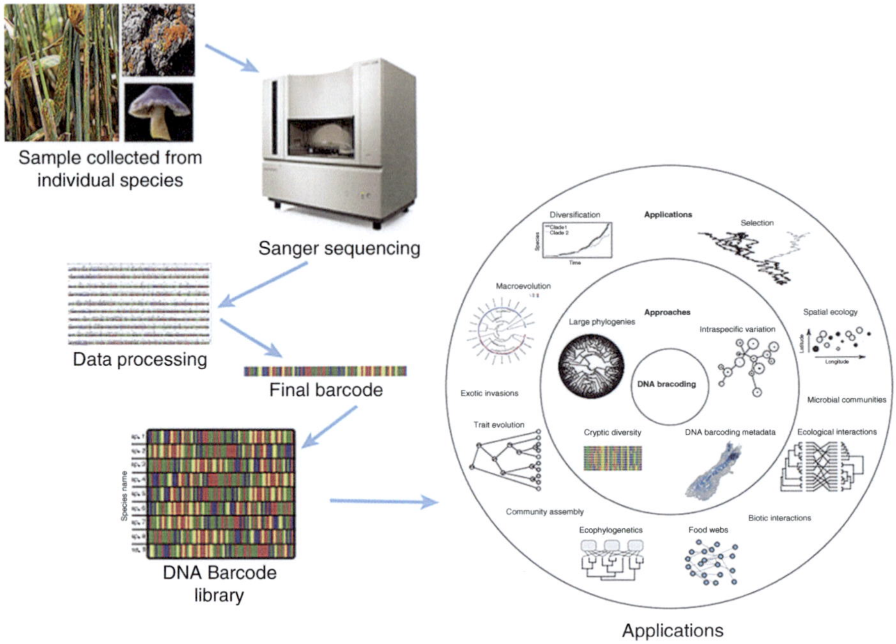

Fig. 5.1 Workflow for generating DNA barcodes and application of DNA barcoding

barcode is listed in Table 5.2. After amplifying with desirable primer, the amplified sequence (amplicon) is submitted for sequencing in one or both directions. The sequencing results are then used for similarity in sequence databases as listed in Table 5.3. Nearest neighbour algorithm is usually used to assign species identity on the basis of maximum sequence similarity in known sequence Basic Local Alignment Search Tool (BLAST) is a common matching tool, provided through NCBI that searches for correspondence between a query sequence and a sequence library. However, some barcodes will be entirely new, and identification may rely on placing the unknown species in a phylogenetic tree with near relatives.

5.5 Selection of an Ideal DNA Barcode Marker

Several criteria have been subjected to determination of an ideal DNA barcode, such as a short standard fragment, universally accepted, having adequate variations among species and conserve within a species, exhibiting a high species resolving power (Hollingsworth et al. 2011). The appropriate intra- and inter-specific sequence variations (i.e. divergent among species but relatively stable within a species) and ease in sequence acquisition are the two most important aspects in evaluating feasibility of a DNA fragment as a barcode marker.

Table 5.2 Major gene/locus selected for DNA barcoding of fungal plant pathogens

Region (locus)	Primer sequence (5′-3′)	Orientation	Reference
Internal transcribed spacer (ITS) region	ITS1: TCCGTAGGTGAACCTGCGG	Forward	White et al. (1990)
	ITS4: TCCTCCGCTTATTGATATGC	Reverse	White et al. (1990)
	ITS1: TCCGTAGGTGAACCTGCGG	Forward	White et al. (1990)
	ITS2: GCTGCGTTCTTCATCGATGC	Reverse	White et al. (1990)
	ITS3: GCATCGATGAAGAACGCAGC	Forward	White et al. (1990)
	ITS4: TCCTCCGCTTATTGATATGC (R)	Reverse	White et al. (1990)
	ITS5: GGAAGTAAAAGTCGTAACAAGG	Forward	White et al. (1990)
	ITS4: TCCTCCGCTTATTGATATGC	Reverse	White et al. (1990)
Nuclear large sub unit (nLSU)	LROR: ACCCGCTGAACTTAAGC	Forward	Vilgalys and Hester (1990)
	LR5: TCCTGAGGGAAACTTCG	Reverse	Vilgalys and Hester (1990)
	LSU1Fd: GRATCAGGTAGGRATACCCG	Forward	Crous et al. (2009)
	LR5: TCCTGAGGGAAACTTCG	Reverse	Vilgalys and Hester (1990)
Translation elongation factor 1-alpha (TEF1α)	EF1α1018F: GAYTTCATCAAGAACATGAT	Forward	Rehner and Buckley (2005)
	EF1α1620R: GACGTTGAADCCRACRTTGTC	Reverse	Rehner and Buckley (2005)
	EF1-728F: CATCGAGAAGTTCGAGAAGG	Forward	Carbone and Kohn (1999)
	EF-2: GGARGTACCAGTSATCATGTT	Reverse	O'Donnell et al. (1998b)
β-tubulin (Btub)	Btub 2Fd: GTBCACCTYCARACCGGYCARTG	Forward	Groenewald et al. (2013)
	Btub4Rd: CCRGAYTGRCCRAARACRAAGTTGTC	Reverse	Groenewald et al. (2013)
	T1:AACATGCGTGAGATTGTAAGT	Forward	O'Donnell and Cigelnik (1997)

(continued)

Table 5.2 (continued)

Region (locus)	Primer sequence (5′-3′)	Orientation	Reference
	β-Sandy-R:GCRCGNGGVACRTACTTGTT	Reverse	Stukenbrock et al. (2012)
	Bsens:ATCACWCACTCICTIGGTGGTGG	Forward	Lesage-Meessen et al. (2011)
	Brev:CATGAAGAARTGIAGACGIGGG	Reverse	Lesage-Meessen et al. (2011)
RNA polymerase II second largest subunit	fRPB2-5F: GAYGAYMGWGATCAYTTYGG	Forward	Liu et al. (1999)
	fRPB2-414R: ACMANNCCCCARTGNGWRTTRTG	Reverse	Quaedvlieg et al. (2011)
Calmodulin	CAL-235F: TTCAAGGAGGCCTTCTCCCTCTT	Forward	Quaedvlieg et al. (2012a, b)
	CAL2Rd: TGRTCNGCCTCDCGGATCATCTC	Reverse	Groenewald et al. (2013)
Actin (ACT)	ACT-512F: ATGTGCAAGGCCGGTTTCGC	Forward	Carbone and Kohn (1999)
	ACT2Rd: ARRTCRCGDCCRGCCATGTC	Reverse	Groenewald et al. (2013)
	CA5R: GTGAACAATGGATGGACCAGATTCGTCG	Forward	Daniel et al. (2001)
	CA14: AACTGGGATGACATGGAGAAGATCTGGC	Reverse	Daniel & Meyer (2003)
60S ribosomal protein L10a (60S L10)	60S-506F:GHGACAAGCGTTTCTCNGG	Forward	Stielow et al. 2015
	60S-908R:CTTVAVYTGGAACTTGATGGT	Reverse	Stielow et al. 2015
Phosphoglycerate kinase (PGK)	PGK-533F GTYGAYTTCAAYGTYCC	Forward	Stielow et al. 2015
	PGK-533R ACACCDGGDGGRCCGTTCCA	Reverse	Stielow et al. 2015
DNA topoisomerase I (TOPI)	TOP1-501F: ACTGCCAAGGTTTTCCGTACHTACAACGC	Forward	Stielow et al. 2015
	TOP1-501R: CCAGTCCTCGTCAACWGACTTRATRGCCCA	Reverse	Stielow et al. (2015)
Hypothetical protein LNS2	LNS2-468F: GGCCATGTGCTGAACATGATCGGHCG WGAYTGGAC	Forward	Stielow et al. (2015)
	LNS2-468R: CGGTTGCCRAAKCCRGCATAGAAKGG	Reverse	Stielow et al. (2015)

(continued)

Table 5.2 (continued)

Region (locus)	Primer sequence (5′-3′)	Orientation	Reference
Histone 3	H3-1a: ACTAAGCAGACCGCCCGCAGG	Forward	Glass and Donaldson (1995)
	H3-1b: GCGGGCGAGCTGGATGTCCTT	Reverse	Glass and Donaldson (1995)
Histone 4	H4-1a: GCTATCCGCCGTCTCGCT	Forward	Glass and Donaldson (1995)
	H4-1b: GGTACGGCCCTGGCGCTT	Reverse	Glass and Donaldson (1995)
Plasma membrane ATPase	A1a: TTCCTCGGTTTCTTCGTCGGTCCC	Forward	Glass and Donaldson (1995)
	A1b: CTAGTCAGACGAGAATGGCCGCTC	Reverse	Glass and Donaldson (1995)
Phosphate permease	pho4-1a: GCTGCCCTTGATGCTTGG	Forward	Glass and Donaldson (1995)
	pho4-1b: CAGGGACCAGCAGGGCGA	Reverse	Glass and Donaldson (1995)
Chitin synthase (CHS)	CHS-79F: TGG AAG AAC CAT CTG TGA GAG TTG	Forward	Carbone and Kohn (1999)
	CHS-354R: TGG GGC AAG GAT GCT TGG AAG AAG	Reverse	Carbone and Kohn (1999)
Ras-related protein gene (Ypt1)	Ypt1F: CGACCATYGGYGTKGACT TT	Forward	Chen and Roxby (1996)
	Ypt5R:GCAGCTTGTTSACGTTCT CR	Reverse	Schena and Cooke (2006)
Mating type locus gene (MAT1-2)	Mat1M72F: ACG GCA AAC GGC TCA GGG AGT G	Forward	Crouch et al. (2009a, b)
	Mat1M72R: AAT GCC GAG TCC CAC GAG GTT CG	Reverse	Crouch et al. (2009a, b)
Manganese superoxide dismutase gene (SOD2)	SOD507F: ATG GCA GCC TTT CCG TTG AGA TAC	Forward	Crouch et al. (2006)
	SOD507R: AGT TGA CAT GAA GCC ACC TAC AGC	Reverse	Crouch et al. (2006)

(continued)

Table 5.2 (continued)

Region (locus)	Primer sequence (5′-3′)	Orientation	Reference
	SOD625F: GCC CAC AGT ACA TAT TGC CTA AGC	Forward	Crouch et al. (2006)
	SOD625R: TCA TCC CGG GAG CCA GAA AAC CT	Reverse	Crouch et al. (2006)
	SODglo2-F: CAG ATC ATG GAG CTG CAC CA	Forward	Moriwaki and Tsukiboshi (2009)
	SODglo2-R: TAG TAC GCG TGC TCG GAC AT	Reverse	Moriwaki and Tsukiboshi (2009)
Glyceraldehyde-3-phosphate dehydrogenase gene (GAPDH)	GDF: GCC GTC AAC GAC CCC TTC ATT GA	Forward	Templeton et al. (1992)
	GDR: GGG TGG AGT CGT ACT TGA GCA TGT	Reverse	Templeton et al. (1992)
	GDF1: GCC GTC AAC GAC CCC TTC ATT GA	Forward	Guerber et al. (2003)
	GDR1: GGG TGG AGT CGT ACT TGA GCA TGT	Reverse	Guerber et al. (2003)
	Gpd1-LM: ATT GGC CGC ATC GTC TTC CGC AA	Forward	Myllys et al. (2002)
	Gpd2-LM: CCC ACT CGT TGT CGT ACC	Reverse	Myllys et al. (2002)
Glutamine synthetase gene (GS)	GSF1: ATG GCC GAG TAC ATC TGG	Forward	Stephenson et al. (1997)
	GSR1 GAA CCG TCG AAG TTC CAG	Reverse	Stephenson et al. (1997)
	GSF3: GCC GGT GGA GGA ACC GTC G	Forward	Weir et al. (2012a, b)
	GSR2: GAA CCG TCG AAG TTC CAC	Reverse	Weir et al. (2012a, b)
Apurinic DNA lyase gene (APN2)	Apn1W1F: ATG GAG CAC AAA AAC GAA CA	Forward	Crouch et al. (2009b)
	Apn1W1R: GCG GAG CAG AGG ATG TAG TC	Reverse	Crouch et al. (2009b)

Table 5.3 Major databases available for extracting sequence and structure information of DNA barcodes for fungi

Database	Description	Website/url address
Barcode of life Datasystem (BOLD)	A web platform that provides an integrated environment for the assembly and use of DNA barcode data.	http://www.barcodinglife.org
Canadian Centre for DNA Barcoding (CCDB)	Birthplace of DNA Barcoding and offers unparalleled access tospecies identifications for over 180,000 animals, plants and molds	http://www.dnabarcoding.ca
Consortium for the Barcode of life (CBOL)	Established in 2004 through support from the Alfred P. Sloan Foundation, CBOL promotes barcoding through working groups, networks, workshops, conferences, outreach, and training.	http://www.barcodeoflife.org
European consortium for the Barcode of life (ECBOL)	Established within EDIT work package 3.4 "DNA barcoding". ECBOL functions as an information and coordination hub for taxonomists in Europe.	http://www.ecbol.org
International Barcode of life (iBOL)	Uses sequence diversity in short, standardized gene regions-DNA barcodes. IBOL's main mission is extending the geographic and taxonomic coverage of the barcode reference library (BOLD).	http://www.ibol.org/
Quarantine Barcode of life (QBOL)	Developing DNA barcoding to identify quarantine organisms in support of plant health.	http://www.qbol.org
Assembling the fungal tree of life (AFTOL)	First database to catalogue fungal subcellular characters and dedicated to significantly enhancing our understanding of the evolution of the kingdom fungi.	http://aftol.org/
ITS database for ectomycorrhizal fungi (UNITE)	An rDNA sequence database to provide a stable platform for sequence identification of ectomycorrhizal and basidiomycetes. It is equipped with a BLAST interface for rapid similarity searches against the records of the database.	http://unite.ut.ee
Mycobank	An online database, documenting new mycological names and combinations.	www.mycobank.org
Maarj AM	Web-based database for studies on the diversity of arbuscular mycorrhizal fungi (Glomeromycota).	http://maarjam.botany.ut.ee/
Q- bank	Comprehensive database on quarantine plant pest and diseases	http://www.q-bank.eu/

(continued)

Table 5.3 (continued)

Database	Description	Website/url address
Fusarium MLST database	Developed to facilitate identification of agriculturally and medically important fusaria by conducting nucleotide BLAST queries of these dedicated DNA sequence databases via the internet	http://www.cbs.knaw.nl/Fusarium
Fusarium-ID	Enables users to explore the diversity of *Fusarium* and accurately identify new isolates based on their sequence similarity to previously characterized species.	http://isolate.fusariumdb.org/index.php

5.6 DNA Barcode Markers

5.6.1 Ribosomal RNA Genes

Ribosomal RNA genes are the most commonly used loci in molecular systematic studies of fungi (Lutzoni et al. 2004). They are of primary importance because various regions of the rRNA cistron are frequently targeted for more than two decades for fungal diagnostic and phylogenetics (Begerow et al. 2010). The eukaryotic rRNA cistron consists of the 18S small subunit (SSU), the 5.8S, and the 28S large subunit (LSU) genes transcribed as a unit by RNA polymerase I. Post transcriptional processes split the cistron, removing two internal transcribed spacers. These two spacers, including 5.8S gene, are usually referred as the ITS region (Naidoo et al. 2013). The 18S nuclear ribosomal small subunit rRNA gene (SSU) is commonly used in phylogenetics because of its conserve nature. The ribosomal DNA (rRNA genes) acquired the characteristics which are appropriate for the recognition of microorganisms at the species level. These rDNA are highly stable and exhibit a mosaic of conserved and diverse regions within the genome (Hibbett 1992). They also occur in multiple copies up to 200 copies per haploid genome (Yao et al. 1992; Bruns et al. 1991) arranged in tandem repeats with each repeat consisting of the 18S small subunit (SSU), 5.8S, and 28S large subunit (LSU) genes. The sequence length of ITS for taxonomy is the complete ITS region, 500–600 bp (ITS1-5.8S-ITS2), depending on the genera. For LSU, it may be D1 and D2 (300–400 bp) or the complete length (~1400 bp) and for environmental sequence it may be 100–300 bp (Sharma et al. 2015).

The internal transcribed spacer (ITS) region of nuclear encoded ribosomal DNA (rDNA) has been proposed as one of the primary fungal barcoding loci (Schoch et al. 2012). ITS sequencing data is easily obtained and good starting point to identify genera and sometimes species. If a genus or species is not represented in curated database such as Q-bank, GenBank blast could be used to supplement these curated databases. Structurally, ITS region is comprised of the internal transcribed spacer 1 (ITS1), 5.8S rRNA gene, and the internal transcribed spacer 2 (ITS2) with the greatest sequence variation in the ITS1 and ITS2 regions. Several studies have examined the implications of ITS region for identification of fungal plant pathogen using high throughput sequencing (Sharma et al. 2015; Feau et al. 2009). The ITS locus is easily amplified and gives a good species resolution in most of fungal groups. However, lack of sufficient ITS interspecies variation within some genera of *Mycosphaerella*-like fungi (e.g. *Septoria*, *Cercospora* and *Pseudocercospora* etc.) and *Fusarium* has been reported (Quaedvlieg et al. 2013; Balajee et al. 2009; Hunter et al. 2006; Verkley et al. 2004). Despite these problems, several ITS rDNA reference sequences are available in the Assembling the Fungal Tree of Life (AFTOL) and GenBank sequence databases and tools have been developed to facilitate the use of ITS for fungal biodiversity assessment, ecological studies, diagnostics and monitoring. Several reports indicated the use of ITS as the primary locus, and if necessary using a secondary locus following a molecular decision protocol (Sharma et al. 2015). At present, this is the most widely accepted and stable marker for reliable identification of fungal plant pathogens.

5.6.2 *Cytochrome* c *Oxidase Subunit I (COI,* COX1*)*

The cytochrome c oxidase is a mitochondrial protein, located in the inner mitochondrial membrane, and is a key enzyme in the electron transport chain. Therefore, it plays a central role in the metabolism of eukaryotic aerobic organisms. It consists of several subunits, and the catalytic cytochrome c oxidase subunit 1 is encoded in the mitochondrial genome (Seifert et al. 2007; Hebert et al. 2004). COI has proven useful in phylogenetic studies of the oomycete genus *Phytophthora* (Kroon et al. 2004; Martin and Tooley 2003), and the success of COI barcoding in red algae (Saunders 2005) made it a very intriguing prospect for barcoding of all oomycetes due to their algal ancestry. Because COI is a protein-coding region, alignment of COI sequences is simple and devoid of gaps if introns are absent. With the use of primers that amplify the 5′ end of COI, accurate species delimitation has been achieved with sequences of ~650 base pairs (bp) or less (Meusnier et al. 2008). Recently, Robideau et al. (2011) reported that both ITS and COI are useful barcodes for accurate identification of many oomycetes. Intraspecific variation of COI is at par with that of ITS, although ITS does provide greater interspecific variation than COI (Seifert 2009). The benefit of COI as barcoding marker is ease in sequencing and aligning a relatively short fragment which has uniform length and can be amplified with degenerate primers throughout the entire oomycete class. This advantage over ITS is especially evident in the downy mildew genera *Basidiophora*, *Plasmopara*, *Plasmoverna* and relatives, which contain insertions in the ITS2 resulting in ITS sequences often longer than 2 kb (Thines 2007), raising difficulties to amplify, sequence and align the complete ITS region. However, a thorough analysis of the COI marker in *Fusarium* revealed a significant number of multiple copies, various numbers of introns and highlighted the problems pertaining to primer design and the need for nested primers to amplify the entire region (Gilmore et al. 2009). Looking at other mitochondrial genes as candidates for barcode regions, Santamaria et al. (2009) demonstrated that introns are very common in mitochondrial genome of Ascomycota and suggested NADH dehydrogenase 6 (nad6) as a possible barcode marker. Based on these studies it is clear that COI will difficult to use as a universal marker for fungal barcoding.

5.6.3 *β-Tubulin (*Btub*)*

Beta-tubulin genes are found in all eukaryotes and have been used for phylogenetic analysis in fungi from the entire kingdom to the species level (Thon and Royse 1999). The beta-tubulin gene sequences contain 3.5-fold more phylogenetic information than the small sub-unit (SSU) rRNA gene, thus it has been reported an ideal marker for analysis of deep level phylogenies and for complex species groups.

5.6.4 *Translation Elongation Factor 1-Alpha (EF-1α)*

The translation elongation factor 1-a (TEF) gene encodes an essential part of the protein translation machinery. This gene has high phylogenetic utility because it is (i) highly informative at the species level especially in case of *Fusarium*; (ii) non-orthologous copies of the gene have not been detected in the genus; and (iii) universal primers have been designed that work across the phylogenetic breadth of the genus (Geiser et al. 2004). This gene was first used as a phylogenetic marker to infer species- and generic-level relationships among Lepidoptera (Mitchell et al. 1997; Cho et al. 1995). Primers were first developed in the fungi to investigate lineages within the *F. oxysporum* complex (O'Donnell et al. 1998a). The ef1 and ef2 primers were designed based on sites shared in exons and can be applied to a number of filamentous ascomycetes. These primers amplify ~700 bp region of TEF, flanking three introns that total over half of the amplicon's length, in all known fusaria. This gene appears to be consistently single-copy in *Fusarium*, and it shows a high level of sequence polymorphism among closely related species, even in comparison to the intron-rich portions of protein-coding genes such as calmodulin, beta-tubulin and histone H3. For these reasons, TEF has become the marker of choice as a single-locus identification tool in *Fusarium*. Nitschke et al. (2009) used this marker for reliable identification based on sequence information of the translation elongation factor 1α (TEF-1α) gene for *Fusarium* spp. isolated from sugar beets. In all, 65 isolates from different species (*Fusarium avenaceum*, F. *cerealis*, *F. culmorum*, *F. equiseti*, *F. graminearum*, *F. oxysporum*, *F. proliferatum*, *F. redolens*, *F. solani*, *F. tricinctum* and *F. venenatum*) were obtained from sugar beet at different developmental stages from worldwide. Database sequences for additional species (*F. sporotrichioides*, *F. poae*, *F. torulosum*, *F. hostae*, *F. sambucinum*, *F. subglutinans*, and *F. verticillioides*) isolated from sugar beets in previous studies, were included in the analysis. Molecular sequence analysis of the partial TEF-1α gene fragment revealed sufficient variability to differentiate between the *Fusarium* spp., resulting in species-dependent separation of the isolates analyzed. This inter-specific divergence could be translated into a polymerase chain reaction restriction fragment length polymorphism assay using only two subsequent restriction digests for the differentiation of 17 of 18 species. Recently, Kashyap et al. (2016a, b) used partial TEF-1α gene fragment to analyse the phylogenetic relationship among Indian races of *Fusarium oxysporum* f. sp. *ciceris* from chickpea.

5.6.5 RNA Polymerase II Second Largest Subunit (RPB2)

RNA polymerase II is one of the three RNA polymerases of eukaryotes which catalyse the transcription of messenger RNA. The RNA polymerase II gene (RPB2), which encodes the second largest protein subunit, is suggested to be a single-copy gene in fungi and well conserved. Twelve highly conserved regions of the RPB2 protein subunit have >85% amino acid identity among fungi (James et al. 1991). Therefore, in combination with numerous morphological characteristics, nucleotide sequance and deduced polypeptide sequence data, the partial RPB2 gene sequence has been used in parsimony analysis to detect phylogenetic links and to understand evolution in plant pathogenic fungi (Malkus et al. 2006).

5.6.6 Other Alternative DNA Barcode Markers

A number of key concepts for the selection process for alternative candidate barcodes were described by several workers (Stielow et al. 2015; Robert et al. 2011; Lewis et al. 2011) to search for gene regions in complete fungal genomes under several optimality criteria. Lewis et al. (2011) inferred suitable candidate genes using translated protein sequences (Pfam domains), whereas Robert et al. (2011) focused entirely on nucleic acids. These studies identified gene sections and novel primer pairs that functioned *in silico* for the genomes available, either as universal fungal primers, or as primers targeting phyla or classes. Furthermore, Stielow et al. (2015) also tested and compared amplification efficiency of two nuclear ribosomal regions (ITS and LSU, D1–D2 domains of 26/28S), the 5′ primed end of β-tubulin2 (TUB2) and γ-actin (ACT), the section '6–7' of the second largest subunit of the RNA-polymerase II gene (RPB2), the commonly used intermediate section of translation elongation factor 1-α (TEF1α) (Sasikumar et al. 2012) corresponding to the section 983–1567 bp (in the rust *Puccinia graminis*), two novel universal high fidelity TEF1-α primer alternatives covering approximately the same region, three sections of the newly identified candidate region and fungus-specific gene, Histone 3 (Mesapogu et al. 2011), translation elongation factor 3 (TEF3) (Belfield and Tuite 2006; Greganova et al. 2011), and one of the small ribosomal proteins required for tRNA transfer, the 60S L10 (L1) (Brodersen and Nissen 2005), phosphoglycerate kinase (PGK), DNA topoisomerase I (TOPI), Lipin/Ned1/Smp2 (LNS2), Indole-3-glycerol phosphate synthase (IGPS), Phosphoribosylaminoimidazolecarboxylase

PurE domain (PurE), Peptide methionine sulphoxidereductase (Msr) and Vacuolar ATPase (vATP) (Stielow et al. 2015). Both TOPI and PGK show prospects for the Ascomycota, while TOPI and LNS2 are attractive for the Pucciniomycotina, for which universal primers for ribosomal subunits often fail. Besides this, some other genes such as glyceraldehyde-3-phosphate dehydrogenase (GAPDH) and glutamine synthetase (GS), either singly or in combinations can be used to reliably distinguish most taxa or will need to be developed as secondary barcodes for species level identification, which is important because fungal plant pathogens are of biosecurity significance (Weir et al. 2012). From all these studies, it is clear that several gene sections are accessible to universal primers yielding a single PCR-product. Barcode gap and multi-dimensional scaling analysis revealed that some of the tested candidate markers have universal properties providing adequate intra- and inter-specific variation that make them attractive barcodes for fungal species identification and diagnosis.

5.7 Bioinformatics Tools for DNA Barcoding

A wide range of programs are available for sequence data analysis. The three commonly used methods for phylogenetic analysis are maximum parsimony (MP), maximum likelihood (ML) (Hillis et al. 1994; Felsenstein 1985), and Bayesian inference (BI). Among these, maximum likelihood is the most discriminative. The maximum likelihood (ML) method evaluates an evolutionary hypothesis in terms of the probability that the proposed model and the hypothesized history would give rise to the observed data set properly. The topology with the highest maximum probability or likelihood is then chosen. This method may have lower variance than other methods and is thus least affected by sampling error and differential rates of evolution. It can statistically evaluate different tree topologies and use all of the sequence information. The maximum parsimony algorithm (Swofford et al. 1996; Farris 1970) searches for the minimum number of genetic events (*e.g.* nucleotide substitutions) to infer the shortest possible tree (i.e., the maximally parsimonious tree). Often the analysis generates multiple equally most parsimonious trees. When evolutionary rates are drastically different among the species analyzed, results from parsimony analysis can be misleading (Felsenstein 1978). Parsimony analysis is most often performed with the computer program PAUP* 4.0 (Swofford 2002a, b), and MEGA (Tamura et al. 2011). The Bayesian phylogenetic inference is model-based method and was proposed as an alternative to maximum likelihood (Ali et al. 2014). The computer program MrBayes 3.0 (Huelsenbeck and Ronquist 2001) performs Bayesian estimation of phylogeny based on the posterior probability distribution of trees, which is approximated using a simulation technique called Markov chain Monte Carlo (or MCMC). MrBayes can combine information from different data partitions or subsets evolving under different stochastic evolutionary models. This allows the user to analyze heterogeneous data sets consisting of different data types, including morphology and nucleotides. Bayesian inference has facilitated the exploration of parameter-rich evolutionary models (Table 5.4).

Table 5.4 Bioinformatic tools and phylogenetic analysis softwares used for DNA barcoding

Tool(s)	Description	Reference
Bayesian evolutionary analysis sampling trees (BEAST)	Used to analyze nucleotide and amino acid sequences, as well as morphological data	Drummond and Rambaut (2007)
BioEdit	A fairly comprehensive sequence alignment and analysis tool and supports a wide array of file types and offers a simple interface for local BLAST searches	Hall (1999)
ClustalX	A windows interface for the Clustal W multiple sequence alignment program. It provides an integrated environment for performing multiple sequence and profile alignments and analyzing the results. This program allows to create neighbor joining trees with bootstrapping	Larkin et al. (2007)
ClustalW	Multiple sequence alignment. This provides one with a number of options for data presentation, homology matrices [BLOSUM (Henikoff), PAM (Dayhoff) or GONNET, and presentation of phylogenetic trees (neighbor-joining, Phylip or distance).	Larkin et al. (2007)
DNA for windows	A compact, easy to use DNA analysis program, ideal for small-scale sequencing projects	http://www.dna-software.co.uk/
Geneious	Provides an automatically-updating library of genomic and genetic data; for organizing and visualizing data. It provides a fully integrated, visually-advanced toolset for: Sequence alignment and phylogenetics; sequence analysis including BLAST; protein structure viewing, NCBI, EMBL, Pubmed auto-find, etc.	Kearse et al. (2012)
MAFFT	A multiple sequence alignment program for unix-like operating systems. It offers a range of multiple alignment methods, L-INS-i (accurate; for alignment of <∼200 sequences), FFT-NS-2 (fast; for alignment of <∼10,000 sequences), etc.	
FigTree	A graphical viewer of phylogenetic trees to display summarized and annotated trees produced by BEAST	http://tree.bio.ed.ac.uk/software/figtree/
Format converter v2.2.5	This program takes as input a sequence or sequences (*e.g.*, an alignment) in an unspecified format and converts the sequence(s) to a different user-specified format	https://www.hiv.lanl.gov/content/sequence/HIV/COMPENDIUM/compendium.html
Genetic algorithm for rapid likelihood inference (GARLI)	A program that uses genetic algorithms to search for maximum likelihood trees. It includes the GTR + Γ model and special cases and can analyze nucleotide, amino acid and codon sequences. A parallel version is also available	Cummings (2014)
Hypothesis testing using phylogenies (HyPhy)	A maximum likelihood program for fitting models of molecular evolution. It implements a high-level language that the user can use to specify models and to set up likelihood ratio tests	Pond et al. (2005)

(continued)

Table 5.4 (continued)

Tool(s)	Description	Reference
ITS2 database	An exhaustive dataset of internal transcribed spacer 2 sequences from NCBI GenBank accurately reannotated. Following an annotation by profile hidden Markov models (HMMs), the secondary structure of each sequence is predicted. Also provides several tools to process ITS2 sequences, including annotation, structural prediction, motif detection and BLAST search on the combined sequence–structure information. Moreover, it integrates trimmed versions of 4SALE and ProfDistS for multiple sequence–structure alignment calculation and neighbor joining tree reconstruction. Together they form a coherent analysis pipeline from an initial set of sequences to a phylogeny based on sequence and secondary structure	Merget et al. (2002)
Molecular evolutionary genetic analysis (MEGA)	A windows-based program with a full graphical user interface that can be run under mac OSX or Linux using windows emulators. It includes distance, parsimony and likelihood methods of phylogeny reconstruction, although its strength lies in the distance methods. It incorporates the alignment program Clustal W and can retrieve data from GenBank	Tamura et al. (2013)
MRBAYES	A Bayesian MCMC program for phylogenetic inference. It includes all of the models of nucleotide, amino acid and codon substitution developed for likelihood analysis	Huelsenbeck and Ronquist (2001)
Modeltest	Uses hierarchical likelihood ratio tests (hLRT) to compare the fit of the nested GTR (general time reversible) family of nucleotide substitution models. Additionally, it calculates the Akaike information criterion estimate associated with the likelihood scores	Posada and Crandall (1998)
OligoCalc	On line tool for to find length, melting temperature, %GC content and molecular weight of DNA sequence	Kibbe (2007)
Phylogenetic analysis by maximum likelihood (PAML)	A collection of programs for estimating parameters and testing hypotheses using likelihood. It is mostly used for tests of positive selection, ancestral reconstruction and molecular clock dating. It is not appropriate for tree searches	Yang (2007)
Phylogeny.fr	A simple to use web service dedicated to reconstructing and analysing phylogenetic relationships between molecular sequences. It includes multiple alignment (MUSCLE, T-Coffee, ClustalW, ProbCons), phylogeny (PhyML, MrBayes, TNT, BioNJ), tree viewer (Drawgram, Drawtree, ATV) and utility programs (*e.g.* Gblocks to eliminate poorly aligned positions and divergent regions)	Dereeper et al. (2008)

(continued)

Table 5.4 (continued)

Tool(s)	Description	Reference
PHYLIP	A package of programs for phylogenetic inference by distance, parsimony and likelihood methods	Felsenstein (1995)
PhyML	A fast program for searching for the maximum likelihood trees using nucleotide or protein sequence data	Guindon et al. (2009)
PAUP	PAUP* has many options and close compatibility with MacClade. It includes parsimony, distance matrix, invariants, and maximum likelihood methods and many indices and statistical tests	Swofford (2002a, b)
ProfDistS	Distance based phylogeny on sequence–structure alignments.	Wolf et al. (2008)
MacClade	A computer program for phylogenetic analysis. Its analytical strength is in studies of character evolution. It also provides many tools for entering and editing data and phylogenies, and for producing tree diagrams and charts	Maddison and Maddison (1992)
Neighbor-joining	For reconstructing phylogenetic trees from evolutionary distance data	Saitou and Nei (1987)
RAxML	A fast program for searching for the maximum likelihood trees under the GTR model using nucleotide or amino acid sequences. The parallel versions are particularly powerful	Stamatakis (2014)
Readseq	A tool for converting between common sequence file formats, particularly useful for those using various phylogenetic analysis tools	Gilbert (2003)
4SALE	A tool for synchronous RNA sequence and secondary structure alignment and editing	Seibel et al. (2006)
Sequencher	The premier DNA sequence analysis software for sanger and NGS datasets	www.genecodes.com
Tree analysis using new technology (TNT)	A fast parsimony program intended for very large data sets	Goloboff et al. (2008)
TreeView	Provides a simple way to view the contents of a NEXUS, PHYLIP, or other format tree file	Page (2002)

5.8 Progress of DNA Barcoding in Taxonomy, Diagnosis and Monitoring of Fungal Plant Pathogens

There is strong community-wide interest in applying molecular techniques for identifying and monitoring of plant pathogen species, but selection of universal standardized region(s) of the genome has not been finalized. A single marker, the ribosomal DNA internal transcribed spacer region, has frequently been suggested as the standard for fungi (Schoch et al. 2012). The ITS region is easy to amplify for DNA sequencing in most species with the use of universal eukaryotic PCR (polymerase chain reaction) primers (Robideau et al. 2011; White et al. 1990). ITS has been successfully used for species discrimination of *Melampsora* (Feau et al.

2009), *Puccinia horiana* (Pedley 2009; Alaei et al. 2007) and *Puccinia helianthi* (Gao et al. 2016). However, Crouch et al. (2009a) observed high error rate and frequency of misidentification (86%) based on ITS sequence similarity comparison within the *Colletotrichum graminicola* species complex. Dita et al. (2014), using primers developed from intergenic spacers (IGS) of the ribosomal genes, developed a real time PCR diagnostic method able to differentiate isolates of *Foc*TR4, *Foc* R1 and Foc R2 races present in Cuba. Similarly, Crouch and Szabo (2011) reported real-time polymerase chain reaction assay that discriminates between *Puccinia polysora* and *P. sorghi* using sequences of the rDNA internal transcribed spacer region determined for *P. polysora* and *P. sorghi*. Further, a DNA array containing 172 oligonucleotides complementary to specific diagnostic regions of internal transcribed spacers (ITS) of more than 100 species was developed for identification and detection of *Pythium* species (Tambong et al. 2006). The developed DNA array is a reliable tool for identification and detection of the majority of *Pythium* species in environmental samples and also beneficial for epidemiological and ecological studies. However, due to the apparent lack of functional constraint on this untranslated region of rDNA, alignment of ITS sequences is hampered by large amounts of insertions and deletions, which can be an issue for accurate comparisons. Indels in the ITS can even be observed within a single strain due to differences in alleles or differences among the multiple copies of the ITS, making direct sequencing of PCR products impossible (Kageyama et al. 2007). In some species of downy mildews, excessive length due to long insertions can raise difficulties when sequencing the complete ITS region. There are also certain cases where the ITS sequences of formally described species are extremely similar, particularly when they are evolutionarily closely related such as *Phytophthora infestans*, *Phytophthora phaseoli*, *Phytophthora ipomoeae*, *Phytophthora* sp. 'andina' and *Phytophthora mirabilis* (Gomez-Alpizar et al. 2008), which are 99.9% similar in ITS sequence (Kroon et al. 2004). Due to these limitations of the ITS region, the use of another region or multiple loci for this purpose may provide more clarity to the molecular depictions of plant pathogen taxonomy. For *Aspergillus* species diagnosis, β-tubulin gene is suggested as the potential DNA barcode locus against COI, ITS and IGS (Geiser et al. 2007). Nuclear 28S rDNA and β -tubulin genes have discriminative capability for *Fusarium* species (O'Donnell and Cigelnik 1997). Translation elongation factor 1 α gene (EF-1α) is also a reliable barcode for *Fusarium* (Geiser et al. 2004). Similarly, Cai et al. (2009) performed comparison of intra- and inter-species variation for six potential barcode (GPDH, CAL, ACT, ITS, CHS, and EF1α) genes for *Colletotrichum gloeosporioides sensu lato*. They concluded that within *C. gloeosporioides sensu lato* GPDH, CAL, and ACT are good candidates for barcodes; whilst ITS, CHS, and EF1α are poor candidates. A similar set of six genes (ITS, ACT, CHS, GPDH, histone 3 and β-tubulin) was used by Damm et al. (2009) for studying *Colletotrichum* species with curved conidia from herbaceous hosts, which included six different clades. The distinction of the species was best with both GPDH and histone 3 genes, which were superior to ITS, ACT, CHS and β-tubulin. Similarly, a membrane-based oligonucleotide array was developed by Chen et al.

(2013) that can detect *Phytophthora* spp. using three DNA regions [internal transcribed spacer (ITS), the 5′ end of cytochrome c oxidase 1 gene (cox1), and the intergenic region between cytochrome c oxidase 2 gene (cox2) and cox1 (cox2-1 spacer)]. Each sequence data set contained ~250 sequences representing 98 described and 15 undescribed species of *Phytophthora*. The developed assay has the potential to detect 82 described and 8 undescribed *Phytophthora* spp., including several quarantine pathogens such as *Phytophtho raramorum*. They also showed that a DNA array containing signature oligonucleotides designed from multiple genomic regions provided robustness and redundancy for the detection and differentiation of closely related taxon groups. This array has the potential to be used as a routine diagnostic tool for *Phytophthora* spp. from complex environmental samples without culturing.

Martínez-Espinoza et al. (2003) used mating type genes to specifically detect *Ustilago maydis* in maize cultivars. Similarly, Sampietro et al. (2010) used partial sequence of TEF1-α gene to identify the fumonisin and trichothecene-producing species in *Fusarium* isolates from diverse regions of Argentina. To enhance the specificity of diagnostic assay, a combination of multiple diagnostic regions is recommended. Several researchers have followed this multi-locus diagnostic strategy, for instance, Collado-Romero et al. (2008) studied the evolutionary relationships among *Verticillium dahliae* vegetative compatibility groups by AFLP fingerprints and sequence analysis of actin, β-tubulin, calmodulin, histone 3 genes, the ITS region, and a *V. dahlia* specific sequence. Dixon et al. (2009) demonstrated the host specialisation and phylogenetic diversity of *Corynespora cassiicola* using the ITS region, actin gene and two random hypervariable loci. Glienke et al. (2011) performed sequence analysis of the ITS region and partial TEF-1α, actin andglyceraldehyde-3-phosphate dehydrogenase (GPDH) genes to study the genetic diversity of *Phyllosticta* spp. allowing differentiation of pathogenic and non pathogenic species and describing two new *Phyllosticta* species. Inderbitzin et al. (2010) reported a high species diversity in Botryosphaeriaceae species by performing phylogenetic analysis based on six loci, including the ITS region and TEF-1α, GPDH, a heat shock protein, histone-3, and β-tubulin genes. Shimomoto et al. (2011) used RAPDs and sequences from β-tubulin, TEF-1α, calmodulin and actin genes to detect pathogenic and genetic variation among isolates of *Corynespora cassiicola*. Bilodeau et al. (2010) developed a multiplex assay for genus- and species-specific detection of *Phytophthora* based on differences in mitochondrial gene order. The system was based on the high copy sequences of the mitochondrial DNA utilizing gene orders (atp9 and nad9 genes) that were highly conserved in the genus *Phytophthor*a but different in the related genus *Pythium* and plants. Recently, a DNA macroarray based on reverse dot-blot hybridization containing 102 oligonucleotides complementary to portions of the β-tubulin region was developed by Úrbez-Torres et al. (2015) for detection of young vine decline (YVD) fungi. Specificity of the array was evaluated against 138 pure fungal cultures representing 72 different taxa from nine genera, including 37 YVD species and has the potential to be used in commercial diagnostics by the grapevine industry to determine the health status of the planting material.

DNA barcoding is also emerged as a promising technology for the development of international quarantine pathogen detection standards. Recently, this concept has been adopted by official and phytosanitary organizations. In Europe, the Quarantine Barcoding of Life initiative of 20 scientific organizations was financed by the European Union (EU) 7th Framework Program with aims to develop DNA barcodes for the identification of quarantine organisms in the genera of *Monilinia*, *Ceratocystis*, *Melampsora*, *Puccinia*, *Thecaphora*, and *Mycosphaerella* (Quaedvlieg et al. 2012). A national scientific project toward quarantine organism barcoding is also underway in China including fungi as important pathogenstargeting 125 quarantine fungal pathoges including *Phytophthora*, *Fusarium*, *Mycosphaerella* and *Diaporthe* etc. (Gao and Zhang 2013). Ultimately, these initiatives will lead to establish international detection standards based on DNA barcoding, which will prove the superiority of DNA barcoding as a taxonomic tool and meet the needs of routine quarantine diagnostic practices for international trade.

5.9 Limitations of DNA Barcoding

DNA-based identification of fungal plant pathogens relies on discriminating intra- or inter-specific genetic variation. The ranges of these types of variation are unknown and may differ between taxa. It may be difficult to resolve recently diverged species or new species that have arisen through hybridization, mutation and selection pressure etc. So far, except ITS, there is no universal gene for DNA barcoding that is conserved in all domains of fungal life and exhibits enough sequence divergence for species discrimination. The validity of DNA barcoding therefore depends on establishing reference sequences from taxonomically validated and authentic specimens. This is likely to be a complex process that will involve cooperation among a diverse group of researchers and institutions. In general, barcode sequences are short (~500–1000 bp) in length and this limits their application in resolving deep branches in phylogenies. Some controversy exists over the value of DNA barcoding, largely because of the perception that this new identification tool would diminish rather than enhance conventional morphology-based taxonomy, and identity solely based on the genetic divergence could result in incorrect species recognition (Ali et al. 2014).

5.10 DNA Metabarcoding

The massive sequencing of polymerase chain reaction (PCR) amplicons from specific barcode genes represents a powerful culture-independent technique to investigate the whole microbial diversity and determine the relative quantity of microbial community members in environmental samples (Korpelainen et al. 2016). This technique is also known as amplicon metagenomics or metabarcoding. In the case of fungal communities, this approach is exclusively based on the amplification of the ITS regions of the ribosomal DNA (rDNA). Thus, DNA metabarcoding represents an unprecedented opportunity for the study and monitoring of biodiversity in a wide array of environmental conditions over time (Orgiazzi et al. 2015). This molecular approach is based on the assumption that each Operational Taxonomic Units (OTUs) can be unequivocally identified through a specific DNA barcode. The general strategy include several steps like extracting DNA from soil, plant or sample of any environment; amplifying a specific DNA sequence chosen for its taxonomic value; sequencing the corresponding DNA amplicons; analysing the sequences using proper pipelines, and finally determining the taxonomic diversity of the analysed environmental sample or identifying the organism from which DNA has been extracted. For the first time, the term 'DNA metabarcoding' was used by Taberlet et al. (2012) to describe high-throughput multispecies identification using whole or typically degraded DNA extracted from an environment sample or from bulk samples of entire organisms. Metabarcoding differs from metagenomics in several ways. Metagenomics used to the study of all genomes within a particular ecosystem, whereas metabarcoding aims to study a subset of gene(s). From methodology point of view, metagenomics involves preparation of shotgun libraries for sequencing, while metabarcoding relies on amplicon sequencing. Metagenomics approach generally used to get deep insights about the interactions between species within an ecosystem at taxonomic and functional level. Currently, meta-barcoding approach is used to detect and characterize species diversity in the ecosystem with better coverage to identify rare taxa within an ecosystem (Pavan-Kumar et al. 2015). The most critical step in the metabarcoding analysis is to assign taxon to OTU (Ji et al. 2013). This can be achieved by comparing each sequence to a reference database that is a subset of public databases (e.g. NCBI GenBank, EMBL, SILVIA and BOLD etc.) or a set of sequences specifically produced for the study. The comparison for sequence similarity between queried sequence and reference database sequence can be performed through BLAST search. Recently, metabarcoding method based on genus-specific primers and 454 Pyrosequencer have been utilized by Prigigallo et al. (2016) to investigate the genetic diversity of *Phytophthora* spp. in soil and root samples of potted ornamentals and fruit tree species. Abdelfattah et al. (2015) also used metabarcoding approach to decipher the fungal diversity associated with olive leaves, flowers and fruits at different phenological stages. They revealed a rich fungal community with 195 different OTUs up to genus level in most of the cases. Ascomycota was the dominating phyla representing 93.6% of the total number of detected sequences followed by unidentified fungi (3.6%) and Basidiomycota (2.8%). A higher level of diversity was revealed for leaves compared to flowers and fruits. Among plant pathogens the genus *Colletotrichum* represented by three species (*C. godetiae* syn. *C. clavatum*, *C. acutatum* s.s and *C. karstii*) was the most abundant on ripe fruits but it was also detected in other organs. *Pseudocercospora cladosporioides* was detected with a high frequency in all leaf samples and to a less extent in ripe fruits. A much lower relative frequency was revealed for *Spilocaea oleagina* and for other putative pathogens including *Fusarium* spp., *Neofusicoccum* spp., and *Alternaria* spp. Among non-pathogen taxa, *Aureobasidium pullulans*, the species complex of *Cladosporium cladosporioides* and *Devriesia* spp. were the most represented. Overall, the study highlights the existence of a complex fungal consortium including both phytopathogenic and potentially antagonistic microorganisms that can have a significant impact on olive productions. Although, metabarcoding is a very promising approach for diagnosis and monitoring of fungal communities including plant pathogens, there are some limitations and constraints that should need to be taken into consideration. Metabarcoding studies include amplicon sequencing, therefore, parameters like PCR efficiency, primer tags and sequencing efficacy need to be considered to avoid errors (Korpelainen et al. 2016; Berry et al. 2011). To circumvent primer tag bias, a two stage PCR where template DNA is first amplified using untagged primers and subsequently by tagged primers during the last few PCR cycles has been useful (Hajibabaei et al. 2011). Another constraint is lack of comprehensively cured reference databases for certain fungi for assigning taxon to the OTUs. Future studies are needed to improve sampling strategies and to understand the relationship between sequence reads and species density (Thomsen and Willerslev 2015; Herder et al. 2014; Taberlet et al. 2012). Moreover, the integration of knowledge from ecology, taxonomy and evolution is essential for addressing any questions related to diagnosis and monitoring of fungal plant pathogens in environmental samples using metabarcoding.

5.10.1 Bio-barcode

The concept of the bio-barcode is unique and represents a potential alternative to the PCR technique. The bio-barcode assay is an ultrasensitive method of amplification and detection of nucleic acids. This assay is characterized by their ultra high sensitivity and multiplexing capability for the detection of oligonucleotides (Kashyap et al. 2016a). In case of oligonucleotide targets, oligo-gold nanoparticles (AuNP) probes are hybridized to oligonucleotide- functionalized magnetic microparticle (MMP) probes using the target sequence as a linker. These complexes are then separated magnetically for subsequent release of the oligonucleotides from the Oligo-AuNP probes. These released biobarcodes are quantitatively analyzed by the scanometric assay (Khiyami et al. 2014). The assay is also promising by allowing the quick detection of nucleic acids at high zeptomolar levels (Kashyap et al. 2017; Nam et al. 2004). Schwenkbier et al. (2015) developed a helicase-dependent isothermal amplification (HDA) in combination with on-chip hybridization for the detection of *Phytophthora* species. The developed chip allows efficient amplification of the yeast GTP-binding protein (Ypt1) target gene region at one constant temperature in a miniaturized heating device. The assay's specificity was determined by on-chip DNA hybridization and subsequent silver nanoparticle deposition. The silver deposits serve as stable endpoint signals that enable the visual as well as the electrical readout (Schwenkbier et al. 2015). These innovations have a very promising future for on-site application of the bio-barcodes for a reliable and rapid detection of several kinds of fungal pathogenic communities in a large agricultural field.

5.11 Conclusions

DNA barcoding has great potential for use in the identification and monitoring of fungal plant pathogens. Despite some drawbacks, DNA barcoding, succeeded in distinguishing fungal plant pathogens from a range of taxa and to reveal cryptic. However, it is known that species identification based on a single DNA sequence may produce some erroneous outcomes. Therefore, efforts should be made to complement the barcoding region that is currently in use. As the advantages and limitations of barcoding become apparent, it is clear that taxonomic approaches integrating DNA sequencing, morphology and ecological studies will achieve maximum efficiency at species or even race identification of fungal plant pathogens. Accurate species identification is the essence of any molecular diagnostic system, and potential genomic regions should be tested rigorously prior to adoption in a standardized system. So far, no single barcode can be used universally for fungal plant disease diagnosis, the range of possible barcode based technologies is increasing, and low-cost multiplex barcode based testing methods may be available in near future. It is necessary to standardise or optimise the diagnostic parameters, such as sensitivity, specificity, error rates, to validate barcode system for species or race specific detection and diagnostic methods. With the advent of better and cheaper sequencing technology, advancement in bioinformatic analysis tools, three is immense possibility that the barcode sequences *per se* and their ever-increasing taxonomic coverage could become an unprecedented resource for taxonomy, systematic, diagnosis and monitoring of fungal pathogens.

References

Abdelfattah A, Li Destri Nicosia MG, Cacciola SO, Droby S, Schena L. Metabarcoding analysis of fungal diversity in the phyllosphere and carposphere of olive (*Olea europaea*). PLoS One. 2015;110(7):e0131069. doi:10.1371/journal.pone.0131069.

Alaei H, Baeyena S, Maes M, Höfte M, Heungens K. Molecular detection of *Puccinia horiana* in *Chrysanthemum* x *morifolium* through conventional and real-time PCR. J Microbiol Methods. 2009;76:136–45.

Ali MA, Gyulai G, Hidvégi N, Kerti B, et al. The changing epitome of species identification-DNA barcoding. Saudi J Biol Sci. 2014;21(3):204–31.

Aroca A, Raposo R. PCR-based strategy to detect and identify species of *Phaeoacremonium* causing grapevine diseases. Appl Environ Microbiol. 2007;73:2911–8. doi:10.1128/AEM.02176-06.

Aroca A, Raposo R, Lunello P. A biomarker for the identification of four *Phaeoacremonium* species using the β-tubulin gene as the target sequence. Appl Microbiol Biotechnol. 2008;80:1131–40.

Atkins SD, Clark IM. Fungal molecular diagnostics: a mini review. J Appl Genet. 2004;45(1):3–15.

Balajee SA, Borman AM, Brandt ME, et al. Sequence-based identification of *Aspergillus*, *Fusarium*, and Mucorales species in the clinical mycology laboratory: where are we and where should we go from here? J Clin Microbiol. 2009;47:877–84.

Begerow D, Nilsson H, Unterseher M, Maier W. Current state and perspectives of fungal DNA barcoding and rapid identification procedures. Appl Microbiol Biotechnol. 2010;87:99–108.

Belfield GP, Tuite MF. Translation elongation factor 3: a fungus-specific translation factor? Mol Microbiol. 2006;9:411–8.

Berry D, Mahfoudh KB, Wagner M, Loy A. Barcoded primers used in multiplex amplicon pyrosequencing bias amplification. Appl Environ Microbiol. 2011;77:7846–9.

Bilodeau GJ, Koike ST, Uribe P, Martin FN. Development of an assay for rapid detection and quantification of *Verticillium dahliae* in soil. Phytopathology. 2012;102:331–43.

Bilodeau GJ, Martin FN, Coffey MD, Blomquist CL. Development of a multiplex assay for genus- and species-specific detection of *Phytophthora* based on differences in mitochondrial gene order. Phytopathology. 2014;104(7):733–48. doi:10.1094/PHYTO-09-13-0263-R.

Brodersen DE, Nissen P. The social life of ribosomal proteins. FEBS J. 2005;272:2098–108.

Bruns TD, White TJ, Talyor JW. Fungal molecular systematics. Annu Rev Ecol Syst. 1991;22:525–64.

Budge GE, Shaw MW, Colyer A, Pietravalle S, Boonham N. Molecular tools to investigate *Rhizoctonia solani* distribution in soil. Plant Pathol. 2009;58:1071–80.

Cai L, Hyde KD, Taylor PWJ, Weir BS, Waller J, et al. A polyphasic approach for studying *Colletotrichum*. Fungal Divers. 2009;39:183–204.

Carbone I, Kohn LM. A method for designing primer sets for speciation studies in filamentous ascomycetes. Mycologia. 1999;91:553–6.

Ceasar AS, Ignacimuthu S. Genetic engineering of crop plants for fungal resistance: role of antifungal genes. Biotechnol Lett. 2012;34:995–1002. doi:10.1007/s10529-012-0871-1.

Chen Y, Roxby R. Characterization of a *Phytophthora infestans* gene involved in vesicle transport. Gene. 1996;181:89–94.

Chen W, Djama ZR, Coffey MD, Martin FN, et al. Membrane-based oligonucleotide array developed from multiple markers for the detection of many *Phytophthora* species. Phytopathology. 2013;103:43–54. doi:10.1094/PHYTO-04-12-0092-R.

Cho S, Mitchell A, Regier JC, Mitter C, Poole RW, Friedlander TP, Zhao S. A highly conserved gene for low level phylogenetics: elongation factor-1 alpha recovers morphology-based tree for heliothine moths. Mol Biol Evol. 1995;12:650–6.

Collado-Romero M, Mercado-Blanco J, Olivares-García C, Jiménez-Díaz RM. Phylogenetic analysis of *Verticillium dahliae* vegetative compatibility groups. Phytopathology. 2008;98:1019–28.

Crouch JA, Clarke BB, Hillman BI. Unraveling evolutionary relationships among the divergent lineages of *Colletotrichum* causing anthracnose disease in turfgrass and corn. Mycology. 2006;96(1):46–60. doi:10.1094/PHYTO-96-0046.

Crouch JA, Szabo LJ. Real-time PCR detection and discrimination of the southern and common corn rust pathogens *Puccinia polysora* and *Puccinia sorghi*. Plant Dis. 2011;95:624–32.

Crouch JA, Clarke BB, Hillman BI. What is the value of ITS sequence data in *Colletotrichum* systematics and species diagnosis? A case study using the falcate-spored graminicolous *Colletotrichum* group. Mycologia. 2009a;101:648–56.

Crouch JA, Tredway LP, Clarke BB, Hillman BI. Phylogenetic and population genetic divergence correspond with habitat for the pathogen *Colletotrichum cereale* and allied taxa across diverse grass communities. Mol Ecol. 2009b;18(1):123–35.

Crous PW, Groenewald JZ, Summerell BA, Wingfield BD, Wingfield MJ. Co-occurring species of Teratosphaeria on eucalyptus. Persoonia. 2009;22:38–48.

Crous PW, Hawksworth DL, Wingfield MJ. Identifying and naming plant-pathogenic fungi: past, present, and future. Annu Rev Phytopathol. 2015;53:247–67. doi:10.1146/annurev-phyto-080614-120245.

Cummings MP. GARLI (Genetic Algorithm for Rapid Likelihood Inference). Dictionary of Bioinformatics and Computational Biology. 2014. doi: 10.1002/9780471650126.dob0912.

Damm U, Woudenberg JHC, Cannon PF, Crous PW. *Colletotrichum* species with curved conidia from herbaceous hosts. Fungal Divers. 2009;39:45–87.

Daniel HM, Meyer W. Evaluation of ribosomal RNA and actin gene sequences for the identification of ascomycetous yeasts. Int J Food Microbiol. 2003;86:61–78.

Daniel HM, Sorrell TC, Meyer W. Partial sequence analysis of the actin gene and its potential for studying the phylogeny of Candida species and their teleomorphs. Int J Syst Evol Microbiol. 2001;51:1593–606.

Davey JW, Hohenlohe PA, Etter PD, Boone JQ, et al. Genome-wide genetic marker discovery and genotyping using next-generation sequencing. Nat Rev Genet. 2011;12:499–510.

de Beer ZW, Duong TA, Barnes I, Wingfield BD, Wingfield MJ. Redefining *Ceratocystis* and allied genera. Stud Mycol. 2014;79:187–219.

De Boer SH, López MM. New grower-friendly methods for plant pathogen monitoring. Annu Rev Phytopathol. 2012;50:197–218.

Dereeper A, Guignon V, Blanc G, Audic S, Buffet S, Chevenet F, Dufayard JF, Guindon S, Lefort V, Lescot M, Claverie JM, Gascuel O. Phylogeny.Fr: robust phylogenetic analysis for the non-specialist. Nucleic Acids Res. 2008;36(Web Server):W465–9.

Dita MA, Waalwijk C, Buddenhagen IW, Souza Jr MT, Kema GHJ. A molecular diagnostic for tropical race 4 of the banana fusarium wilt pathogen. Plant Pathol. 2010;59:348–57.

Dixon LJ, Schlub RL, Pernezny K, Datnoff LE. Host specialization and phylogenetic diversity of *Corynespora cassiicola*. Phytopathology. 2009;99(9):1015–27.

Drummond AJ, Rambaut A. BEAST: Bayesian evolutionary analysis by sampling trees. BMC Evol Biol. 2007;7:214. doi:10.1186/1471-2148-7-214.

Duressa D, Rauscher G, Koike ST, Mou B, Hayes RJ, Maruthachalam K, Subbarao KV, Klosterman SJ. A real-time PCR assay for detection and quantification of *Verticillium dahliae* in spinach seed. Phytopathology. 2012;102:443–51.

Fang Y, Ramasamy RP. Current and prospective methods for plant disease detection. Biosensors. 2015;4:537–61. doi:10.3390/bios5030537.

Farris JS. Methods for computing Wagner trees. Syst Zool. 1970;19:83–92.

Feau N, Vialle A, Allaire M, Maier W, Hamelin RC. DNA barcoding in the rust genus Chrysomyxa and its implications for the phylogeny of the genus. Mycologia. 2011;103(6):1250–66.

Feau N, Vialle A, Allaire M, Tanguay P, Joly DL, Frey P, Callan BE, Hamelin RC. Fungal pathogen (mis-) identifications: a case study with DNA barcodes on Melampsora rusts of aspen and white poplar. Mycol Res. 2009;113:713–24. doi:10.1016/j.mycres.2009.02.007.

Felsenstein J. Cases in which parsimony and compatibility methods will be positively misleading. Syst Zool. 1978;27:401–10.

Felsenstein J. Confidence limits on phylogenies: an approach using the bootstrap. Evolution. 1985;39:783–91.

Felsenstein J. Phylip (phylogeny inference package). Seattle: Department of Genetics, University of Washington; 1995.

Fraaije BA, Lowell DJ, Coelho JM, Baldwin S, Hollomon DW. PCR-based assay to assess wheat varietal resistance to blotch (*Septoria tritici* and *Stagonospora nodorum*) and rust (*Puccinia striiformis* and *Puccinia recondita*) diseases. Eur J Plant Pathol. 2001;107:905–17.

Frederick RD, Snyder CL, Peterson GL, Bonde MR. Polymerase chain reaction assays for the detection and discrimination of the soybean rust pathogens *Phakopsora pachyrhizi* and *P. meibomiae*. Phytopathology. 2002;92:217–27.

Gao R, Zhang G. Potential of DNA barcoding for detecting quarantine fungi. Phytopathology. 2013;103(11):1103–7.

Geiser DM, Jiménez-Gasco M, Kang S, Makalowska I, Veeraraghavan N, et al. FUSARIUM-ID v. 1.0: a DNA sequence database for identifying *Fusarium*. Eur J Plant Pathol. 2004;110:473–9.

Geiser DM, Klich MA, Frisvad JC, Peterson SW, Varga J, Samson RA. The current status of species recognition and identification in *Aspergillus*. Stud Mycol. 2007;59:1–10. doi:10.3114/sim.2007.59.01.

Gilbert D. Sequence file format conversion with command-line readseq. Curr Protoc Bioinformatics. 2003;00:1E:A.1E.1–A.1E.4.

Gilmore SR, Gräfenhahn T, Louis-Seize G, Seifert KA. Multiple copies of cytochrome oxidase 1 in species of the fungal genus *Fusarium*. Mol Ecol Resour. 2009;9:90–8.

Glass NL, Donaldson GC. Development of primer sets designed for use with the PCR to amplify conserved genes from filamentous ascomycetes. Appl Environ Microbiol. 1995;61(4):1323–30.

Glienke C, Pereira OL, Stringari D, Fabris J, Kava-Cordeiro V, Galli-Terasawa L, Cunnington J, Shivas RG, Groenewald JZ, Crous PW. Endophytic and pathogenic *Phyllosticta* species, with reference to those associated with citrus black spot. Persoonia. 2011;26:47–56.

Glynn NC, Dixon LJ, Castlebury LA, Szabo LJ, Comstock JC. PCR assays for the sugarcane rust pathogens *Puccinia kuehnii* and *P. melanocephala* and detection of a SNP associated with geographical distribution in *P. kuehnii*. Plant Pathol. 2010;59:703–11.

Goloboff PA, Farris JS, Nixon KC. TNT, a free program for phylogenetic analysis. Cladistics. 2008;24(5):774–86.

Gomez-Alpizar L, CH H, Oliva R, Forbes G, Ristaino JB. Phylogenetic relationships of *Phytophthora andina,* a new species from the highlands of Ecuador that is closely related to the Irish potato famine pathogen *Phytophthora infestans*. Mycologia. 2008;100:590–602.

Greganova E, Altmann M, Buetikofer P. Unique modifications of translation elongation factors. FEBS J. 2011;278:2613–24.

Groenewald JZ, Nakashima C, Nishikawa J, Shin HD, Park JH, Jama AN, Groenewald M, Braun U, Crous PW. Species concepts in Cercospora: spotting the weeds among the roses. Stud Mycol. 2013;75(1):115–70. doi:10.3114/sim0012.

Guerber JC, Liu B, Correll JC, Johnston PR. Characterization of diversity in *Colletotrichum acutatum* sensu lato by sequence analysis of two gene introns, mtDNA and intron RFLPs, and mating compatibility. Mycologia. 2003;95:872–95.

Guindon S, Delsuc F, Dufayard JF, Gascuel O. Estimating maximum likelihood phylogenies with PhyML. Methods Mol Biol. 2009;537:113–37. doi:10.1007/978-1-59745-251-9_6.

Guo D, Jing J, Hu W, Navi SS, Jing L. Internal transcribed spacer sequence analysis of *Puccinia helianthi* Schw. And its application in detection of sunflower rust. J Phytopathol. 2016;164(2):141–6. doi:10.1111/jph.12404.

Guo Y, Li W, Sun H, Ning W, Hanshou Y, Chen H. Detection and quantification of *Rhizoctonia cerealis* in soil using real-time PCR. J Gen Plant Pathol. 2012;78:247–54.

Hajibabaei M, Shokralla S, Zhou X, Singer GAC, Baird DJ. Environmental barcoding: a next-generation sequencing approach for biomonitoring applications using river benthos. PLoS One. 2011;6:e17497.

Hall TA. BioEdit: a user-friendly biological sequence alignment editor and analysis program for windows 95/98/NT. Nucleic Acids Symp Ser. 1999;41:95–8.

Hatsch D, Phalip V, Jeltsch JM. Use of genes encoding cellobiohydrolase-C and topoisomerase II as targets for phylogenetic analysis and identification of *Fusarium*. Res Microbiol. 2004;155(4):290–6.

Hebert PDN, Stoeckle MY, Zemlak TS, Francis CM. Identification of birds through DNA barcodes. PLoS Biol. 2004;2(10):e312. doi:10.1371/journal.pbio.0020312.

Herder JE, Valentini A, Bellemain E, Dejean T, van Delft CWJJ, et al. Environmental DNA-A review of the possible applications for the detection of invasive species. Stitching RAVON, Nijmegen. Report 2013–2014; 2014.

Hibbett DS. Ribosomal RNA and fungal systematics. Trans Mycol Soc. 1992;33:533–56.

Hillis DM, Huelsenbeck JP, Swofford DL. Hobgoblin of phylogenetics ? Nature. 1994;369:363–4.

Hollingsworth PM, Graham SW, Little DP. Choosing and using a plant DNA barcode. PLoS One. 2011;6:e19254.

Hong S-Y, Kang M-R, Cho E-J, Kim H-K, Yun S-H. Specific PCR detection of four quarantine *Fusarium* species in Korea. Plant Pathol J. 2010;26:409–16.

Huelsenbeck JP, Ronquist F. MrBayes: Bayesian inference of phylogenetic trees. Bioinformatics. 2001;17(8):754–5.

Hunter GC, Wingfield BD, Crous PW, Wingfield MJ. A multi-gene phylogeny for species of *Mycosphaerella* occurring on eucalyptus leaves. Stud Mycol. 2006;55:147–61.

Inderbitzin P, Bostock RM, Trouillas FP, Michailides TJ. A six locus phylogeny reveals high species diversity in Botryosphaeriaceae from California almond. Mycologia. 2010;102:1350–68.

James P, Whelen S, Hall BD. The RET1 gene of yeast encodes the second-largest subunit of RNA polymerase III. J Biol Chem. 1991;266:5616–24.

Ji Y, Ashton L, Pedley SM, Edwards DP, Tang Y, Nakamura A, Kitching R, Dolman PM, Woodcock P, Edwards FA, Larsen TH, Hsu WW, Benedick S, Hamer KC, Bruce C, Wang X, Levi T, Lott M, Emerson BC, Yu DW. Reliable, verifiable and efficient monitoring of biodiversity via metabarcoding. Ecol Lett. 2013;16:1245–57.

Joly S, Davies TJ, Archambault A, Bruneau A, et al. Ecology in the age of DNA barcoding: the resource, the promise and the challenges ahead. Mol Ecol Resour. 2014;14:221–32. doi:10.1111/1755-0998.12173.

Kageyama K, Senda M, Asano T, Suga H, Ishiguro K. Intra-isolate heterogeneity of the ITS region of rDNA in *Pythium helicoides*. Mycol Res. 2007;111:416–23.

Kang S, Mansfield MA, Park B, Geiser DM, Ivors KL, Coffey MD, Grünwald NJ, Martin FN, Lévesque CA, Blair JE. The promise and pitfalls of sequence-based identification of plant pathogenic fungi and oomycetes. Phytopathology. 2010;100:732–7.

Kashyap PL, Kumar S, Gurjar MS, Singh A, et al. Phytopathogenomics in plant disease management: a paradigm shift. In: Prasad D, Ray DP, editors. Biotechnological approaches in crop protection. New Delhi: Biotech Book publsihers; 2013. p. 241–62.

Kashyap PL, Kumar S, Srivastava AK. Nanodiagnostics for plant pathogens. Environ Chem Lett. 2017;15:7–13. doi:10.1007/s10311-016-0580-4.

Kashyap PL, Rai P, Sharma S, Chakdar H, Kumar S, Pandiyan K, Srivastava AK. Nanotechnology for the detection and diagnosis of plant pathogens. In: Ranjan S, et al., editors. Nanoscience in Food and Agriculture 2, sustainable Agriculture reviews 21. Cham: Springer International Publishing; 2016a. doi:10.1007/978-3-319-39306-3_8.

Kashyap PL, Rai S, Kumar S, Srivastava AK. Genetic diversity, mating types and phylogenetic analysis of Indian races of Fusarium oxysporum f. Sp. ciceris from chickpea. Arch Phytopathol Plant Protect. 2016b;49:533–53. doi:10.1080/03235408.2016.1243024.

Kashyap PL, Rai S, Kumar S, Srivastava AK, Anandaraj M, Sharma AK. Mating type genes and genetic markers to decipher intraspecific variability among *Fusarium udum* isolates from pigeonpea. J Basic Microbiol. 2015;55:846–56. doi:10.1002/jobm.201400483.

Kashyap PL, Sanghera GS, Wani SH, Shafi W, et al. Genes of microorganisms: paving way to tailor next generation fungal disease resistant crop plants. Not Sci Biol. 2011;3:147–57.

Katoh K, Misawa K, Kuma K, Miyataa T. MAFFT: a novel method for rapid multiple sequence alignment based on fast Fourier transform. Nucleic Acids Res. 2002;30(14):3059–66.

Kearse M, Moir R, Wilson A, Stones-Havas S, Cheung M, Sturrock S, Buxton S, Cooper A, Markowitz S, Duran C, Thierer T, Ashton B, Meintjes P, Drummond A. Geneious Basic: an integrated and extendable desktop software platform for the organization and analysis of sequence data. Bioinformatics. 2012;28(12):1647–9. doi:10.1093/bioinformatics/bts199.

Khiyami MA, Almoammar H, Awad YM, Alghuthaym MA, et al. Plant pathogen nanodiagnostic techniques: forthcoming changes? Biotechnol Biotechnol Equip. 2014;28(5):775–85. doi:10.1080/13102818.2014.960739.

Kibbe WA. OligoCalc: an online oligonucleotide properties calculator. Nucleic Acids Res. 2007;35(Web Server issue):W43–6. doi:10.1093/nar/gkm234.

Knutsen AK, Torp M, Holst-Jensen A. Phylogenetic analyses of the *Fusarium poae*, *F. sporotrichioides* and *F. langsethiae* species complex based on partial sequences of the translation elongation factor-1 alpha gene. Int J Food Microbiol. 2004;95:287–95.

Korpelainen H, Pietiläinen M, Huotari T. Effective detection of indoor fungi by metabarcoding. Ann Microbiol. 2016;66:495–8. doi:10.1007/s13213-015-1118-x.

Kress WJ, García-Robledo C, Uriarte M, Erickson DL. DNA barcodes for ecology, evolution, and conservation. Trends Ecol Evol. 2015;30(1):25–35. doi:10.1016/j.tree.2014.10.008.

Kristensen R, Torp M, Kosiak B, Holst-Jensen A. Phylogeny and toxigenic potential is correlated in *Fusarium* species as revealed by partial translation elongation factor 1 alpha gene sequences. Mycol Res. 2005;109:173–86.

Kroon LP, Bakker FT, van den Bosch GB, Bonants PJ, Flier WG. Phylogenetic analysis of *Phytophthora* species based on mitochondrial and nuclear DNA sequences. Fungal Genet Biol. 2004;41:766–82.

Kumar S, Singh R, Kashyap PL, Srivastava AK. Rapid detection and quantification of *Alternaria solani* in tomato. Sci Hortic. 2013;151:184–9.

Larkin MA, Blackshields G, Brown NP, Chenna R, McGettigan PA, McWilliam H, Valentin F, Wallace IM, Wilm A, Lopez R, Thompson JD, Gibson TJ, Higgins DG. Clustal W and Clustal X version 2.0. Bioinformatics. 2007;23:2947–8.

Lewis CT, Bilkhu S, Robert V, et al. Identification of fungal DNA barcode targets and PCR primers based on Pfam protein families and taxonomic hierarchy. Open Appl Inform J. 2011;5:30–44.
Lievens B, Houterman PM, Rep M. Effector gene screening allows unambiguous identification of *Fusarium oxysporum* f. Sp. *lycopersici* races and discrimination from other formae speciales. FEMS Microbiol Lett. 2009;300:201–15.
Liu YJ, Whelen S, Hall BD. Phylogenetic relationships among ascomycetes: evidence from an RNA polymerse II subunit. Mol Biol Evol. 1999;16:1799–808.
Lutzoni F, Kauff F, Cox CJ, McLaughlin D, et al. Assembling the fungal tree of life: progress, classification, and evolution of subcellular traits. Am J Bot. 2004;91:1446–80.
Maddison WP, Maddison DR. MacClade version 3: analysis of phylogeny and character evolution. Sunderland: Sinauer Associates; 1992.
Malkus A, Linda CP, Sabina MZ, Kuang-ren C, Jonathan S, et al. RNA polymerase II gene (RPB2) encoding the second largest protein subunit in *Phaeosphaeria nodorum* and *P. avenaria*. MycolRes. 2006;110:1152–64.
Mann SK, Kashyap PL, Sanghera GS, Singh G, Singh S. RNA interference: an eco-friendly tool for plant disease management. Transgenic Plant J. 2008;2:110–26.
Martin FN, Tooley PW. Phylogenetic relationships among *Phytophthora* species inferred from sequence analysis of mitochondrially encoded cytochrome oxidase I and II genes. Mycologia. 2003;95:269–84.
Martínez-Espinoza AD, León-Ramírez CG, Singh N, Ruiz-Herrera J. Use of PCR to detect infection of differentially susceptible maize cultivars using *Ustilago maydis* strains of variable virulence. Int Microbiol. 2003;6(2):117–20.
McCartney HA, Foster SJ, Fraaije BA, Ward E. Molecular diagnostics for fungal plant pathogens. Pest Manag Sci. 2003;59:129–42.
Mehl HL, Epstein L. Identification of *Fusarium solani* f. Sp. *cucurbitae* race 1 and race 2 with PCR and production of disease-free pumpkin seeds. Plant Dis. 2007;91:1288–92.
Merget B, Koetschan C, Hackl T, Förster F, Dandekar T, Müller T, Schultz J, Wolf M. The ITS2 database. J Vis Exp. 2012;61:3806. doi:10.3791/3806.
Mesapogu S, Babu BK, Bakshi A, Reddy SS, Saxena S. Rapid detection and quantification of *Fusarium udum* in soil and plant samples using real-time PCR. J Plant Pathol Microbiol. 2011;2:107. doi:10.4172/2157-7471.1000107.
Meusnier I, Singer GA, Landry JF, Hickey DA, Hebert PD, Hajibabaei M. A universal DNA minibarcode for biodiversity analysis. BMC Genomics. 2008;9:214. doi:10.1186/1471-2164-9-214.
Miller SA, Beed FD, Harmon CL. Plant disease diagnostic capabilities and networks. Annu Rev Phytopathol. 2009;l47:15–38.
Mitchell A, Cho S, Regier JC, Mitter C, Poole RW, et al. Phylogenetic utility of elongation factor-1 alpha in Noctuoidea (Insecta: Lepidoptera): the limits of synonymous substitution. Mol Biol Evol. 1997;14:381–90.
Moriwaki J, Tsukiboshi T. *Colletotrichum echinochloae*, a new species on Japanese barnyard millet (*Echinochloa utilis*). Mycoscience. 2009;50:273–80. doi:10.1007/s10267-009-0485-1.
Mostert L, Groenewald JZ, Summerbell RC, Gams W, Crous PW. Taxonomy and pathology of *Togninia* (Diaporthales) and its *Phaeoacremonium* anamorphs. Stud Mycol. 2006;54:1–115.
Mulè G, Susca A, Stea G, Moretti A. Specific detection of the toxigenic species *Fusarium proliferatum* and *F. oxysporum* from asparagus plants using primers based on calmodulin gene sequences. FEMS Microbiol Lett. 2004;230:235–40.
Myburg H, Gryzenhout M, Wingfield BD, Wingfield MJ. β-tubulin and histone H3 gene sequences distinguish *Cryphonectria cubensis* from South Africa, Asia, and South America. Can J Bot. 2002;80:590–6.
Myllys L, Stenroos S, Thell A. New genes for phylogenetic studies of lichenized fungi: glyceraldehyde-3-phosphate dehydrogenase and beta-tubulin genes. Lichenologist. 2002;34(3):237–46.
Naidoo K, Steenkamp ET, Coetzee MPA, Wingfield MJ, Wingfield BD. Concerted evolution in the ribosomal RNA cistron. PLoS One. 2013;8(3):e59355. doi:10.1371/journal.pone.0059355.
Nam JM, Stoeva SI, Mirkin CA. Bio-bar-code-based DNA detection with PCR-like sensitivity. J Am Chem Soc. 2004;126(19):5932–3.

Nguyen HD, Seifert KA. Description and DNA barcoding of three new species of *Leohumicola* from South Africa and the United States. Persoonia. 2008;21:57–69. doi:10.3767/003158508X361334.

Nitschke E, Nihlgard M, Varrelmann M. Differentiation of eleven *Fusarium* spp. isolated from sugar beet, applying restriction fragment analysis of polymerase chain reaction–amplified translation elongation factor 1α gene fragment. Phytopathology. 2009;99:921–9.

O'Donnell K, Cigelnik E, Nirenberg HI. Molecular systematics and phylogeography of the *Gibberella fujikoroi* species complex. Mycologia. 1998a;90:465–93.

O'Donnell K, Kistler HC, Cigelnik E, Ploetz RC. Multiple evolutionary origins of the fungus causing Panama disease of banana: concordant evidence from nuclear and mitochondrial gene genealogies. Proc Natl Acad Sci U S A. 1998b;95:2044–9.

O'Donnell K, Cigelnik E. Two divergent intragenomic rDNA ITS2 types within a monophyletic lineage of the fungus *Fusarium* are non orthologous. Mol Phylogenet Evol. 1997;7(1):103–16.

O'Donnell K, Rooney AP, Proctor RH, Brown DW, McCormick SP, et al. Phylogenetic analysis of RPB1 and RPB2 support a middle cretaceous origin for a clade comprising all agriculturally and medically important fusaria. Fungal Genet Biol. 2013;52:20–31. doi:10.1016/j.fgb.2012.12.004.

Okubara PA, Harrison LA, Gatch EW, Vandemark G, Schroeder KL, du Toit LJ. Development and evaluation of a TaqMan real-time PCR assay for *Fusarium oxysporum* f. Sp. *spinaciae*. Plant Dis. 2013;97:927–37.

Okubara PA, Schroeder KL, Paulitz TC. Identification and quantification of *Rhizoctonia solani* and *R. oryzae* using real-time polymerase chain reaction. Phytopathology. 2008;98:837–47. doi:10.1094/PHYTO-98-7-0837.

Orgiazzi A, Dunbar MB, Panagos P, Groot GA, Lemanceau P. Soil biodiversity and DNA barcodes: opportunities and challenges. Soil Biol Biochem. 2015;80:244–50.

Packer L, Gibbs J, Sheffield C, Hanner R. DNA barcoding and the mediocrity of morphology. Mol Ecol Resour. 2009;9:42–50.

Page RD. Visualizing phylogenetic trees using TreeView. Curr Protoc Bioinformatics. 2002;00:6.2:6.2.1–6.2.15. doi: 10.1002/0471250953.bi0602s01.

Pannecoucque J, Hofte M. Detection of rDNA ITS polymorphism in *Rhizoctonia solani* AG 2-1 isolates. Mycologia. 2009;101:26–33.

Pavan-Kumar A, Gireesh-Babu P, Lakra WS. DNA metabarcoding: a new approach for rapid biodiversity assessment. J Cell Sci Mol Biol. 2015;2(1):111.

Pedley KF. PCR-based assays for the detection of *Puccinia horiana* on chrysanthemums. Plant Dis. 2009;93:1252–8.

Pond SL, Frost SD, Muse SV. HyPhy: hypothesis testing using phylogenies. Bioinformatics. 2005;21(5):676–9.

Posada D, Crandall KA. MODELTEST: testing the model of DNA substitution. Bioinformatics. 1998;14(9):817–8.

Pouzoulet J, Mailhac N, Couderc C, Besson X, Daydé J, Lummerzheim M, Jacques A. A method to detect and quantify *Phaeomoniella chlamydospora* and *Phaeoacremonium aleophilum* DNA in grapevine-wood samples. Appl Microbiol Biotechnol. 2013;97:10163–75. doi:10.1007/s00253-013-5299-6.

Prigigallo MI, Abdelfattah A, Cacciola SO, Faedda R, Sanzani SM, Cooke DEL, Schena L. Metabarcoding analysis of *Phytophthora* diversity using genus-specific primers and 454 pyrosequencing. Phytopathology. 2016;106(3):305–13. doi:10.1094/PHYTO-07-15-0167-R.

Quaedvlieg W, Groenewald JZ, de Jesus Y-MM, Crous PW. DNA barcoding of *Mycosphaerella* species of quarantine importance to Europe. Persoonia. 2012;29:101–15. doi:10.3767/003158512X661282.

Quaedvlieg W, Kema GHJ, Groenewald JZ, Verkley GJM, Seifbarghi S, Razavi M, Mirzadi Gohari A, Mehrabi R, Crous PW. Zymoseptoria gen. Nov.: a new genus to accommodate Septoria-like species occurring on graminicolous hosts. Persoonia. 2011;26:57–69.

Quaedvlieg W, Verkley GJM, Shin H-D, Barreto RW, Alfenas AC, Swart WJ, Groenewald JZ, Crous PW. Sizing up Septoria. Stud Mycol. 2013;75:307–90.

Rai S, Kashyap PL, Kumar S, Srivastava AK, Ramteke PW. Comparative analysis of microsatellites in five different antagonistic *Trichoderma* species for diversity assessment. World J Microbiol Biotechnol. 2016a;32:8.

Rai S, Kashyap PL, Kumar S, Srivastava AK, Ramteke PW. Identification, characterization and phylogenetic analysis of antifungal *Trichoderma* from tomato rhizosphere. SpringerPlus. 2016b;5:1939. doi:10.1186/s40064-016-3657-4.

Rehner SA, Buckley E. A Beauveria phylogeny inferred from nuclear ITS and EF1-alpha sequences: evidence for cryptic diversification and links to Cordyceps teleomorphs. Mycologia. 2005;97:84–9810. 3852/mycologia.97.1.84

Riit T, Tedersoo L, Drenkhan R, Runno-Paurson E, Kokko H, Anslan S. Oomycete-specific ITS primers for identification and metabarcoding. MycoKeys. 2016;14:17–30. Doi.org/10.3897/mycokeys.14.9244

Robert V, Szoke S, Eberhardt U, et al. The quest for a general and reliable fungal DNA barcode. Open Appl Inform J. 2011;5:45–61.

Robideau GP, De Cock AWAM, Coffey MD, Voglmayr H, et al. DNA barcoding of oomycetes with cytochrome c oxidase subunit I and internal transcribed spacer. Mol Ecol Resour. 2011;11:1002–11.

Romero MP, Wise KA. Development of molecular assays for detection of *Stenocarpella maydis* and *Stenocarpella macrospora* in corn. Plant Dis. 2015;99:761–9.

Saitou N, Nei M. The neighbor-joining method: a new method for reconstructing phylogenetic trees. Mol Biol Evol. 1987;4(4):406–25.

Sampietro DA, Marín P, Iglesias J, Presello DA, Vattuone MA, Catalan CA, Gonzalez Jaen MT. A molecular based strategy for rapid diagnosis of toxigenic *Fusarium* species associated to cereal grains from Argentina. Fungal Biol. 2010;114:74–81. doi:10.1016/j.mycres.2009.10.008.

Santamaria M, Vicario S, Pappadà G, Scioscia G, Scazzocchio C, Saccone C. Towards barcode markers in fungi: an intron map of Ascomycota mitochondria. BMC Bioinformatics. 2009;10:S15.

Sasikumar AN, Perez WB, Goss Kinzy T. The many roles of eukaryotic elongation factor 1 complex. Wiley Interdisciplinary Rev RNA. 2012;3:543–55.

Saunders GW. Applying DNA barcoding to red macroalgae: a preliminary appraisal holds promise for future applications. Phil Trans Royal Soci B BiolSci. 2005;360:1879–88.

Schena L, Cooke DL. Assessing the potential of regions of the nuclear and mitochondrial genome to develop a "molecular tool box" for the detection and characterization of Phytophthora species. J Microbiol Methods. 2006;67(1):70–85.

Schoch CL, Seifert KA, Huhndorf S, Robert V, Spouge JL. Nuclear ribosomal internal transcribed spacer (ITS) region as a universal DNA barcode marker for fungi. PNAS. 2012;109(16):6241–6. doi:10.1073/pnas.1117018109.

Schroeder KL, Okubara PA, Tambong JT, Lévesque CA, Paulitz TC. Identification and quantification of pathogenic *Pythium* spp. from soils in eastern Washington using real-time polymerase chain reaction. Phytopathology. 2006;96:637–47.

Schubert K, Groenewald JZ, Braun U, Dijksterhuis J, Starink M, et al. Biodiversity in the Cladosporium herbarum Complex (Davidiellaceae, Capnodiales), with standardisation of methods for Cladosporium taxonomy and diagnostics. Stud Mycol. 2007;58:105–56.

Schwenkbier L, Pollok S, König S, Urban M, et al. Towards on-site testing of *Phytophthora* species. Anal Methods. 2015;7:211–7.

Seibel PN, Müller T, Dandekar T, Schultz J, Wolf M. 4SALE – a tool for synchronous RNA sequence and secondary structure alignment and editing. BMC Bioinformatics. 2006;7:498. doi:10.1186/1471-2105-7-498.

Seifert KA. Progress towards DNA barcoding of fungi. Mol Ecol Resour. 2009;9:83–9.

Seifert KA, Samson RA, Dewaard JR, et al. Prospects for fungus identification using CO1 DNA barcodes, with *Penicillium* as a test case. PNAS. 2007;104:3901–6.

Sharma R, Polkade AV, Shouche YS. Species concept' in microbial taxonomy and systematic. Curr Sci. 2015;108:1804–14.

Sharma S, Rai P, Rai S, Srivastava M, Kashyap PL, Sharma A, Kumar S. Genomic revolution in crop disease diagnosis: a review. In: Singh SS, editor. Plants and microbes in an ever changing environment. Nova Science Publishers; 2017. p. 257–93.

Shimomoto Y, Sato T, Hojo H, Morita Y, Takeuchi S, et al. Pathogenic and genetic variation among isolates of *Corynespora cassiicola* in Japan. Plant Pathol. 2011;60:253–60.

Shishido M, Sato K, Yoshida N, Tsukui R, Usami T. PCR-based assays to detect and quantify *Phomopsis sclerotioides* in plants and soil. J General Plant Pathol. 2010;76:21–30.

Singh R, Kumar S, Kashyap PL, Srivastava AK, Mishra S, et al. Identification and characterization of microsatellite from *Alternaria brassicicola* to assess cross-species transferability and utility as a diagnostic marker. Mol Biotechnol. 2013;56:1049–59.

Stamatakis A. RAxML version 8: a tool for phylogenetic analysis and post-analysis of large phylogenies. Bioinformatics. 2014;30(9):1312–3. doi:10.1093/bioinformatics/btu033.

Steenkamp ET, Wingfield BD, Coutinho TA, Wingfield MJ, Marasas WFO. Differentiation of *Fusarium subglutinans* f. Sp. *pini* by histone gene sequence data. Appl Environ Microbiol. 1999;65(8):3401–6.

Steenkamp ET, Wingfield BD, Coutinho TA, Zeller KA, Wingfield MJ, Marasas WFO, Leslie JF. PCR-Based identification of MAT-1 and MAT-2 in the *Gibberella fujikuroi* species complex. Appl Environ Microbiol. 2000;66(10):4378–82.

Stephenson SA, Green J, Manners J, Maclean DJ. Cloning and characterisation of glutamine synthetase from *Colletotrichum gloeosporioides* and demonstration of elevated expression during pathogenesis on *Stylosanthes guianensis*. Curr Genet. 1997;31:447–54. doi:10.1007/s002940050228.

Stielow JB, Lévesque CA, Seifert KA, Meyer W, Irinyi L, et al. One fungus, which genes? Development and assessment of universal primers for potential secondary fungal DNA barcodes. Persoonia. 2015;35:242–63.

Sundelin T, Collinge DB, Lübeck M. A cultivation independent, PCR-based protocol for the direct identification of plant pathogens in infected plant material. Eur J Plant Pathol. 2009;123(4):473–6.

Swofford DL. PAUP*. Phylogenetic analysis using parsimony (*and other methods). Sunderland: Sinauer Associates; 2002a.

Swofford DL. PAUP: phylogenetic analysis using maximum parsimony (and other method). Version 4.0b10. Sunderland: Sinauer; 2002b.

Swofford DL, Olsen GJ, Waddell PJ, Hillis DM. Phylogenetic inference. In: Hillis DM, Moritz C, Mable BK, editors. Molecular systematics. 2nd ed. Sunderland: Sinauer; 1996. p. 407–514.

Taberlet P, Coissac E, Pompanon F, Brochmann C, Willerslev E. Towards next-generation biodiversity assessment using DNA metabarcoding. Mol Ecol. 2012;21:2045–50.

Tambong JT, de Cock AW, Tinker NA, Lévesque CA. Oligonucleotide array for identification and detection of *Pythium* species. Appl Environ Microbiol. 2006;72:2691–706.

Tamura K, Peterson D, Peterson N, Stecher G, Nei M, Kumar S. MEGA5: molecular evolutionary genetics analysis using maximum likelihood, evolutionary distance, and maximum parsimony methods. Mol Biol Evol. 2011;10:2731–9.

Tamura K, Stecher G, Peterson D, Filipski A, Kumar S. MEGA6: molecular evolutionary genetics analysis version 6.0. Mol Biol Evol. 2013;30(12):2725–9. doi:10.1093/molbev/mst197.

Templeton MD, Rikkerink EHA, Solon SL, Crowhurst RN. Cloning and molecular characterization of the glyceraldehyde-3-phosphate dehydrogenase-encoding gene and cDNA from the plant pathogenic fungus *Glomerella cingulata*. Gene. 1992;122(1):225–30.

Thines M. Characterization and phylogeny of repeated elements giving rise to exceptional length of ITS2 in several downy mildew genera (Peronosporaceae). Fungal Genet Biol. 2007;44:199–207.

Thomsen PF, Willerslev E. Environmental DNA – an emerging tool in conservation for monitoring past and present biodiversity. Biol Conserv. 2015;183:4–18.

Thon MR, Royse DJ. Evidence for two independent lineages of shiitake of the americas (*Lentinula boryana*) based on rDNA and beta tubulin gene sequences. Mol Phylogenet Evol. 1999;13:520–4.

Úrbez-Torres JR, Haag P, Bowen P, Lowery T, O'Gorman DT. Development of a DNA macroarray for the detection and identification of fungal pathogens causing decline of young grapevines. Phytopathology. 2015;105:1373–88.

Verkley GJM, Starink-Willemse M, van IA, Abeln ECA. Phylogenetic analyses of *Septoria* species based on the ITS and LSU-D2 regions of nuclear ribosomal DNA. Mycologia. 2004;96:558–71.

Vilgalys R, Hester M. Rapid genetic identification and mapping of enzymatically amplified ribosomal DNA from several Cryptococcus species. J Bacteriol. 1990;172:4239–46.

Weir BS, Johnston PR, Damm U. The *Colletotrichum gloeosporioides* species complex. Stud Mycol. 2012;73(1):115–80. doi:10.3114/sim0011.

White TJ, Bruns T, Lee S, Taylor J. Amplification and direct sequencing of fungal ribosomal RNA genes for phylogenetics. In: Innis MA, Gelfand DH, Sninksky JJ, White TJ, editors. PCR protocols: a guide to method and amplifications. San Diego, CA: Academic; 1990. p. 315–22.

Wolf M, Ruderisch B, Dandekar T, Schultz J, Müller T. ProfDistS: (profile-) distance based phylogeny on sequence--structure alignments. Bioinformatics. 2008;24(20):2401–2. doi:10.1093/bioinformatics/btn453.

Xu J. Fungal DNA barcoding. Genome. 2016;59(11):913–32. doi:10.1139/gen-2016-0046.

Yadav MK, Babu BK, Saxena AK, Singh BP, Singh K, et al. Real-time PCR assay based on topoisomerase- II Gene for detection of *Fusarium udum*. Mycopathologia. 2011;171:373–81.

Yang Z. PAML 4: phylogenetic analysis by maximum likelihood. Mol Biol Evol. 2007;24(8):1586–91.

Yao C, Frederiksen RA, Magill CW. Length heterogeneity in ITS2 and the methylation status of CCGG and GCGC sites in the rRNA genes of the genus *Peronosclerospora*. Curr Genet. 1992;22:415–20.

Zhao P, Luo J, Zhuang W-Y. Practice towards DNA barcoding of the nectriaceous fungi. Fungal Divers. 2011;46:183–91. doi:10.1007/s13225-010-0064-y.

In the original version of Chapter 5, illustrations 1–10 had permission issues. In the current version, Chapter 5 has been replaced with a new, revised chapter with one original figure replacing the previous 10 illustrations.

Chapter 6
New Generation Markers for Fingerprinting and Structural Analysis of Fungal Community

Hillol Chakdar, Atul Singha, and Pratik Satya

6.1 Introduction

Fungi represent an essential and a major component of global biological diversity and inhabit almost every habitat on the earth playing a key role in the function and dynamics of terrestrial ecological systems, by affecting the structure of biotic communities through various types of interactions. Apart from their pivotal roles in geochemical cycling fungi are also a major part of terrestrial food webs. Fungal hyphae account for the highest fraction of soil biomass and serve as the primary carbon source in many soil food webs (Wardle 2002). These eukaryotes are the most common and important plant pathogens, causing substantial crop loss and influencing the population structure and dynamics of natural plant communities.

The key role of fungi in most of the life sustaining processes on earth is inseparably related to their enormous diversity. As of now, there are approximately 0.1 million described species although it is estimated that 0.8–5.1 million fungi exist on the Earth (Blackwell 2011). Despite their ubiquity, enormous diversity and absolute ecological importance, the studies of fungal communities is impeded due to lack to identify the fungal species and communities. Morphometric methods are not ideal for many reasons like cultures are time consuming, are biased toward fast-growing fungi, preclude many biotrophic fungi, can only be done on fresh materials, and in

H. Chakdar
Microbial technology Unit II, National Bureau of Agriculturally Important Microorganisms, Maunath Bhanjan, Uttar Pradesh 275103, India

A. Singha
Division of Crop Production, ICAR-Central Institute for Subtropical Horticulture, Lucknow, Uttar Pradesh 227107, India

P. Satya (✉)
Crop Improvement Division, ICAR-Central Research Institute for Jute and Allied fibres, Barrackpore, Kolkata 700120, India
e-mail: pscrijaf@gmail.com

B.P. Singh, V.K. Gupta (eds.), *Molecular Markers in Mycology*, Fungal Biology,
DOI 10.1007/978-3-319-34106-4_6

some cases are based on flawed morphological species concepts. Though macroscopic structures like fruiting bodies and microscopic spores and filaments have long been used as diagnostic features for identification, these features are significantly influenced by environmental conditions and many a times harbour cryptic complexes and do not represent the species abundance accurately. Application of polymerase chain reaction (PCR) and development of fungal-specific primers for amplification of the internal transcribed spacer (ITS) region of the ribosomal RNA genes opened the way for fast and accurate identification of fungal species and analyzing structure and function of fungal communities (Gardes and Bruns 1993). Initial studies demonstrated the usefulness of the molecular approach to fungal community ecology (Gardes and Bruns 1996). In the past few decades, boom of easy, cheap, high-throughput molecular methods and techniques have enabled researchers to know the population structure and functional roles of fungi/fungal community in a given ecological niche. Providing a glimpse of the initial progresses made in the molecular identification of fungi, we will concentrate on the recent developments for identification and diversity analysis in fungi, particularly using diverse new generation marker systems and next generation sequencing techniques.

6.2 Molecular Identification of Fungi and Community Analysis: Past Developments

Development of DNA fingerprinting techniques, use of specific primers and probes have enabled accurate identification of fungi and empowered to resolve differentiation among closely related strains. Traditional methods for fungal identification mostly relied on isolation and culturing; laboratory identification by robust morphological features like fruiting bodies, spores etc., disease symptoms and biochemical tests which are insufficient to resolve beyond species level. Although these methods are still routinely used, there is an increasing move towards molecular identification of fungi due to their fast, accurate and reproducible results.

6.2.1 Species Identification

With increasing knowledge about DNA, genes and genomes, fungal taxons are increasingly being defined by sequence variation in DNA. Ideal DNA-based identification method utilizes sequence variations at multiple loci, as no single locus could be unequivocally reliable for species recognition (Taylor et al. 2000). However, in ecological arena, multi-locus information of species is difficult to obtain; and this approach is problematic when all the sequences are drawn from a common environmental pool like soil or plant material. For these reasons, a single locus viz. the internal transcribed spacer (ITS) region of the nuclear ribosomal RNA gene, has become widely used for species level identification. This multi-copy, tripartite, and approximately 550 bp segment combines the advantages of resolution at various scales (ITS1: rapidly evolving, 5.8S : very conserved, ITS2 : moderately rapid to rapidly evolving; Hillis and Dixon 1991; Hershkovitz and Lewis 1996) with the ease of amplification of a multi-copy region at the species level. Transcribed spacer regions in the rDNA operon have been successfully used to identify distinct lineages of fungal species and resolve relationships between closely related species (Witthuhn et al. 1998;). Nucleotide variations in ITS region have been shown to be sufficient to distinguish different species in many different genera such as the Ascomycetes (Bridge and Arora 1998) or the Oomycetes (Cooke et al. 2000). However, ITS-based characterization often is found insufficient for sub-species characterization and identification of fungal strains.

6.2.2 Sub-species and Strain Characterization

Among the first generation DNA markers, RAPD (Random Amplified Polymorphic DNA) has been extensively used for intra-specific characterization of microorganisms including fungi over the past twenty years despite inherent problem of low reproducibility. For distinguishing strains and developing fingerprints for molecular epidemiology, RAPDs has been highly effective. The marker has been widely employed to study intra-specific variation within species such as *Fusarium oxysporum* f. sp. *ciceri* (Bayraktar et al. 2008), *Macrophomina phaseolina* (Babu et al. 2010) and *Bipolaris oryzae* (Kandan et al. 2015). Universally primed PCR (UP-PCR), another technique very similar to RAPD has considerably higher reproducibility than RAPD and has been used to discriminate isolates of *Trichoderma*, *L. lecanii* (Mitina et al. 2007) and *Fusicladium effusum* (Bock et al. 2014). A variant of the UP-PCR technique is UP-PCR product cross hybridization assay which augments investigation of sequence similarity (homology) of UP-PCR products (Bulat et al. 1998). This allows grouping of strains into hybridization groups which can be used to separate the strains into genetic entities with high genetic similarity. PCR–RFLP (PCR-Restriction Fragment Length Polymorphism) has also been widely used for assessing strain relatedness. The discriminatory power of this technique depends mainly on careful selection of restriction enzymes.

AFLP (Amplification Fragment Length Polymorphism) is a powerful method for fingerprinting strains and for generating a large number of dominant markers for genetic analyses (Vos et al. 1995). Being a hybrid of restriction digestion and PCR, AFLP generates large number of fragments in a single reaction and is highly reproducible. Different enzymes and/or selective extension nucleotides can be used to create new sets of markers. This technique has been used for intraspecific genetic variation in isolates of *Cladosporium fulvum* (Majer et al. 1996), *Pyrenopeziza brassicae* (Majer et al. 1996), *Duddingtonia flagrans* (Ahren et al. 2004), and *Fusarium oxysporum* (Bogale et al. 2006).

DNA markers based on hypervariable tandemly repeated units of di-, tri- or multiple nucleotides are known as microsatellite or SSR (Simple Sequence Repeats) markers. Microsatellites can be located by probing a genomic library with simple repeated sequences or by searching databases of gene sequences. Due to its widespread applications in fungal identification and population studies, a number of curated database of microsatellites/SSR for many fungal genomes are available (Table 6.1). SSR markers have been identified and characterized for various fungal spcies, i.e., *Beauveria bassiana* (Rehner and Buckley 2003), *Metarhizium anisopliae* (Enkerli et al. 2005; Oulevey et al. 2009) and *Paecilomyces fumosoroseus* (Dalleau-Clouet et al. 2005). SSR markers have also been exploited in numerous plant pathogenic fungi also. Karaoglu et al. (2005) suggested that there are at least 11,000 SSRs with lengths of 10 bp or more in *M. grisea*. Microsatelite Lenght Polymorphism (MLP) typing is a SSR based marker system that exploits the high variability in the repeat number of microsatellite sequences, defined as tandemly repetitive stretches of two to six nucleotides. MLP is easy to perform, rapid and is amenable for automation and high throughput. Overall, MLP is one of the most discriminative methods for fungi identification. Combining several markers located on different chromosomes in the same typing system allows more accurate characterization of *C. albicans* strains by MLP analysis. Multiplex PCR systems co-amplifying many microsatellite markers are possible when primers are labelled with different dyes allowing rapid and highly resolutive strains typing (Sampaio et al. 2005). Recently, inclusion of high-resolution melting (HRM) technology to microsatellite marker

Table 6.1 SSR/microsatellite database for fungi

Name of the database	Fungi	Reference/website
Yeast_Msat	*Saccharomyces cerevisiae*	http://intranet.pasteur.edu.uy/yeast/search.php
Database of Molecular Mycology Lab	*Aspergillus nidulans, Encephalitozoon cuniculi, Fusarium graminearum, Magnaporthe grisea Neurospora crassa S. cerevisiae Schizosacchaaromyces pombe Ustilago maydis*	www.mycologylab.org/DefaultInfo.aspx?Page=SSR Karagolu et al. (2005)
Fungal Simple Sequence Repeat Database		http://capssr.s.ac.kn
EuMicroSatDb	*Asprgillus fumigates, A. oryzae, Candida glbrata, Cryptococcus neoformans, Eremothecium gossypii, S. cerevisiae, Schizosacchaaromyces pombe*	http://ipu.ac.in/usbt/EuMicroSatdb.htm Aishwarya et al. (2007)
Fusarium SSR Db	*14 different Fusarium* spp.	http://webapp.cabgrid.res.in
UgMicroSatDb	*Dictiostelium discoideum, Giberella moniliformis, Magnaporthe grisea, Neurospora crassa, Fibrobasidiella neoformans, Phytophthora infestans*	Aishwarya et al. (2008) http://ipu.ac.in/usbl/ugmisatdb.html

typing of *C. albicans* strains increased discriminatory power of MLP analysis (Ben Abdeljelil et al. 2012).

The ERIC, BOX and REP-PCR approaches are fast and highly reproducible microbial characterization methods, which involve PCR amplification using primers corresponding to the Enterobacterial Repetitive Intergenic Consensus (ERIC) and Repetitive Extragenic Palindromic (REP) elements distributed over the genome. These repetitive sequences were initially described in *Escherichia coli, Streptococcus pneumoniae* and *Salmonella typhimurium* (Stern et al. 1984; Martin et al. 1992; Versalovic et al. 1991). Although rep-PCR primers were developed for repetitive elements in prokaryotic genomes, these primers have been successfully applied in the fingerprinting of eukaryotic genomes (van Belkum et al. 1998). Repetitive element based PCR fingerprinting has been used for identification of *Fusarium* species (Healy et al. 2005), *Histoplasma capsulatum*, *Blastomyces dermatitidis*, *Coccidioides* spp. (Pounder et al. 2006), and species from Botryosphariaceae family (Alves et al. 2007).

DNA microarray has also been used for high throughput identification of multiple microorganisms. There are three major classes of DNA microarray, functional gene arrays (FGAs), phylogenetic oligonucleotide arrays (POAs) and community genome arrays (CGAs), classified according to the nature of gene used in microarray construction. ITS based microarrays are the most common example of POAs. ITS based array has been developed for several fungi species (Fukushima et al. 2003; Leinberger et al. 2005; Hong et al. 2010). Hsiao et al. (2005) also used ITS-1 and ITS-2 sequences for developing a DNA microarray for identification of 64 species of clinically important filamentous fungi. A microarray system with fluorescent probes targeted to the ITS-2 region was developed for the identification of 20 common species belonging to the genera *Aspergillus, Fonsecaea, Phialophora, Cladosporium, Sporothrix, Mucor* and the dermatophytes (Huang et al. 2006). Hong et al. (2010) developed an 18S rRNA probe based microarray to detect and identify several fungi genus of order Eurotiales.

6.3 Analysis of Fungal Community

To understand the role of fungal species and community in ecological processes, it is very important for the ecologists to understand their identity, prevalence and population dynamics. Till date very limited information is available on the role of fungal community in any studied ecosystem. The enormity of fungal diversity is probably undermined both due to scanty global exploration and research effort, and difficulties in culturing a large number of fungal species. Although culture-based approaches have told us most of what we presently know about fungal ecology, methods used to isolate fungi tend to select for species able to grow on particular media, and are therefore quite limited. Most fungi isolated in this way are fast-growing species adapted to high substrate levels and may not necessarily be the dominant organisms of the community. There has been widespread concern within

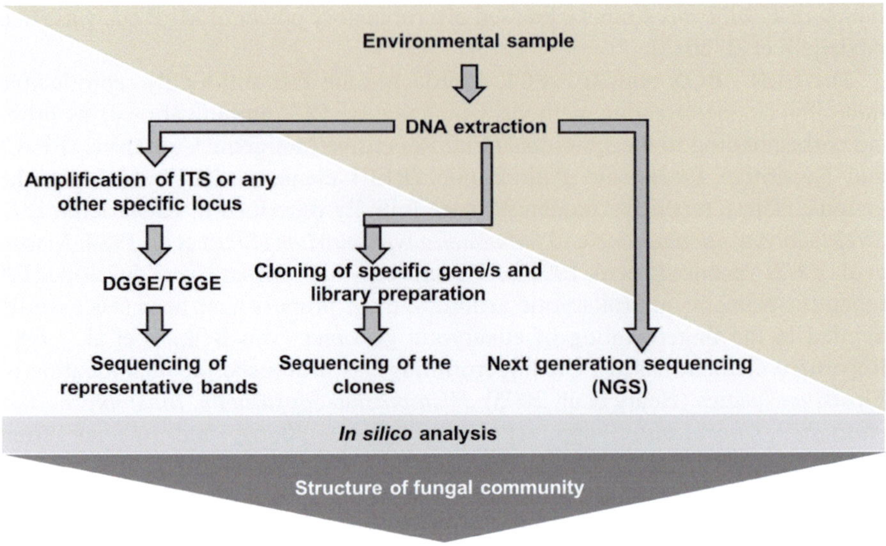

Fig 6.1 Schematic outline for culture independent community analysis in fungi

microbial ecology as a whole that cultivation based approaches have given a biased view of many microbial ecosystems. To overcome these problems, microbial ecologists have turned in recent years towards culture independent analysis of community structures.

Culture-independent analysis of community structures determines presence and prevalance of specific genotypes and helps to compare structures of microbial communities in complex biological samples (Kirk et al. 2004). For this approach specific marker genes are amplified from complex DNA sample and subsequently resolved by use of various analytical procedures (Fig. 6.1). A large number of different primers has been designed during last few decades that are used for community structure analyses of different taxonomic groups, i.e., at various taxonomic levels starting from genus up to the fungal kingdom (Smit et al. 1999; Yergeau et al. 2005; Oliveira et al. 2009). The small subunit of rRNA gene and the ITS region of the rRNA gene cluster have been the main target loci for analyses of fungal community structures though another locus, EF1-a has been targeted as well (Yergeau et al. 2005).

Internal Transcribed Spacer (ITS) region in fungi are highly variable in sequence composition and also vary in length between species and hence, suitable for profiling a number of ribotypes present in a community. Amplicons of different sizes can be separated conventionally on a polyacrylamide gel, a technique known as ribosomal intergenic spacer analysis (RISA), or automatically on a sequencer (automated ribosomal intergenic spacer analysis, or ARISA). Each ribotype will form one discrete band (on a gel) or peak (in a sequencer profile), revealing the number of species present. The intensity of the band or size of the peak height relative to the overall sample can be used as a crude estimate of the abundance of certain ribotypes in the community. For example, ARISA has been used to profile fungal communities from soil (Ranjard et al. 2001) and salt marsh (Torzilli et al. 2006).

Denaturing or Temperature gradient gel electrophoresis (DGGE or TGGE) are the most commonly used electrophoretic separations based on sequence composition (Muyzer et al. 1993). Upon amplification with suitable primer (e.g., ITS), the amplified sequences are electrophoreses on a polyacralamide gel either containing an increasing gradient of chemical denaturants like Urea or formamide in case of DGGE, or using a temperature gradient (TGGE). As the amplicons migrate down the gel, they denature from double stranded to single stranded form. These techniques benefit from the ability to analyse and compare numerous samples on a single gel and allow a rapid, simultaneous comparison in a population. The development of specific software packages has substantially improved the DGGE and TGGE based fungal community characterization (Fromin et al. 2002). TGGE and DGGE have been used successfully to profile fungal communities from environments as diverse as wheat rhizosphere (Smit et al. 1999), maize rhizosphere (Gomes et al. 2003), forest soil (Agnelli et al. 2004), leaves of *Magnolia liliifera* (Duong et al. 2006) and municipal composts (Bonito et al. 2010).

Terminal RFLP (t-RFLP) is one popular variant of classical RFLP which has been applied to investigate fungal diversity in agricultural land that was turned into fallow fields (Klamer and Hedlund 2004) and many other fungal community analyses (Liu et al. 1997; Schwarzenbach et al. 2007). T-RFLP is a very powerful technique, particularly for the identification of target organisms in a community if robust database is available.

Another approach that has been extensively used in various studies is to shotgun clone and sequence entire PCR amplicons of marker regions, and to subsequently identify sequences by performing similarity searches in public databases. Despite being laborious and potentially costly, the analysis of clone libraries produced from environmental DNA is a useful technique and is complementary to the other community profiling techniques mentioned above. This approach has been applied to identify and compare fungal communities, e.g., in plant roots (Vandenkoornhuyse et al. 2002), plant rhizospheres (Smit et al. 1999), and in various types of soils including agricultural (Lynch and Thorn 2006), tundra (Schadt et al. 2003), and forest (O'Brien et al. 2005) soils. In these studies, analyses have been performed on different taxonomic levels using fungus (Smit et al. 1999; Schadt et al. 2003) or basidiomycete specific primers (Lynch and Thorn 2006).

6.4 Population Genetic Analysis of Fungi

Fungi exhibit a wide array of reproductive strategies rendering direct impact on their population biology. An important factor for many plant-pathogenic fungi is the relative abundance of sexual and asexual reproduction within populations. Asexual reproduction leads to clonal lineages that may exhibit a limited spectrum of virulence within each lineage (Zeigler et al. 1995). Unlike diploid Eukaryotes, most dominant markers can be easily analysed for population genetic studies in haploid fungi, though it can be a problem with many Basidiomycetes and Oomycetes that are heterokaryons, diploid or polyploid. In diploid fungal species, the interpretation on the number of loci and the number of alleles per locus for dominant markers can be tedious. In sexually reproducing fungal species, crosses could be constructed and meiotic progenies be analysed. But for diploid species with no known sexual cycle, the interpretation of complex fingerprinting patterns is highly problematic. Co-dominant, single copy genetic markers are best suited for allelic assignments for strains of diploid species. Knowledge of the spatial distribution of genotypes within populations can provide information about dispersal potential within fields (Chen et al. 1994; Kohli et al. 1995). If measures of genotype diversity indicate towards sexual reproduction, the hypothesis of random mating can be tested by measures of gametic disequilibrium (Chen and McDonald 1996).

Comparisons among populations can also be useful for defining population boundaries. In particular, plant pathologists would like to know whether quarantine measures are effective. If they are not, it would be useful to be able to determine the source of introduction and paths of migration among populations. Brewer and Milgroom (2010) used multi-locus (ITS/IGS, TUB2 & EF-1) markers for population structure and phylogeographic studies on 146 isolates of *Erisyphe necator* collected from the eastern US, Europe, Australia, and the western US. Phylogeographic analyses show that the two genetic groups in Europe represent two separate introductions and that the genetic groups may be derived from eastern US ancestors.

The type of marker suitable for population genetic analysis depends on many factors, including the purpose of the study, ploidy level of the fungi and the mating system of the species. According to Fisher's fundamental theorem of natural selection the evolutionary potential of a population is proportional to the amount of genetic diversity in that population. However, Fisher's theorem applies only to genes that are known to have an effect on fitness. Neutral DNA markers will not necessarily reflect the amount of genetic variation present in genes that condition fitness. The genic markers thus have special applications in population genetic analysis.

6.5 New Generation Markers for Fungal Diversity and Community Analysis

6.5.1 Genic Markers

Polymorphic restriction profile of particular gene or genomic region has been extensively used for studying genetic variability in fungal genera to species level. The internal transcribed spacer region of the rRNA gene cluster has been the main target for this type of analyses and it has been applied for fungi like *Pandora neoaphidis* (Rohel et al. 1997), *Conidiobolus* spp. (Tymon et al. 2004), *Zoophthora radicans* (Guzman-Franco et al. 2008), *Rhizoctonia solani* (Kiliçoğlu and Özkoç 2010),

Penicillium spp. (Diguta et al. 2011), *Aspergillus* sp. (Diba et al. 2014). Restriction profiling of ribosomal large subunit gene has also been deployed to assess genetic variations and relations among *B. brongniartii* isolates (Neuve'glise et al. 1997) and within and/or among the genera *Entomophthora*, *Eryniopsis* and *Entomophaga* (Jensen and Eilenberg 2001; Hajek et al. 2003). PCR–RFLP approach has also been extended to genes like pathogenicity related Pr1 protease gene and chitinase genes to determine strain relatedness and population structure in *B. bassiana* (Wang et al. 2003) and *M. Anisopliae* (Leal et al. 1997; Enkerli et al. 2009). PCR-RFLP of other genes like DNA Topoisomerase II gene has also been used for identification of dermatophytes (Kamiya et al. 2004).

In comparison to multi-copy genes, protein coding genes occurring as single copies in genomes are advantageous for quantitative comparison of taxon abundances. Furthermore, coding genes often contain introns with sufficient power for discrimination among species. But a disadvantage with many protein coding genes is that they occur in gene families where gene duplications within genome often have taken place within the same period of time as speciation, making the identification of gene orthologs complex and problematic (Lindah and Taylor 2004; Bödeker et al. 2014). Woobling at third base makes it further difficult to design primers that cover all possible sequence variants, even when highly conserved functional domains are targeted. Genes encoding cytochrome oxidases, histones, actin, β-tubulins etc. including mating type locus have been used for identification and characterization of fungi used (O'Donnell et al. 2004; James et al. 2006; Keeling and Inagaki 2004). Although there is now wide acceptance for the need of standardization of gene regions used in taxonomic studies (Caterino et al. 2000), it has proven more difficult to determine which particular gene regions and sequence length are most informative (Roe and Sperling 2007). Initially, a 658-bp segment of mitochondrial DNA from the cytochrome c oxidase gene subunit 1 (*CO1*) was proposed as the single standard DNA barcoding region (Hebert et al. 2003). Seifert et al. (2007) sequenced the barcode region of *CO1* for 360 strains from one group of the mold genus *Penicillium* using PCR primers designed from mitochondrial genomes for the ascomycete family Trichocomaceae. *CO1* provided species-specific barcodes for about 66 % of species, superior to ITS (about 25 %). Similar results were obtained for seven species of the soil-borne genus, *Leohumicola* (Nguyen and Seifert 2008), where *CO1* and ITS provided similar species resolution. *CO1* functions well as a barcode in the fungus-like Oomycete genera *Pythium* and *Phytophthora* (Seifert 2009). Pre-barcoding phylogenetic studies of *Phytophthora* by Martin and Tooley (2003) used *CO1*and *CO2* genes as markers. O'Donnell et al. (2000) used the DNA sequences of the nuclear rDNA large subunit, mitochondrial small subunit, and β-tubulin to develop a phylogeny that includes 36 taxa in the *G. fujikuroi* species complex. Utilizing 21 fungal genomes, Aguileta et al. (2008) identified 246 single-copy orthologous gene clusters in an optimally performing gene set. Protein-coding genes such as *β-tubulin*, *tef-1α*, *actin, chitinases*, *chitin synthases*, *RNA polymerases* and others were not among the best performing genes found by Aguileta et al. (2008). In addition, *β-tubulin*, *tef-1α*, *chitinases*, and *dehydrogenases* were not in the cross-genome single-copy ortholog set, most likely due to the presence of paralogs. Schmitt et al. (2009) developed primers sets for two single-copy genes *Mcm7* and *Tsr1*, which Aguileta et al. (2008) determined as the most phylogenetically informative markers in their optimally performing gene set. Although these genes show promise for taxonomic resolution from species to class levels, species amplified by Schmitt et al. (2009) belonged exclusively to the lichen-forming classes Eurotiomycetes, Lecanoromycetes, and Lichinomycetes. Primer sequences, however, were compared with published *Mcm7* and *Tsr1* sequences for representative taxa suggesting that the primers would work outside the tested groups (Schmitt et al. 2009). Raja et al. (2011) confirmed phylogenetic utility of *Mcm7* from species to class levels in the Dothideomycetes, Eurotiomycetes, Geoglossomycetes, Lecanoromycetes, Leotiomycetes, and Sordariomycetes.

Self-incompatible (heterothallic) fungi in the phylum Ascomycota contain a single mating-type locus (*MAT1*) with one of two possible specificities. For two heterothallic individuals to be compatible and mate successfully, each must possess a different mating specificity at *MAT1* locus. In most filamentous ascomycetes, the mating types are named *MAT1-1* and *MAT1-2* (Yun et al. 2000). The regions flanking the idiomorphs (alternate sequences of *MAT1* locus) are homologous for isolates of both mating types and the gene order in the flanking regions is conserved among most ascomycete fungi (Debuchy and Turgeon 2006). The idiomorphs *MAT1-1* and *MAT1-2* are characterized by conserved motifs in their encoded proteins. Idiomorphs may contain genes in addition to *MAT1-1-1* or *MAT1-2-1*; for example, some species of ascomycetes have *MAT1-1* idiomorphs with *MAT1-1-3*, which is also characterized as having an HMG domain, although it belongs to a different phylogenetic grouping and is not as conserved as in *MAT1-2-1* (Debuchy and Turgeon 2006). Identification of mating-type genes was useful for comparative studies on the structure and evolution of the *MAT1* locus within the Erysiphales, and among other ascomycetes. Still, use of mating type loci for species level phylogenetics has been limited compared to ITS (Barve et al. 2003; Turgeon 1998). O'Donnell et al. (2004) used mating type locus for resolving phylogenetic relationship in *Fusarium graminearum* clade and nine phylogenetically distinct species could be resolved.

6.5.2 Gene Targeted Start Codon Based Markers

Start codon based marker is a very recent addition to the armory of the molecular marker technology. Molecular markers from the transcribed region offers potential for genotyping as they reveal polymorphisms directly related to gene function. A novel marker system called Start Codon Targeted Polymorphism (SCoT) was described by Collard and Mackill (2009a), based on the observation that the short conserved regions of plant genes are surrounded by the ATG translation start codon (Sawant et al. 1999). Like RAPD, the technique uses single primer. The primer is designed to anneal to the flanking regions of the ATG initiation codon on both DNA strands (Fig. 6.2). The generated amplicons are possibly distributed within gene regions that contain genes on both the DNA strands. It is also possible that

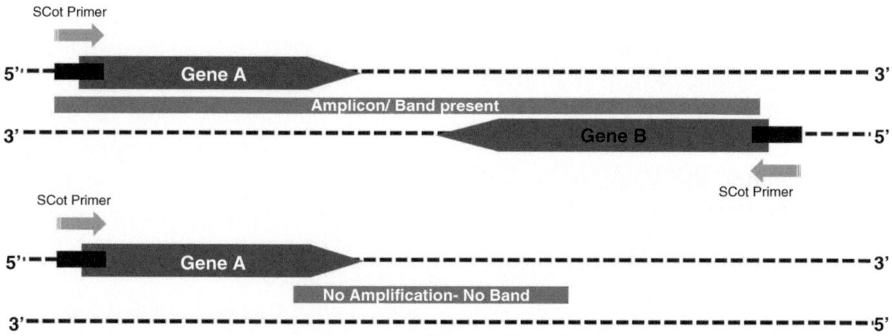

Fig 6.2 Principle of SCoT analysis

pseudogenes and (genes within) transposable elements may be used as primer binding sites by SCoT polymorphism technique. As the distance between the primer binding sites on the template may vary considerably, a relatively longer extension time of the thermal cycle is required. SCoT markers are generally reproducible, and it is suggested that primer length and annealing temperature are not the sole factors determining reproducibility (Gorji et al. 2011). They are dominant markers, however, a number of co-dominant markers are also generated during amplification, and thus SCoT can be used for genetic diversity analysis (Collard and Mackill 2009b). This marker has been validated initially through study of genetic diversity among rice varieties (Collard and Mackill 2009b) and further extended to many other plants (Guo et al. 2012; Satya et al. 2015). Recently, SCoT markers are also gaining importance in mycological research. Zhao et al. (2013) used SCoT analysis for determining genetic variability and population structure of *Pleurotus eryngii* var. *tuoliensis*. Nath et al. (2013) used SCoT marker for analysis of *Phytophthora colocasiae* and reported its high efficiency in evaluating the genetic diversity.

6.5.3 Single Nucleotide Ploymorphism (SNP)

Single nucleotide polymorphisms (SNPs) are among the fastest-developing categories of DNA markers because of high frequency, low mutation rate and ease of scoring and data sharing for a variety of biological analyses (Brumfield et al. 2003). There are several common methods for SNPs analysis, such as DNA sequencing, single-strand conformation polymorphism PCR (SSCP-PCR) (Piddock et al. 1998), the mismatch amplification mutation assay (MAMA-PCR) (Zirnstein et al. 1999), the Light Cycler-based PCR hybridization mutation assay (Walker et al. 2001), denaturing high performance liquid chromatography (DHPLC) (Eaves et al. 2002), TaqMan-MGB (Minor Groove Binder) probes (Saito et al. 2003), allele-specific PCR (ASPCR) (Ma and Michailides 2004), and primer-introduced restriction analysis PCR (PIRA-PCR) (Luo et al. 2009). ASPCR is a good method, though it frequently requires extensive optimization. TaqMan-MGB is a precise method for detecting single nucleotide mutations but relative high cost has limited its use. PCR coupled with SSCP has been found to be effective for SNP identification arbuscular mycorrhizal strains (Simon et al. 1993). Kjøller and Rosendahl (2000) showed that specific SSCP patterns in *Glomus mossae* and *G. caledonium* could be used to detect these two fungi in plant roots. SSCP has also been used to determine genetic diversity among the single spore isolates of *G. intraradices* (Jansa et al. 2002). SSCP of the rRNA genes (18S or 28S) has been used to identify a wide variety of fungi, but to our knowledge it has not been assessed on more polymorphic genes to fingerprint fungal strains of the same species (Hui et al. 2000).

Heteroduplex mobility assay (HMA) is another emerging technique for identification of SNPs and small differences in DNA sequences. After PCR amplification of DNA regions of interest, the DNA is denatured by heat and allowed to anneal, forming homoduplexes when the sequences are identical, and heteroduplexes when there is a difference in one or more bases. Huang et al. (2010, 2011) used HMA for detection and phylogenetic analysis of different *Colletotichum* spp. responsible for Anthracnose disease. A heteroduplex mobility assay with a panel of reference fragments (HPA) has been used to assess intraspecific variation in a group of aflatoxin-producing *Aspergillus* spp. (Kumeda and Asao 2001). HMA has been used for differentiation of yeasts like *Candida* spp. (Ramos et al. 2006) and *Saccharomyces* species (Ramos et al. 2001). Kristensen et al. (2007) have developed SNP based multiplexing assay for identification of trichlothecene and moniliformin producing *Fusarium*. A few studies have also utilized genic SNPs for fungal characterization. For example, Fournier et al. (2010) developed SNP based assay with genic regions like RNA polymerase II (*RPB1, RBPII*), *β-tubulin* (BTUB) gene, elongation factor 1a-like (*EFL*), large subunit (LSU) and small subunit (SSU) of rDNA genes together with the internal transcribed spacer (ITS) to discriminate 15 *Pandora neoaphidis* isolates.

6.6 Next Generation Sequencing (NGS)

The traditional DNA-sequencing method (Sanger et al. 1977) can only sequence individual specimens and, therefore, is not adequate for processing complex environmental samples, especially for large-scale studies. These samples contain mixtures of DNA from numerous individuals beyond the scope of the ability of conventional sequencing (Hajibabaei et al. 2011). Recovering DNA sequences from the thousands of specimens present in an environmental bulk sample requires the ability to read DNA from multiple templates at a go; something that next-generation sequencing technologies do effectively at cheaper costs. Next generation sequencing platforms use electrophoresis-free systems to decipher DNA sequences by monitoring each sequencing step *in situ*. By removing the requirement for electrophoretic fractionation of reaction products, next-generation sequencing methodologies can produce large amounts of sequence data in a single run. Large scale or next generation sequencing approaches provide the capacity to generate thousands of sequences, which may allow obtaining maximal resolution and species representation (Ronaghi et al. 1998; Margulies et al. 2005). These technologies allow shotgun sequencing of single genomes (Margulies et al. 2005), massive parallel sequencing of PCR amplified target regions like rRNA genes or ITS regions from metagenomic samples (Christen 2008; Petrosino et al. 2009), and whole genome shotgun sequencing of environmental DNA/RNA samples (Tringe and Rubin 2005; Petrosino et al. 2009), where partial genomes of diverse organisms are sequenced simultaneously.

6.6.1 NGS Workflow and Platforms

The NGS workflow is well described in a number of reviews (Haridas et al. 2011; Solomon et al. 2014). The workflow starts with sample collection followed by DNA or RNA isolation (Fig. 6.3). The quality of the nucleic acid is a critical factor for success of NGS. Another factor is to determine the depth of coverage of NGS, which depends on the purpose of the study. Low genome coverage is sufficient for ITS based taxonomic identification using NGS. But for whole transcriptome sequencing high depth of sequencing is required. Similarly, library fragment size, assembly program and k-mer vary according to the purpose of the experiment. Various library preparation protocols for NGS are available suited to need and instrument used for analysis. Initially, the isolated nucleic acid is fragmented to

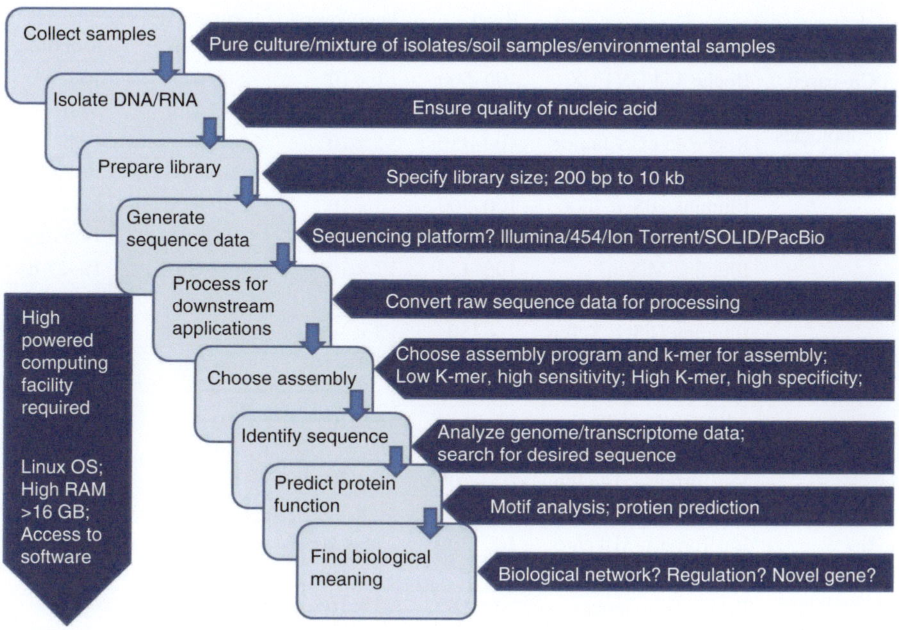

Fig 6.3 Guideline for next generation sequencing in fungi

suitable size and ligated (tagged) with specific adapters. The adapters are designed to contain elements for immobilization of the sequence on a surface such as beads or slides. A size selection step excludes undesirable fragment size and free adapters in the pool. The prepared library is then subjected to PCR to generate sufficient amount of DNA for sequencing of the library.

Table 6.2 (modified from Shokralla et al. 2012) provides a comparative description of currently available NGS technologies. The 454 sequencing technology (454 Life Sciences, Roche) is one of the first used next generation technologies. In this technology, DNA fragments are ligated to flanking adapters to form amplicons and are bound to polystyrene beads during emulsion PCR (emPCR). Each of these beads carry thousands of copies of different and unique DNA amplicon which are deposited into wells of a PicoTiter Plate (PTP). Successful nucleotide sequence reads of DNA bound to one to two million of these beads result from pyrosequencing. Unlike 454, Illumina uses different methods of template preparation, sequencing and detection. The basic strategy for Illumina is sequencing by synthesis (SBS) technology i.e., tracking the addition of fluorescently labelled nucleotides as the DNA chain is being copied. The Ion Torrent technology is another such PCR based technology which uses semiconductor technology to detect base incorporation in terms of H^+ release during base addition. Other NGS technologies (PacBio & Helicos Heliscope) use single molecule sequencing (SMS) technology. Single-

Table 6.2 Comparison of currently available next-generation sequencing technologies

Category	Platform	Read length (bp)	Max. number of reads/run	Sequencing output/run	Run time
PCR-based NGS technologies	Roche 454 GS FLX	400–500	1×10^6	≤500 Mb	10 h
	Roche 454 GS FLX+	600–800	1×10^6	≤700 Mb	23 h
	Roche 454 GS Junior	400–450	1×10^5	~35 Mb	10 h
	Illumina HiSeq 2000	100–200	6×10^9	≤540–600 Gb	11 day
	Illumina HiSeq 1000	100–200	3×10^9	≤270–300 Gb	8.5 day
	Illumina GAIIx	50–75	6.4×10^8	≤95Gb	14.5 day
	Illumina MiSeq	100–150	7×10^6	≤1–2 Gb	19–27 h
	Illumina HiSeq 2500	2 × 125–150	2×10^8	≤1 Tb	1–6 day
	Illumina HiSeq X Ten	2 × 150	3×10^8	1.6–1.8 Tb	<3 day
	AB SOLiD 5500 system	35–75	2.4×10^9	~100 Gb	4 day
	AB SOLiD 5500 xl system	35–75	6×10^9	~250 Gb	7–8 day
	Ion Torrent -314 chip	100–200	1×10^6	≥10 Mb	3.5 h
	Ion Torrent -316 chip	100–200	6×10^6	≥100 Mb	4.7 h
	Ion Torrent -318 chip	100–200	11×10^6	≥1 Gb	5.5 h
SMS technologies	Helicos HeliScope	30–35	1×10^9	~20–28 Gb	≤1d
	Pacific Biosciences system	≥1500	5×10^4	~60–75 Mb	0.5 h

molecule sequencing systems carry out sequencing-by-synthesis using laser excitation to generate a fluorescent signal from labelled nucleotides, which is detected using a camera. Although these technologies have higher error rate, random distribution of errors and use of error correction software improves the performance of these technologies.

6.6.2 Whole Genome Sequencing Using NGS Technology

De novo whole genome sequencing using NGS platforms has been a very attractive choice for the biologists to look inside the genomes of fungi with very little known genomic information. With the availability of ultra-high-throughput sequencing, one strategy is to just sequence whole genomes of several target organisms within an environment sample, also known as metagenome analysis, and identify SNPs by comparing these total data sets. Table 6.3 presents a list of ascomycetes fungi for which permanent genome draft has been prepared using various NGS technology since 2010. Using Ion Torrent Technology, Leboldus et al. (2015) discovered total of 5783 and 2373 unique ‘sequence tags’ containing 16441 and 9992 single nucleotide polymorphisms (SNPs) from natural populations of *Pyrenophora teres* f. *maculata* and *Sphaerulina musiva*, respectively.

Another application of NGS in whole genome analysis is re-sequencing of fungal strains for which genome sequence is already available. Ojeda et al. (2014) discovered 9531–17266 SNP in *Leptographium longiclavatum*, a mountain pine beetle-associated symbiotic fungus, using whole-genome resequencing. Using the whole genome and SNP data, they developed genotyping assay using the Illumina ®Sequenom iPLEX Gold and reported that this approach for population genomics studies in other mountain pine beetle fungal symbionts and other fungal non-model species. In another study, McDonald et al. (2015) re-sequenced 13 isolates of *Zymoseptoria triticeae*, a fungal pathogen of wheat to identify presence-absence variation in comparison with the reference strain *Z. triticeae* IPO 323 and outlined several possibilities for comparison of strains using re-sequencing. However, whole genome sequencing may not be the best choice for many eukaryotes with large genomes and is less informative for poorly assembled new genomes (where linkage between the sequenced fragments is unknown). Sequencing large genome is also costly, which multiplies if more samples have to be sequenced.

6.6.3 NGS for Single Locus and Multi-locus Sequence Typing

Mycologists have initially used NGS for amplification of specific regions of genome, such as ribosomal DNA (rDNA) sequence from a large number of samples or from an environmental sample for characterization. As stated previously, the rDNA genes are excellent choice for taxonomic identification in microbial community. Since the gene

Table 6.3 Genome sequencing of agriculturally important fungi (2010–2014) using NGS platform

Organism name	NCBI Bioproject Accession No.	NGS platform	Sequencing depth	Total seq. Length (bp)
Alternaria arborescens EGS 39–128	PRJNA78243	Illumina GAii	90x	3,38,89,384
Bipolaris maydis ATCC 48331	PRJNA83117	454 &Illumina	67.5x	3,29,29,167
Blumeria graminis f. sp. *hordei* A6	PRJNA186717	454 &Illumina HiSeq	70x	4,74,65,002
Blumeria graminis f. sp. *hordei* K1	PRJNA186718	454 &Illumina HiSeq	70x	5,37,86,025
Ceratocystis fimbriata C1421	PRJNA67151	454-GS-FLX-Titanium	20x	2,35,89,795
Cercospora canescens BHU	PRJNA183604	Illumina	12x	3,39,67,224
Colletotrichum gloeosporioides Nara gc5	PRJNA171218	Illumina HiSeq	37x	5,56,07,143
Colletotrichum orbiculare MAFF 240422	PRJNA171217	454/Illumina Gaiix	55x	9,00,86,138
Debaryomyces hansenii MTCC 234	PRJNA77881	Illumina GAii	152x	1,14,62,343
Diplodia pinea CMW39103	PRJNA242796	Illumina MiSeq	10x	3,52,41,951
Fusarium fujikuroi B14	PRJNA171493	Illumina HiSeq	75.8x	4,38,12,727
Fusarium oxysporum f. sp. *conglutinans* race 2 54008	PRJNA73543	Illumina	124x	5,35,75,352
Fusarium oxysporum f. sp. *cubense* race 1	PRJNA174274	Illumina GAiix	83x	4,76,57,417
Fusarium oxysporum f. sp. *cubense* race 4	PRJNA174275	Illumina GAiix	87x	5,29,26,277
Fusarium oxysporum f. sp. *cubense* tropical race 4 54006	PRJNA73539	Illumina	208x	4,65,53,780
Fusarium oxysporum f. sp. *lycopersic*i MN25	PRJNA72769	Illumina	210x	4,86,37,398
Fusarium oxysporum f. sp. *melonis* 26406	PRJNA73541	Illumina	157x	5,40,34,280
Fusarium oxysporum f. sp. *pisi* HDV247	PRJNA72771	Illumina	242x	5,51,88,216
Fusarium oxysporum f. sp. *radicis-lycopersici* 26381	PRJNA73535	Illumina	188x	4,93,59,289
Fusarium oxysporum f. sp. *raphani* 54005	PRJNA73545	Illumina	145x	5,34,99,362
Fusarium oxysporum f. sp. *vasinfectum* 25433	PRJNA73537	Illumina	165x	5,29,14,414
Fusarium oxysporum Fo47	PRJNA67069	Illumina	138x	4,96,64,628

Fusarium oxysporum Fo5176	PRJNA68027	454-GS-FLX-Titanium	8x	5,47,67,602
Fusarium oxysporum FOSC 3-a	PRJNA67067	Illumina	195x	4,79,06,303
Fusarium pseudograminearum CS3096	PRJNA66583	Illumina GA/GAiix	180x	3,69,73,259
Guignardia citricarpa CGMCC3.14348	PRJNA188924	454/Illumina Gaiix	300x	3,20,07,210
Magnaporthe oryzae 4091-5-8	PRJNA184380	Illumina GAiix	68x	3,75,19,115
Magnaporthe oryzae KJ201	PRJNA179498	Illumina GAiix	61x	4,50,96,509
Mycosphaerella graminicola STIR04 1.1.1	PRJNA63041	Illumina	45x	3,09,57,533
Mycosphaerella graminicola STIR04 1.1.2	PRJNA63043	Illumina	45x	3,11,36,334
Mycosphaerella graminicola STIR04 3.11.1	PRJNA46489	Illumina	45x	3,34,56,794
Mycosphaerella graminicola STIR04 3.3.2	PRJNA63047	Illumina	45x	3,28,56,836
Mycosphaerella graminicola STIR04 5.9.1	PRJNA63039	Illumina	45x	3,24,96,248
Mycosphaerella graminicola STIR04 A48b	PRJNA63029	Illumina	45x	3,17,94,146
Mycosphaerella populicola P02.02b	PRJNA81737	454-GS-FLX-Titanium	18x	3,31,88,813
Mycosphaerella populorum SO2202	PRJNA51781	454/Illumina	35x	2,93,52,103
Pyrenophora teres f. *teres* 0–1	PRJNA50389	Illumina	20x	3,35,83,335
Sclerotinia homoeocarpa LT30	PRJNA167556	454	8x	2,97,25,518
Sphaeropsis sapinea CMW190	PRJNA215898	Illumina	15x	3,60,53,350
Zymoseptoria ardabiliae STIR04_3.13.1	PRJNA63045	Illumina	45x	3,31,90,438
Zymoseptoria pseudotritici STIR04_2.2.1	PRJNA63031	Illumina	45x	3,16,23,993
Zymoseptoria pseudotritici STIR04_4.3.1	PRJNA63035	Illumina	45x	3,20,97,292
Zymoseptoria pseudotritici STIR04_5.3	PRJNA63037	Illumina	45x	3,27,96,701
Zymoseptoria tritici STIR04_A26b	PRJNA62705	Illumina	45x	3,26,41,306

is easily amplified and sequenced using universal primers it has been an easy choice for microbiologists. Besides availability of a huge ITS database allows accurate matching. NGS based ITS sequencing of microbial community has been attempted in a number of studies (Lindah et al. 2013), but ITS based classification is less reliable in fungi than in bacteria. Major concerns of using rDNA are high variability of rDNA within species and sequence similarity of the ITS regions across species, which may differ by a few bases. Combination of ITS and other loci such as chloroplast *rbcL* has been used for identification of fungal symbionts in subtropical forests (Toju et al. 2014). Multi-locus sequence typing (MLST) uses multiple sequence loci for molecular identification of fungi, which is currently considered the most effective solution for microbial taxonomy. Using NGS, a next generation MLST (NGMLST) has been developed by Chen et al. (2015), which utilized multiplex PCR and Pacific Biosciences (PacBio) circular consensus sequencing (CCS) technology to type nine consensus unlinked loci in *Cryptococcus neoformans*, a human fungal pathogen. The protocol was highly efficient by generating 32,932 barcode sequences, 818 CCS reads and identified 306 alleles in 34 reference sequences. They also compared the sequences generated by PacBio CCS with previously Sanger sequenced loci and found 99.98 % similarity, which confirms the robustness of the NGMLST technology.

6.6.4 Restriction Site-Associated DNA Sequencing

Of the emerging NGS technologies, Restriction site-associated DNA sequencing (RADSeq) can identify and score thousands of genetic markers randomly distributed across the target genome, from a group of individuals using Illumina technology. RADSeq can be used to carry out population genetic studies on species with no, or limited, existing sequence data, and has several advantages over previous methods for marker discovery. RADSeq can be used on crosses of any design, and in wild populations, enabling not only genotyping and SNP discovery, but also more complex analyses such as quantitative genetic and phylogeographic studies. RAD analyses incorporated into high-throughput sequencing (i.e., on the Illumina platform; Baird et al. 2008). For this, the genomic DNA is fragmented to much smaller than the average distance between restriction sites by random shearing, which is followed by preparation of the sheared ends and ligation of the second adapter, and amplification of specific fragments that contain both adapters using PCR. The first adapter contains a short DNA sequence barcode. Different DNA samples can be prepared with different barcodes to allow for sample tracking when multiple samples are sequenced in the same reaction (Hohenlohe et al. 2010; Baird et al. 2008). These RAD tags can then be subjected to high-throughput sequencing for more efficient RAD mapping. Cromie et al. (2013) used RADSeq to decipher genomic diversity and population structure of 262 strains of *S. cerevisiae* from diverse geographical origin. Hebert et al. (2013) also used RADSeq characterization in boxwood blight fungus *Calonectaria pseudonaviculata* to develop a whole genome-scale suite of variable genomic markers in *C. pseudonaviculata*, estimate general worldwide genetic diversity, and determine the source of the boxwood blight fungus in North America.

6.6.5 RNA-seq

RNA-seq, or RNA sequencing of the transcriptome is another exciting area of application of NGS technologies. Since the same genome can have different transcriptome composition depending on internal factors (juvenile vs. adult, vegetative vs. sexual, mycelia vs. spore etc.) or external elements (stress, nutrient, environmental condition, culture etc.), RNA-seq offers identification of transcriptional landscape, identification of differentially expressing genes, identification of genes associated with pathways, finding alternate splice sites and understanding RNA editing process (Filichkin et al. 2009; Zhao et al. 2013). It has been also applied to identify transcriptionally active regions (TARs), which are RNA regions outside the known genes. Such TARs have been identified from *Candida albicans* (Bruno et al. 2010), *C. glabrata* (Linde et al. 2015), *Aspergillus flavus* (Lin et al. 2013) *Aspergillus niger* (Novodvorska et al. 2013) and *Aspergillus oryzae* (Wang et al. 2010).

RNA-seq is also extensively employed for understanding genetic basis of host-microbe interaction. A host-pathogen interaction analysis between rice and *Magnaporthe oryzae* identified 240 candidate fungal genes involved in the process of initiation of fungal infection in rice (Kawahara et al. 2012). Many of these genes encoded for cell wall degrading enzymes like chitinase, cutinase, xylanase, proteins involved in appressoria and hyphae formation and signal transduction proteins. Similarly, 16084 rice transcripts were identified of which 23 genes were highly upregulated with a fold change of more than 150.

Not only fungal transcriptome analysis, but plant transcriptome study may also explain the role of fungi in plant physiological processes. Beneficial fungi like mycorrhiza help to increase yield of many crops mimicking fertilizer application response. Comparative RNA-seq of tomato plants supplied with mycorrhiza or fertilizer showed that mycorrhizal association affects ripening of tomato fruit by influencing a core set of genes that function as nutrient transporters (Zouari et al. 2014). This association also benefits the plant by activating wide array of defence genes in plants, thereby explaining the role of mycorrhiza in plant protection.

6.6.6 NGS for Fungal Community Analysis

NGS is a promising tool for fungal richness analysis in a community or environment such as analysis of gut microbial diversity, soil microbial diversity, endophyte diversity, diversity of saprophytic fungal community, or identification fungal symbionts associated with forest trees. The 454 pyrosequencing has been used to analyze the fungal diversity of bovine rumen by Fouts et al. (2012). Using Ion Torrent technology, Brown et al. (2013) performed ITS based amplicon sequencing to study soil fungal community responses to long-term prescribed fire treatments in a loblolly pine forest on the Piedmont of Georgia. Voříšková et al. (2014) used pyrosequencing to determine the seasonal dynamics of fungal communities in Oak forest. Community analysis of fungi by 454 pyrosequencing identified 575 different operational taxonomic units (OTU) inhabiting in pinewoods in Japan (Fukasawa and Matsuoka 2015). On the other hand, Womack et al. (2015) used Illumina HiSeq 2000 based analysis of atmosphere over amazon rainforest to characterize active and total fungal communities.

6.6.7 Future Challenges for Mycologists

Present NGS sequencing technology is not full-proof; error in sequencing is an undeniable component of NGS. Besides, NGS technology generates huge amount of sequence data which needs significant expertise and knowledge in computational biology for downstream processing. In most cases, the data needs expert handling for assembly and annotation. Thus, development of high performance computing facility is a prerequisite for bioinformatic analysis of the data. Due to its high input cost and expertise needed for interpretation of the data, application of NGS is limited to fund availability. For small scale analysis, marker based studies are still considered to be more economic. It will be unwise to invest huge sum in NGS analysis in routine screening of pathogens unless if alternative low-cost technologies are available.

Using NGS for fungal community analysis is much more problematic as samples are heterogeneous. If reference sequence is available, the error rate will be low, but for de novo assembly misreading of overlaps of short reads could lead to erroneous results. To overcome these challenges, a number of bioinformatic tools have been developed that can assemble short reads more efficiently, such as Trinity (Grabherr et al. 2011) and ABYSS (Simpson et al. 2009). It expected that new developments in NGS technologies in the next few years will enable sequencing of longer reads which will be helpful to develop better assembly for further downstream analysis at lower cost.

6.7 Conclusions

Rapid progress in the area of DNA marker development has enabled mycologists to choose from vast array of technologies suitable from low-scale diagnostics to high-scale community analysis or in-depth genomic studies. Despite arrival of new technologies, traditional low cost markers are quite effective for low scale studies and case by case investigation. However with increase in the volume and complexity of the study, new generation markers systems do offer better and more accurate results allowing the depth of knowledge. Advances in high volume computing and mathematical development for designing high quality algorithms for sequence assembly, sequence identification, functional prediction and marker design have helped tremendously in new generation marker development, identification of large number of SSR and SNP markers and a burst in next generation sequencing technology development. Within a few years, these new generation marker systems have helped to dissect complex problems generating huge database and identifying vast array of up- and down-regulated genes in fungi and its hosts. The NGS technologies are highly efficient for high volume community analysis allowing association analysis of plant and fungi at different peripheries such as pathogenic, mycorrhizal, saprophytic or neutral association. What hinders the large scale use of these new markers systems is the cost involved in analysis, facilities required for high volume bioinformation management and expertise required for handling and interpreting large datasets. Although the number of mycologists trained for bioinformatic analysis of NGS data is on the rise, this may be the most important limiting factor in coming days, as the cost of computing facilities and sequencing are expected to trim down with development of new technologies.

References

Agnelli A, Ascher J, Corti G, Ceccherini MT, Nannipieri P, Pietramellara G. Distribution of microbial communities in a forest soil profile investigated by microbial biomass, soil respiration and DGGE of total and extracellular DNA. Soil Biol Biochem. 2004;36(5):859–68.

Aguileta G, Marthey S, Chiapello H, Lebrun MH, Rodolphe F, Fournier E, Giraud T. Assessing the performance of single-copy genes for recovering robust phylogenies. Syst Biol. 2008;57(4):613–27.

Ahren D, Faedo M, Rajashekar B, Tunlid A. Low genetic diversity among isolates of the nematode-trapping fungus *Duddingtonia flagrans*: evidence for recent worldwide dispersion from a single common ancestor. Mycol Res. 2004;108(10):1205–14.

Aishwarya V, Grover A, Sharma, PC. EuMicroSat db: A database for microsatellites in the sequenced genomes of eukaryotes. BMC Genomics 2007; 8(1):1.

Aishwarya V, Sharma, PC. UgMicroSatdb: database for mining microsatellites from unigenes. Nucleic Acids research 2008; 36:D53–D56.

Alves A, Phillips AJ, Henriques I, Correia A. Rapid differentiation of species of Botryosphaeriaceae by PCR fingerprinting. Res Microbiol. 2007;158(2):112–21.

Babu BK, Reddy SS, Yadav MK, Sukumar M, Mishra V, Saxena AK, Arora DK. Genetic diversity of *Macrophomina phaseolina* isolates from certain agro-climatic regions of India by using RAPD markers. Indian J Microbiol. 2010;50(2):199–204.

Baird NA, Etter PD, Atwood TS, Currey MC, Shiver AL, et al. Rapid SNP discovery and genetic mapping using sequenced RAD markers. PLoS One. 2008;3:e3376.

Barve MP, Arie T, Salimath SS, Muehlbauer FJ, Peever TL. Cloning and characterization of the mating type (MAT) locus from *Ascochyta rabiei* (teleomorph: *Didymella rabiei*) and a MAT phylogeny of legume-associated *Ascochyta* spp. Fungal Genet Biol. 2003;39(2):151–67.

Bayraktar H, Dolar FS, Maden S. Use of RAPD and ISSR markers in detection of genetic variation and population structure among *Fusarium oxysporum* fspciceris isolates on chickpea in Turkey. J Phytopathol. 2008;156(3):146–54.

Ben Abdeljelil J, Saghrouni F, Cabaret O, Boukadida J, Bretagne S, Ben Saïd M. Temporal similarity between *Candida albicans* genotypes in a tunisian neonatal intensive care unit suggests several nosocomial cross-contamination episodes. DNA Cell Biol. 2012;31(7):1161–6.

Blackwell M. The fungi: 1, 2, 3 51 million species? Am J Bot. 2011;98:426–38.

Bock CH, Endalew TT, Biswas BK, Yadav AK, Sitther V, Hotchkiss MW, Wood BW. A comparison of UP-PCR and RAPD markers to study genetic diversity of Fusicladium effusum (GWinter), cause of pecan scab. Forest Pathol. 2014;44(4):266–75.

Bödeker I, Clemmensen KE, Boer W, Martin F, Olson Å, Lindahl BD. Ectomycorrhizal *Cortinarius* species participate in enzymatic oxidation of humus in northern forest ecosystems. New Phytologist. 2014;203(1):245–56.

Bogale M, Wingfield BD, Wingfield MJ, Steenkamp ET. Characterization of *Fusarium oxysporum* isolates from Ethiopia using AFLP, SSR and DNA sequence analyses. Fungal Diversity. 2006;23:51–66.

Bonito G, Isikhuemhen OS, Vilgalys R. Identification of fungi associated with municipal compost using DNA-based techniques. Biores Technol. 2010;101(3):1021–7.

Brewer MT, Milgroom MG. Phylogeography and population structure of the grape powdery mildew fungus, *Erysiphe necator*, from diverse *Vitis* species. BMC Evol Biol. 2010;10(1):268.

Bridge PD, Arora DK. Interpretation of PCR methods for species definition. In: Bridge PD, Arora DK, Reddy CA, Elander RP, editors. Applications of PCR in mycology. Wallingford: CAB International; 1998. p. 63–84.

Brown SP, Callaham MA, Oliver AK, Jumpponen A. Deep Ion Torrent sequencing identifies soil fungal community shifts after frequent prescribed fires in a southeastern US forest ecosystem. FEMS Microbiol Ecol. 2013;86(3):557–66.

Brumfield RT, Beerli P, Nickerson DA, Edwards SV. The utility of single nucleotide polymorphisms in inferences of population history. Trends Ecol Evol. 2003;18(5):249–56.

Bruno VM, Wang Z, Marjani SL, Euskirchen GM, Martin J, Sherlock G, Snyder M. Comprehensive annotation of the transcriptome of the human fungal pathogen *Candida albicans* using RNA-seq. Genome Res. 2010;20:1451–8.

Bulat SA, Lübeck M, Mironenko N, Funck Jensen D, Stephensen Lübeck P. UP-PCR analysis and ITS1 ribotyping of strains of *Trichoderma* and *Gliocladium*. Mycol Res. 1998;102(08):933–43.

Caterino MS, Cho S, Sperling FA. The current state of insect molecular systematics: a thriving Tower of Babel. Ann Rev Entomol. 2000;45(1):1–54.

Chen RS, McDonald BA. Sexual reproduction plays a major role in the genetic structure of populations of the fungus *Mycosphaerella graminicola*. Genetics. 1996;142:1119–26.

Chen RS, Boeger JM, McDonald BA. Genetic stability in a population of a plant pathogenic fungus over time. Mol Ecol. 1994;3:209–18.

Chen Y, Frazzitta AE, Litvintseva AP, Fang C, Mitchell TG, Springer DJ, Ding Y, Yuan G, Perfect JR. Next generation multilocus sequence typing (NGMLST) and the analytical software program MLSTEZ enable efficient, cost-effective, high-throughput, multilocus sequencing typing. Fungal Genet Biol. 2015;75:64–71.

Christen R. Global sequencing: a review of current molecular data and new methods available to assess microbial diversity. Microbes Environ. 2008;23(4):253–68.

Collard BC, Mackill DJ. Start codon targeted (SCoT) polymorphism: a simple, novel DNA marker technique for generating gene-targeted markers in plants. PMB Rep. 2009a;27(1):86–93.

Collard BCY, Mackill DJ. Conserved DNA-derived polymorphism (CDDP): a simple and novel method for generating DNA markers in plants. PMB Rep. 2009b;27(4):558–62.

Cooke DEL, Drenth A, Duncan JM, Wagels G, Brasier CM. A molecular phylogeny of *Phytophthora* and related oomycetes. Fungal Genet Biol. 2000;30:17–32.

Cromie GA, Hyma KE, Ludlow CL, Garmendia-Torres C, Gilbert TL, May P, Fay JC. Genomic sequence diversity and population structure of *Saccharomyces cerevisiae* assessed by RAD-seq. G3 Genes Genomes Genetics. 2013;3(12):2163–71.

Dalleau-Clouet C, Gauthier N, Risterucci A, Bon MC, Fargues J. Isolation and characterization of microsatellite loci in the entomopathogenic *fungus Paecilomyces fumosoroseus*. Mol Ecol Notes. 2005;5:496–8.

Debuchy R, Turgeon BG. Mating-type structure, evolution, and function in Euascomycetes. In: Growth, differentiation and sexuality. Berlin/Heidelberg: Springer; 2006. p. 293–323.

Diba K, Mirhendi H, Kordbacheh P. Development of RFLP-PCR method for the identification of medically important *Aspergillus* species using single restriction enzyme MwoI. Braz J Microbiol. 2014;45(2):503–7.

Diguta CF, Vincent B, Guilloux-Benatier M, Alexandre H, Rousseaux S. PCR ITS-RFLP: a useful method for identifying filamentous fungi isolates on grapes. Food Microbiol. 2011; 28(6):1145–54.

Duong LM, Jeewon R, Lumyong S, Hyde KD. DGGE coupled with ribosomal DNA gene phylogenies reveal uncharacterized fungal phylotypes. Fungal Diversity. 2006;23(1):121–38.

Eaves DJ, Liebana E, Woodward MJ, Piddock LJ. Detection of gyrA mutations in quinolone-resistant *Salmonella enterica* by denaturing high performance liquid chromatography. J Clin Microbiol. 2002;40:4121–5.

Enkerli J, Kolliker R, Keller S, Widmer F. Isolation and characterization of microsatellite markers from the entomopathogenic fungus *Metarhizium anisopliae*. Mol Ecol Notes. 2005;5:384–6.

Enkerli J, Ghormade V, Oulevey C, Widmer F. PCR-RFLP analysis of chitinase genes enables efficient genotyping of *Metarhizium anisopliae var anisopliae*. J Invertebr Pathol. 2009;102:185–8.

Filichkin SA, Priest HD, Givan SA, Shen R, Bryant DW, Fox SE, Wong WK, Mockler TC. Genome-wide mapping of alternative splicing in *Arabidopsis thaliana*. Genome Res. 2009;20:45–58.

Fournier A, Widmer F, Enkerli J. Development of a single-nucleotide polymorphism (SNP) assay for genotyping of *Pandora neoaphidis*. Fungal Biol. 2010;114(5):498–506.

Fouts DE, Szpakowski S, Purushe J, Torralba M, Waterman RC, MacNeil MD, Nelson KE. Next generation sequencing to define prokaryotic and fungal diversity in the bovine rumen. PLoS One. 2012;7(11):e48289.

Fromin N, Hamelin J, Tarnawski S, Roesti D, Jourdain-Miserez K, Forestier N, et al. Statistical analysis of denaturing gel electrophoresis (DGE) fingerprinting patterns. Environ Microbiol. 2002;4:634–43.

Fukasawa Y, Matsuoka S. Communities of wood-inhabiting fungi in dead pine logs along a geographical gradient in Japan. Fungal Ecol. 2015;18:75–82.

Fukushima M, Kakinuma K, Hayashi H, Nagai H, Ito K, Kawaguchi R. Detection and identification of Mycobacterium species isolates by DNA microarray. J Clin Microbiol. 2003;41:2605–15.

Gardes M, Bruns TD. ITS primers with enhanced specificity for basidiomycetes-application to the identification of mycorrhizae and rusts. Mol Ecol. 1993;2:113–8.

Gardes M, Bruns TD. Community structure of ectomycorrhizal fungi in a *Pinus muricata* forest: above- and below-ground views. Can J Bot. 1996;74:1572–83.

Gomes NCM, Fagbola O, Costa R, Rumjanek NG, Buchner A, Mendona-Hagler L, Smalla K. Dynamics of fungal communities in bulk and maize rhizosphere soil in the tropics. Appl Environ Microbiol. 2003;69(7):3758–66.

Gorji AM, Poczai P, Polgar Z, Taller J. Efficiency of arbitrarily amplified dominant markers (SCoT, ISSR and RAPD) for diagnostic fingerprinting in tetraploid potato. Am J Potato Res. 2011;88:226–37.

Grabherr MG, Haas BJ, Yassour M, Levin JZ, et al. Full-length transcriptome assembly from RNA-seq data without a reference genome. Nat Biotechnol. 2011;29:644–52.

Guo DL, Zhang JY, Liu CH. Genetic diversity in some grape varieties revealed by SCoT analyses. Mol Biol Rep. 2012;39(5):5307–13.

Guzman-Franco AW, Atkins SD, Alderson PG, Pell JK. Development of species-specific diagnostic primers for *Zoophthora radicans* and Pandora *blunckii;* two co-occurring fungal pathogens of the diamondback moth, *Plutella xylostella*. Mycol Res. 2008;112:1227–40.

Hajek AE, Jensen AB, Thomsen L, Hodge KT, Eilenberg J. PCR-RFLP is used to investigate relations among species in the entomopathogenic genera *Eryniopsis* and *Entomophaga*. Mycologia. 2003;95:262–8.

Hajibabaei M, Shokralla S, Zhou X, Singer GA, Baird DJ. Environmental barcoding: a next-generation sequencing approach for biomonitoring applications using river benthos. PLoS One. 2011;6(4), e17497.

Haridas S, Breuill C, Bohlmann J, Hsiang T. A biologist's guide to de novo genome assembly using next-generation sequence data: a test with fungal genomes. J Microbiol Methods. 2011;86(3):368–75.

Healy M, Reece K, Walton D, Huong J, Frye S, Raad II, Kontoyiannis DP. Use of the DiversiLab System for species and strain differentiation of *Fusarium* species isolates. J Clin Microbiol. 2005;43(10):5278–80.

Hebert PD, Ratnasingham S, de Waard JR. Barcoding animal life: cytochrome c oxidase subunit 1 divergences among closely related species. Proc R Soc Lond B Biol Sci. 2003;270 Suppl 1:S96–9.

Hebert JB, Marra RE, Hillman BI, Crouch JA. Population genomic RAD-Seq characterization of the boxwood blight fungus, *Calonectria pseudonaviculata*. APS-MSA Joint Meet, Austin, Texas, USA, 2013.

Hershkovitz MA, Lewis LA. Deep-level diagnostic value of the rDNA-ITS region. Mol Biol Evol. 1996;13:1276–95.

Hillis DM, Dixon MT. Ribosomal DNA: molecular evolution and phylogenetic inference. Q Rev Biol. 1991;66:411–53.

Hohenlohe PA, Bassham S, Etter PD, Stiffler N, Johnson EA, Cresko WA. Population genomics of parallel adaptation in three spine stickleback using sequenced RAD tags. PLoS Genet. 2010;6(2), e1000862.

Hong JW, Park JY, Fomina M, Gadd GM. Development and optimization of an 18S rRNA-based oligonucleotide microarray for the fungal order Eurotiales. J Appl Microbiol. 2010;108:985–97.

Hsiao CR, Huang L, Bouchara JP, Barton R, Li HC, Chang TC. Identification of medically important molds by an oligonucleotide array. J Clin Microbiol. 2005;43:3760–8.

Huang A, Li JW, Shen ZQ, Wang XW, Jin M. High-throughput identification of clinical pathogenic fungi by hybridization to an oligonucleotide microarray. J Clin Microbiol. 2006;44:3299–305.

Huang TP, Yeh Y, Tzeng DDS. Heteroduplex mobility assay for identification and phylogenetic analysis of anthracnose fungi. J Phytopathol. 2010;158(1):46–55.

Huang TP, Yeh Y, Tzeng DDS. Barcode-like heteroduplex DNA pattern as an aid for rapid identification of anthracnose fungi. New Biotechnol. 2011;28(1):72–8.

Hui M, Ip M, Chan PK, Chin ML, Cheng AF. Rapid identification of medically important *Candida* to species level by polymerase chain reaction and single-strand conformational polymorphism. Diagn Microbiol Infect Dis. 2000;38(2):95–9.

James TY, Kauff F, Schoch CL, Matheny PB, Hofstetter V, Cox CJ, Spotts RA. Reconstructing the early evolution of Fungi using a six-gene phylogeny. Nature. 2006;443(7113):818–22.

Jansa J, Mozafar A, Banke S, McDonald BA, Frossard E. Intra-and intersporal diversity of ITS rDNA sequences in *Glomus intraradices* assessed by cloning and sequencing, and by SSCP analysis. Mycol Res. 2002;106(06):670–81.

Jensen AB, Eilenberg J. Genetic variation within the insect-pathogenic genus *Entomophthora*, focusing on the *E. muscae* complex, using PCR-RFLP of the ITS II and the LSU rDNA. Mycol Res. 2001;105:307–12.

Kamiya A, Kikuchi A, Tomita Y, Kanbe T. PCR and PCR-RFLP techniques targeting the DNA topoisomerase II gene for rapid clinical diagnosis of the etiologic agent of dermatophytosis. J Dermatol Sci. 2004;34:35–48.

Kandan A, Jameel A, Singh B, Dixit D, Chand D, Roy A, Rajkumar S, Agarwal PC. Molecular diversity of *Bipolaris oryzae* infecting *Oryza sativa* in India. Phytoparasitica. 2015;43(1):5–14.

Karaoglu H, Lee CMY, Meyer W. Survey of simple sequence repeats in completed fungal genomes. Mol Biol Evol. 2005;22(3):639–49.

Kawahara Y, Oono Y, Kanamori H, Matsumoto T, Itoh T, Minami E. Simultaneous RNA-Seq analysis of a mixed transcriptome of rice and blast fungus interaction. PLoS One. 2012;7(11), e49423.

Keeling PJ, Inagaki Y. A class of eukaryotic GTPase with a punctate distribution suggesting multiple functional replacements of translation elongation factor 1α. Proc Natl Acad Sci U S A. 2004;101(43):15380–5.

Kiliçoğlu MÇ, Özkoç İ. Molecular characterization of *Rhizoctonia solani* AG4 using PCR-RFLP of the rDNA-ITS region.Turk. J Biol. 2010;34(3):261–9.

Kirk JL, Beaudette LA, Hart M, Moutoglis P, Klironomos JN, Lee H, Trevors JT. Methods of studying soil microbial diversity. J Microbiol Methods. 2004;58(2):169–88.

Kjøller R, Rosendahl S. Detection of arbuscular mycorrhizal fungi (Glomales) in roots by nested PCR and SSCP (single stranded conformation polymorphism). Plant Soil. 2000;226(2):189–96.

Klamer M, Hedlund K. Fungal diversity in set-aide agricultural soil investigated using terminal-restriction fragment length polymorphism. Soil Biol Biochem. 2004;36(6):983–8.

Kohli Y, Brunner LJ, Yoell H, Milgroom MG, Anderson JB, Morrall RAA, Kohn LM. Clonal dispersal and spatial mixing in populations of the plant pathogenic fungus, *Sclerotinia sclerotiorum*. Mol Ecol. 1995;4:69–77.

Kristensen R, Gauthier G, Berdal KG, Hamels S, Remacle J, Holst-Jensen A. DNA microarray to detect and identify trichothecene-and moniliformin-producing *Fusarium* species. J Appl Microbiol. 2007;102(4):1060–70.

Kumeda Y, Asao T. Heteroduplex panel analysis, a novel method for genetic identification of *Aspergillus* section flavi strains. Appl Environ Microbiol. 2001;67(9):4084–90.

Leal SCM, Bertioli DJ, Butt TM, Carder JH, Burrows PR, Peberdy YF. Amplification and restriction endonuclease digestion of the Pr1 gene for the detection and characterization of *Metarhizium* strains. Mycol Res. 1997;101:257–65.

Leboldus JM, Kinzer K, Richards J, Ya Z, Yan C, Friesen TL, Brueggeman R. Genotype-by-sequencing of the plant-pathogenic fungi *Pyrenophora teres* and *Sphaerulina musiva* utilizing ion torrent sequence technology. Mol Plant Pathol. 2015;16:623–32.

Leinberger DM, Schumacher U, Autenrrieth IB, Bachmann TT. Development of a DNA microarray for detection and identification of fungal pathogens involved in invasive mycoses. J Clin Microbiol. 2005;43:4943–53.

Lin JQ, Zhao XX, Zhi QQ, Zhao M, He ZM. Transcriptomic profiling of *Aspergillus flavus* in response to 5-azacytidine. Fungal Genet Biol. 2013;56:78–86.

Lindah BD, Taylor AF. Occurrence of N-acetylhexosaminidase-encoding genes in ectomycorrhizal basidiomycetes. New Phytol. 2004;164(1):193–9.

Lindah BD, Nilsson RH, Tedersoo L, Abarenkov K, Carlsen T, et al. Fungal community analysis by high-throughput sequencing of amplified markers – a user's guide. New Phytol. 2013;199:288–99.

Linde J, Duggan S, Weber M, Horn F, Sieber P, Hellwig D, Riege K, Marz M, Martin R, Guthke R, Kurzai O. Defining the transcriptomic landscape of *Candida glabrata* by RNA-Seq. Nucleic Acids Res. 2015. doi:10.1093/nar/gku1357.

Liu WT, Marsh TL, Cheng H, Forney LJ. Characterization of microbial diversity by determining terminal restriction fragment length polymorphisms of genes encoding 16S rRNA. Appl Environ Microbiol. 1997;63(11):4516–22.

Luo QQ, Xu JQ, Hou YP, Chen CJ, Wang JX, Zhou MG. PIRA-PCR for detection of *Fusarium graminearum* genotypes with moderate resistance to carbendazim. Plant Pathol. 2009;58(5):882–7.

Lynch MD, Thorn RG. Diversity of basidiomycetes in Michigan agricultural soils. Appl Environ Microbiol. 2006;72(11):7050–6.

Ma Z and Michailides TJ. An allele-specific PCR assay for detecting azoxystrobin-resistant Alternaria isolates from pistachio in California. J Phytopathol 2004; 152:118–21.

Majer D, Mithen R, Lewis BG, Vos P, Oliver RP. The use of AFLP fingerprinting for the detection of genetic variation in fungi. Mycol Res. 1996;100(9):1107–11.

Margulies M, Egholm M, Altman WE, Attiya S, Bader JS, Bemben LA, Volkmer GA. Genome sequencing in microfabricated high-density picolitre reactors. Nature. 2005;437(7057):376–80.

Martin FN, Tooley PW. Phylogenetic relationships among *Phytophthora* species inferred from sequence analysis of mitochondrially encoded cytochrome oxidase I and II genes. Mycologia. 2003;95(2):269–84.

Martin B, Humbert O, Camara M, Guenzi E, Walker J, Mitchell T, Claverys JP. A highly conserved repeated DNA element located in the chromosome of *Streptococcus pneumonia*. Nucleic Acids Res. 1992;20(13):3479–83.

McDonald MC, Williams AH, Milgate A, Pattemore JA, Solomon PS, Hane JK. Next-generation re-sequencing as a tool for rapid bioinformatic screening of presence and absence of genes and accessory chromosomes across isolates of *Zymoseptoria tritici*. Fungal Genet Biol. 2015;79:71–5.

Mitina GV, Mikhailova LA, Yli-Mattila T. RAPD-PCR, UP-PCR and rDNA sequence analyses of the entomopathogenic fungus *Verticillium lecanii* and its pathogenicity towards insects and phytopathogenic fungi. Arch Phytopathol Pflanzenschutz. 2007;41:113–28.

Muyzer G, De Waal EC, Uitterlinden AG. Profiling of complex microbial populations by denaturing gradient gel electrophoresis analysis of polymerase chain reaction-amplified genes coding for 16S rRNA. Appl Environ Microbiol. 1993;59(3):695–700.

Nath VS, Hegde VM, Jeeva ML, Misra RS, Veena SS, Raj M, Nair SSD. Genetic diversity of phytophthora colocasiae causing Taro Leaf Blight: analysis using Start Codon Targeted (SCoT) polymorphism. J Root Crops. 2013;39:168–77.

Neuve'glise C, Brygoo Y, Riba G. 28S rDNA group-I introns: a powerful tool for identifying strains of *Beauveria brongniartii*. Mol Ecol. 1997;6:373–81.

Nguyen HDT, Seifert KA. Description and DNA barcoding of three new species of *Leohumicola* from South Africa and the United States. Persoonia Mole Phylogeny Evol Fungi. 2008;21:57.

Novodvorska M, Hayer K, Pullan ST, Wilson R, Blythe MJ, Stam H, Stratford M, Archer DB. Trancriptional landscape of *Aspergillus niger* at breaking of conidial dormancy revealed by RNA-sequencing. BMC Genomics. 2013;14:246.

O'Brien HE, Parrent JL, Jackson JA, Moncalvo JM, Vilgalys R. Fungal community analysis by large-scale sequencing of environmental samples. Appl Environ Microbiol. 2005;71(9): 5544–50.

O'Donnell K, Nirenberg HI, Aoki T, Cigelnik E. A multigene phylogeny of the *Gibberella fujikuroi* species complex: detection of additional phylogenetically distinct species. Mycoscience. 2000;41(1):61–78.

O'Donnell K, Ward TJ, Geiser DM, Kistler HC, Aoki T. Genealogical concordance between the mating type locus and seven other nuclear genes supports formal recognition of nine phylogenetically distinct species within the *Fusarium graminearum* clade. Fungal Genet Biol. 2004;41(6):600–23.

Ojeda DI, Dhillon B, Tsui CK, Hamelin RC. Single-nucleotide polymorphism discovery in *Leptographium longiclavatum*, a mountain pine beetle-associated symbiotic fungus, using whole-genome resequencing. Mol Ecol Res. 2014;14(2):401–10.

Oliveira CA, Sá NM, Gomes EA, Marriel IE, Scotti MR, Guimarães CT, Alves VM. Assessment of the mycorrhizal community in the rhizosphere of maize (*Zea mays* L) genotypes contrasting for phosphorus efficiency in the acid savannas of Brazil using denaturing gradient gel electrophoresis (DGGE). Appl Soil Ecol. 2009;41(3):249–58.

Oulevey C, Widmer F, Ko¨lliker R, Enkerli J. An optimized microsatellite marker set for detection of Metarhizium anisopliae genotype diversity on field and regional scales. Mycol Res. 2009;113:1016–24.

Petrosino JF, Highlander S, Luna RA, Gibbs RA, Versalovic J. Metagenomic pyrosequencing and microbial identification. Clin Chem. 2009;55(5):856–66.

Piddock LJ, Ricci V, McLaren I, Griggs DJ. Role of mutation in the gyrA and parC genes of nalidixic-acid-resistant *Salmonella* serotypes isolated from animals in the United Kingdom. J Antimicrobial Chemo. 1998;41(6):635–41.

Pounder JI, Hansen D, Woods GL. Identification of *Histoplasma capsulatum, Blastomyces dermatitidis*, and *Coccidioides* species by repetitive-sequence-based PCR. J Clin Microbiol. 2006;44(8):2977–82.

Raja H, Schoch C, Hustad V, Shearer C, Miller A. Testing the phylogenetic utility of MCM7 in the Ascomycota. MycoKeys. 2011;1:63–94.

Ramos JP, Valente P, De Souza RA, Rosa CA, Leoncini O. Heteroduplex mobility assay of the D1/D2 region of the 26S rDNA for differentiation of *Saccharomyces* species. Lett Appl Microbiol. 2001;33(3):206–10.

Ramos JP, Rosa CA, Carvalho EM, Leoncini O, Valente P. Heteroduplex mobility assay of the 26S rDNA D1/D2 region for differentiation of clinically relevant *Candida* species. Anton van Leeuwenhoek. 2006;89(1):39–44.

Ranjard L, Poly F, Lata JC, Mougel C, Thioulouse J, Nazaret S. Characterization of bacterial and fungal soil communities by automated ribosomal intergenic spacer analysis fingerprints: biological and methodological variability. Appl Environ Microbiol. 2001;67(10):4479–87.

Rehner SA, Buckley EP. Isolation and characterization of microsatellite loci from the entomopathogenic fungus *Beauveria bassiana* (Ascomycota: Hypocreales). Mol Ecol Notes. 2003;3: 409–11.
Roe AD, Sperling FA. Patterns of evolution of mitochondrial cytochrome c oxidase I and II DNA and implications for DNA barcoding. Mol Phylogenet Evol. 2007;44(1):325–45.
Rohel E, Couteaudier Y, Papierok B, Cavelier N, Dedryver CA. Ribosomal internal transcribed spacer size variation correlated with RAPD-PCR pattern polymorphisms in the entomopathogenic fungus *Erynia neoaphidis* and some closely related species. Mycol Res. 1997;101:573–9.
Ronaghi M, Uhlén M, Nyrén P. A sequencing method based on real-time pyrophosphate. Science. 1998;281(5375):363–5.
Saito K, Miyake S, Moriya H, Yamazaki M, Itoh F, Imai K, Miyamoto A. Detection of the four sequence variations of MDR1 gene using TaqMan® MGB probe based real-time PCR and haplotype analysis in healthy Japanese subjects. Clin Biochem. 2003;36(7):511–8.
Sampaio P, Gusmão L, Correia A, Alves C, Rodrigues AG, Pina-Vaz C, Pais C. New microsatellite multiplex PCR for Candida albicans strain typing reveals microevolutionary changes. J Clin Microbiol. 2005;43(8):3869–76.
Sanger F, Nicklen S, Coulson AR. DNA sequencing with chain-terminating inhibitors. Proc Natl Acad Sci U S A. 1977;74(12):5463–7.
Satya P, Karan M, Jana S, Mitra S, Sharma A, Karmakar PG, Ray DP. Start codon targeted (SCoT) polymorphism reveals genetic diversity in wild and domesticated populations of ramie (*Boehmeria nivea* LGaudich), a premium textile fiber producing species. Meta Gene. 2015;3:62–70.
Sawant SV, Singh PK, Gupta SK, Madnala R, Tuli R. Conserved nucleotide sequences in highly expressed genes in plants. J Genet. 1999;78:123–31.
Schadt CW, Martin AP, Lipson DA, Schmidt SK. Seasonal dynamics of previously unknown fungal lineages in tundra soils. Science. 2003;301(5638):1359–61.
Schmitt I, Crespo A, Divakar PK, Fankhauser JD, Herman-Sackett E, Kalb K, Lumbsch HT. New primers for promising single-copy genes in fungal phylogenetics and systematic. Persoonia. 2009;23:35.
Schwarzenbach K, Enkerli J, Widmer F. Objective criteria to assess representativity of soil fungal community profiles. J Microbiol Methods. 2007;68(2):358–66.
Seifert KA. Progress towards DNA barcoding of fungi. Mol Ecol Res. 2009;9(s1):83–9.
Seifert KA, Samson RA, Houbraken J, Lévesque CA, Moncalvo JM, Louis-Seize G, Hebert PD. Prospects for fungus identification using CO1 DNA barcodes, with *Penicillium* as a test case. Proc Natl Acad Sci U S A. 2007;104(10):3901–6.
Shokralla S, Spall JL, Gibson JF, Hajibabaei M. Next-generation sequencing technologies for environmental DNA research. Mol Ecol. 2012;21(8):1794–805.
Simon L, Levesque RC, Lalonde M. Identification of endomycorrhizal fungi colonizing roots by fluorescent single-strand conformation polymorphism-polymerase chain reaction. Appl Environ Microbiol. 1993;59(12):4211–5.
Simpson JT, Wong K, Jackman SD, Schein JE, Jones SJM, Birol I. ABySS: a parallel assembler for short read sequence data. Genome Res. 2009;19:1117–23.
Smit E, Leeflang P, Glandorf B, van Elsas JD, Wernars K. Analysis of fungal diversity in the wheat rhizosphere by sequencing of cloned PCR-amplified genes encoding 18S rRNA and temperature gradient gel electrophoresis. Appl Environ Microbiol. 1999;65:2614–21.
Solomon KV, Haitjema CH, Thompson DA, O'Malley MA. Extracting data from the muck: deriving biological insight from complex microbial communities and non-model organisms with next generation sequencing. Curr Opin Biotechnol. 2014;28:103–10.
Stern MJ, Ames GF, Smith NH, Robinson EC, Higgins CF. Repetitive extragenic palindromic sequences: a major component of the bacterial genome. Cell. 1984;37:1015–26.
Taylor JW, Jacobson DJ, Kroken S, Kasuga T, Geiser DM, Hibbett DS, Fisher MC. Phylogenetic species recognition and species concepts in fungi. Fungal Genet Biol. 2000;31:21–32.
Toju H, Sato H, Tanabe AS. Diversity and spatial structure of belowground plant–fungal symbiosis in a mixed subtropical forest of ectomycorrhizal and arbuscular mycorrhizal plants. PLoS One. 2014;9(1), e86566.

Torzilli AP, Sikaroodi M, Chalkley D, Gillevet PM. A comparison of fungal communities from four salt marsh plants using automated ribosomal intergenic spacer analysis (ARISA). Mycologia. 2006;98(5):690–8.

Tringe SG, Rubin EM. Metagenomics: DNA sequencing of environmental samples. Nat Rev Genet. 2005;6(11):805–14.

Turgeon BG. Application of mating type gene technology to problems in fungal biology. Ann Rev Phytopathol. 1998;36(1):115–37.

Tymon AM, Shah PA, Pell JK. PCR-based molecular discrimination of *Pandora neoaphidis* isolates from related entomopathogenic fungi and development of species specific diagnostic primers. Mycol Res. 2004;108:419–33.

van Belkum, Alex, Stewart Scherer, Loek Van Alphen, and Henri Verbrugh. "Short-sequence DNA repeats in prokaryotic genomes." Microbiology and Molecular Biology Reviews 62, no. 2 (1998):275–93.

Vandenkoornhuyse P, Baldauf SL, Leyval C, Straczek J, Young JPW. Extensive fungal diversity in plant roots. Science. 2002;295:2051.

Versalovic J, Koeuth T, Lupski JR. Distribution of repetitive DNA sequences in eubacteria and application to fingerprinting of bacterial genomes. Nucleic Acids Res. 1991;19:6823–31.

Voříšková J, Brabcová V, Cajthaml T, Baldrian P. Seasonal dynamics of fungal communities in a temperate oak forest soil. New Phytol. 2014;201(1):269–78.

Vos P, Hogers R, Bleeker M, Reijans M, Van de Lee T, Hornes M, Zabeau M. AFLP: a new technique for DNA fingerprinting. Nucleic Acids Res. 1995;23(21):4407–14.

Walker RA, Saunders N, Lawson AJ, Lindsay EA, Dassama M, Ward LR et al. Use of a LightCycler gyrA mutation assay for rapid identification of mutations conferring decreased susceptibility to ciprofloxacin in multiresistant Salmonella enterica serotype Typhimurium DT104 isolates. J Clin Microbiol 2001; 39(4):1443–48.

Wang CS, Shah FA, Patel N, Li ZZ, Butt TM. Molecular investigation on strain genetic relatedness and population structure of *Beauveria bassiana*. Environ Microbiol. 2003;5:908–15.

Wang B, Guo G, Wang C, Lin Y, Wang X, Zhao M, Guo Y, He M, Zhang Y, Pan L. Survey of the transcriptome of *Aspergillus oryzae* via massively parallel mRNA sequencing. Nucleic Acids Res. 2010;38:5075–87.

Wardle DA. Communities and ecosystems: linking the above ground and below ground components. Princeton): Princeton University Press; 2002.

Witthuhn RC, Wingfield BD, Wingfield MJ, Wolfaard M, Harrington TC. Monophyly of the conifer species in the *Ceratocystis coerulescens* complex using DNA sequence data. Mycologia. 1998;90:96–101.

Womack AM, Artaxo PE, Ishida FY, Mueller RC, Saleska SR, Wiedemann KT, Green JL. Characterization of active and total fungal communities in the atmosphere over the Amazon rainforest. Biogeosci Dis. 2015;12:7177–207.

Yergeau E, Filion M, Vujanovic V, St-Arnaud M. A PCR-denaturing gradient gel electrophoresis approach to assess *Fusarium* diversity in asparagus. J Microbiol Methods. 2005;60(2):143–54.

Yun SH, Arie T, Kaneko I, Yoder OC, Turgeon BG. Molecular organization of mating type loci in heterothallic, homothallic, and asexual Gibberella/Fusarium species. Fungal Genet Biol 2000; 31:7–20.

Zeigler RS, Cuoc LX, Scott RP, Bernardo MA, Chen DH, Valent B, Nelson RJ. The relationship between lineage and virulence in *Pyricularia grisea* in the Philippines. Phytopathology. 1995;85:443–51.

Zhao C, Waalwijk C, de Wit PJGM, Tang D, van der Lee T. RNA-Seq analysis reveals new gene models and alternative splicing in the fungal pathogen *Fusarium graminearum*. BMC Genomics. 2013;14:21. doi:10.1186/1471-2164-14-21.

Zirnstein G, Li Y, Swaminathan B, Angulo F. Ciprofloxacin resistance in *Campylobacter jejuni* isolates: detection of gyrA resistance mutations by Mismatch Amplification Mutation Assay PCR and DNA sequence analysis. J Clin Microbiol. 1999;37(10):3276–80.

Zouari I, Salvioli A, Chialva M, Novero M, Miozzi L, Tenore GC, Bagnaresi P, Bonfante P. From root to fruit: RNA-Seq analysis shows that arbuscular mycorrhizal symbiosis may affect tomato fruit metabolism. BMC Genomics. 2014;15:221.

Chapter 7
Genic Molecular Markers in Fungi: Availability and Utility for Bioprospection

Anirudha Chattopadhyay, Kapil Kumar Tiwari, Kul Bhushan, and Dharmendra Pratap

7.1 Introduction

Fungi are one of the most important groups of organism on this earth. They play a vital role in the biosphere in various ways with either beneficial or harmful impact on the other organism. Despite their extraordinary impacts on the mankind, their taxonomic information is available to a lesser extent relative to other kingdoms. Due to the huge diversity and high morphological and biological similarities within the fungal kingdom, the taxonomic identification becomes cumbersome and furthermore, produces complexity to distinguish them individually within a single lowest rank of a taxon or below. On the basis of visual morphological examination, the identification and characterization of fungi up to the species level or its below rank is highly selective and unmanageable (Bridge et al. 2004). Therefore, a proper understanding of their physiological divergence remains untouched. Hence, the available genomic information could be used to disclose their various morphological and physiological complexity. The development of various

The original version of this chapter was revised. A correction to this chapter can be found at DOI https://doi.org/10.1007/978-3-319-34106-4_18

A. Chattopadhyay
Department of Plant Pathology, C.P. College of Agriculture, S.D. Agricultural University, S.K. Nagar, Gujarat, India

K.K. Tiwari
Department of Plant Molecular Biology & Biotechnology, C.P. College of Agriculture, S.D. Agricultural University, S.K. Nagar, Gujarat, India

K. Bhushan
Division of Plant Pathology, Indian Agricultural Research Institute, New Delhi, India
e-mail: lsr_bhushan@rediffmail.com

D. Pratap (✉)
Department of Genetics & Plant Breeding, Ch. Charan Singh University, Meerut, UP, India
e-mail: pratapbiotech@gmail.com

B.P. Singh, V.K. Gupta (eds.), *Molecular Markers in Mycology*, Fungal Biology,
DOI 10.1007/978-3-319-34106-4_7

molecular techniques for genetic analysis can ameliorate our knowledge of fungal genetics and interpretation of the structural and functional aspects of the fungal genome. Among those techniques, molecular markers can be used to link the functional behaviour with the morphological and physiological characteristic. Therefore, genome-wide identification and utilization of genetic markers may be looked as the effective and simplest way to decipher the obscured events of fungal biology, their phylogenetic and evolutionary study, population genetics and epidemiology.

With the recent advancement in molecular biology and genomics, a vast amount of genetic data has been generated and prevailed to identify various molecular markers. These molecular markers can be derived from both the variable and conserved regions of the nuclear and mitochondrial genome of fungi (Bridge et al. 2004). The different molecular markers have different property and criteria to work and therefore, they have their different forms of application. Some of them are used to define populations to entire levels beginning from an individual isolate upwards (Bridge et al. 2004) while some others could be used to correlate various morphological and physiological characteristics of the targeted gene of fungi, which may be of commercial importance. In the recent past, the molecular markers derived from universal gene sequences i.e. internal transcribe spacer (ITS) region. The 'universal' ITS primers amplify the ITS region containing two variable non-coding regions that are nested within the rDNA repeat between highly conserved small subunit (5.8S) and large subunit of rRNA genes. The ITS region is frequently highly variable among morphologically distinct as well as indistinct fungal species of *Fusarium*, *Aspergillus*, *Arthrobotrys*, and *Monacrosporium*, etc. (Gardes and Bruns 1991, 1993; Gardes et al. 1991; Baura et al. 1992; Meyer et al. 2005; Zhang et al. 2011). Therefore, the ITS region is aimed to design the specific primers that can be used as a universal marker in a range of fungal genera without any prior genomic information. But the primers derived from ITS region are merely not enough to resolve the difference between the closely related fungal strains or isolates (Carter et al. 2004). Hence, for many applications, hybridization and PCR amplification based markers are popularized. They are either hybridizing to or amplifying from repetitive sequences.

The genetic markers like Restriction Fragment Length Polymorphism (RFLP) and Random Amplified Polymorphic DNA (RAPD) were introduced to differentiate and identify fungi at the intraspecific level. Despite their desire for the differentiation of closely related fungal isolates or strains based on their phenotypic traits like virulence or aggressiveness or other physiological characters, they were not considered as popularized markers, due to their complicated methodology, difficult interpretation and low reproducibility. Therefore, the development of species-specific genic markers is essential for establishing the correct taxonomic classification and proper phenotypic characterization. The rapid development of sequencing technology with reduced cost helps to generate a flood of genome sequence data from a plethora of fungi that have agricultural, medical, and industrial significance. The availability of whole-genome sequences has also fundamentally changed the methods of the identification and application of genetic markers for mapping certain phenotypic trait. The huge repository of genome sequences made it easy to develop genic markers for fungi that are either gene targeted or functional in nature and associated with the characterization of the polymorphisms. These genic markers are located either in regulatory or in functional site and are directly or indirectly correlated in defining the phenotypic trait variation (Varshney et al. 2007). Hence, 'genic' molecular markers can be employed for the detection and identification of fungal strains and can be used to decipher their biological as well as functional diversity in order to achieve marker-assisted selection for trait improvement, comparative genome mapping, and to explore fungal adaptation to different environments and many more applications.

7.2 Molecular Markers Used for Genotyping of Fungi

Molecular marker is defined as a segment of DNA sequence that is associated with a part of the genome carrying gene coding for a certain trait. These markers can be exploited based on their naturally occurring polymorphisms in DNA sequences which are responsible for genomic variation and linked to structural and functional attributes of fungi. These markers can be used to build up genetics maps providing valuable insights into the evolutionary history and to evaluate the effect of genomic variation in the expression of particular traits. Therefore, these markers must be polymorphic and should have high reproducibility, and should provide moderately high genotyping throughput in low cost (Kesawat and Das 2009). It is extremely difficult to find a molecular marker which would meet all those criteria. For fungal genotyping, a marker system can be used that would fulfil at least some of the above criteria. Depending on the type of study to be undertaken, technically different types of molecular markers are used in mycology that are differ from each other with respect to their principle and functionality. Based on the genotyping techniques used, these markers can be categorised into two sections: (1) DNA Sequence typing based markers, and (2) DNA Fingerprinting based markers.

7.2.1 DNA Sequence Based Barcode Marker

7.2.1.1 Universal DNA Barcode Markers

These markers are designed based on highly conserved genome sequences. They can amplify the specific target sequence of the genome from any fungal species without prior genomic information. They can be used to detect different fungal species due to interspecific variation. The most significant illustration is the amplification of ribosomal DNA (rDNA) gene cluster. Ribosomal DNA Gene Cluster is the DNA sequences consisting ribosomal RNA (rRNA) encoding genes arranged as tandem repeats and form a gene cluster. The basic unit consists the three genes for the small ribosomal subunit, the 5.8S subunit, and the large subunit ribosomal subunits separated by two internally transcribed spacers (ITS) (White et al. 1990). Whereas the each basic unit within gene cluster are separated by an intergenic spacer

(IGS) and many contain the 5S subunit gene in many fungi (Hillis and Dixon 1991). The ribosomal RNA gene cluster is found in both nuclear and mitochondrial genome of fungi and consisting both highly conserved and variable regions. The conserved regions include the large subunit (LSU) and small subunit (SSU) genes that have been exploited to study the many phylogenetic relationships among distantly related fungi due to their interspecific genetic variations (Gaudet et al. 1989; Bowman et al. 1992a, b; Bruns et al. 1992; Hibbett 1992; Lindahl et al. 2013). Due to the inadequate species-level resolution in some taxonomic groups, these nuclear ribosomal subunits are considered to be slightly inferior to the ITS. Both, spacer regions, i.e. internal transcribed spacers (ITS), and intergenic spacers (IGS) are representing considerably more variable regions than the subunit sequences, and most useful to study molecular systematics and relationships at the species level within a single genus, and even within species (e.g., to identify geographic races) (Buchko and Klassen 1990; Spreadbury et al. 1990; Nazar et al. 1991; Anderson and Stasorski 1992; Baura et al. 1992; Kim et al. 1992; Arora et al. 1996; Singh et al. 2006). Due to the highest probability of successful identification for the broadest range of fungi up to interspecific level, nuclear ribosomal ITS region is considered as the universal DNA barcode marker for fungi (Schoch et al. 2012). The ITS based molecular identification of fungi is very popular because of easy PCR amplification of relatively shorter sized ITS region (500–800 bp) using universal ITS primers (White et al. 1990) even from low quantity DNA (Gardes and Bruns 1993) and higher degree of sequence variability in ITS region among distinct species make it easy to estimate genetic distances for systematic and phylogenetic analysis (Baura et al. 1992; Bruns et al. 1992; Gardes and Bruns 1993; Williams et al. 1995). Hence, it is used as potent barcode marker for fungal taxonomy and genomic study.

Although, ITS is most commonly used marker for fungal identification, even then, it has certain limitations, *viz.*, (1) due to the identical ITS sequences in different species of genus complexes, identification of several species is limited. For example, identification of critical mycotoxigenic, industrial, and medically important species of *Aspergillus* sp. only by using ITS is very difficult and requires additional markers to solve this problem (Geiser et al. 2007) and (2) the failure of the universal ITS primers to work in some fungal groups is also documented (Seifert 2008; Voigt and Kirk 2011; Nilsson et al. 2008). In spite of them, ITS region was selected to designed species-specific probes due to their occurrence in multiple copy number and higher sequence variability between related fungal species,

7.2.1.2 Partially Universal DNA Barcode Markers

The DNA sequences of some other protein coding genes are also used as an alternative to ITS marker for phylogenetic analyses or species identification. Generally, they have highly conserved coding sequences between distantly related fungal groups but may contain short introns that can be extremely variable in the insertion position and number (Edelmann and Staben 1994). This sequence variation in introns can be used to differentiate closely related fungal species and sub-species and may be used as potential fungal barcode marker. Protein coding housekeeping nuclear genes like largest subunit of RNA polymerase II (Stockinger et al. 2014) β-tubulin (Mostert et al. 2006; Aroca et al. 2008) translation elongation factor 1 alpha (TEF1α) (Voigt and Wostemeyer 2001; Geiser et al. 2004; Knutsen et al. 2004; Kristensen et al. 2005). calmodulin (Mule et al. 2004) avirulence genes (Lievens et al. 2009) actin (Voigt and Wostemeyer 2001; Daniel and Meyer 2003) mating type genes (Wallace and Covert 2000; Foster et al. 2002) and mitochondrial genes like multicopy cox I and cox II and their intergenic region (Martin and Tooley 2003; Seifert et al. 2007; Nguyen and Seifert 2008) are widely tested. These house-keeping gene sequence typing based markers are superior over that of ribosomal genes to differentiate inter- and intra-species and to resolve relationships at various taxonomic levels (Schoch et al. 2009). But there are certain limitations for using them as 'universal' barcode marker. Due to the restricted application in the wide taxonomic range (Schoch et al. 2012) and lack of significant level of variability between fungal species for certain genes, these markers have limited use in a practical sense. Therefore, they can be considered as 'partially universal' DNA barcode marker.

7.2.2 Fingerprinting Based Molecular Markers

DNA fingerprinting is the technique used to identify DNA banding patterns in gel electrophoresis, which will appear due to variation in the length of DNA sequence or variation in primer binding site. This variation appears due to the polymorphism in the DNA sequence which is detected either by restriction digestion of nucleic acid sequence or by amplifying the target sequence using a molecular marker. It is performed to compare the nucleotide sequences of DNA fragments collected from different sources. In fungi, the DNA fingerprinting is performed to study the phylogenetic relationship, population biology, for the detection of fungi at the lowest taxonomic level, and even for strain differentiation within the same species with different host range, virulence, compatibility group or mating type (Capote et al. 2012). In this method, the molecular markers used to amplify the DNA segment can bind to the target site of the genome which may be random or specific. This molecular marker may either be randomly distributed on the genome or located in a specific region which may be coding or non-coding sequences.

7.2.2.1 Random Molecular Markers

In terms of molecular biology, 'random' molecular marker is used where genomic information is not required and locus of the amplicons is absolutely unknown. This category of markers includes random amplification of polymorphic DNA (RAPD), amplified fragment length polymorphism (AFLP), and Inter-simple sequence repeats (ISSRs). It is the class of molecular markers which is derived from anonymous regions of the genome that may or may not be polymorphic within a gene. Random Molecular Markers (RMMs) are derived at random from polymorphic sites in the genome (Andersen and Lubberstedt 2003). It is used to screen random regions of the fungal genome for identifying species-specific sequences when conserved genes have not enough variation to successfully identify species (McCartney et al. 2003). These markers help in the total genome profiling.

7.2.2.2 Locus Specific Markers

In this category of the marker, either genic or non-genic positions amplify using specific primer or probe for the detection of polymorphism. These molecular markers can either be derived from genomic DNA libraries, SSR library (Tiwari and Chandra 2010). EST library, whole genome and transcriptome assembly. Based on locus position, these markers are of two types, genic and non-genic molecular markers.

7.2.2.3 Non-genic Molecular Markers

If the marker position located in other than the functional region of the genome, it can be anticipated as non-genic marker. It mostly includes Restriction Fragment Length Polymorphism (RFLP), Simple Sequence Repeats (SSRs), Single Nucleotide Polymorphisms (SNPs), etc that have a non-genic origin.

7.2.2.4 Genic Molecular Markers

When a genetic marker is developed directly from the parts of functional genes, *i.e*,, coding and non-coding gene sequences or the genic regions, with or without any known function, such markers are called as 'genic' molecular markers (GMMs). They are used to detect polymorphism within genes and can be distinguished from gene-linked markers on the basis of the polymorphic site. Gene linked markers are developed from regions which are tightly linked to the gene of interest (Kage et al. 2015). whereas genic markers are constructed from coding or non-coding region (including introns, UTRs, promoters and regulators) of any gene with or without phenotypic trait variation, (Aggarwal et al. 2007; Salgotra et al. 2014). Therefore, they may or may not be functional (Andersen and Lubberstedt 2003). Based on the functional characterization of polymorphism, these GMMs can be categorised as gene-targeted markers and functional markers (Andersen and Lubberstedt 2003). In gene-targeted markers (GTMs), polymorphism placed within genes which is independent of phenotypic trait variation (Andersen and Lubberstedt 2003; Schmitt et al. 2006; Aggarwal et al. 2007). Whereas, functional markers (FMs) are derived from the polymorphic site of candidate genes and directly or indirectly associated with phenotypic trait variation (Andersen and Lubberstedt 2003; Salgotra et al. 2014).

These GMMs are utilized to identify polymorphisms in genic regions. Similar to RDMs, GMMs can detect both length and sequence polymorphism in the genome, but specifically in genic regions. Being derived from the conserved genomic region, they are highly robust and have interspecific/intergeneric transferability (Varshney et al. 2007). Sometimes they can also be used for functional validation of candidate genes when derived from the functional marker (a special class of GMMs). Therefore, these genic markers are better over random molecular markers for functional genetic diversity estimation, construction of transcript map, marker-assisted selection, trait association analysis and comparative as well as functional genomics study. The schematic diagram of the development of genic marker is shown in Fig. 7.1.

7.3 Types of Genic Molecular Markers

7.3.1 Genic RFLP Markers

It is a hybridization-based marker developed to detect polymorphism at the DNA sequence level. In this technique, organisms can be differentiated based on the cleavage pattern of their DNA. If two organisms differ in the distance between sites of cleavage for a particular restriction endonuclease, the length of the fragments

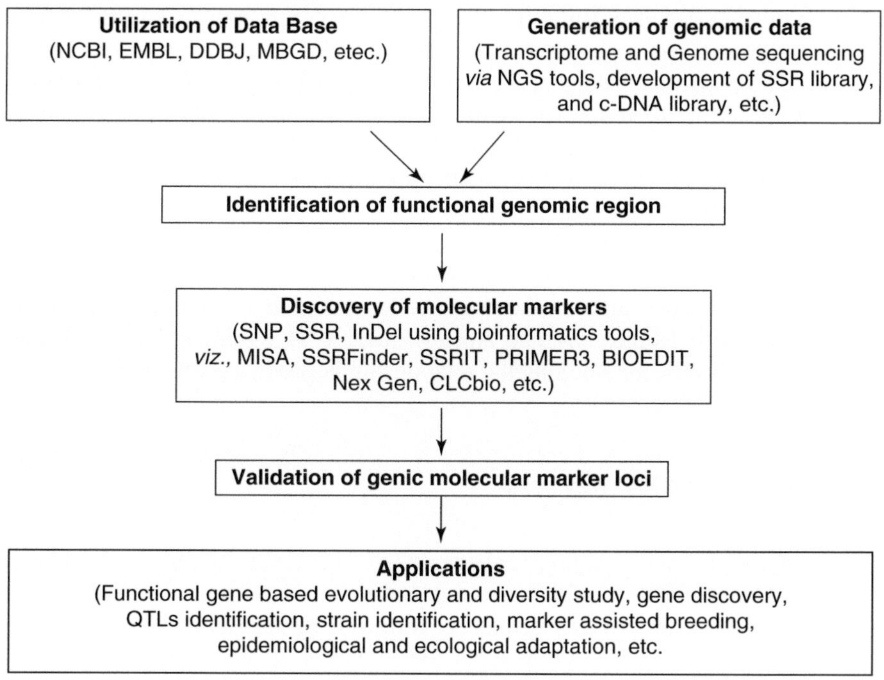

Fig. 7.1 Schematic presentation of the development of genic molecular markers

produced will vary when the DNA is digested with a restriction endonuclease resulting in differential DNA fragment profile on gel electrophoresis. This polymorphic fingerprinting pattern can be used to differentiate fungal species and strains from one another. The polymorphism is detected by hybridizing a chemically labelled DNA probe to a Southern blot. Polymorphisms are the result of point mutations in the restriction sequence (single nucleotide polymorphism) or insertion/deletions between restriction sites and unequal crossing over (Kumar et al. 2009). If such mutations or insertion/deletions occur in the genic region, *i.e.,* coding as well as noncoding sequences of a gene, the designed marker can be called as genic RFLP marker. This type of polymorphism is directly or indirectly influencing the expression of a gene leading to the phenotypic trait variation.

7.3.2 Genic STS Markers

It is a relatively short DNA sequence (200–500 bp) which is unique and not similar to other part of the genome. Thus, it can be detected in the presence of all other genomic sequences. Sequence-tagged-site (STS) markers of arbitrarily selected genes can easily be amplified by PCR using a pair of primers that flank the STS and are based on the sequence of either RFLP probe (Blake et al. 1996) or AFLP fragments. Sometimes, STS contain repetitive elements/sequences that may appear elsewhere in the genome also. But as long as the sequences at both ends of the site are unique and conserved, researchers can unambiguously identify this portion of the genome. STS markers are considered to be robust and highly efficient in functionality when located in the transcribed part of the genome, especially gene sequences/ESTs/unigenes. These STS are considered 'genic' marker, directly linked to some specific traits. A genic STS marker usually amplifies very specific region, *i.e.,* transcribed part of the genome and produce single band during fingerprinting showing less polymorphism. This problem can be solved by cleaving STS amplified PCR products which generate polymorphism within target genomic region and led to the development of new marker technology, called Cleaved Amplified Polymorphic Sequence (CAPS) (Konieczny and Ausubel 1993). STS primers have a broader range of applications, for studying the relationship between various species.

7.3.3 Genic CAPS Markers

Cleaved Amplified Polymorphic Sequence (CAPS) is a versatile marker technique which combines PCR amplification and restriction digestion. It involves PCR amplification of a target DNA sequence, followed by restriction digestion using endonuclease (Konieczny and Ausubel 1993; Jarvis et al. 1994; Michaels and Amasino 1998). Here, the polymorphisms are detected by restriction digestion of PCR amplicons and compared in fingerprinting gel. Cleaved amplified polymorphic sequence (CAPS) technology is very suitable to use as genic or functional markers for targeting polymorphisms within a certain gene that is directly responsible for the phenotype trait variation. This variation is due to the allelic variation within coding and/or non-coding sequence of gene generated either by point mutation or small insertion/deletions. Thus, cleaved amplified polymorphic sequence (CAPS) technology is particularly suitable for the development of genic and functional markers. Due to its higher reproducibility, accuracy and low cost, it has been widely used for identification of microorganisms and functional genetic diversity studies based on their certain important trait. CAPS markers can also be used to detect a commercially important strain of fungi.

7.3.4 Genic SSR Markers

It is a tandem array of short DNA sequences, containing 2–6 bp motif repeated number of times (Chambers and MacAvoy 2000). They are dispersed throughout the eukaryotic as well as prokaryotic genome including fungi (Litt and Luty 1989; Carter et al. 2004; Karaoglu et al. 2005). These di, tri, tetra, pen and hexa-nucleotide motifs are usually repeated at least five times. The SSR length polymorphism is due to the length variation in the number of repeat units of a microsatellite. This variation is created because of DNA polymerase slippage during DNA replication (Tachida and Iizuka 1992). or slipped strand mispairing and unequal crossing over (Matsuoka et al. 2002). These polymorphisms can be identified by constructing PCR primers for the flanking region of SSRs which is unique and found to be conserved within the species, sometimes in higher taxa also. PCR primers designed to an SSR within one species can amplify a corresponding locus in related species, thus enabling comparative genomic analysis of closely related species as well phylogenetic and population biology study (Xia et al. 2007; Hwang et al. 2009; Islam et al. 2012). These markers are distributed in the transcribed and non-transcribed regions in the eukaryotic genome (Toth et al. 2000; Katti et al. 2001; Morgante et al. 2002). When microsatellites loci are present in the protein-coding and non-coding sequences (intron, UTRs, promoter and even regulators) of a gene, such SSRs can be called as genic SSR marker. Usually, SSRs are more abundant in noncoding regions than in exons (Hancock 1995). However, certain trinucleotides and hexanucleotides repeats are found to be more abundant in coding regions than in noncoding regions of higher eukaryotic genomes (Metzgar et al. 2000; Toth et al. 2000; Subramanian et al. 2003a, b). The SSRs present in promoter regions may influence transcriptional activity of the gene (Kashi and Soller 1997). They may also influence protein-protein interactions (Gerber et al. 1994; Perutz et al. 1994) and protein-DNA interaction. Thus, some SSRs play an important role in gene regulation and gene expression (Li et al. 2002; Huang et al. 2003; Hefferon 2004; Kunzler et al. 1995). The primers for genic-SSR marker can be designed from the most conserved region of the gene thus show higher interspecific transferability. Hence, it can be used as a potent 'functional genic marker' to study the fungal population.

7.3.5 *Genic SNP Markers*

Single nucleotide polymorphisms are single nucleotide differences in the DNA sequences of individuals in a population that are arises due to transversions (C/G, A/T, C/A and T/G), transitions (C/T or G/A) and insertions/deletions, also known as indels. It creates genetic variation within the genome of any organism. This variation can easily be identified and can be used for the mapping of heritable traits (Chagne et al. 2007). The SNPs are widely distributed within the genome of fungus (Carter et al. 2004). The high heritability of such SNPs makes them an excellent indicator of genetic diversity and phylogenetic evolution of fungal species. SNPs that are mostly prevalent in the non-coding regions of the genome are selectively neutral to change in amino acid sequence. If located in the genic region, *i.e.,* the coding and non-coding sequences of a gene, the SNP can either be synonymous type without alteration of amino acid sequence or non-synonymous type resulting in an amino acid sequence change and functional phenotypic modification, which can be used as criterion for designing marker (Agarwal et al. 2008). The tracking of such single nucleotide change in the genic region is a powerful marker for fingerprinting of the fungal population. Mostly, genic SNPs are positioned in the untranslated regions (UTRs) of a gene than the coding sequences (CDSs). Due to less selective pressure, UTRs are more prone to sequence variation than the CDSs (Li et al. 2014). Expressed sequence tag (EST)-derived single-nucleotide-polymorphism (SNP) can be used as promising molecular markers for species identification in which microsatellites are difficult to isolate and which have low levels of polymorphism (Delmotte et al. 2011). When 'genic' SNPs and InDels are mined from the EST data base, they can be considered to be an indispensable functional marker, if the particular gene is responsible for phenotypic variations.

7.4 Development of Genic Molecular Markers

The rise in the importance of agricultural, medicinal and industrially important fungi in commercial aspects, structural and functional genomics studies becomes of great significance for the identification of suitable fungal strains, their characterization and genetic improvement that have many beneficial downstream applications. Genic molecular markers give more insight on the characterization of a phenotypic trait. The majority of the molecular markers developed and used in the fungal genomic study are randomly derived from the genomic DNA, *i.e.,* either the transcribed or the non-transcribed part of the genome, without any functional information. In contrast, GMMs developed from coding sequences like ESTs or fully characterized genes with known function. These genic molecular markers can be developed in the following way,

7.4.1 *Development of Genic Molecular Markers by Sequencing Technique*

7.4.1.1 Development of Genic Molecular Marker by Sanger Sequencing

At present, a vast amount of partially utilized functional information of genes is available in the public available genome database. This information can be used for identification and development of functional molecular markers. Sanger sequencing is extremely useful and robust tool for determination of targeted DNA sequence up to 1–1.5 kb. This method generally used detection of SNPs and InDel in targeted genomic region of candidate genes. Sanger-based sequencing technologies are still used for targeted amplification of genomic regions due to fidelity in their procedure despite the availability of other platform. Addition to targeted or gene-based sequencing, Sanger-based technology is also used for the development of cDNA, genomic, and SSR library. Generated data may be further utilized the detection of polymorphism with a wide range of molecular marker namely SSR, SNP, CAPS, STS.

7.4.1.2 Development of Genome-Wide Markers Using Next Generation Sequencing Technologies

In the last decade, many next generation technologies got emerged as powerful tools for high-throughput genome and transcriptome sequencing. Out of various platform available commercially, only a few of them are so popular (454/FLX sequencing, Solexa/Illumina, Ione torrent etc.) due to their capacity, quality, and cost effective property, while some of the platform have been out dated (ABI SOLiD: Sequencing by ligation) or less popular with scientific community world wide. In next generation sequencing methods, genomic DNA is randomly sheared and immobilized on solid support by using a microscopic bead. Subsequently, fragmented individual DNA molecules amplified using emulsion or bridge PCR. After amplification sequencing chemistry is applied directly to the solid support and based on fluorescence or chemiluminescence nucleotide of fragmented DNA is determined. These technologies used for both, denovo sequencing as well as re-sequencing of the genome. After the introduction of these technologies, a large amount of data has been generated in plant species, species of microbes, animals etc. and these data are available in publically available data base like NCBI (SRA). By the comparison of genomic data within species and between species large amount functional molecular marker like SNPs, InDel, SSRs etc. can be generated by the integration of transcriptomic data. In many of the plant species like rice, wheat, maize, etc. large amount of functional markers have been generated, while in fungal species only a few of the reports are available.

7.4.2 Development of Expression Based (ESTs) Genic Molecular Marker

The genic, as well as functional molecular markers, have been successfully used in high throughput expression analysis to reveal the transcriptome profiling. The EST-based approaches have been enormously recognized to discover functional marker as it excludes non-functional region of the genome. For the identification of suppression, the subtractive hybridization experiment directly correlates with the function of the desired character, if EST library represents the suppression transcriptome. Expression level polymorphism (ELP) also can be used as a functional marker for characterization of the diverse isolate. If parental genotypes produce the same amount of mRNA but contain polymorphism within DNA will result in differential hybridization in the microarray experiment. These types of polymorphism have been termed as single feature polymorphism (SFP).

7.4.3 Development of Genic Molecular Markers by In Silico Technique

By utilizing bioinformatics tools, information accumulated in the public data base can be utilized for the development of functional marker. Data base like NCBI, EMBL, DDBJ etc. possesses large scale information in the form of candidate gene, EST, Microarray, Genome and transcriptome assembly. By the implementation of bioinformatics tools like MISA, SSR finder, Clustal W and others, data can be analysed for the detection of functional molecular markers.

7.5 Availability and Utility of Genic Markers in Fungi

Identification of genic molecular marker in the genome sequences of fungal species is one of the best ways to characterise there phenotypic trait. Availability of a large amount of whole genome sequence information increases the opportunity to identify the genic molecular markers by utilising the data base. But the report of only a few genic molecular markers exists to date. Among them, genic SSRs and SNPs are most commonly used to resolve various important mycological aspects discussed below.

7.5.1 Functional Diversity Study

Characterization of variation within natural populations is crucial for effective conservation and exploitation of bio-resources. Molecular markers have been proven to be the most useful tool for assessment of these variations. Fungi are the second most diverse organism after insects on this earth (Rossman et al. 1998). This biodiversity accounts for its huge morphological, ecological and functional diversity within the population and is also important in respect to natural conservation and utilization for the mankind. Therefore, studies of population genetics of each single species of fungi are gaining importance. To make inferences about population structure on a finer scale within a single species, it is often necessary to use markers that are polymorphic. Random molecular markers are most commonly used to resolve this issue. But, these anonymous markers are locus nonspecific, and unsuitable high throughput genotyping. During the last decade, some unique loci specific and highly polymorphic markers like microsatellites or simple sequence repeats (SSRs), and single-nucleotide polymorphisms (SNPs) have become the best choice for population genetic studies (Dutech et al. 2007). In most of the cases, these markers are located in the non-functional part of the genome and can introduce serious error in population genetic studies (Dutech et al. 2007). When SSRs and SNPs are located in the coding region and non-coding parts of any gene, they may have a significant role in gene expression and gene function. Usually, these genes are conserved and little variation within the sequence is very important to regulate gene function as well as phenotypic trait variations. Therefore, evaluation of biodiversity using GMMs is gaining importance due to the variation in the functional part of gene and reduced the biases for characterizing genetic variation.

Genic SSRs and SNPs are most commonly used to characterise the population dynamics of any fungi based on functional significance. While using the genic-SSR markers for functional diversity studies, the length of SSR repeats in the gene with known function can be tested to establish the close association with biological function and phenotypic trait variation (Ayers et al. 1997). The variation in the length of SSRs in the coding sequence of a gene or its non-coding parts (UTRs, Promoter, and regulator sequence) may directly or indirectly influence the gene expression and gene function with phenotypic variation. This was evident in several mushroom fungi to characterise qualitative and quantitative traits (Table 7.1). Similarly, the use of genic SNPs for genetic diversity studies may correlate the single nucleotide polymorphism within coding regions of the gene with the phenotypic variation of many plant pathogenic fungi (Table 7.1) and provided opportunities to examine functional diversity in relation to their adaptation to a new environment.

7.5.2 Strain Identification

Conventional (phenotypic) strain typing of fungi is problematic due to a lack of stable characteristics distinguishing between isolates. Most of the fungal strains show uniformity in both morphological and biological appearance, although physiological variations within a population do exist. Therefore, identification of genetically related strains and variants within a species is the key to exploring the microbial diversity that information can be utilized for their agricultural, medical and industrial application. The commercial strains of many industrially important fungi are supposed to be genetically very similar due to their origin from a same hereditary line. Many highly polymorphic molecular markers with the whole-genome coverage, accuracy and reproducibility have been exploited for fungal strain typing. But for high throughput analysis and more amenable functional characterisation, the markers located in gene coding regions are proven to be promising. Next generation sequencing technology makes it affordable to conduct robust assays to identify simple sequence repeats (SSRs) and single nucleotide polymorphisms (SNPs) for higher resolution. But, SSR and SNP mining relying on whole genome sequence comparisons are not economically and feasible for identifying commercial important fungal strains. Moreover, a large number of SNPs and SSRs located in the non-coding region of genomes have limited application in functional mapping of fungal strains. The discovery of SNPs and SSRs markers located in transcribed regions has become a common application in fungi because of a large number of ESTs available in databases (Table 7.1). EST-derived SSRs and SNPs can be for higher resolution fungal genotyping and strain differentiation based on functional significance. The correlation between the virulence phenotyping with molecular genotyping of fungal isolates was established by using such functional markers. Identification of fungicide resistance strains of many plant pathogenic fungi is also becoming easier with SNPs based detection located on functional gene conferring resistance.

7.5.3 Marker Assisted Beeding Programme

Marker-assisted selection is the easiest and cheapest method for the screening and breeding of commercially important strains of fungi. The use of random molecular markers in marker assisted selection depends on the persistence of linkage between a marker and target loci. The tight linkage between a marker and trait located within the same cross is important for the unbiased selection process. Owing to frequent recombination in fungi, undesirable crossing over between RDMs and gene may result in a false positive selection. Therefore, phenotypic validation is always essential in RDM assisted selection process to confirm the presence of the target gene(s) and to avoid undesirable selection. This validation can be circumvented by using genic molecular markers (GMMs). Because GMMs are derived either directly from functional motifs or from a noncoding sequence of a gene. Due to complete linkage with the favourable target locus allele, marker-assisted selection based on these functional markers would be more efficient than RDM-based selection (Andersen and Lubberstedt 2003). Now days, the discovery of GMMs providing an attractive choice for marker assisted breeding for many commercially important fungal species, especially mushroom. At present, in fungal kingdom, adequate evidence are not available, but in near future GMM based

Table 7.1 Availability and utility of genic molecular markers in different fungal species

Marker	Source	Fungal species	Functional utility/application	References
CAPS/ PCR-RFLP	Converted from anSNP in rDNA intergenic spacer region 1 and eburicol 14-α demethylase gene	*Erysiphe necator*	To identify different genetic groups of E. necator.	Araya et al.(2014)
	Converted from SNP incytochrome *b* gene	*Blumeria graminis* f. sp. *hordei*	Conferring resistance to strobilurin group of fungicide.	Baumler et al. (2003)
	Restritcion digesation of PCR amplicon of succinate dehydro-genase subunit B, and D	*Alternaria solani*	To screen and characterize resistant strains of pathogen to succinate dehydrogenase inhibitors (boscalid, fluopyram and penthiopyrad)	Miles et al. (2013)
	Point mutations in the fungal mitochondrial cytochrome band the β-tubulin gene.	*Cercospora beticola*	To monitor fungicide resistance within pathogen population.	Rosenzweig et al. (2015)

(continued)

Table 7.1 (continued)

Marker	Source	Fungal species	Functional utility/application	References
SSR	EST data base	*Puccinia coronata f.sp. lolii.*	Determination of functional diversity within pathogen population	Dracatos et al. (2006)
	EST data base	*Neotyphodium* sp. and *Epichloe* sp.	Discrimination between species & genetic variation within and between endophytic fungal species to correlate with variation for endophyte-derived agronomic traits	van Zijll de Jong et al. (2003)
	EST data base	*Phellinus linteus*	To identify putative unigenes involved in sterol biosynthesis and helps in marker-assisted breeding and gene mapping	Huang et al., (2015)
	EST data base	*Ervsiphe necator* (syn.*Uncinula necator*)	To study the population biology and to detect genetic differentiation between fungal isolates based on their origin and aggressiveness.	Frenkel et al.(2012)
	Transcriptome sequence	*Auricularia polytricha*	To study functional genetic diversity and to construct a transcript map	Zhou et al. (2014).
	EST data base	*Agaricus subrufescens*	For gene discovery and molecular marker development	Foulongne-Oriol et al. (2014)
	EST data base	*Pleurotus ostreatus, Lentinula edodes, Agaricus bisporus*	For comparative genetic studies of different mushroom species	Polat et al. (2010)
	Whole genome sequence	*Pleurotus ostreatus*	Functional analysis of the subset of genes involve in evolution, environmental adaptation, etc.	Qu et al. (2016).
	Whole genome sequence & EST data base	*Laccaria bicolor*	To identify protein-coding genes contained SSRs in their exons and characterise altered gene phenotypes based on SSR polymorphism	Labbe et al. (2011)
	Whole genome sequence	*Aspergillus nidulans, Magnaporthe grisea, Neurospora crassa*	To decipher the difference of different types of microsatellite repeats and the number of motifs in protein coding regions of three fungi to decipher their functional genomics	Li et al. (2009).
	EST data base	*Lentinula edodes*	To study genetic diversity and to assisted breeding of edible fungi	Yan et al.(2010)
	EST data base	*Pleurotus ostreatus*	To characterise genes associated with both qualitative and quantitative traits within a primary core collection of fungi	Yao et al. (2012)
	EST data base	*Lentinus edodes*	To study genetic diversity	Wang et al. (2010)
	EST data base	*Agrocybe aegerita*	To decipher genetic differences between strains and to compare the analysis of agronomic traits.	Wang (2012)
	EST data base	*Agrocybe cylindracea*	To study the genetic diversity and comparative genomics of edible mushroom.	Wang et al. (2012)
	EST library	*Puccinia triticina*	To correlate virulence with genetic variability and population genetic study	Wang et al. (2010a)
	EST sequence	*P. graminis* f. sp. *tritici*	For characterization of population structures and for gene mapping	Zhong et al. (2009)
	EST database	*Fusarium graminearum*	To understand the evolution and genetic diversity	Singh et al. (2011)
	EST sequences	*Polyporus umbellatus*	Helps in genetic diversity study and taxonomic, molecular breeding, and functional gene analysis	Zhang et al. (2015)

SNP	EST and transcriptome data base	*Fusarium oxysporu - mf*.sp. *lycopersici*, *F. oxy*.f.sp. *cucume –rium*, &*F. oxy*. f.sp.*melonis*	To study genetics, population history, ecology and evolution of different *Fusarium* species	Rai and Maurya, (2014)
	cDNA library/ transcriptome data	*Podosphaera plantaginis*	To reveal genetic diversity in the pathogen meta-population and importance of mixed-genotype infection within a single host as the main driver of pathogen evolution	Tollenaere et al. (2012)
	DNAsequence of six protein coding geno-mic region	*Pandora neoaphidis*	To detect intraspecific variations and to understand population dynamics of the fungi with better approach to monitor potential biological control strains for aphid management.	Fournier et al. (2010)
	Genome and transcriptome data	*Leptographium longiclavatum*	For population structure, genetic diversity and the identification of candidate adaptive SNPs involved in host defence detoxification and pathogenicity	Ojeda et al. (2014)
	EST resource	*Neotyphodium* sp. and *Epichloe* sp.	To study complex genetic traits and to understand genome evolution	Spangenberg et al. (2005)
	Genomic DNA library	*Lentinula edodes*	For the identification of different shiitake lines and role SNP polymorphism in affecting phenotypic expression of laccase gene.	Kim et al. (2015)
	EST library	*Plasmopara viticola*	To study population genetics and evolution mechanism	Delmotte et al. (2011)
	DNA sequence of β-tubulin gene	*Erysiphe necator*	For epidemiological and population genetic study.	Amrani and Corio-Costet (2006)

marker-assisted selection can be exploited for qualitative and quantitative trait improvement of fungi.

7.5.4 Evolutionary Study

After the Assembling the Fungal Tree of Life (AFTOL) project, a significant progress has been achieved in the taxonomic and evolutionary aspect of fungi. Plenty of molecular data were used to resolve deep branches of the phylogenetic tree of fungal life up to species and below species rank. A multi-locus sequence typing (MLST) approach based on six genes (nu-SSU, -LSU, and 5.8S rRNA, *rpb*1, *rpb*2 and *tef*1) was adopted to address the evolutionary relationship in every major clade of fungi. But, use of sequence information of these genes could underestimate the real diversity at strain/race level. Thus, information on a number of simple sequence repeat motif sequences in the coding sequence of the functional genes like β-tubulin would be required for better understanding of fungal evolutionary dynamics. Due to their high mutability, SSRs are thought to play an active role in the genome evolution (Tautz et al. 1986). Usually, the number of repeat sequence in the coding sequence of any eukaryotic genome is correlated with a higher level of evolution and this was evident from many plant species. Although there is some evidence of the relative abundance of SSRs differing with the genome size of fungi and is not consistent with the fungal evolution (Morgante et al. 2002; Hancock 1996, 2002). But analysis of relative abundance, relative density, length and frequency of different simple sequence repeat motifs on the genome organization of fungal species may shed more light on the fungal evolution.

7.5.5 Epidemiological & Ecological Adaptation

Fungi are distributed in a variety of ecological environments either as a saprophyte or as a plant parasite. In each ecological niche, there always have some changing climatic scenario which imposes some selection pressure on the existing population of fungi and ultimately leads to the evolution of new variants and subsequent adaptation in the new ecological situation. The evolutionary process is commonly evident within fungal pathogen populations in the intensified agro-ecosystem than wild ecosystems due to increasing directional selection and rapid adaptation of the new virulent pathotype. The DNA-based molecular markers are a powerful tool to predict this evolution and adaptation of fungal pathogens. Usually, genic molecular markers are better than random molecular markers in this regard. Both the genic SSRs and genic SNPs can be used to characterise the epidemiological and ecological adaptation of fungi. In the case of genic-SSR, the size variation of repeat motif numbers in the natural population of fungi suggests their evolutionary consequences. The tri-nucleotide SSRs in the exonic region are translated into amino-acid repeats, which possibly contribute to the biological function of the protein. Contraction and expansion of these exonic tri-nucleotide SSRs may play a significant role in the evolution of gene function and facilitate adaptation to new environments (Li et al. 2004; Kashi and King 2006).

7.5.6 Mapping and Tagging of Gene(s) & QTLs

A genetic map is a partial representation of the genome, showing the relative position and distances (in centiMorgan) between markers and genes along a chromosome. It insights into the genome organization of fungi. It can be utilised to map genes or quantitative trait loci (QTL) for a trait of interest like virulence genes in pathogenic species and agronomically relevant trait in cultivated edible mushrooms (Foulongne-Oriol 2012; Chattopadhyay et al. 2014) and also facilitate whole genome sequencing process. The genetic map can be constructed either by biparental population mapping or association mapping of the unrelated natural population. QTL mapping using biparental mapping population is popular for different mushroom fungi. Whereas, association mapping is done by using the natural population of many fungi like *Heterobasidion annosum* (Dalman et al. 2013). *Magnaporthe oryzae* (Yoshida et al. 2009). The robustness of genetic map depends upon the linkage disequilibrium between phenotyping data with genotyping information. The genome-wide availability of high-density marker can be used to generate unbiased genomic information. Random molecular markers, developed from genomic libraries, can belong to either the transcribed region or the non-transcribed region of the genome without any functional information. By contrast, genic molecular markers (both, gene targeted marker and functional marker) often have known or 'putative' functions or affect the function of the gene in which they reside and can be used for functional genotyping of fungal populations. It is believed that the distribution of GMMs in the genetic map mirrors the distribution of genes along the genetic map (Varshney et al. 2005). Therefore, use of GMMs is thought to be very effective both for 'genome-wide association mapping' and candidate gene based mapping of the unstructured fungal population.

7.5.7 Other Applications

The genic molecular markers can also be used to resolve the issues of various other mycological aspects like rapid detection of mutant within existing strains/isolate of fungi. To get genetically pure culture of any desired fungi for commercial application. The high throughput detection of toxic genic species of mushroom can also be easy by using functional gene-based GMMs.

7.6 Future Prospects and Conclusion

Although the fungal genomic study has reached some milestone but in reality, the majority of the research is focussed on the structural aspect. Use of genic molecular marker, especially the functional markers would be the best tool to decipher the functional genomic of fungi. These markers directly represent the genes responsible for traits of interest and thus help in easy selection. But the low level of polymorphism of the GMMs restricts their use. In spite of that, being the easy and cost-effective approach, GMMs are suitable to assay the functional biodiversity within a fungal population, evolution and adaptation of fungi in a new ecological niche and many other applications. The higher interspecific transferability enriches their wider application. With the rapid development of next-generation sequencing technology, the whole genome sequence as well as EST sequence data of many fungal species has already been generated and many are underway. Along with that, huge application of *in silico* genomics tool made it possible to have a comparative study of fungal genomics. It also helps in easy identification and development of GMMs/FMs. Therefore, application of GMMs/FMs in the field of mycology can be accelerated in coming years to get the higher resolution based fungal identification system, fine mapping of genes/QTLs, rapid marker assisted selection of useful fungal strains and correct assessment of fungal evolution and their phylogenetic representation. Besides, modern 'omics' approach like genomics/expression genetics, transcriptomics, proteomics and metabolomics, etc., development and utilization of allele-specific GMMs will be very important tools for the newly identified candidate genes and their functional characterisation, which ultimately give a better insight to the functional genomics of fungi.

References

Agarwal M, Shrivastava N, Padh H. Advances in molecular marker techniques and their applications in plant sciences. Plant Cell Rep. 2008;27:617–31.

Aggarwal RK, Hendre PS, Varshney RK, Bhat PR, Krishna KV, Singh L. Identification, characterization and utilization of EST-derived genetic microsatellite markers for genome analyses of coffee and related species. Theor Appl Genet. 2007;114:359–72.

Amrani L, Corio-Costet MF. A single nucleotide polymorphism in the β-tubulin gene distinguishing two genotypes of *Erysiphe necator* expressing different symptoms on grapevine. Plant Pathol. 2006;55(4):505–12.

Andersen JR, Lubberstedt T. Functional markers in plants. Trends Plant Sci. 2003;8:554–60.

Anderson JB, Stasorski E. Molecular phylogeny of Northern Hemisphere species of *Armillaria*. Mycologia. 1992;84:505–76.

Araya C, Rosales IM, Mendez MA, Delmotte F. Identification and geographic distribution of genetic groups of *Erysiphe necator* in Chilean vineyards. Vitis. 2014;53(3):163–5.

Aroca A, Raposo R, Lunello P. A biomarker for the identification of four *Phaeoacremonium* species using the beta-tubulin gene as the target sequence. Appl Microbiol Biotechnol. 2008; 80(6):1131–40.

Arora DK, Hirsch PR, Kerry BR. PCR-based molecular discrimination of *Verticillium chlamydosporum* isolates. Mycol Res. 1996;100:801–9.

Ayers NM, McClung AM, Larkin PD, Bligh HFJ, Jones CA, Park WD. Microsatellites and a single nucleotide polymorphism differentiate apparent amylose classes in an extended pedigree of US rice germplasm. Theor Appl Genet. 1997;94:773–81.

Baumler S, Felsenstein FG, Schwarz G. CAPS and DHPLC analysis of a single nucleotide polymorphism in the Cytochrome b gene conferring resistance to Strobilurins in field isolates of *Blumeria graminis f. sp. hordei*. J Phytopathol. 2003;151(3):149–52.

Baura G, Szaro TM, Bruns TD. *Gatrosuillus laricinus* is a repeat derivative of *Suillus grevellei* : molecular evidence. Mycologia. 1992;84:592–7.

Blake TK, Kadyrzhanova D, Shepherd KW, Islam A, Langridge PL, McDonald CL, Erpelding J, Larson S, Blake NK, Talbert LE. STS-PCR markers appropriate for wheat-barley introgression. Theor Appl Genet. 1996;82:715–21.

Bowman BH, Taylor JW, White TJ. Molecular evolution of the fungi: human pathogens. Mol Biol Evol. 1992a;9:893–904.

Bowman BH, Taylor JW, Browilee AG, Lee J, Lu S, White TJ. Molecular evolution of the fungi relationship of basidiomycetes, ascomycetes and chytridiomycetes. Mol Biol Evol. 1992b;9:285–96.

Bruns TD, Vilgalys R, Barns SM, Gonzalez D, Hibbert DS, Lane DJ, Simon L, Stickel S, Szaro TM, Weisburg WG, Sogin ML. Evolutionary relationships within fungi analysis of nuclear small subunit rRNA sequences. Mol Phylogeny Evol. 1992;1:231–41.

Buchko J, Klassen GR. Detection of length heterogeneity in the ribosomal DNA of *Pythium ultimum* by PCR amplification of the intergenic region. Curr Genet. 1990;18:203–5.

Capote N, Pastrana AM, Aguado A, Sanchez-Torres P. Molecular tools for detection of plant pathogenic fungi and fungicide resistance. In: Cumagun CJ, editor, Plant pathology. 2012. p. 151–202. InTech Europe. www.intechopen.com.

Carter DA, Tran-Dinh N, Marrar RE, Vera RE. The development of genetic markers from fungal genome initiatives. In: Arora DK, Khachatourians GG, editors. Fungal genomics-Applied Mycology & Biotechnology, vol. 4. Amsterdam: Elsevier; 2004. p. 1–27.

Chagne D, Batley J, Edwards D, Forster JW. Single nucleotide polymorphisms genotyping in plants. In: Oraguzie NC, Rikkerink EHA, Gardiner SE, De Silva HN, editors. Association mapping in plants. New York: Springer; 2007. p. 77–94.

Chambers GK, MacAvoy ES. Microsatellites: consensus and controversy. Compar Biochem Physiol (Part B). 2000;126:455–76.

Chattopadhyay A, Kadappa HS, Meena SC. Recent molecular genetics and biotechnological approaches in mushroom research. In: Chandra R, editor. Microbial biodiversity in sustainable agriculture. New Delhi: Daya Publishing House, Astral International (P) Ltd; 2014. p. 323–43.

Dalman K, Himmelstrand K, Olson A, Lind M, Brandstrom-Durling M, Stenlid J. A genome-wide association study identifies genomic regions for virulence in the non-model organism *Heterobasidion annosum s.s.* PLoS One. 2013;8(1):e53525. doi:10.1371/journal.pone.0053525.

Daniel HM, Meyer W. Evaluation of ribosomal RNA and actin gene sequences for the identification of ascomycetous yeasts. Int J Food Microbiol. 2003;86:61–78.

Delmotte F, Machefer V, Giresse X, Richard-Cervera S, Latorse MP, Beffa R. Characterization of single-nucleotide-polymorphism markers for *Plasmopara viticola*, the causal agent of grapevine downy mildew. Appl Environ Microbiol. 2011;77(21):7861–3.

Dracatos PM, Dumsday JL, Olle RS, Cogan NO, Dobrowolski MP, Fujimori M, Roderick H, Stewart AV, Smith KF, Forster JW. Development and characterization of EST-SSR markers for the crown rust pathogen of ryegrass (*Puccinia coronata f.sp. lolii*). Genome. 2006;49(6):572–83.

Dutech C, Enjalbert J, Fournier E, Delmotte F, Barres B, Carlier J, Tharreau D, Giraud T. Challenges of microsatellite isolation in fungi. Fungal Genet Biol. 2007;44(10):933–49.

Edelmann SE, Staben C. A statistical analysis of sequence features within genes from *Neurospora crassa*. Exp Mycol. 1994;18:70–81.

Foster SJ, Ashby AM, Fitt BDL. Improved PCR-based assays for pre-symptomatic diagnosis of light leaf spot and determination of mating type of *Pyrenopeziza brassicae* on winter oilseed rape. Eur J Plant Pathol. 2002;108(4):379–83.
Foulongne-Oriol M. Genetic linkage mapping in fungi: current state, applications, and future trends. Appl Microbiol Biotech. 2012;95(4):891–904.
Foulongne-Oriol M, Lapalu N, Ferandon C, Spataro C, Ferrer N, Amselem J, Savoie JM. The first set of expressed sequence tags (EST) from the medicinal mushroom *Agaricus subrufescens* delivers resource for gene discovery and marker development. Appl Microbiol Biotechnol. 2014;98(18):7879–92.
Fournier A, Widmer F, Enkerli J. Development of a single-nucleotide polymorphism (SNP) assay for genotyping of *Pandora neoaphidis*. Fungal Biol. 2010;114:498–506.
Frenkel O, Portillo I, Brewer MT, Peros JP, Cadle-Davidson L, Milgroom MG. Development of microsatellite markers from the transcriptome of *Erysiphe necator* for analysing population structure in North America and Europe. Plant Pathol. 2012;61:106–19.
Gardes M, Bruns TD. Rapid characterization of ectomycorrhizae using RFLP pattern of their PCR amplified-ITS. Mycol Soc Newsletter. 1991;41:14.
Gardes M, Bruns TD. ITS primers with enhanced specificity for basidiomycetes application to the identification of mycorrhizae and rusts. Mol Ecol. 1993;2:113–8.
Gardes M, White TJ, Fortin JA, Bruns TD, Taylor JW. Identification of indigenous and introduced symbiotic fungi in ectomycorrhizae by amplification of nuclear and mitochondrial ribosomal DNA. Can J Bot. 1991;69:180–90.
Gaudet J, Julien J, Lafay JF, Brygoo Y. Phylogeny of some Fusarium species, as determined by large subunits rRNA sequence comparison. Mol Biol Evol. 1989;6:227–42.
Geiser DM, Jimenez-Gasco M, Kang S, Makalowska I, Veeraraghavan N, Ward TJ, Zhang N, Kuldau GA, O'Donnell K. Fusarium-ID v. 1.0: A DNA sequence database for identifying *Fusarium*. Eur J Plant Pathol. 2004;110(5-6):473–9.
Geiser DM, Klich MA, Frisvad JC, Peterson SW, Varga J, Samson RA. The current status of species recognition and identification in *Aspergillus*. Stud Mycol. 2007;59:1–10.
Gerber HP, Seipel K, Georgiev O, Hofferer M, Hug M, Rusconi S, Schaffner W. Transcriptional activation modulated by homopolymeric glutamine and proline stretches. Science. 1994; 263:808–11.
Hancock JM. The contribution of slippage-like processes to genome evolution. J Mol Evol. 1995;41:1038–47.
Hancock JM. Simple sequences and the expanding genome. Bioessays. 1996;18:421–5.
Hancock JM. Genome size and the accumulation of simple sequence repeats: implications of new data from genome sequencing projects. Genetica. 2002;115(1):93–103.
Hefferon TW. A variable dinucleotide repeat in the CFTR gene contributes to phenotype diversity by forming RNA secondary structures that alter splicing. Proc Natl Acad Sci U S A. 2004; 101:3504–9.
Hibbett DS. Ribosomal RNA and fungal systematics. Trans Mycol Soc Jpn. 1992;33:533–56.
Hillis DM, Dixon MT. Ribosomal DNA: molecular evolution and phylogenetic inference. Quart Rev Biol. 1991;66:411–53.
Huang TS, Lee CC, Chang AC, Lin S, Chao CC, Jou YS, Chu YW, Wu CW, Whang-Peng J. Shortening of microsatellite deoxy(CA) repeats involved in GL331-induced down-regulation of matrix metalloproteinase-9 gene expression. Biochem Biophys Res Commun. 2003; 300:901–7.
Huang Y, Wu X, Jian D, Zhan Y, Fan G. De novo transcriptome analysis of a medicinal fungi *Phellinus linteus* and identification of SSR markers. Biotechnol Biotechnol Equip. 2015; 29(2):395–403.
Hwang TY, Sayama T, Takahashi M, Takada Y, Nakamoto Y, Funatsuki H, Hisano H, Sasamoto S, Sato S, Tabata S, Kono I, Hoshi M, Hanawa M, Yano C, Xia Z, Harada K, Kitamura K, Ishimoto M. High-density integrated linkage map based on SSR markers in soybean. DNA Res. 2009;16(4):213–25.

Islam MN, Molla MR, Rohman MM, Hasanuzzaman M, Islam SMN, Rahman L. DNA fingerprinting and genotyping of cotton varieties using SSR markers. Not Bot Horti Agrobo. 2012;40:261–5.

Jarvis P, Lister C, Szabo V, Dean C. Integration of CAPS markers into the RFLP map generated using recombinant inbred lines of *Arabidopsis thaliana*. Plant Mol Bio. 1994;24:685–7.

Kage U, Kumar A, Dhokane D, Karre S, Kushalappa AC. Functional molecular markers for crop improvement. Crit Rev Biotechnol. 2015;16:1–14.

Karaoglu H, Lee CY, Meyer W. Survey of simple sequence repeats in completed fungal genomes. Mol Biol Evol. 2005;22(3):639–49.

Kashi Y, King DG. Simple sequence repeats as advantageous mutators in evolution. Trends Genet. 2006;22(5):253–9.

Kashi Y, King D, Soller M. Simple sequence repeats as a source of quantitative genetic variation. Trends Genet. 1997;13:74–8.

Katti MV, Ranjekar PK, Gupta VS. Differential distribution of simple sequence repeats in eukaryotic genome sequences. Mol Biol Evol. 2001;18:1161–7.

Kesawat MS, Das BK. Molecular markers. Its application in crop improvement. J Crop Sci Biotechnol. 2009;12(4):169–81.

Kim KH, Ka KH, Kang JH, Kim S, Lee JW, Jeon BK, Yun JK, Park SR, Lee HJ. Identification of single nucleotide polymorphism markers in the Laccase gene of shiitake mushrooms (*Lentinula edodes*). Mycobiology. 2015;43(1):75–80.

Kim WK, Zuerucha T, Klassen GR. A region of heterogeneity adjacent to 5S ribosomal RNA genes of cereal rusts. Curr Genet. 1992;22:101–5.

Knutsen AK, Torp M, Holst-Jensen A. Phylogenetic analyses of the *Fusarium poae, Fusarium sporotrichioides* and *Fusarium langsethiae* species complex based on partial sequences of the translation elongation factor-1 alpha gene. Int J Food Microbiol. 2004;95(3):287–95.

Konieczny A, Ausubel FM. A procedure for mapping Arabidopsis mutations using co-dominant ecotype-specific PCR-based markers. Plant J. 1993;4:403–10.

Kristensen R, Torp M, Kosiak B, Holst-Jensen A. Phylogeny and toxigenic potential is correlated in *Fusarium* species as revealed by partial translation elongation factor 1 alpha gene sequences. Mycol Res. 2005;109(2):173–86.

Kumar P, Gupta VK, Mishra AK, Modi DR, Pandey BK. Potential of molecular markers in plant biotechnology. Plant Omics J. 2009;2:141–62.

Kunzler P, Matsuo K, Schaffner W. Pathological, physiological, and evolutionary aspects of short unstable DNA repeats in the human genome. Biol Chem Hoppe Seyler. 1995;6:201–11.

Labbe J, Murat C, Morin E, Le Tacon F, Martin F. Survey and analysis of simple sequence repeats in the *Laccaria bicolor* genome, with development of microsatellite markers. Curr Genet. 2011;57(2):75–88.

Li CY, Liu L, Yang J, Li JB, Su Y, Zhang Y, Wang YY, Zhu YY. Genome-wide analysis of microsatellite sequence in seven filamentous fungi. Interdiscip Sci Comput Life Sci. 2009;1(2):141–50.

Li X, Gao W, Guo H, Zhang X, Fang DD, Zhongxu L. Development of EST-based SNP and InDel markers and their utilization in tetraploid cotton genetic mapping. BMC Genomics. 2014; 15:1046.

Li YC, Korol AB, Fahima T, Beiles A, Nevo E. Microsatellites: genomic distribution, putative functions and mutational mechanisms: a review. Mol Ecol. 2002;11:2453–65.

Li YC, Korol AB, Fahima T, Nevo E. Microsatellites within genes: structure, function, and evolution. Mol Biol Evol. 2004;21(6):991–1007.

Lievens B, Van Baarlen P, Verreth C, Van Kerckhove S, Rep M, Thomma BPHJ. Evolutionary relationships between *Fusarium oxysporum f. sp. lycopersici* and *F. oxysporum f. sp. radicis-lycopersici* isolates inferred from mating type, elongation factor-1 alpha and exopolygalacturonase sequences. Mycol Res. 2009;113:1181–91.

Lindahl BD, Nilsson RH, Tedersoo L, Abarenkov K, Carlsen T, Kjoller R, Koljalg U, Pennanen T, Rosendahl S, Stenlid J, Kauserud H. Fungal community analysis by high-throughput sequencing of amplified markers-a user's guide. New Phytol. 2013;199(1):288–99.

Litt M, Luty JM. A hypervariable microsatellite revealed by *in vitro* amplification of a dinucleotide repeat within the cardiac muscle actin gene. Am J Hum Genet. 1989;44:397–401.

Martin FN, Tooley PW. Phylogenetic relationships among *Phytophthora* species inferred from sequence analysis of mitochondrially encoded cytochrome oxidase I and II genes. Mycologia. 2003;95(2):269–84.

Matsuoka Y, Mitchell SE, Kresovich S, Goodman M, Doebley J. Microsatellites in Zea-variability, patterns of mutations, and use for evolutionary studies. Theor Appl Genet. 2002;104:436–50.

McCartney HA, Foster SJ, Fraaije BA, Ward E. Molecular diagnostics for fungal plant pathogens. Pest Manag Sci. 2003;59(2):129–42.

Metzgar D, Bytof J, Wills C. Selection against frameshift mutations limits microsatellite expansion in coding DNA. Genome Res. 2000;10:72–80.

Meyer SLF, Carta LK, Rehner SA. Morphological variability and molecular phylogeny of the nematophagous fungus *Monacrosporium drechsleri*. Mycologia. 2005;97:405–15.

Michaels SD, Amasino RM. A robust method for detecting single-nucleotide changes as polymorphic markers by PCR. Plant J. 1998;14:381–5.

Miles TD, Miles LA, Fairchild KL, Wharton PS. Screening and characterization of resistance to succinate dehydrogenase inhibitors in *Alternaria solani*. Plant Pathol. 2013;63(1):155–64.

Morgante M, Hanafey M, Powell W. Microsatellites are preferentially associated with nonrepetitive DNA in plant genomes. Nat Genet. 2002;30:194–200.

Mostert L, Groenewald JZ, Summerbell RC, Gams W, Crous PW. Taxonomy and pathology of *Togninia* (Diaporthales) and its *Phaeoacremonium* anamorphs. Stud Mycol. 2006;54:1–113.

Nguyen HDT, Seifert KA. Description and DNA barcoding of three new species of *Leohumicola* from South Africa and the United States. Persoonia. 2008;21:57–69.

Nilsson RH, Kristiansson E, Ryberg M, Hallenberg N, Larsson KH. Intraspecific ITS variability in the kingdom Fungi as expressed in the international sequence databases and its implications for molecular species identification. Evol Bioinform Online. 2008;4:193–201.

Nazar RN, Hu X, Schmidt J, Culham D, Robb J. Potential use of PCR-amplified ribosomal intergenic sequences in the detection and differentiation of Verticillium wilt pathogens. Physiol Mol Plant Pathol. 1991;39:1–11.

Ojeda DI, Dhillon B, Tsui CKM, Hamelin RC. Single-nucleotide polymorphism discovery in *Leptographium longiclavatum* a mountain pine beetle-associated symbiotic fungus, using whole-genome resequencing. Mol Ecol Resour. 2014;14:401–10.

Bridge PD, Singh T, Arora DK. The application of molecular markers in the epidemiology of plant pathogenic fungi. In: Arora DK, Bridge PD, Bhatnagar D, editors. Fungal biotechnology in agricultural, food, and environmental applications. New York: Marcel Dekke, INC; 2004. p. 57–68.

Perutz MF, Johnson T, Suzuki M, Finch JT. Glutamine repeats as polar zippers: their possible role in inherited neurodegenerative diseases. Proc Natl Acad Sci. 1994;91:5355–8.

Polat E, Ince AG, Karaca M, Naci Onus A. Mining and utilization of mushroom ESTs for microsatellites. Conserv Genet. 2010;11:1123–6.

Qu J, Huang C, Zhang J. Genome-wide functional analysis of SSR for an edible mushroom *Pleurotus ostreatus*. Gene. 2016;575:524–530.

Rai S, Maurya DK. Computational analysis of single-nucleotide polymorphisms and insertions/deletions in coding regions of three *formae speciales* of *Fusarium oxysporum*. J Sci Res. 2014;58:51–8.

Rosenzweig N, Hanson LE, Clark G, Franc GD, Stump WL, Jiang QW, Stewart J, Kirk WW. Use of PCR-RFLP analysis to monitor fungicide resistance in *Cercospora beticola* populations from sugarbeet (*Beta vulgaris*) in Michigan United States. Plant Dis. 2015;99(3):355–62.

Rossman AY, Tulloss RE, O'Dell TE, Thorn RG. Protocols for an all taxa biodiversity inventory of fungi in a Costa Rican conservation area. Boone: Parkway Publishers, Inc; 1998. p. 195.

Salgotra RK, Gupta BB, Stewart Jr CN. From genomics to functional markers in the era of next-generation sequencing. Biotechnol Lett. 2014;36:417–26.

Schmitt BA, Costa JH, de Melo DF. AOX –A functional marker for efficient cell reprogramming under stress. Trends Plant Sci. 2006;11:281–7.

Schoch CL, Seifert KA, Huhndorf S, Robert V, Spouge JL, Levesque CA, Chen W, Fungal Barcoding Consortium, Fungal Barcoding Consortium Author List. Nuclear ribosomal internal transcribed spacer (ITS) region as a universal DNA barcode marker for fungi. Proc Natl Acad Sci U S A. 2012;109(16):6241–6.

Schoch CL, Sung GH, Lopez-Giraldez F, Townsend JP, Miadlikowska J, Hofstetter V, Robbertse B, et al. The Ascomycota tree of life: a phylum-wide phylogeny clarifies the origin and evolution of fundamental reproductive and ecological traits. Syst Biol. 2009;58:224–39.

Seifert KA. The all-fungi barcoding campaign (FunBOL). Persoonia. 2008;20:106.

Seifert KA, Samson RA, deWaard JR, Houbraken J, Lévesque CA, Moncalvo JM, Louis-Seize G, Hebert PDN. Prospects for fungus identification using CO1 DNA barcodes, with *Penicillium* as a test case. Proc Natl Acad Sci U S A. 2007;104(10):3901–6.

Singh BP, Saikia R, Yadav M, Singh R, Chauhan VS, Arora DK. Molecular characterization of *Fusarium oxysporum* f. sp.*ciceri* causing wilt of chickpea. Afr J Biotechnol. 2006;5: 497–502.

Singh R, Sheoran S, Sharma P, Chatrath R. Analysis of simple sequence repeats (SSRs) dynamics in fungus *Fusarium graminearum*. Bioinformation. 2011;5(10):402–4.

Spangenberg GC, Felitti SA, Shields K, Ramsperger M, Tian P, Ong EK, Singh D, Logan E, Edwards D. Gene discovery and microarray-based transcriptome analysis of the grass-endophyte association. In: Roberts CA, West CP, Spiers DE, editors. *Neotyphodium* in cool-season grasses. Iowa,USA: Blackwell Publishing Professional; 2005. p. 103–21.

Spreadbury CL, Bainbridge BW, Cohen J. Restriction fragment length polymorphisms in isolates of *Aspergillus fumigatus* probed with part of the intergenic spacers region from the ribosomal RNA genes complex of *Aspergillus nidulans*. J Gen Microbiol. 1990:136:1991–1994.

Stockinger H, Peyret-Guzzon M, Koegel S, Bouffaud ML, Redecker D. The largest subunit of RNA Polymerase II as a new marker gene to study assemblages of arbuscular mycorrhizal fungi in the field. PLoS One. 2014;9(10), e107783.

Subramanian S, Mishra RK, Singh L. Genome-wide analysis of microsatellite repeats in humans: their abundance and density in specific genomic regions. Genome Biol. 2003a;4:R13.

Subramanian S, Madgula VM, George R, Mishra RK, Pundit MW, Kumar CS, Singh L. Triplet repeats in human genome: distribution and their association with genes and other genomic regions. Bioinformatics. 2003b;19:549–52.

Tachida H, Iizuka M. Persistence of repeated sequences that evolve by replication slippage. Genetics. 1992;131:471–8.

Tautz D, Trick M, Dover GA. Cryptic simplicity in DNA is a major source of genetic variation. Nature. 1986;322:652–6.

Tiwari KK, Chandra A. Use of degenerate primers in rapid generation of microsatellite markers in *Panicum maximum*. J Environ Biol. 2010;31(6):965–8.

Tollenaere C, Susi H, Nokso-Koivisto J, Koskinen P, Tack A, Auvinen P, Paulin L, Frilander MJ, Lehtonen R, Laine AL. SNP design from 454 sequencing of *Podosphaera plantaginis* transcriptome reveals a genetically diverse pathogen metapopulation with high levels of mixed-genotype infection. PLoS One. 2012;7(12), e52492.

Toth G, Gaspari Z, Jurka J. Microsatellites in different eukaryotic genomes: survey and analysis. Genome Res. 2000;10(7):967–81.

Van Zijll De Jong E, Guthridge KM, Spangenberg GC, Forster JW. Development and characterization of EST derived simple sequence repeat (SSR) markers for pasture grass endophytes. Genome. 2003;46(2):277–90.

Varshney RK, Graner A, Sorrells ME. Genic microsatellite markers in plants: features and applications. Trends Biotechnol. 2005;23(1):48–55.

Varshney RK, Mahendar T, Aggarwal RK, Borner A. Genetic molecular markers in plants: development and applications. In: Varshney RK, Tuberosa R, et al., editors. Genomics-assisted crop improvement: genomics approaches and platforms. Berlin: Springer; 2007. p. 13–29.

Voigt K, Kirk PM. Recent developments in the taxonomic affiliation and phylogenetic positioning of fungi: Impact in applied microbiology and environmental biotechnology. Appl Microbiol Biotechnol. 2011;90:41–57.

Voigt K, Wostemeyer J. Phylogeny and origin of 82 zygomycetes from all 54 genera of the Mucorales and Mortierellales based on combined analysis of actin and translation elongation factor EF-1alpha genes. Gene. 2001;270(1-2):113–20.

Wallace MM, Covert SF. Molecular mating type assay for *Fusarium circinatum*. Appl Environ Microbiol. 2000;66(12):5506–8.

Wang X, Mulock B, Guus B, Mccallum B. Development of EST-derived simple sequence repeat markers for wheat leaf rust fungus, *Puccinia triticina* Eriks. Can J Plant Pathol. 2010a; 32(1):98–107.

Wang XF, Mao HL, Ouyang L, Ge XQ. EST-SSR markers in *Agrocybe cylindracea* and its application. Mycosystema. 2012;31(05):800–6.

Wang YF, Zhao YH, Wang AY, Wang QQ, Liu LD, Zhou RL, Yu XF. Analysis of SSR information in EST resource of *Lentinus edodes*. Food Sci. 2010b;31(1):137–40.

Wang ZF. Agronomy traits analysis and EST-SSR marker application in *Agrocybe aegerita*. Nanchang University Master's thesis, China. 2012.

White TJ, Bruns T, Lee S, Taylor JW. Amplification and direct sequencing of fungal ribosomal RNA genes for phylogenetics. In: Innis MA, Gelgard DH, Sninsky JJ, White TJ, editors. PCR protocols: a guide to methods and applications. New York: Academic; 1990. p. 315–22.

Williams DW, Wilson MJ, Lewis MAG, Potts AJC. Identification of Candida species by PCR and restriction fragment length polymorphism analysis of intergenic spacer regions of ribosomal DNA. J Clin Microbiol. 1995;33:2476–9.

Xia Z, Tsubokura Y, Hoshi M, Hanawa M, Yano C, Okamura K, Ahmed TA, Anai T, Watanabe S, et al. An integrated high-density linkage map of soybean with RFLP, SSR, STS and AFLP markers using a single F2 population. DNA Res. 2007;14:257–69.

Yan LC, Yi LN, Qiong ZY. Development of expressed sequence tag-simple sequence repeat (EST-SSR) markers in *Lentinula edodes*. Acta Edulis Fungi. 2010;17(2):1–6.

Yao Q, Zhang X-M, Gong Z-Y, Gao XX, Jin LI, Han JD, Ren PF, Wan LZ, Ren HX. Development of EST-SSR Markers in *Pleurotus ostreatus* and construction of primary core collection. Mushroom Sci. 2012;18:180–8.

Yoshida K, Saitoh H, Fujisawa S, Kanzaki H, Matsumura H, Yoshida K, Tosa Y, Chuma I, Takano Y, Win J, Kamoun S, Terauchi R. Association genetics reveals three novel avirulence genes from the rice blast fungal pathogen *Magnaporthe oryzae*. Plant Cell. 2009;21:1573–91.

Zhang Y, Chen Y, Wang R, Zeng A, Deyholos MK, Shu J, Guo H. Development of microsatellite markers derived from expressed sequence tags of Polyporales for genetic diversity analysis of endangered *Polyporus umbellatus*. BioMed Research International ID 941357. 2015. (http://dx.doi.org/10.1155/2015/941357).

Zhang Y, Yu ZF, Xu J, Zhang KQ. Divergence and dispersal of the nematodetrapping fungus *Arthrobotrys oligospora* from China. Environ Microbiol Rep. 2011;3:763–73.

Zhong S, Leng Y, Friesen TL, Faris JD, Szabo LJ. Development and characterization of expressed sequence tag-derived microsatellite markers for the wheat stem rust fungus *Puccinia graminis f. sp. tritici*. Phytopathology. 2009;99:282–9.

Zhou Y, Chen L, Fan X, Bian Y. De Novo assembly of *Auricularia polytricha* transcriptome using Illumina sequencing for gene discovery and SSR marker identification. PLoS One. 2014;9(3), e91740.

Chapter 8
Molecular Markers for the Identification and Diversity Analysis of Arbuscular Mycorrhizal Fungi (AMF)

R. Krishnamoorthy, N. Premalatha, M. Karthik, R. Anandham, M. Senthilkumar, N.O. Gopal, G. Selvakumar, and Tongmin Sa

8.1 Introduction

Fungal phylum Glomeromycota is a monophyletic group which is conceivably separated from the same ancestor as the Ascomycota and Basidiomycota (Schüssler et al. 2001). Glomeromycota consists of arbuscular mycorrhizal fungi (AMF), which form symbiotic association with more than 80 % of the land plants and the endocytobiotic *Geosiphon pyriformis*. AMF are widely distributed organisms in soil with the widespread mycelia network, but no sexual state has been observed so far. AMF cannot be easily identified as they do not have conspicuous above ground fruiting body and they cannot grown in the absence of plant roots. From the 97,000 of fungal species described so far only 214 species are belong to Glomeromycota, this may be due to the difficulties in the identification and the requirement of specific growth condition.

R. Krishnamoorthy • N. Premalatha • R. Anandham (✉) • N.O. Gopal
Department of Agricultural Microbiology, Agricultural College and Research Institute, Tamil Nadu Agricultural University, Madurai, Tamil Nadu 625 104, India
e-mail: anandhamranga@gmail.com

M. Karthik • M. Senthilkumar
Department of Agricultural Microbiology, Tamil Nadu Agricultural University, Coimbatore, Tamil Nadu 641 003, India

G. Selvakumar • T. Sa (✉)
Department of Agricultural Chemistry, Chungbuk National University, Cheongju 361 763, Republic of Korea
e-mail: tomsa@chungbuk.ac.kr

B.P. Singh, V.K. Gupta (eds.), *Molecular Markers in Mycology*, Fungal Biology,
DOI 10.1007/978-3-319-34106-4_8

AMF spore morphological characters are considered as an important tool for identification and diversity analysis (Del Val et al. 1999). However, some AMF species shared similar spore wall characters and morphology varies with different stages of spore formation (Long et al. 2010). Many markers are widely used for the identification of AMF and diversity analysis, which lead to obtain more precious information. Markers can be at phenotypic level (by Morphological markers), difference in proteins (by Biochemical markers) or difference in the nucleotide sequence of DNA (by Molecular markers). Many molecular markers have been developed during last two decades (Verbruggen et al. 2012; Holland et al. 2014) and molecular markers are more reliable than phenotypic marker. Nonetheless, each molecular marker has its own pros and cons, using more than one molecular marker may give reliable information on AMF community structure rather than using single marker (Dickie and Fitzjohn 2007).

Molecular markers such as PCR-Restriction Fragment Length Polymorphism (PCR-RFLP), Single Stranded Conformation Polymorphism (SSCP), Denaturing Gradient Gel Electrophoresis (DGGE), Temperature Gradient Gel Electrophoresis (TGGE), Terminal-Restriction Fragment Length Polymorphism (T-RFLP), Amplified Fragment Length Polymorphism (AFLP), Random Amplification of Polymorphic DNA (RAPD), Variable Number Tandem Repeat (VNTR), Simple Sequence Repeat (SSR), Single Nucleotide Polymorphism (SNP), Short Tandem Repeat (STR), Single Feature Polymorphism (SFP), Diversity Arrays Technology (DArT) are widely used molecular markers for the identification and diversity analysis of microorganisms. In this Chapter we have summarized the molecular markers which are frequently used for the identification and diversity analysis of AMF present in an ecosystem.

8.2 Arbuscular Mycorrhizal Fungi (AMF)

Mycorrhiza referred as the symbiotic association between plant roots and fungus. The term mycorrhiza was coined by Frank (1885) which is derived from two Greek words '*mycos*' (meaning fungus) and '*rhiza*' (meaning root). So far seven different type of mycorrhiza are described (arbuscular, ecto, ectendo-, arbutoid, monotropoid, ericoid and orchidaceous mycorrhiza), in which arbuscular mycorrhiza and ectomycorrhiza are the most abundant and widely studied (Allen et al. 2003). Arbuscular mycorrhizal fungi (AMF) are obligate biotrophs, which form symbiotic association with more than 80 % of the land plants (Smith and Read 1997). Figure 8.1 illustrates the AMF colonization and formation of vesicles and arbuscules in plant roots. AMF spores feed germinating presymbiontic hyphae through the catabolism of storage lipids, just for a few days. Perception of the host root by the presymbiontic hypha is mediated by plant exudates, which contains stringolactones as a signaling molecule (Akiyama et al. 2005).

Stringolactones stimulate AMF metabolism and hypha branching and in response to the plant exudates mycorrhizal hypha secrets exudates contains Myc

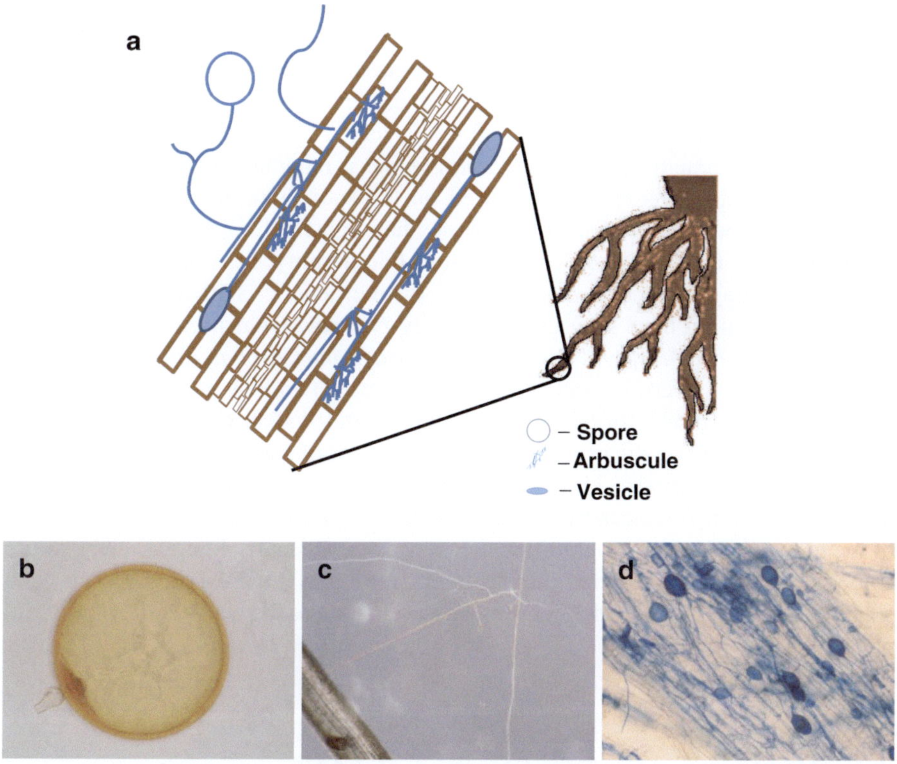

Fig. 8.1 Illustration of AMF root colonization, vesicle, arbuscule and spore formation in roots (**a**), AMF spore (**b**), *in vitro* growth of AMF hypha (**c**), Maize root colonization by AMF (**d**)

factors (Maillet et al. 2011). Plant roots prepare to accommodate the hyphae after receiving the signals from fungal Myc factor. Fungal exudates trigger calcium spiking in the roots through the activation of common symbiotic pathway. The contact between roots and fungus is followed by the adhesion of hyphopodium to the root surface. This triggers the cytoplasm to aggregate and form pre penetration apparatus (PPA) in the epidermal cell and underlying cortical cell. Then intracellular hypha follows the PPA from epidermis to inner cortical cell and develops arbuscule inside the plant cell (Oldroyd 2013). Arbuscules are tree like structure, which involved in nutrient exchange between plant and fungus. There are two different types of arbuscules are formed in plant roots that is depends on the AMF species and host plant. The two types are *Arum* and *Paris*, the *Arum* type is characterized by the formation of intercellular hypha between the cortical cells and the intracellular arbuscules with in the cortical cells. Whereas *Paris* type is characterized by, the fungi form extensive intracellular hyphae coils and arbusculate coils in the root cortex not in intracellular spaces (van Aarle et al. 2005).

8.3 AMF Phylogenetic Position and Classification

Gerdemann and Trappe (1974) proposes first classification of AMF, in that AMF is separated in to 4 different genera under one order *Endogonales*, this provided sound basic in taxonomy. In 1990, Morton and Benny added two more genera to the previously described genera and the order *Endogonales* is renamed as *Glomerales*. These first two classifications are purely based on the spore morphological characters. Later development of molecular techniques has lightened on AMF classification. Schuβler et al. (2001) created new phylum Glomeromycota for AMF under the kingdom of fungi, which is mainly based on the genetic characters. This rDNA based classification have nine genera namely *Glomus*, *Gigaspora*, *Scutellospora*, *Acaulospora*, *Entrophospora*, *Diversispora*, *Paraglomus*, *Archaeospora*, *Giasiphon*. Five years later Sieverding and Oehl (2006) erected two new genera (*Kuklospora* and *Intraspora*) from *Entrophosphora*. Schußler and Walker (2010) proposed new classification based on the near full length sequences of small sub units of rDNA, it contains 18 genera under 11 families. Unfortunately, this classification is recognized as an incomplete because of the unavailability of the living materials for some of the previously described glomeromycota species. Recently International Culture Collection of (Vesicular) Arbuscular Mycorrhizal Fungi (INVAM) updated the classification with 11 family and 25 genera, however some of the genera information are not available.

8.4 Markers for the Identification of AMF

8.4.1 Morphological Markers

Arbuscular mycorrhizal fungi taxonomy was morphologically driven up to the end of last millennium (Oehl et al. 2011). Morphological characters of spores are hierarchically organized (primary, secondary and tertiary characters) with primary characters being defined by the spore wall, germinal walls and germination structure. Species level variation is mostly confined to variation in spore wall layers (Morton et al. 1995). Morphological characters are mainly used for taxonomical classification during 1990 (Morton and Benny 1990). A morphological threshold is defined as the point at which it is impossible to distinguish a strain as a different species (Sokolski et al. 2011). Problems associated with identification of different taxa based on spore morphological markers. Molecular markers are developed, after 20th century, the diversity study and AMF classification are mainly focused on DNA based techniques (Schuβler et al. 2001). Molecular techniques are used to clearly distinguish morphologically more similar spores and species level diversity analysis. Eventhough there are much molecular techniques for diversity studies are developed, still spore taxonomic characters are playing a vital role in AMF diversity analysis (Sturmer and Siqueira 2011). Morphological markers were used to differentiate *Archaeosporaceae*,

Gigasporaceae, *Acaulosporaceae* and *Glomeraceae* isolated from Brazilian forest (Sturmer and Siqueira 2011). Table 8.1 representing the phenotypic characters used for the differentiation of various genus of AMF.

8.4.2 Biochemical Markers

Despite the importance of AMF in soil ecosystem, quantifying their biomass and identification in the field is hindered by various factors. As a consequence, biochemical markers such as ergosterol, chitin and glomalin-related soil protein (GRSP) have been considered for studying AMF (Rosier et al. 2006). Due to the production of ergosterol and chitin by many soil organisms, their usage as AMF marker is somewhat limited. Hence, GRSP produced by mycorrhiza has the potential to serve as a specific biomarker for AMF identification. Several studies were used GRSP for the identification of AMF in different ecosystem (Lovelock et al. 2004; Krivtsov et al. 2004; Krishnamoorthy et al. 2014).

Lipids are used for the identification of microorganisms for last several decades. Due to the different enzymes involved in lipid metabolism, fatty acids display an abundant structural diversity and biological specificity, which provides an integrated and quantitative measurement of community structure in different ecosystem. In particular, neutral lipid fatty acid (NLFA) and phospholipid fatty acid (PLFA) are used as a biochemical marker for the identification of AMF (Ngosong et al. 2012). Lipid biomarker has been used for last two decades for the quantification of AMF biomass in the soil (Olsson et al. 1995; Larsen et al. 1998; Van Aarle and Olsson 2003). Olsson et al. (1995) was the first person to use fatty acid 16:1ω5 for qualification of AMF in soil and plant roots. PLFA fraction of 16:1ω5 was used for assessing viable fungal hyphal biomass and NLFA was used to determine spore storage lipids. Eventhough 16:1ω5 PLFA is proposed to be the origin of AMF, other soil inhabiting bacteria also contains this compound and influences on the biomass estimation. Because of that use of 16:1ω5 PLFA in the field condition is unsuitable and should be restricted to controlled laboratory conditions. However, 16:1ω5 NLFA is not only present in AMF spores but also forms the transport vehicle for carbon between intraradical and extraradical mycelium (Bago et al. 2002). These clearly indicates that the 16:1ω5 NLFA is a good marker to assess AMF biomass in soil compared to 16:1ω5 PLFA. Nonetheless, PLFA biomarker 16:1ω5 was also used for predicting AMF population in arable soils (Beauregard et al. 2010).

8.4.3 Molecular Markers

Genetic marker technology has advanced rapidly compared to other methods. Morphological and biochemical markers are being replaced by DNA based methods of increasing complexity. Early markers were few in number and very

Table 8.1 Morphological markers for the identification of different AMF genus

Family[a]	Genus	Presence of mycorrhizal structures[b]	Spore formation	No. of spore walls	Spore germination
Glomeraceae 	*Glomus*	V, A, H	Glomoid	1	Germ tube (gt) through hypha
	Funneliformis (former *Glomus* Group Aa, '*Glomus mosseae* clade')	V, A, H	Glomoid	1	gt hrough hypha
	Rhizophagus (former *Glomus* Group Ab, '*Glomus intraradices* clade')	V, A, H	Glomoid	1	gt through hypha
	Sclerocystis (basal in former *Glomus* Group Ab)	V, A, H	Glomoid	1	gt through hypha
	Septoglomus	V, A, H	Glomoid	1	gt through hypha
Pacisporaceae 	*Pacispora*	V, A, H	Pacisporoid	2	gt through wall; multiply lobed germ structure
Acaulosporaceae 	*Acaulospora* (including the former *Kuklospora*)	V, A, H	Acaulosporoid	3	gt through wall; mono-(to multiply) lobed, hyaline germ shield (=orb)

Diversisporaceae 	*Corymbiglomus*	NA	NA	NA	NA
	Diversispora (former *Glomus* Group C)	V, A, H	Glomoid	1	gt through hypha
	Otospora	V, A, H	Acaulosporoid	1	NA
	Redeckera	V, A, H	Glomoid	1	gt through hypha
	Tricispora	V, A, H	Entrophosporoid	1	NA
Gigasporaceae 	*Gigaspora*	A, H	Gigasporoid	1	gt through wall; germ warts on inner spore wall layer
	Racocetra	A, H	Scutellosporoid	2	gt through wall; multiply lobed, hyaline germ shield
	Scutellospora	A, H	Scutellosporoid	3	gt through wall; bi-lobed, hyaline, violin-shaped germ shield
	Cetraspora	A, H	Scutellosporoid	3	gt through wall; multiply lobed, hyaline germ shield
	Dentiscutata	A, H	Scutellosporoid	3	gt through wall; brown germ shield with multiple small compartments
	Intraornatospora	NA	NA	NA	NA
	Paradentiscutata	NA	NA	NA	NA
Claroideoglomeraceae	*Claroideoglomus* (former *Glomus* Group B, '*Glomus claroideum* clade')	V, A, H	Glomoid	1	gt through hypha

(continued)

Table 8.1 (continued)

Family[a]	Genus	Presence of mycorrhizal structures[b]	Spore formation	No. of spore walls	Spore germination
Paraglomaceae 	*Paraglomus*	A, H	Glomoid	1	gt through hypha
Archaeosporaceae 	*Archaeospora* (including the former *Intraspora*)	A, H	Bimorph: Acaulo- & Glomo-archaeosporoid	2	gt through wall; germ trunk & gt through hypha
Ambisporaceae	*Ambispora*	V, A, H	Bimorph: Acaulo- & Glomo-ambisporoid	3	Multiply-lobed germ structure & gt through hypha
Geosiphonaceae 	*Geosiphon*	Associated with cyanobacteria	Glomoid	1	gt through hypha
Sacculosporaceae	Sacculospora	V, A, H	Entrophosporoid	3	NA

Modified from Oehl et al. (2011); INVAM
[a]Family with spore structure included
[b]*V* vesicles, *A* arbuscles, *H* hyphae, *NA* Information not available

difficult to assay. However, the development of the technology provides simple methods to assess all genetic variation in the genome. Molecular markers have evolved through several phases. Early methods using non-DNA based methods were replaced by DNA based methods as the technologies for DNA analysis improved. Early DNA hybridization based methods were replaced rapidly following the development of polymerase chain reactions (PCR). Table 8.2 summarized the molecular marker used for the identification and diversity analysis of AMF.

8.4.3.1 Properties Desirable for Ideal DNA Marker

- It must be polymorphic as it is polymorphism that is measured for genetic identification and diversity studies
- A marker should be evenly and frequently distributed throughout the genome
- It should be easy, fast and cheap to detect
- DNA fragment should not possess horizontal gene transferring nature
- DNA fragment should be highly conserved in nature
- Easy and fast assay
- High reproducibility

It is very difficult to find a molecular marker, which would meet all the above criteria. A wide range of molecular methods is available that detects the polymorphism at the DNA level. Depending upon the type of study, to be undertaken, a marker system can be identified which would fulfill at least a few of the above mentioned characters (Weising et al. 1995; Kumar et al. 2009).

8.4.3.2 Nuclear Ribosomal Genes as a Marker

Ribosomal RNA gene is considered as the best marker for studying the phylogenetic relationship among the AMF, it is universal and composed of highly conserved as well as variable domains. In all AMF the ribosome consists of two subunits, the small subunit (SSU) contains a single RNA species (18S rRNA gene) and the large subunit (LSU) contains three RNA species (5S, 5.8S and 28S rRNA gene). SSU and LSU are separated from each other by the Internal Transcribed Spacers (ITS) (Fig. 8.2). The 5S subunit is not normally located within the ribosomal tandem repeat. The ITS, SSU and LSU evolve at different rates; ITS are variable region that change more frequently than the SSU and LSU. This high degree of polymorphism of ITS region can be explained by the relatively low evolutionary pressure on these ITS regions, which are not included in the structure of mature ribosome, but spliced during rRNA maturation. For this reason, the ITS is widely used marker for AMF identification.

Table 8.2 Molecular markers used in the identification of arbuscular mycorrhizal fungi

Molecular markers	DNA fragment targeted	Primer used	AMF family identified	Reference
DGGE	SSU rDNA	AML1/AML2; G101/NS31	*Glomeraceae*	Long et al. (2010)
DGGE	SSU rDNA	NS1 and NS41; Mixture of AM1, AM2, AM3	*Glomeraceae* *Gigasporaceae* *Archaeosporaceae*	Hassan et al. (2011)
DGGE	SSU rDNA	GeoA2 and Geo11; AM1 and NS31; NS31-GC and Glol	*Glomeraceae*	Jiao et al. (2011)
DGGE	SSU rDNA	Nested PCR GeoA2/Geo11, AM1/NS31 and Glo1/NS31-GC	*Glomeraceae*	Krishnamoorthy et al. (2014)
DGGE	SSU rDNA	Nested PCR GeoA2/Geo11, AM1/NS31 and Glo1/NS31-GC	*Glomeraceae*	Krishnamoorthy et al. (2015)
TGGE	SSU rDNA	MH2/MH4 and NS31/AM1	*Glomeraceae*	Sonjak et al. (2009)
PCR-RFLP	ITS	ITS1 and ITS4	*Glomeraceae*	Martin-Laurent et al. (1996)
PCR-RFLP	ITS	ITS1 and ITS4	*Glomeraceae* *Gigasporaceae*	Redecker et al. (1997)
PCR-RFLP	SSU rDNA	LR1 and FLR2	*Glomeraceae*	Jacquot et al. (2000)
PCR-RFLP	SSU rDNA	NS31/AM1; ARCH1311/NS8	*Glomeraceae* *Gigasporaceae*	Vallino et al. (2006)
PCR-RFLP	SSU rDNA	AM1/NS31	*Glomeraceae*, *Gigasporaceae*, *Archaeosporaceae*	Schreiner and Mihara (2009)
T-RFLP	LSU rDNA	FLR3 and FLR4	*Glomeraceae*, *Gigasporaceae*, *Acaulosporaceae*	Mummey and Rillig (2007)
T-RFLP	LSU rDNA	FLR1 and FLR2	*Glomeraceae*, *Gigasporaceae*	Mirás-Avalos et al. (2011)
T-RFLP	LSU rDNA	LR1–FLR2 and FLR3–FLR4	*Glomeraceae* *Gigasporaceae*	Verbruggen et al. (2012)
T-RFLP	LSU rDNA	LR1/FLR2 and FLR3/FLR4	*Glomeraceae* *Gigasporaceae*	Krishnamoorthy et al. (2014)
T-RFLP	LSU rDNA	LR1/FLR2 and FLR3/FLR4	*Glomeraceae* *Gigasporaceae*	Krishnamoorthy et al. (2015)
SSCP	SSU rDNA	Taxon specific primers VALETC, VAGLO, VAACAU, and VAGIGA	*Glomeraceae* *Gigasporaceae* *Acaulosporaceae*	Simon et al. (1993)

Table 8.2 (continued)

Molecular markers	DNA fragment targeted	Primer used	AMF family identified	Reference
SSCP	28S rDNA	LSU-Primers	*Glomeraceae*	Kjoller and Rosendahl (2000)
SSCP	SSU rDNA	VANS1	*Glomeraceae*	Redecker (2002)
SSCP	ITS	Glomus-specific ITS primer	*Glomeraceae*	Kjoller and Rosendahl (2003)
RAPD	Genomic DNA	OPA-02 and OPA-04, OPA-18 and P124, OPA-18 and P124	*Glomeraceae* *Gigasporaceae* *Acaulosporaceae*	Wyss and Bonfante (1993)
RAPD	Genomic DNA	ITS1 and ITS4	*Glomeraceae*	Gomez-Leyva et al. (2008)

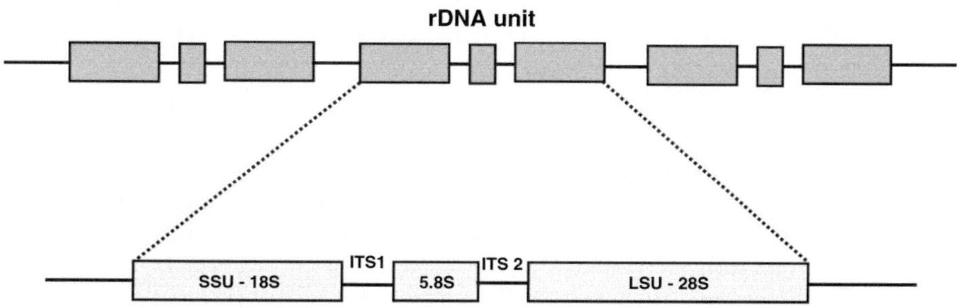

Fig. 8.2 Ribosomal DNA and other marker genes (SSU, LSU and ITS) used for identification arbuscular mycorrhizal fungi

SSU

In prokaryotes the SSU (16S rDNA) has been the prime target for phylogenetic and community analysis. Therefore, it may seem natural to use the corresponding SSU (18S rDNA) for AMF. However, in fungi the SSU is more conserved compared to the prokaryotes. Initially SSU rDNA region was analyzed for species differentiation of AMF. Unfortunately, it was shown that analysis of this region does not allow resolution of closely related AMF species. For example two closely related AMF species *Funneliformis caledonium* and *Funneliformis geosporum* exhibited a difference of only 2–3 bp in SSU region (Rosendahl 2008). New primer has been developed which targets SSU of rDNA and facilitates the effective detection of AMF directly from field samples (Lee et al. 2008). The sequencing of SSU region can reveal high variability of taxon richness and composition between particular ecosystems. The phylogenetic analysis based on the SSU rDNA of Azcomycota and Zygomycota phylum showed only 52 % success rate for species identification (Molitor et al. 2009).

Lanfranco et al. (2001) used SSU to distingue *Gigaspora rosea* from *Gigaspora margarita* and *Gigaspora gigantea*. Based on the comprehensive SSU rDNA analysis

AMF can be separated into a monophyletic clade which is not related to Zygomycetous fungi but shares common ancestry with the Ascomycota and Basidiomycota clade (Schuβler et al. 2001). Recently Kruger et al. (2011) has used SSU for the identification of new AMF species *Acaulospora brasiliensis* and *Acaulospora alpine*.

LSU

Another widely used part of DNA region for AMF species identification is the 5′ region of the LSU. Using an ~ 800 bp of LSU region resulted in more resolved species identification than using ITS region. However the identification of some members of the *Gigasporaceae* was still not successful using LSU region. Geue and Hock (2004) used the 5′ end of the large subunit of the ribosomal RNA gene for the identification of *Acaulospora longula* and *Glomus* subgroup in the plant roots of grassland. New primer was constructed (FLR3 and FLR4) for the amplification of LSU region of AMF present in the roots of *Agrostis capillaris* and *Lolium perenne* (Gollotte et al. 2004).

ITS

AMF has a distribution of ITS fragment lengths concentrated between 550 and 650 bp which found in 96.4% of in silico analyzed AMF sequences (Patreze et al. 2009). These authors recovered 8 AMF genera and 31 species from 422 ITS sequences. ITS regions located in rRNA genes between the 18S and 5.8S coding regions (ITS1) and between the 5.8S and 28S coding regions (ITS2). The subregions ITS1 and ITS2 show high evolution rate and they are typically species specific (Bruns and Shefferson 2004). In addition, higher copies (more than 250) of ITS per cell resulted ITS region as good target to sequence. The taxonomic discrepancies between ITS1 and ITS2 was analyzed separately and also in relation to the full ITS region (Nilsson et al. 2009). It is suggested to use the ITS2 region, because it is as variable and long as ITS1, but ITS2 has a major number of accesses at International Nucleotide Sequence Databases to perform comparisons. Although the ITS sequencing allows species identification, the number of samples required for environmental studies may be higher. The ITS region was used to access the genetic diversity of geographical isolates of *Funneliformis mosseae* (Avio et al. 2009). ITS region was used for the identification of *Glomus* and *Archaeospora* genus in the plants grown in the fly ash pond (Babu and Reddy 2011).

8.4.3.3 Random Amplified Polymorphic DNA (RAPD)

RAPD is based on the amplification of the genomic DNA with a single primer of arbitrary sequences of nine or ten bases of length, which hybridize with sufficient affinity to chromosomal DNA sequences at low annealing temperatures so that they can be used to initiate the amplification of AMF genome region. The amplification is followed by agarose gel electrophoresis, which yields a band pattern that should be characteristic of the particular AMF strain. RAPD is a useful diagnostic tool to analyze isolates of individual species and possibly to distinguish among species (Wyss and Bonfante 1993). RAPD analysis was used to differentiate the *Glomus*, *Gigaspora* and *Acaulospora* genera of AMF (Gomez-Leyva et al. 2008). RAPD have all the advantages of a PCR based marker, with the additional benefit that primers are commercially available and do not require prior knowledge of the target DNA sequence. In addition, RAPD are the ease with which a large number of loci and individuals can be screened. Presence of paralogous PCR product and low reproductively due to low annealing temperature used in the PCR amplification limits the use of RAPD markers.

8.4.3.4 Restriction Fragment Length Polymorphism (PCR-RFLP)

Fragment in a genome where the distance between two restriction sites varies among different species/strains. These fragments are identified by using restriction enzyme digests of chromosomal DNA, and the use of southern blotting for the identification of specific fragment. RFLP markers were the first DNA based genetic markers developed by Botstein et al. (1980). In RFLP analysis total DNA is digested with restriction endonucleases, which resulted in large pool of restriction fragments of different sizes. Since 1980s many restriction endonucleases have been discovered that cleave DNA at specific recognition sites of varying length and sequence. Nonetheless, only few restriction enzymes (*Hin*dIII, *Eco*RI, *Bam*HI) are widely used because they generally provides the best size distribution of DNA fragments and are inexpensive. Restriction enzymes recognition sites are found throughout the genome (both coding and non-coding regions) and cut into the fragments. The restriction fragments are then separated by size on an agarose gel using electrophoresis. Separated fragments on the gel can be visualized by UV after stained with ethidium bromide. Due to the higher number of fragments after digestion, it is very difficult to see all the bands. To overcome this problem, the fragmented DNA is transferred and chemically bound to a nylon membrane by the process called southern blotting. After the fragments are transferred to a nylon membrane the desirable DNA fragment can be visualized by hybridizing the fragment with radioactive or fluorescently labeled DNA probe.

PCR-RFLP was successfully employed to distinguish AMF species from the DNA isolated from spores (Sanders et al. 1995), however when applied to field sample this technique may generate polymorphism in non-target organism. Differentiation was able to make between *F. mosseae* and other *Glomeraceae* family AMF (Avio et al. 2009) by using *Hinf*I restriction enzyme.

Similarly, RFLP analysis was performed to differentiate the AMF fragments obtained from the Oregon vineyards (Schreiner and Mihara 2009). To our knowledge, there are no studies using solely PCR-RFLP to characterize AMF from an ecosystem. Recent report from van Diepen et al. (2011) used this method to select clone's representatives of each AMF type and re-amplified and sequenced.

8.4.3.5 Terminal Restriction Fragment Length Polymorphism (T-RFLP)

Molecular approaches for diversity analysis minimized data variation in the morphological characters that hamper traditional taxonomy and have revealed a considerable unknown AMF diversity in soil (Rosendahl 2008). T-RFLP analysis works based on the restriction digestion of fluorescently labeled PCR product. The digests are separated in electrophoresis and detected in automated sequencer. The results represent the community profile of particular organism in the community. The oligonucleotides are fluorescently labeled that enables generation of fingerprinting microbial community. One or both the primes in PCR are fluorescently labeled at 5′ end. Sequence heterogeneity between rDNA of different species results in different terminal restriction fragment size when PCR amplifications are digested with restriction enzymes (Mummey and Rillig 2007). The choice of primers, target region and restriction enzymes are critical for T-RFLP analysis. Generally internal transcribed spacer (ITS), small subunit (SSU) or large sub units (LSU) are targets to identify the particular fungal community in soil. For AMF, primers ITS4/ITS5 are used for the amplification of ITS region (Lekberg et al. 2007) and AM1/NS31 are used to amplify SSU (Uibopuu et al. 2009). However, primers FLR3/FLR4 are widely used by many authors, which target the LSU of AMF (Koch et al. 2011). Next important thing is the choice of restriction enzymes, which may affect the efficiency of the result. A finest restriction enzyme would result in a wide range of fragments size and show little variation within the species. Even though there are many number of restriction enzymes available in the market *AluI* and *MboI* are widely used by many authors (Barto et al. 2011). The resulting terminal restriction fragments are sized on a sequencing gel or by capillary method. Each fragments may be represent as a single species, because different organisms contain different restriction sites. The complete results give the community profile of the AMF in a particular area. In addition to this, T-RFLP can be used to identify the possible ribotype for the restriction fragments obtained by digestion.

8.4.3.6 Denaturing Gradient Gel Electrophoresis (DGGE) and Temperature Gradient Gel Electrophoresis (TGGE)

Both DGGE and TGGE are the similar marker used for the identification of AMF community structure. These two techniques were first developed to detect point mutation in DNA sequences. Denaturing gradient gel electrophoresis separates nucleic acid fragments by their sequence composition not by their size (Hovig et al. 1991). In this method, chemical denaturants like urea and formamide are used. Gel composed of linear gradient of denaturant is used to separate the amplified product composed of different sequences. This technique was originally developed to detect point mutation on DNA sequences. Based on the G+C content of PCR amplified DNA, sequences of organisms are separated in different gradients of urea and formamide (Kirk et al. 2004). Three hydrogen bonds between G and C require more denaturant to break the bond compared to the two bonds between A and T. As DNA separates, the migration becomes slowdown in the gel. High GC sequences (GC clamp) are attached to the PCR primer to prevent complete strand separation which results in sharp band in the gel (Sheffield et al. 1989). If a mixer of PCR product with different DNA composition is amplified from sample, then DGGE separate them during electrophoresis. The resulting band profile represents the community structure and the approximate populations (represented by each band) and their abundance (represented by band intensity) in the community. TGGE also follow the same method as DGGE except the gradient is temperature instead of chemical gradients.

Use of nested PCR with AM1-NS31 primer set and Glo1-NS31 yielded a high resolution AMF band profile to soil samples including in a polluted soils (Krishnamoorthy et al. 2014, 2015). These two methods have the advantages of being consistent, reproducible, quick and to some extent inexpensive. Multiple samples can also be analyzed simultaneously, making it possible to compare different AMF samples. Steps involved in DGGE and T-RFLP are presented in Fig. 8.3.

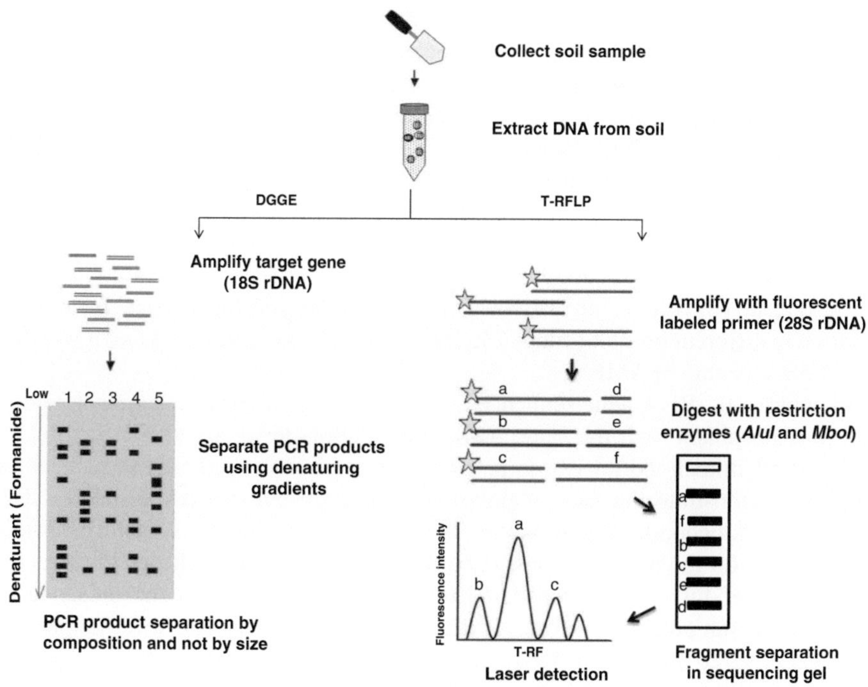

Fig. 8.3 Illustration of DGGE and T-RFLP methodologies (Modified from Nakatsu 2007)

8.4.3.7 Single Strand Conformation Polymorphism (SSCP)

Like DGGE/TGGE, SSCP also depends on the electrophoretic separation based on the differences in DNA sequences and this was originally developed to detect polymorphisms or point mutations in DNA sequences. Taxon specific primers were developed for identifying different AMF genus using SSCP (Simon et al. 1993). In order to detect AMF species, nested PCR based on the sequence differences in the gene coding for the larger ribosomal subunit can be coupled to a method known as SSCP and the differences among AMF species are visualized in polyacrylamide gel with no denaturing conditions. Kjoller and Rosendahl (2000) used SSCP to detect four different species of *Glomeraceae* family in root tissues of four plant species. A study conducted by Alguacil et al. (2010) in arid gypsophilous plant communities SSCP was used to select clones. In such work, representatives of different SSCP pattern were selected for sequencing while the remaining was classified based on SSCP pattern. These authors identified nine fungal types: six belonged to the genus *Glomus* and three to *Acaulospora*. The SSCP method is a very sensitive and reproducible technique that has the potential to be applied successfully in studies in order to analyze the AMF diversity.

8.5 Mitochondrial rDNA as a Marker

Mitochondrial DNA (mtDNA) is non-nuclear DNA which is located in the cell organelles cytoplasm called mitochondria. mtDNA is physically separated from the rest of the cell DNA and so it is relatively easy to isolate for the fungal spores/hypha. Mitochondrial DNA has a long history as a molecular marker that spreads into the era PCR development (Bruns et al. 1989). In metazoan population studies mitochondrial genes have played an important role due to the high variability and complete absence of recombination make the organelle genomes as a unique tool for population studies. Nonetheless some reports show the recombination in fungal mitochondria (Saville et al. 1998). Different modes of inheritance have been reported in different group of fungi (Xu 2005), but it is currently not known whether one of them occurs in AMF.

In contrast to rRNA gene sequences, mtDNA of the AMF were demonstrated to lack polymorphism with the same AMF strains (Borstler et al. 2008). Raab et al. (2005) used mitochondrial ribosomal RNA large subunit (mtLSU) sequences of AMF to demonstrate the lack of polymorphism between two AMF species. The sequenced mitochondrial genome of *Rhizophagus intraradices* confirmed this homogeneity for the whole genome (Lee and Young 2009). In addition, the increasing number of fully sequenced AMF mitochondria highlighted the presence of intergenic region potentially useful for the development of isolate specific markers (Beaudet et al. 2013; de la Providencia et al. 2013).

Although the potential of mitochondrial DNA (mtDNA) as a highly informative genetic marker has been recognized, minimum number of studies have so far been carried out considering mtDNA markers for AMF species, mainly belonging to *Rhizophagus* genera (Borriello et al. 2014). Recently Borriello et al. (2014) developed largest mitochondrial gene coding region for the Cytochrome c Oxidase I (COI), sequence dataset of Glomeromycota. This dataset represents a promising resource for further investigations on mtDNA-based marker for the phylogenetic analysis of Glomeromycota.

8.6 Protein Encoding Gene Marker

Protein-encoding genes usually occur as single copies in genomes, which may be advantageous for phylogenetic analysis. β-tubulin gene might serve as a marker for fungal growth and found to be differently expressed in the symbiotic state of AMF. Moreover, β-tubulin gene sequences have already been characterized for many AMF taxa (Corradi et al. 2004). Eventhough tubulin encoding gene shows a strong separation between fast evolving (Basidiomycota, Ascomycota) and slowly evolving fungi lineages (Glomeromycota, Chytridiomycota) it still has the problem in phylogenetic reconstruction.

The elongation factor 1 (EF 1) and alpha gene (*tef*) are usually present in a single copy and encodes the translation elongation factor that controls the rate and consistency of protein synthesis (Baldauf 1999). Similarly, the actin gene (*act*) encodes actin which also found in a single copy in the majority of the fungi. These *tef* and *act* genes have been used to study the evolutionary relationship between zygomycota and other fungal group (Voigt and Wostemeyer 2003). Therefore use of EF1 and actin genes as the marker for identifying the AMF and finding the relationship between AMF and other fungal group.

Two largest subunits of the RNA polymerase II known as *rpb1* and *rpb2* offers an excellent possibilities for phylogenetic studies. The *rpb1* gene coding protein phylogeny of AMF strongly supports the monophyletic Glomeromycota, however this is not effective for the symbiomycotan clade (comprising Glomeromycetes, Ascomycetes and Basidiomycetes) (Redecker and Raab 2006). All AMF families contain an intron at the same location in their *rpb1* gene and this region is seems to be ideal molecular marker for AMF identification using restriction analysis. This *rpb1* gene stood out for several reasons (1) it has several variable regions useful for species discrimination (2) according to present knowledge it is a single-copy gene in fungi, avoiding problems with paralogues. These properties render this marker very attractive for large-scale diversity surveys of the Glomeromycota. Recently, Stockinger et al. (2014) proved that *rpb1* gene as a valuable tool to study AMF communities under field conditions which is highly likely to facilitate community analyses and contribute to a better understanding of the field ecology of this symbiosis.

Eventhough the protein encoding genes act as an efficient marker for the identification of AMF, some of the protein encoding genes (*rpb1* and actin genes) indicates that the Glomeromycota is closely related to Zygomycota. In future, protein encoding genes may be the promising marker for the phylogenetic relationship among the members of Glomeromycota phylum. However, the sequences of nuclear encoded rDNA are widely used in molecular phylogenetic identification as they can be amplified from small quantities of DNA, due to the high copy number of rRNA genes present in AMF nucleus. Protein encoding genes are less frequently used due to the following reasons (1) difficulties in designing primer across a broad range of taxa (2) heterozygous loci requires cloning (3) paralogy and gene families complicates the use of protein encoding genes as a marker for AMF (Thiery 2010).

8.7 AMF Functional Traits as the Marker

Functional traits of organisms in a community have long been used in examining the dynamics of ecosystem processing. AMF are functionally varied at several levels of systematic integration and vary in many functional characters, including nutrient uptake capabilities, carbohydrate requirements and fungal growth patterns. However, it is not known to what extent current morphological and rDNA-based phylogenetic groupings of AMF parallel with functional characters. It has been proposed to include functional traits of AMF groupings in ecological studies to provide greater understanding into their impacts on ecosystem processes (Gamper et al. 2010).

AMF are mostly characterized using nuclear rRNA, internal transcribed spacers and the 5.8S gene. Eventhough these DNA fragments are useful for AMF identification, these fragments are inappropriate for species identification due to a high degree of sequence variation. Sometimes an intersporal variation in rDNA is much higher than an interspecific variation and sequence overlap across the AMF species are the major problem with the correct identification of species. In addition to protein encoding genes for tubulin, actin and P type H^+ ATPase has already been investigated but with inadequate achievements.

Phosphate transporter genes of the *Glomus* are considered as a marker for AMF identification (Benedetto et al. 2005). In addition, Sokolski et al. (2011) demonstrated that the phosphate transporter gene of AMF allow clear differentiation between the morphologically defined *Glomus* species, they proposed that they can be used as a routine tool for species identification and particularly to distinguish between closely related species. Necessary steps has to be taken in the development and implementation of genetic trait based markers includes (1) construction of nucleotide sequence databases of functional traits genes from broad AMF taxonomic range, (2) development of effective AMF gene specific PCR amplification system to examine sequence diversity in environmental samples, (3) development and optimization of reliable AMF mRNA extraction procedures for analyzing root and soil samples (4) development of suitable high throughput tools to detect and quantify AMF mRNAs (Gamper et al. 2010).

8.8 Conclusion and Future Prospects

The lack of defined biological materials from different AMF lineages represents a practical problem to the establishment of sequence databases, which is necessary for the development of functional gene markers. Further development of efficient AMF pure culture method and free exchange of AMF strains among the scientists are the crucial for the development of markers for AMF identification. Many open questions regarding AMF growth and development could be answered by studying defined pure cultures and isolates, or some times by more in depth characterization of field materials. It is very surprising that many of the fully described AMF species biological materials are seems not available at all, except the voucher specimens. And lack of AMF sequence information available in the databases and the possibility of primer bias towards certain AMF genus are likely to prevent the use of single primer pair with universal application to study AMF diversity. In addition, the high genetic variability of the rRNA genes within the single spores of AMF, new alternative gene marker are needed for the better resolution of closely related species. The future is going to witness the tremendous increases in the development of markers for AMF, which hopefully will help resolve many issues being challenged today.

Acknowledgements This work was partially supported by Tamil Nadu Agricultural University, Coimbatore and Basic Science Research Program through the National Research Foundation of Korea (NRF) funded by the Ministry of Education, Science and Technology (2015R1A2A1A05001885). RK was supported by Department of Science and Technology, Science and Engineering Research Board (DST-SERB) through their National Post Doctoral Fellowship (PDF/2015/000456).

References

Akiyama K, Matsuzaki K, Hayashi H. Plant sesquiterpenes induce hyphal branching in arbuscular mycorrhizal fungi. Nature. 2005;435:824–7.

Alguacil M d M, Lozano Z, Campoy MJ, Roldán A. Phosphorus fertilization management modifies the biodiversity of AM fungi in a tropical savanna forage system. Soil Biol Biochem. 2010;42:1114–22.

Allen MF, Swenson W, Querejeta JI, Egerton-Warburton LM, Treseder KK. Ecology of mycorrhizae: a conceptual framework for complex interactions among plants and fungi. Ann Rev Phytopathol. 2003;41:271–303.

Avio L, Cristani C, Giovannetti SP. Genetic and phenotypic diversity of geographically different isolates of *Glomus mosseae*. Can J Microbiol. 2009;55:242–53.

Babu AG, Reddy MS. Diversity of Arbuscular mycorrhizal fungi associated with plants growing in fly ash pond and their potential role in ecological restoration. Curr Microbiol. 2011;63:273–80.

Bago B, Pfeffer PE, Zipfel W, Lammers P, Shachar-Hill Y. Tracking metabolism and imaging transport in arbuscular mycorrhizal fungi. Plant Soil. 2002;244:189–97.

Baldauf SL. A search for the origins of animals and fungi: comparing and combining molecular data. Am Nat. 1999;154:178–88.

Barto EK, Antunes PM, Stinson K, Koch AM, Klironomos JN, Cipollini D. Differences in arbuscular mycorrhizal fungal communities associated with sugar maple seedlings in and outside of invaded garlic mustard forest patches. Biol Invasions. 2011;13:2755–62.

Beaudet D, Terrat Y, Halary S, de la Providencia IE, Hijri M. Mitochondrial genome rearrangements in glomus species triggered by homologous recombination between distinct mtDNA haplotypes. Genome Biol Evol. 2013;5:1628–43.

Beauregard MS, Hamel C, Nayyar A, St-Arnaud M. Long-Term phosphorus fertilization impacts soil fungal and bacterial diversity but not AM fungal community in Alfalfa. Microb Ecol. 2010;59:379–89.

Benedetto A, Magurno F, Bonfante P, Lanfranco L. Expression profiles of a phosphate transporter gene (GmosPT) from the endomycorrhizal fungus *Glomus mosseae*. Mycorrhiza. 2005;15:620–7.

Borriello R, Bianciotto V, Orgiazzi A, Lumini E, Bergero R. Sequencing and comparison of the mitochondrial COI gene from isolates of Arbuscular Mycorrhizal Fungi belonging to Gigasporaceae and Glomeraceae families. Mol Phylogenet Evol. 2014;75:1–10.

Borstler B, Raab PA, Thiéry O, Morton JB, Redecker D. Genetic diversity of the arbuscular mycorrhizal fungus *Glomus intraradices* as determined by mitochondrial large subunit rRNA gene sequences is considerably higher than previously expected. New Phytol. 2008;180:452–65.

Botstein D, White RL, Skolnick M, Davis RW. Construction of a genetic linkage map in man using restriction fragment length polymorphisms. Am J Hum Genet. 1980;32:314–31.

Bruns TD, Fogel R, White TJ, Palmer JD. Accelerated evolution of a false-truffle from a mushroom ancestor. Nature. 1989;339:140–2.

Bruns TD, Shefferson RP. Evolutionary studies of ectomycorrhizal fungi: recent advances and future directions. Can J Bot. 2004;82:1122–32.

Corradi N, Kuhn G, Sanders IR. Monophyly of b-tubulin and H+-ATPase gene variants in *Glomus intraradices*: consequences for molecular evolutionary studies of AMF genes. Fungal Genet Biol. 2004;41:262–73.

De la Providencia IE, Nadimi M, Beaudet D, Rodriguez Morales G, Hijri M. Detection of a transient mitochondrial DNA heteroplasmy in the progeny of crossed genetically divergent isolates of arbuscular mycorrhizal fungi. New Phytol. 2013;200:211–21.

Del Val C, Barea JM, Azcon-Aguilar C. Diversity of arbuscular mycorrhizal fungus populations in heavy metal contaminated soils. Appl Environ Microbiol. 1999;65:718–23.

Dickie IA, Fitzjohn RG. Using terminal restriction fragment length polymorphism (T-RFLP) to identify mycorrhizal fungi: a methods review. Mycorrhiza. 2007;17:259–70.

Frank AB. Uber di auf werzelsymbiose beruhende Ernahrung gewisser Baume durch unterirdischeplize. Ber Dtsch Bot Ges. 1885;3:128–45.

Gamper HA, Marcel GA, van der Heijden KGA. Molecular trait indicators: moving beyond phylogeny in arbuscular mycorrhizal ecology. New Phytol. 2010;185:67–82.

Gerdemann JW, Trappe JM. Endogonaceae in the pacific Northwest. Mycol Mem. 1974;5:1–76.

Geue H, Hock B. Determination of *Acaulospora longula* and *Glomus* subgroup Aa in plant roots from grassland using new primers against the large subunit ribosomal DNA. Mycol Res. 2004;108:76–83.

Gollotte A, van Tuinen D, Atkinson D. Diversity of arbuscular mycorrhizal fungi colonising roots of the grass species *Agrostis capillaris* and *Lolium perenne* in a field experiment. Mycorrhiza. 2004;14:111–7.

Gomez-Leyva JF, Lara-Reyna J, Hernandez-Cuecas LV, Martinez-Soriano JP. Specific polymerase chain reaction based assay for the identification of the arbuscular mycorrhizal fungus *Glomus intraradices*. J Biol Sci. 2008;8:563–9.

Hassan SED, Boon E, St-Arnaud M, Hijri M. Molecular biodiversity of arbuscular mycorrhizal fungi in trace metal-polluted soils. Mol Ecol. 2011;20:3469–83.

Holland TC, Bowen P, Bogdanoff C, Hart MM. How distinct are arbuscular mycorrhizal fungal communities associating with grapevines? Biol Fertil Soils. 2014;50:667–74.

Hovig E, Smithsorensen B, Brogger A, Borresen AL. Constant denaturant gel-electrophoresis, a modification of denaturing gradient gel-electrophoresis, in mutation detection. Mutat Res. 1991;262:63–71.

Jacquot E, Van Tuinen D, Gianinazzi S, Gianinazzi-Pearson V. Monitoring species of arbuscular mycorrhizal fungi in planta and in soil by nested PCR: Amplification to the study of the impact of sewage sludge. Plant Soil. 2000;226:179–88.

Jiao H, Chen Y, Lin X, Liu R. Diversity of arbuscular mycorrhizal fungi in greenhouse soils continuously planted to watermelon in North China. Mycorrhiza. 2011;21:681–8.

Kirk JL, Beaudette LA, Hart M, Moutoglis P, Klironomos JN, Lee H, et al. Methods of studying soil microbial diversity. J Microbiol Meth. 2004;58:169–88.

Kjoller R, Rosendahl S. Detection of arbuscular mycorrhizal fungi (Glomales) in roots by nested PCR and SSCP (single stranded conformation polymorphism). Plant Soil. 2000;226:189–96.

Kjoller R, Rosendahl S. Molecular diversity of Glomalean (arbuscular mycorrhizal) fungi determined as distinct Glomus specific DNA sequences from roots of field-grown peas. Mycol Res. 2003;105:1027–32.

Koch AM, Antunes PM, Barto EK, Cipollini D, Mummey DL, Klironomos JN. The effects of arbuscular mycorrhizal (AM) fungal and garlic mustard introductions on native AM fungal diversity. Biol Invasions. 2011;13:1627–39.

Krishnamoorthy R, Kim CG, Subramanian P, Kim KY, Selvakumar G, Sa TM. Arbuscular mycorrhizal fungi community structure, abundance and species richness changes in soil by different levels of heavy metal and metalloid concentration. PLoS One. 2015;10(6), e0128784. doi:10.1371/journal.pone.0128784.

Krishnamoorthy R, Kim KY, Kim CG, Sa TM. Changes of arbuscular mycorrhizal traits and community structure with respect to soil salinity in a coastal reclamation land. Soil Biol Biochem. 2014;72:1–10.

Krivtsov V, Griffiths BS, Salmond R, Liddell K, Garside A, Bezginova T, et al. Some aspects of interrelations between fungi and other biota in forest soil. Mycol Res. 2004;108:933–46.

Kruger M, Walker C, Schußler A. *Acaulospora brasiliensis* comb. nov. and *Acaulospora alpine* (Glomeromycota) from upland Scotland: morphology, molecular phylogeny and DNA-based detection in roots. Mycorrhiza. 2011;21:577–87.

Kumar P, Gupta VK, Misra AK, Modi DR, Pandey BK. Potential of molecular markers in plant biotechnology. Plant Omics J. 2009;2(4):141–62.

Lanfranco L, Bianciotto V, Lumini E, Souza M, Morton JB, Bonfante P. A combined morphological and molecular approach to characterize isolates of arbuscular mycorrhizal fungi in *Gigaspora* (Glomales). New Phytol. 2001;152:169–79.

Larsen J, Olsson PA, Jakobsen I. The use of fatty acid signatures to study mycelial interactions between the arbuscular mycorrhizal fungus *Glomus intraradices* and the saprotrophic fungus *Fusarium culmorum* in root-free soil. Mycol Res. 1998;102:1491–6.

Lee J, Lee S, Young PW. Improved PCR primers for the detection and identication of arbuscular mycorrhizal fungi. FEMS Microbiol Ecol. 2008;65:339–49.

Lee J, Young JPW. The mitochondrial genome sequence of the arbuscular mycorrhizal fungus *Glomus intraradices* isolate 494 and implications for the phylogenetic placement of Glomus. New Phytol. 2009;183:200–11.

Lekberg Y, Koide RT, Rohr JR, Aldrichwolfe L, Morton JB. Role of niche restrictions and dispersal in the composition of arbuscular mycorrhizal fungal communities. J Ecol. 2007;95:95–105.

Long LK, Yao Q, Guo J, Yang RH, Huang YH, Zhu HH. Molecular community analysis of arbuscular mycorrhizal fungi associated with five selected plant species from heavy metal polluted soils. Eur J Soil Biol. 2010;46:288–94.

Lovelock CE, Wright SF, Nichols K. Using glomalin as an indicator for arbuscular mycorrhizal hyphal growth: an example from a tropical rain forest soil. Soil Biol Biochem. 2004;36:1009–12.

Maillet F, Poinsot V, Andre O, Pages VP, Haouy A, Gueunier M, et al. Fungal lipochitooligosaccharide symbiotic signals in arbuscular mycorrhiza. Nature. 2011;469:58–63.

Martin-Laurent FA, Dumas-Gaudot E, Franken P, Schlichter U, Antonie JE, Gianinazzi-Pearson V, et al. Differential display reverse transcriptase polymerase chain reaction (DDRT–PCR): a new approach to detect symbiosis-related genes induced in arbuscular mycorrhiza. In: Azcon-Aguilar C, Barea JM. Mycorrhizae in sustainable system from genes to plant development. 1996. p. 195–8.

Mirás-Avalos JM, Antunes PM, Koch A, Khosla K, Klironomos JN, Dunfieldd KE. The influence of tillage on the structure of rhizosphere and root-associated arbuscular mycorrhizal fungal communities. Pedobiol. 2011;54:235–41.

Molitor C, Inthavong B, Sage L, Geremia RA, Mouhamadou B. Potentiality of the cox1 gene in the taxonomic resolution of soil fungi. FEMS Microbiol Lett. 2009;302:76–84.
Morton JB, Benny GL. Revised classification of arbuscular mycorrhizal fungi (Zygomycetes). A new order, Glomales, two new suborders, Glomineae and Gigasporineae, and two new families, Acaulosporaceae and Gigasporaceae, with an emendation of Glomaceae. Mycotaxon. 1990;37:471–91.
Morton JB, Bentivenga SP, Bever JD. Discovery, measurement, and interpretation of diversity in arbuscular endomycorrhizal fungi (Glomales, Zygomycetes). Can J Bot. 1995;73:25–32.
Mummey DL, Rillig MC. Evaluation of LSU rRNA-gene PCR primers for analysis of arbuscular mycorrhizal fungal communities via terminal restriction fragment length polymorphism analysis. J Microbiol Meth. 2007;70:200–4.
Nakatsu CH. Soil microbial community analysis using denaturing gradient gel electrophoresis. Soil Sci Soc Am J. 2007;71:562–71.
Ngosong C, Gabriel E, Ruess L. Use of the signature fatty acid 16:1ω5 as a tool to determine the distribution of arbuscular mycorrhizal fungi in soil. J Lipid. 2012; Article ID 236807. doi:10.1155/2012/236807.
Nilsson RH, Ryberg M, Abarenkov K, Sjökvist E, Kristiansson E. The ITS regions a target for characterization of fungal communities using emerging sequencing technologies. FEMS Microbiol Lett. 2009;296:97–101.
Oehl F, Sieverding E, Palenzuela J, Ineichen K, da Silva GA. Advances in Glomeromycota taxonomy and classification. IMA Fungus. 2011;2:191–9.
Oldroyd GED. Speak, friend, and enter: signalling systems that promote beneficial symbiotic associations in plants. Nat Rev Microbiol. 2013;11:252–63.
Olsson PA, Baath E, Jakobsen I, Soderstrom B. The use of phospholipid and neutral lipid fatty acids to estimate biomass of arbuscular mycorrhizal fungi in soil. Mycol Res. 1995;99:623–9.
Patreze CM, Paulo EN, Martinelli AP, Cardoso EJB, Tsai SM. Characterization of fungal soil communities by F-RISA and arbuscular mycorrhizal fungi from *Araucaria angustifolia* forest soils after replanting and wildfire disturbances. Symbiosis. 2009;48:164–72.
Raab PA, Brennwald A, Redecker D. Mitochondrial large ribosomal subunit sequences are homogeneous within isolates of *Glomus* (arbuscular mycorrhizal fungi, Glomeromycota). Mycol Res. 2005;109:1315–22.
Redecker D, Raab P. Phylogeny of the Glomeromycota (arbuscular mycorrhizal fungi): recent developments and new gene markers. Mycologia. 2006;98:885–95.
Redecker D, Thierfelder H, Walker C, Werner D. Restriction analysis of PCR-amplified internal transcribed spacer of ribosomal DNA as a tool for species identification in different genera of the order Glomales. Appl Environ Microbiol. 1997;63:1756–61.
Redecker D. Molecular identification and phylogeny of arbuscular mycorrhizal fungi. Plant Soil. 2002;244:67–73.
Rosendahl S. Communities, populations and individuals of arbuscular mycorrhizal fungi. New Phytol. 2008;178:253–66.
Rosier CL, Hoye AT, Rillig MC. Glomalin-related soil protein: assessment of current detection and quantification tools. Soil Biol Biochem. 2006;38:2205–11.
Sanders IR, Alt M, Groppe K, Boller T, Wiemken A. Identification of ribosomal DNA polymorphisms among and within spores of the Glomales: application to studies on the genetic diversity of arbuscular mycorrhizal fungal communities. New Phytol. 1995;130:419–27.
Saville BJ, Kohli Y, Anderson JB. mtDNA recombination in a natural population. Proc Natl Acad Sci U S A. 1998;95:1331–5.
Schreiner RP, Mihara KL. The diversity of arbuscular mycorrhizal fungi amplified from grapevine roots (*Vitis vinifera* L.) in Oregon vineyards is seasonally stable and influenced by soil and vine age. Mycologia. 2009;101:599–611.
Schüssler A, Schwarzotta D, Walker C. A new fungal phlyum, the Glomeromycota: phylogeny and evolution. Mycol Res. 2001;105:1413–21.
Schußler A, Walker C. The Glomeromycota: a species list with new families and genera. Arthur Schüßler & Christopher Walker, Gloucester. Published in December 2010 in libraries at The

Royal Botanical Garden Edinburgh, The Royal Botanic Garden Kew, Botanische Staatssammlung Munich, and Oregon State University. 2010. Electronic version freely available online at www.amf-phylogeny.com.
Schußler A, Schwarzott D, Walker C. A new fungal phylum, the Glomeromycota: phylogeny and evolution. Mycol Res. 2001;105:1413–21.
Sheffield VC, Cox DR, Lerman LS, Myers RM. Attachment of a 40-base-pair G + C-rich sequence (GC-clamp) to genomic DNA fragments by the polymerase chain reaction results in improved detection of single-base changes. Proc Natl Acad Sci. 1989;86:232–6.
Sieverding E, Oehl F. Revision of Entrophospora and description of Kuklospora and Intraspora, two new genera in the arbuscular mycorrhizal Glomeromycetes. J Appl Bot Food Qual. 2006;80:69–81.
Simon L, Levesque RC, Lalonde M. Identification of endomycorrhizal fungi colonizing roots by fluorescent single-strand conformation polymorphism-polymerase chain reaction. Appl Environ Microb. 1993;59:4211–5.
Smith SE, Read DJ. Mycorrhizal symbiosis. London: Academic Press; 1997.
Sokolski S, Dalpe Y, Piche Y. Phosphate transporter genes as reliable gene markers for the identification and discrimination of arbuscular mycorrhizal fungi in the genus glomus. Appl Environ Microb. 2011;77:1888–91.
Sonjak S, Beguiristain T, Leyval C, Regvar M. Temporal temperature gradient gel electrophoresis (TTGE) analysis of arbuscular mycorrhizal fungi associated with selected plants from saline and metal polluted environments. Plant Soil. 2009;314:25–34.
Stockinger H, Peyret-Guzzon M, Koegel S, Bouffaud M-L, Redecker D. The largest subunit of RNA polymerase II as a new marker gene to study assemblages of arbuscular mycorrhizal fungi in the field. PLoS One. 2014;9, e107783. doi:10.1371/journal.pone.0107783.
Sturmer SL, Siqueira JO. Species richness and spore abundance of arbuscular mycorrhizal fungi across distinct land uses in Western Brazilian Amazon. Mycorrhiza. 2011;21:255–67.
Thiery O. Molecular markers from the mitochondrial genome of arbuscular mycorrhizal fungi (Glomeromycota): evolutionary dynamics and application. University of Basel, Faculty of Science. PhD thesis 2010. University of Basel, Faculty of Science.
Uibopuu A, Moora M, Saks U, Daniell T, Zobel M, Opik M. Differential effect of arbuscular mycorrhizal fungal communities from ecosystems along management gradient on the growth of forest understorey plant species. Soil Biol Biochem. 2009;41:2141–6.
Vallino M, Massa N, Lumini E, Bianciotto V, Berta G, Bonfante P. Assessment of arbuscular mycorrhizal fungal diversity in roots of Solidago gigantean growing in a polluted soil in Northern Italy. Environ Microb. 2006;8:971–83.
Van Aarle IM, Cavagnaro TR, Smith SE, Smith FA, Dickson S. Metabolic activity of *Glomus intraradices* in *Arum* – and *Paris* -type arbuscular mycorrhizal colonization. New Phytol. 2005;166:611–8.
Van Aarle IM, Olsson PA. Fungal lipid accumulation and development of mycelial structures by two arbuscular mycorrhizal fungi. Appl Environ Microbiol. 2003;69:6762–7.
Van Diepen LTA, Lilleskov EA, Pregitzer KS. Simulated nitrogen deposition affects community structure of arbuscular mycorrhizal fungi in northern hardwood forests. Mol Ecol. 2011;20:799–811.
Verbruggen E, Van der Heijden MGA, Weedon JT, Kowalchuk GA, Roling WF. Community assembly, species richness and nestedness of arbuscular mycorrhizal fungi in agricultural soils. Mol Ecol. 2012;21:2341–53.
Voigt K, Wostemeyer J. Phylogeny and origin of 83 Zygomycetes from all 54 genera of the Mucorales and Mortierellales based on combined analysis of actin and translation elongation factor EF-1 K genes. Gene. 2003;270:113–20.
Weising K, Nybom H, Wolff K, Meyer W. DNA fingerprinting in plants and fungi (ed. Arbor A). Boca Raton: CRC Press; 1995. p. 1–3.
Wyss P, Bonfante P. Amplification of genomic DNA of arbuscular-mycorrhizal (AM) fungi by peR using short arbitrary primers. Mycol Res. 1993;97:1351–7.
Xu J. Evolutionary genetics of fungi. Norfolk: Horizon Bioscience; 2005.

Chapter 9
Detection of Edible Mushroom Species by Using Molecular Markers

Selima Khatun, Aminul Islam, Kamala Gupta, and Bhaskar Gupta

9.1 Introduction

In today's world with burgeoning population, the growth and cultivation of edible mushrooms hold great significance because of their nutritive and therapeutic properties. They have low fat, high fiber and all essential amino acids content and with the exception of iron, contain almost all the important minerals (Sadler 2003). Mushrooms are rich source of neutraceuticals (Çaglarirmak 2007; Elmastas et al. 2007; Khatun et al. 2012, 2015) and their bioactive properties have been proven by the scientific community (Lindequist et al. 2005; Khatun et al. 2013).

Classification system for gilled fungi and their allies (Basidiomycota) is increasingly relying on molecular data. Morphological information has been shown to be of limited value for fungal systematics due to their inherent simplicity, evolutionary convergence, parallelisms, and phenotypic plasticity (Hofstetter et al. 2002). Few

S. Khatun
Department of Botany, Government General Degree College, Singur, Hooghly 712409, India

A. Islam
Department of Research and Development and quality control, Cymbio Pharma Pvt. Ltd., Yelahanka, Venkatala, Bangalore 560064, India

K. Gupta (✉)
Department of Botany, Government General Degree College, Singur, Hooghly 712409, India

Department of Biological Sciences, Presidency University, 86/1 College Street, Kolkata 700073, India
e-mail: kamalagupta@gmail.com

B. Gupta (✉)
Department of Zoology, Government General Degree College, Singur, Hooghly 712409, India

Department of Biological Sciences, Presidency University, 86/1 College Street, Kolkata 700073, India
e-mail: bhaskarzoology@gmail.com

B.P. Singh, V.K. Gupta (eds.), *Molecular Markers in Mycology*, Fungal Biology, DOI 10.1007/978-3-319-34106-4_9 wild mushrooms are deadly poisonous. Morphologically some deadly poisonous mushrooms look like edible species, for example *Volvariella volvacea* (edible) may be confused with *Amanita phalloides*, which is deadly poisonous (Tibuhwa 2013). Therefore, correct mushroom identification is very important to avoid harm that may be a result of eating poisonous mushrooms. In different parts of the world, information on how to recognize and differentiate between edible and non-edible mushroom is largely based on culture traditions, and vernacular naming system's 'folk taxonomy'. Folk taxonomy is distinguished from scientific taxonomy in that it remains within social relations and hence nonuniversal (Tibuhwa 2012, 2013).

Taxonomic and phylogenetic studies of Basidiomycota have been based mainly on the analysis and comparison of morphological characters like the shape, size and color of caps and gills (Lee et al. 2006; Tibuhwa et al. 2012). These methods for identifying mushrooms using morphological characteristics are subtle, hence unreliable (Lian et al. 2008).

More than two decades ago, the magnitude of fungal diversity was estimated conservatively to be at least 1.5 million species (Hawksworth 1991). Of the 1.5 million estimated fungi, 140,000 species produce fruiting bodies of sufficient size and suitable structure to be considered as macrofungi and are popularly called 'mushrooms'. Of these, about 7000 species are considered to posses varying degrees of edibility, and more than 3000 species are regarded as prime edible mushrooms. To date, only 200 of them are experimentally grown, 100 economically cultivated, approximately 60 commercially cultivated, and about 10 have reached an industrial scale of production in many countries (Chang and Mshigeni 2004).

Mushroom mycelia and spores are often microscopic and usually filamentous with very few phenotypic markers that can be used to differentiate between individuals in a population. In Indian context, all edible mushrooms other than the common button mushroom, *Agaricus* are grouped under the specialty mushrooms (Sharma 1997). With current advances in biotechnology, molecular genetic markers have been used for rapid identification and have superseded the traditional method (Froslev et al. 2005; Moreau et al. 2006; Fonseca et al. 2008). Molecular markers including rapid amplified DNA (RAPD) markers, restriction fragment length polymorphic (RFLP) markers and microsatellites have all been employed to discriminate different kinds of organisms including mushrooms (Barroso et al. 2000; Vos et al. 1995).

The objective of the review work is to gain proper identification knowledge of the mushrooms through molecular markers, which may provide direction towards domestication and commercialization of the wild species for economic benefits apart from aiding molecular taxonomy.

9.2 Common Cultivated Mushrooms

Although there are over 300 genera of mushrooms and related fleshy basidiomycetes, only a few species of these fungi are cultivated commercially. This may be due to the fact that many of them are mycorhizal and may not sporulate in the absence of the host. But many saprophytic species have been amenable to cultivation. Some of the more common cultivated species listed here (Table 9.1) are the button mushroom (*Agaricus bisporus*), which was widely cultivated in Europe before being exported to North America by the settlers; the Shiitake mushroom (*Lentinus edodes*), which is grown for centuries in China and other Oriental countries and the oyster mushroom (*Pleurotus ostreatus*), which were collected as wild specimens from forests in Florida and later actively cultivated in several countries around the world (Chakravarty 2011). Also grown are the oriental 'Enoke' or velvet stem mushroom (*Flammulina velutipes*) whose major production is in Japan, the paddy straw mushroom (*Volvariella volvacaea*) and ear fungus (*Auricularia auricula*), which have great medicinal value. Other cultivated mushrooms are the Reishi mushroom (*Ganoderma lucidum*), which is used as an alternative medicine and also as flavouring agent in Japan; the Nameko (*Pholiota nameko*) grown in the Orient and *Tremelia fuciformis* or white jelly fungi that is grown for use as food supplements in Taiwan. Varieties of *A. bisporus* that are grown commercially include the 'crimini' and 'portabello'. Truffles (*Tuber* species) live in close mycorhizal association with roots of specific trees. They are considered a food delicacy and rated as one of the most expensive natural food in the world (Trappe et al. 2007).

9.3 Molecular Markers

9.3.1 Random Amplified Polymorphic DNA (RAPD)

Random Amplified polymorphic DNA (RAPD) has been successfully applied in the determination of genetic diversity in several mushroom breeding materials intended for crossing (Khush et al. 1991). This is because RAPD technique is simple yet efficient and it requires no prior sequence knowledge (Karp 1997). However, the RAPD technique has proved not to be reproducible especially between laboratories as it is highly influenced by experimental conditions (Jones et al. 1997; Virk et al. 2000; Staub and Serquen 1996). The preferential amplification of DNA fragments also masks relatedness between taxa or populations and limit reproducibility (Mueller and Wolfenbarger 1999).

RAPD has some advantages, with efficiency to generate a large number of markers for genomic mapping (Williams et al. 1990; Welsh and Mclelland 1990; Manaf et al. 2006). Without any previous knowledge about the organism genetics, the requirement of small amount of DNA, the quickness, simplicity and reproducibility in the data acquisition, the low cost and accessibility of this technology and the potential automation, with the possibility of being used in programs of quality control as HACCP (Hazard Analysis Critical Control Point) for controlling and monitoring of critical points of contamination in a process, make RAPD a significant molecular tool for studying genetic diversity in plethora of organisms. *Pleurotus ostreatus* and *P. sajorcaju* were characterized genetically by differentiation through RAPD method (Fonseca et al. 2008). The number of amplifications for each primer

Table 9.1 List of few common commercial mushrooms

Common name	Latin name	Origin	Figure
Button mushroom	*Agaricus bisporus*	Europe	
Shiitake mushroom	*Lentinus edodes*	China	
Oyster mushroom	*Pleurotus ostreatus*	Florida	
Velvet stem mushroom	*Flammulina velutipes*	Japan	

Table 9.1 (continued)

Common name	Latin name	Origin	Figure
Paddy straw mushroom	*Volvariella volvacea*	China, Japan	
Ear fungus	*Auricularia auricula*	China, Japan	
Reishi mushroom	*Ganoderma lucidum*	Japan	

(continued)

Table 9.1 (continued)

Common name	Latin name	Origin	Figure
White jelly fungi	*Tremella fuciformis*	Taiwan	
Nomeko mushroom	*Pholiota nameko*	China, Japan	
Truffle	*Tuber aestivum*	Europe, New Zealand, Australia	

From: Chakravarty (2011)

varied from 1 to 12, totaling 87, of which 77 were polymorphic and 10 were monomorphic (present in both species) (Table 9.2). DNA fingerprint patterns might be useful in identifying the species and as an aid to quality control (Cheng et al. 2000). Capelari and Fungaro (2003) analyzed the genetic variability by RAPD of isolates of *Pleurotus cystidiosus* and *P. smithii* which indicated that the criteria used to separate the two species are unsatisfactory and *P. smithii* should be considered a synonym of *P. cystidiosus*. Shnyreva et al. (2003) studied *P. ostreatus* and *Agaricus bisporus*, using molecular markers for differentiation, which allowed differentiating

Table 9.2 Results of the RAPD polymorphism with number of detected bands for each tested primer in oyster mushroom species

Primer	Species	Sequence 3'-5'	G+C (%)	Amplification	Polymorphism	N° Band
AC-04	*P.ostreatus*	ACGGGACCTO	70	Yes	Yes	10
AC-04	*P. sajor-caju*	ACGGGACCTG	70	Yes	Yes	3
AC-09	*P. ostreatus*	AGACGGTACC	60	No	–	–
AC-09	*P. sajor-caju*	AGACGGTACC	60	No	–	–
AD-08	*P. ostreatus*	GGCAGGCAAG	70	Yes	Yes	5
AD-08	*P. sajor-caju*	GGCAGGCAAG	70	Yes	Yes	1
AD-09	*P. ostreatus*	TCGCTTCTCC	60	Yes	Yes	3
AD-09	*P. sajor-caju*	TCGCTTCTCC	60	Yes	Yes	2
B-02	*P. ostreatus*	TGATCCCTGG	60	No	–	–
B-02	*P.sajor-caju*	TGATCCCTGG	60	No	–	–
B-05	*P. ostreatus*	GTAGACCCGT	60	No	–	–
B-05	*P- sajor-caju*	GTAGACCCGT	60	No	–	–
B-07	*P. ostreatus*	CCTTGACGCA	60	No	–	–
B-07	*P. sajor-caju*	CCTTGACGCA	60	No	–	–
B-14	*P. ostreatus*	ACCCCCGAAG	70	Yes	Yes	5
B-14	*P.sajor-caju*	ACCCCCGAAG	70	Yes	Yes	5
B-15	*P. ostreatus*	GGACCCAACC	70	Yes	No	6
B-15	*P. sajor-caju*	GGACCCAACC	70	No	–	–
D-05	*P. ostreatus*	TCTGGTGAGG	60	No	–	–
D-05	*P. sajor-caju*	TCTGGTGAGG	60	No	–	–
D-06	*P. ostreatus*	TTGGCACGGG	70	Yes	Yes	6
D-06	*P. sajor-caju*	TTGGCACGGG	70	No	–	–
D-20	*P. ostreatus*	CTGGGGACTT	60	Yes	Yes	7
D-20	*P. sajor-caju*	CTGGGGACTT	60	Yes	Yes	2
G-02	*P. ostreatus*	GGCACTGAGG	70	Yes	Yes	6
G –02	*P. sajor-caju*	GGCACTGAGG	70	Yes	Yes	2
G –06	*P. ostreatus*	GGTCTACACC	60	No	–	–
G –06	*P. sajor.caju*	GGTCTACACC	60	No	–	–
G –07	*P. ostreatus*	GAACCTGCGG	70	Yes	No	2
G –07	*P. sajor-caju*	GAACCTGCGG	70	No	–	
G –08	*P. ostreatus*	TCACGTCCAC	60	Yes	No	2
G –08	*P. sajor-caju*	TCACGTCCAC	60	No	–	–
G –09	*P. ostreatus*	CCGAGGGGTT	70	No	–	–
G –09	*P. sajor-caju*	CCGAGGGGTT	70	No	–	–
S-03	*P. ostreatus*	CAGAGGTCCC	70	Yes	Yes	12
S-03	*P. sajor-caju*	CAGAGGTCCC	70	No	–	–
S-17	*P.ostreatus*	TGGGGACCAC	70	No	–	–
S-17	*P. sajor-caju*	TGGGGACCAC	70	Yes	Yes	8
Total						

From Fonseca et al. (2008)

groups of genetically similar and distant strains. *P. ostreatus* (Fig. 9.1) showed a higher genetic variation while *A. bisporus* strains showed a higher level of homology. Most taxonomic and phylogenetic studies of Basidiomycota have been based on the analysis of morphological characters. Recently, relationships among species in several genera of Basidiomycota have often been established by amplification of nuclear sequences by Polymerase Chain Reaction (Pringle et al. 2000; Bos 1996). RAPD analysis was first developed to detect polymorphism between organisms, despite the absence of sequence information, to produce genetic markers, and to construct genetic maps (Williams et al. 1990). Genetic diversity of mushrooms have been determined previously using molecular markers especially RAPD (Ravash et al. 2009; Staniaszek et al. 2002) and this technique is used to assess the genetic diversity among 37 *Pleurotus* species of mushrooms and it was observed that this technique provides better discrimination than morphological analysis (Stajic et al. 2005). Sometimes morphological grouping differ compared to molecular/genomic relationship among the species (Pei-Sheng et al. 1999). Therefore, RAPD technique can be used to interpret overall genetic relatedness and dissimilarity (Ravash et al. 2009). RAPD markers have been used to fingerprint commercial and wild strain of *A. bisporus*, while sequence-characterised amplified sequences (Paran and Michelmore 1993) have been used to study the inheritance of cap colour in the course of *Agaricus* breeding programmes (Loftus et al. 2000). Lucia et al. (2001) showed that the use of RAPD markers reveal the high level of genetic homology of commercial strains of *A. bisporus*, and allows, at the same time, to distinguish between them. Molecular marker techniques like RAPD, potentially linked to the agronomic character 'mushroom weight', has been identified by bulked segregant analysis. DNA fingerprinting techniques that recognize DNA fragmentation patterns from a specific DNA sample by randomly primed polymerase chain reaction (PCR), restriction endonuclease treatment, or a combination thereof, have proven their capacity to verify closely related species or strains belonging to the same species (Zervakis et al. 2001; Lopandic et al. 2005). The RAPD analysis has proven to be useful in discrimination, characterization and differentiation of the fungal cultivars (Agarwal et al. 2013).

9.3.2 Internal Transcribed Spaces (ITS) and Nuclear Large Subunit (nLSU) Markers

Many plant biologists have mainly focused their investigation on nucleotide sequences of the ITS located between the nuclear rDNA 18S and 28S subunit genes, and made it possible to determine the relationships between fungal species from the genus *Pleurotus* (Molcalvo et al. 1995). Ribosomal RNA genes exist in genomes as multiple copies arranged in tandem repeat along one or more chromosomes (Fig. 9.2).

Several features of rDNA make it appropriate for systematic and phylogenetic studies. First, this region of the genome is well characterized and conserved. Many

Fig. 9.1 Morphological features of *Pleurotus ostreatus*

primers already are available to amplify regions of the rDNA repeat that would supply sequence data for a wide range of taxa (White et al. 1990). Second, substantial research has been done on rDNA from many fungi because of which ample datasets are available for reference. Additionally, different regions of rDNA evolve at variable rates, which can be used to investigate fungal relationship at different taxonomic levels (Bruns et al. 1991).

Comparative analysis of coding and non-coding regions of ribosomal DNA has become a widely used tool for construction of phylogenetic trees of many organisms including mushrooms. ITS is the proposed standard bar-coding marker for fungi (Schoch et al. 2012). The ITS region is perhaps the most extensively sequenced DNA region in fungi. This region has higher degree of variation than other genetic regions of rDNA (Gardes and Bruns 1993) and are polymorphic thus provide sequence variability that allows distinguishing among different species or strains of mushrooms (Martin et al. 2004). Hussein et al. (2014) used nucleotide sequence data from the ITS region of the nuclear ribosomal and the nLSU gene to characterize selected Tanzanian SWEM (Saprophytic Wild Edible Species). The electropherogram of genomic DNA, PCR products of ITS and LSU regions are shown in Fig. 9.4. Mushroom species belonging to genera *Lentinus*, *Polyporus*, *Panus*, *Pluteus* and *Macrolepiota* form a monophyletic clade composing of four and three distinct subclades for ITS and LSU data set, respectively. This is because they belong to the same class Agaricomycetes. However, in LSU analysis the genera *Lentinus*, *Polyporus* and *Panus*, grouped together with low support while *Pluteus* and *Microlepiota* formed a different subclade with high support. Both ITS and LSU delineated the SWEM taxa to the six genera. However, the obtained support values showed that ITS sequences have the highest possibility of successful

Fig. 9.2 A schematic representation of the location of ITS region (From Daniel 2012)

delineating the studied SWEM to species level than LSU (Tibuhwa et al. 2012). Moreover, these two molecular markers have thus contributed towards understanding the taxonomic ambiguities, which have been prevailing for quite sometimes in genus *Panus*.

Schoch et al. (2012) also noted that the ITS regions have the highest probability of successful identification for the broadest range of fungi, with the most clearly defined barcode gap between inter- and intraspecies variation. Nevertheless, they found nuclear ribosomal large subunit giving superior species resolution in some taxonomic groups, such as the early diverging lineages and the Ascomycete yeasts, but slightly inferior to the ITS. Thus Schoch et al. (2012) observed that for identification of mushrooms and other fungi, a combination of two markers (ITS and LSU) is the best.

The ITS region of the ribosomal DNA (rDNA) has also been widely used for the phylogenetic identification of mushrooms at both the species and genus level (Sanchez-Ballesteros et al. 2000). Different regions of rDNA also evolve at variable rates and this makes them suitable for investigating fungal relationship at different taxonomic levels (Bruns et al. 1991). The ribosomal RNA genes (rDNA) of fungi are located in a single chromosome and are present as repeated subunits of a tandem array of transcribed and non-transcribed stretches of DNA, which appear highly conserved (Wipf et al. 1999). ITS is referred to a piece of non-functional RNA situated between structural ribosomal RNAs on a common precursor transcript. Sequence comparison of the ITS region is widely used in taxonomy and molecular phylogeny because it is easy to amplify even from minute quantities of DNA and has a high degree of variation between closely related species. This ITS region is most widely used to sequence the DNA region in fungi. It has typically been most useful for molecular systematic at the species level and within the species.

Singh et al. (2003) performed molecular characterization of specialty mushroom collected from germplasm accessions from Rajasthan using DNA fingerprinting and rRNA gene sequencing and added two new additions to the Indian Basidiomycetes biodiversity. This is the first kind of report documented from India on molecular characterization of specialty mushrooms. In the study by Sasidhara and Thirunalasundari (2012), the rDNA-ITS fragment sequencing was attempted to identify wild mushrooms, which again is the first report from India as far as wild mushrooms are concerned. Molecular identification results of the ITS sequences of 5.8 rRNA gene and published mushroom records of India validate that *Agaricomycetes* sp. India-01 is a new addition to the Indian Agaricomycetes.

9.3.3 Restriction Fragment Length Polymorphic (RFLP) Markers

RFLPs have been used for analysis of genetic diversity of fungal species because of their specificity and codominant nature (Chyi et al. 1992). However, the RFLP analysis generates relatively small numbers of polymorphisms and is therefore not suitable for studying new or alternative crops such as wild mushrooms where little prior data is available (Pradhan et al. 1992; Lanner et al. 1997). Castle et al. (1987) used RFLP technique for genotyping of *Agaricus brunnescens*. Ma and Luo (2002) used ITS-RFLP in genotype identification in the genus *Pleurorus*. Restriction polymorphisms have been used to discriminate species and strains of fungi as well as other biological taxa. One approach is to digest genomic DNA with a restriction enzyme and directly examine the resulting bands in agarose or polyacylamide gels after electrophoresis. Depending upon the size of the genome and the frequency of restriction recognition sites in the genome, it may be possible to directly compare digests of whole genomic DNAs from different species/strains. In complex genomes such as those in fungi, this direct comparison can only detect differences in high copy number DNA molecules, e.g., the ribosomal DNA genes and mitochondrial DNA. For RFLPs detected by Southern hybridization with probes targeted to single copy DNA markers, the interpretations are straightforward and data can be used in a variety of ways. Single copy RFLP markers are excellent for addressing population and evolutionary genetic questions in diploid or dikaryotic fungi (Xu et al. 1997, 1998).

9.3.4 Amplified Fragment Length Polymorphic (AFLP) Markers

The development of AFLP method has had a significant impact in its relatively short history. AFLP is a powerful method for fingerprinting strains and for generating a large number of dominant markers for the analysis of genetic crosses (Vos et al. 1995) AFLP has several powerful advantages over the other molecular marker methods. Comparatively many more fragments can be generated and analyzed in a simple reaction. It can detect restriction site variations as well as insertions and deletion within a genomic region. Therefore, AFLP can provide an almost limitless set of genetic markers. In addition, the fragments are stable and highly reproducible since they are amplified with two specific primers under stringent conditions.

AFLP is a highly accurate method to detect polymorphisms among individuals, populations, and independently evolving mushroom lineages (Mueller and Wolfenbarger 1999). The visible polymorphism of AFLP fragments is primarily generated through variations in restriction enzymes sites, and the incorporation of PCR allows for rapid and efficient marker generation. AFLP technique has widely been used to study mushroom lineages including *Pleurotus ostreatus* (Zhuo et al.

2006; Zheng et al. 2007). Terefework et al. (2001) analyzed the suitability of AFLP markers to evaluate genetic variability in mushroom lineages at species level. AFLP markers showed that the wild and cultivated *Pleurotus* species were closely related (Daniel 2012). A total of 643 AFLP loci were generated from 84 samples of *Pleurotus* species using five primer combinations (Table 9.3). The primer combinations used in Table 9.3 produced 330 polymorphic loci across all the species accounting for 51 % of the total scorable loci. The number of scorable loci generated by each AFLP primer pair varied from 20 to 228. The number of polymorphic loci for each primer pair varied from 16 to 116. The loci ranged in size from 51 to 497 bp as generated by GeneMapper version 4.1. The number of loci varied for different primer combinations. The primer combination of 5′ Eco+AGG- Mse+CTC 3′ gave the smallest number of both scorable (20) and polymorphic (16) loci, respectively while 5′ Eco+AAC- Mse+CTG 3′ gave the highest number of both scorable (228) and polymorphic (116) loci. Five primer pairs that were used generated an average of 7 scorable and 4 polymorphic loci across all the 84 species studied. A dendrogram based on Nei's genetic distance (D) is illustrated in Fig. 9.3. The dendrogram clustered 71 wild and 13 cultivated *Pleurotus* species into 3 major clades. Clade I and II consisted mainly of wild species (KK, MK, AS) with bootstrap values of 38 % and 67 % respectively. The cultivated species (JK) formed a distinct cluster with a bootstrap support of 66 %. Distribution of the wild species within each cluster did not correspond to their geographical origin.

AFLP markers have proven to be more reliable compared to other molecular tools for genotyping mushroom lineages. AFLP technique has been successfully applied to discriminate the genomes of various mushrooms such as *Pleurotus ostreatus* (Meng et al. 2003), *Tricholoma matsutake* (Chen et al. 2003), *Lentinula edodes* (Zhuo et al. 2006), *Agaricus bisporus* (Gu et al. 2003) and *Ganoderma lucidum* (Zheng et al. 2007).

9.3.5 Retrotransposon Microcrosatellite Amplified Polymorphism (REMAP) Fingerprinting Markers

As described earlier various molecular approaches have proved to be potent tools in the classification of complex fungal taxonomic groups including mushrooms. However, currently the applicability of this technique is limited principally to the

Table 9.3 AFLP primers and polymorphism

Primer pair	Total number of loci	Polymorphic loci	Polymorphism (%)
E-AAC/M-CTC	228	116	52
E-ACA/M-CAT	115	102	89
E-AT/M-CTG	160	40	25
E-AGG/M-CTG	20	16	80

From Daniel (2012)

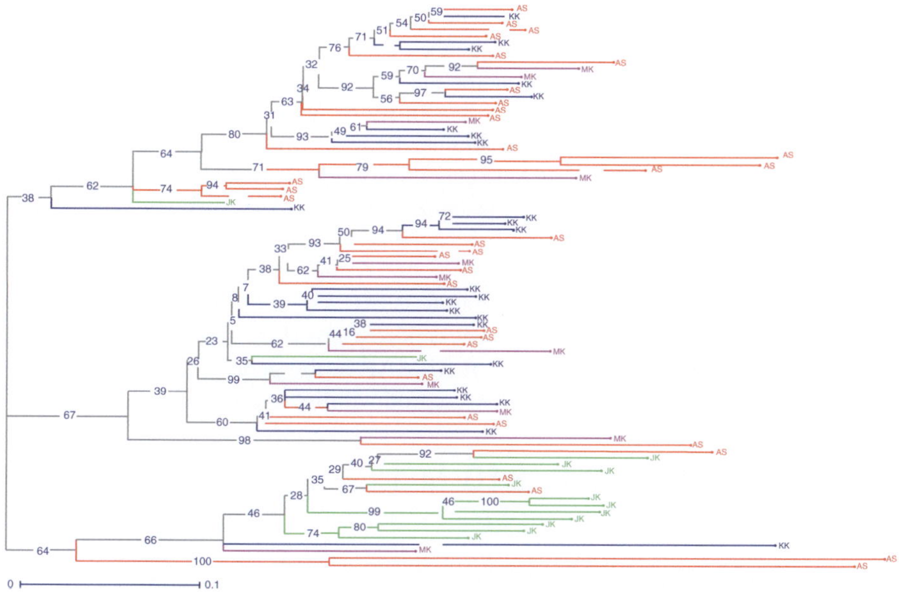

Fig. 9.3 Dendogram of clustering analysis of 84 *Pleurotus species*. KK(blue) AS(red) MK(pink) and JK(green) represents species from Kakamega, Arabuko Sokoke, Mt. Kenya and cultivated species from JKUAT respectively (From Daniel 2012)

classification of fungi at the species level (Garnica et al. 2003; Peintner et al. 2004). REMAP is such a technique, and is premised on the amplification of the DNA region between a long terminal repeat (LTR) of the retrotransposon and simple sequence repeat (SSR) using an outward-facing LTR primer and an SSR specific primer (Kalendar et al. 1999; Kalendar and Schulman 2006). The LTR-SSR regions are recurring sequences along the genomic DNA. The design of primers that anneal these regions enables the primers to amplify DNA fragments of different lengths, and the recurring patterns of these regions differ among various species or even among strains of the same species. Therefore, REMAP is an appealing method for study and application in the classification of mushrooms.

The gypsy-type retroelement marY1 was first identified in *Tricholoma matsutake* (Murata and Yamada 2000) and later in other basidiomycetes species (Murata and Miyazaki 2001; Murata et al. 2001, 2005). This retroelement spans approximately 6 kb and harbors 5′-LTR, gag, prt, pol, and 3′-LTR regions in consecutive order (Murata and Yamada 2000). The intact unit of the marY1 LTR retrotransposon may be contained in the genomic DNA in several to over 1,000 copies, which are equivalent to 0.05–5.5 % of the approximately 34 Mb genomic sequence of each different *Tricholoma* species (Murata and Babasaki 2005). Interestingly, the LTR region alone has been a consistent evidence of markedly higher copy numbers than those observed in the coding region of marY1 (Murata and Babasaki 2005). It was also determined that this LTR region is highly conserved in a variety of higher fungi (Murata et al. 2001). These properties of marY1, and in particular the LTR region of marY1, render it as an excellent candidate for fungal fingerprinting techniques. Le et al. (2008) aligned the mar Y1 sequences of multiple mushroom species and selected the highly conserved region for the cross-species amplification of DNA regions of interest. They assessed the feasibility of utilizing these primers in the fingerprinting of various mushroom strains. REMAP method was applied to 10 mushroom species and the primer set successfully discriminated between different commercial mushroom cultivars of the same strains of 14 *Pleurotus ostreatus* and 16 *P. eryngii* (Fig. 9.4).

Fourteen cultivars of *P. ostreatus* (Fig. 9.5a) and 16 cultivars of *P. eryngii* (Fig. 9.5c) were selected and subjected to REMAP. Variations in the band patterns are more distinctively evident in adjacent dendrograms (Fig. 9.5b, d). These observations strongly advocate for the use of REMAP in the genotyping of mushroom species at the sub-species level. REMAP reproducibility was superior to other pop-

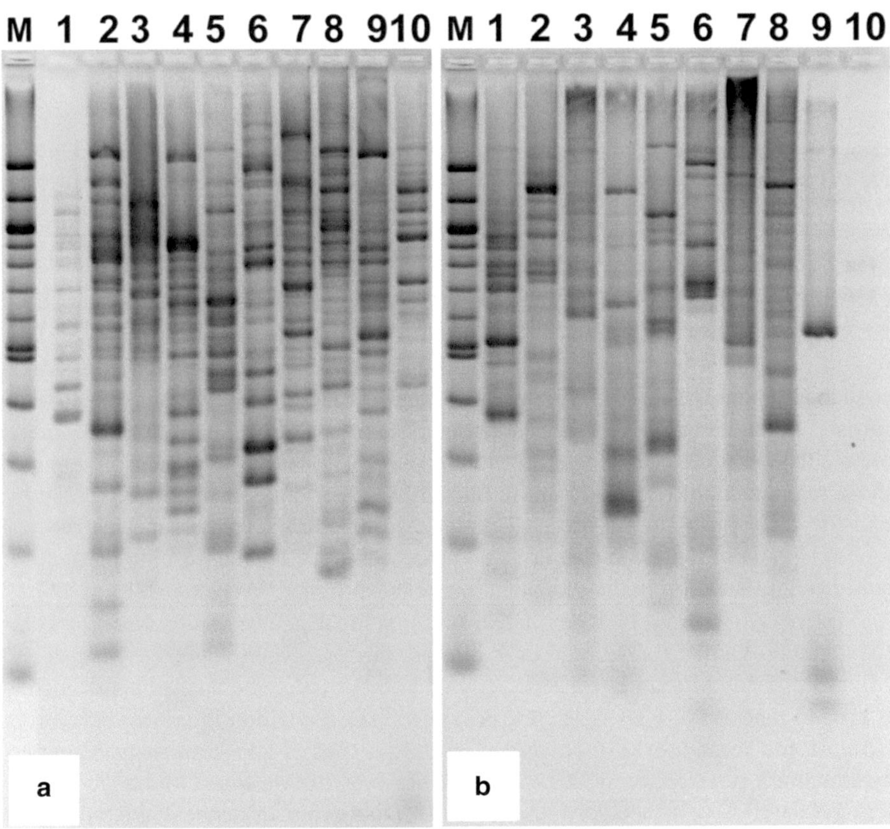

Fig. 9.4 REMAP analyses of 10 mushroom species. Genomic DNA of the mushrooms were subjected to REMAP analysis using the Com3 (**a**) and Com5 (**b**) primer sets. Lanes: *1 Phellinus linteus*; *2 Cordyceps* spp.; *3 Ganoderma lucidum*; *4 Coprinus comatus*; *5Hericium erinaceus*; *6 Flammulina velutipes*; *7 Pleurotus eryngii*; *8 Pleurotus ostreatus*; *9 Lentinula edodes*; *10 Agaricus bisporus* (From Le et al. 2008)

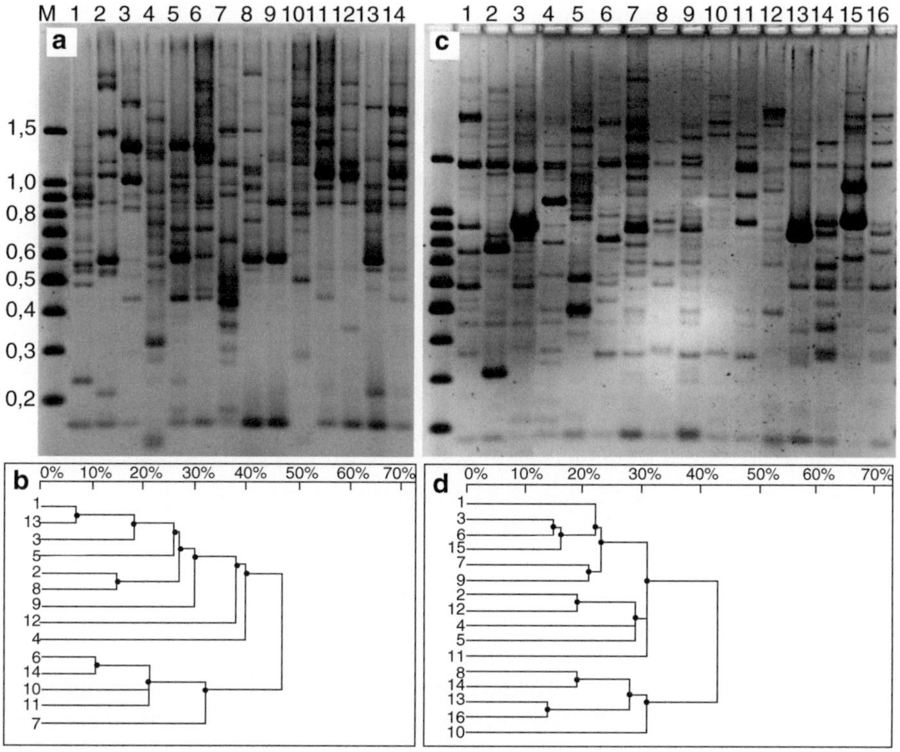

Fig. 9.5 REMAP analyses on the mushroom strains within the same species. Band patterns and dendrograms of 14 *P. ostreatus* strains (**a** and **b**, respectively) and 16 *P. eryngii* strains (**c** and **d**, respectively) are displayed. The 14 *P. ostreatus* strains identified by lane are: *1* IUM1051; *2* IUM1313; *3* IUM1367; *4* IUM1490; *5* IUM1556; *6* IUM1491; *7* IUM557; *8* IUM1677; *9* IUM165; *10* IUM778; *11* IUM1520; *12* IUM1373; *13* IUM1341; *14* IUM1346. Similarly, the 16 *P. eryngii* strains are: *1* IUM2501; 2 IUM2502; *3* IUM2503; *4* IUM2510; *5* IUM1511; *6* IUM2512; *7* IUM2513; *8* IUM2514; *9* IUM2515; *10* IUM2518; *11* IUM2519; *12* IUM2520; *13* IUM2521; *14* IUM2523; *15* IUM2524; *16* IUM2322. The patterns of DNA bands were analyzed and dendrograms were created with Bio-1D++ (From Le et al. 2008)

ular DNA fingerprinting methodologies including the random amplified polymorphic DNA method.

9.3.6 Structural Variation (SV) Markers

A novel type of marker that is based on SV loci has been developed by Feuk et al. (2006). These markers, which are used in genome re-sequencing, are reportedly more specific to individuals than SNPs (Li et al. 2011; Kidd et al. 2008).

In normal wild-type populations, approximately 5 % of the genome is defined as SV and the size is equal to 250–300 genes (Feuk et al. 2006). The SV markers may represent the deletion, duplication, insertion, translocation or inversion of DNA segments in the genome and can profoundly affect the correlation between genetic and physical distance for the same intervals in plants (Dooner and He 2008). Ren et al. (2012) constructed a high-resolution genetic map of anchoring scaffolds in the sequenced watermelon genome. Overall, 953 molecular markers, including 36 SV markers, were used in the linkage analysis, which suggested that SV markers could be applied to genetic map construction.

With respect to the haploid fungi basidiomycete, which typically possesses small genomes and specialized life cycles, SV loci can easily be detected if the genomes of two 'compatible strains' are sequenced and re-sequenced. A number of fruiting fungi have been sequenced or re-sequenced, such as *Laccaria bicolor* (Ohm et al. 2010), *Schizophyllum commune* (Ohm et al. 2010), *Coprinus cinereus* (Stajich et al. 2010), *Ganoderma lucidum* (Chen et al. 2012; Liu et al. 2012), *Volvariella volvacea* (Bao et al. 2013; Chen et al. 2013) and *Flammulina velutipes* (Park et al. 2014; Wang et al. 2015b). It would be useful to identify SV loci and develop SV markers by comparing these genomic sequences and re-sequencing.

A genetic linkage map is a useful tool in gene mapping, molecular breeding for genetic improvement, and genetic dissection of QTL (Guimarães 2007; Zheng et al. 2008). In combination with analysis of the draft genome, these linkage maps can provide a scaffold for assembling a detailed physical map and can promote research in functional genomics (Meksem and Kahl 2005; Harushima et al. 1998). *Volvariella volvacea*, also known as the Chinese straw mushroom (Fig. 9.6), is an important edible fungus that is cultivated extensively across subtropical and tropical East and Southeast Asia (Chang 1978). Although *V. volvacea* has been cultivated for approximately 300 years and its genomic sequence is currently available (Bao et al. 2013; Chen et al. 2013), the number of chromosomes in the *V. volvacea* genome remains unconfirmed. Wang et al. (2015a) reported a strategy to develop new type SV markers, and construct a genetic map of the basidiomycete *V. volvacea*.

This application for SV markers from the genomic sequence may assist with *V. volvacea* genome assembly and genetic research. The construction method for genetic mapping in *V. volvacea* can also be applied to other basidiomycete species. As a novel type of marker, SV markers possess many advantages. First, SV markers can amplify unique polymorphic bands in a mapping population with linkage Groups. The amplified bands are easily detected using electrophoresis, which makes this a simple, time-saving and reproducible marker technique. Second, SV markers are very stable and convenient to use. They can be rapidly developed once the genome of the target organism is re-sequenced or has been sequenced multiple times. Third, the success rate of

SV markers in constructing linkage Groups are higher than those of other markers. Because SV marker primers are designed from the flanking sequences of SV loci, amplified bands are always detectable in all single spores. SV loci and developed markers were used in constructing the high-resolution genetic map of watermelon successfully (Ren et al. 2012).

Although genetic maps for fungi were developed later than those of plants and animals, several linkage maps already exist for certain models or other important basidiomycete fungal species, including *Pleurotus ostreatus* (Larraya et al. 2000),

Fig. 9.6 Morphology of *Volvariella volvacea* (Khatun et al. 2012)

C. cinereus (Muraguchi et al. 2003), *L. bicolor* (Labbe et al. 2008), *Pleurotus pulmonarius* (Okuda et al. 2009), *Agaricus bisporus* (Kerrigan et al. 1993; Moquet et al. 1999; Foulongne-Oriol et al. 2010), *Pleurotus eryngii* (Okuda et al. 2012) (Fig. 9.7) and *Lentinula edodes* (Kwan and Xu 2002; Terashima et al. 2002; Terashima et al. 2006; Miyazaki et al. 2008; Gong et al. 2014), using molecular markers other than SV markers. Several genetic maps of these basidiomycete fungi were combined with QTLs (Moquet et al. 1999), elucidating karyotypes (Tanesaka et al. 2012) and alignment to the whole-genome sequence assemblies (Labbe et al. 2008). Progress in the genetic characterization of *V. volvacea* remains poor, primarily due to a scarcity of information on its life cycle and the wide range of variation among its single-spore isolates (Li and Chang 1979, 1991; Chang et al. 1981, Chang and Li 1991). Wang et al. (2015a) found two compatible monokaryotic strains demonstrating a heterothallic life cycle in *V. volvacea* and constructed the first genetic map of *V. volvacea* using markers developed based on SV loci. The constructed genetic map consisted of 102 SV markers (two markers were not linked: SV011; SV978) that were distributed across ten linkage Groups, accounting for a total length of 411.61 cM.

9.3.7 *Inter Simple Sequence Repeat (ISSR Markers)*

ISSR technique is a PCR based method, involving amplification of DNA segment(s) located in between two identical microsatellite repeat regions oriented in opposite direction. This method utilizes microsatellites (~16–25 bp long) as PCR primers in a single primer PCR reaction targeting multiple genomic loci to amplify mainly the inter-SSR sequences of different sizes. Interestingly, the microsatellite repeats used as primers can be di-nucleotide, tri-nucleotide, tetra- nucleotide or penta-nucleotide (Reddy et al. 2002).

ISSR markers with their whole-genome coverage and distribution, accuracy, reproducibility and robustness have proved to be promising among other markers for genetic diversity analysis of several plant species. One of the most important species of the genus *Agaricus* is the white button mushroom *A. bisporus*, which is widely cultivated in most parts of the world. Commercial strains of white button mushroom are classified based on the morphological characters, which are easily

Fig. 9.7 Morphology of *Pleurotus eryngii*

affected by the environmental factors. Environmental factors along with the close genetic relation between the isolates make their isolation and identification difficult and sometimes impossible. In the past two decades, different molecular markers based on nucleic acid polymorphisms, have been exploited in genetic studies of the edible mushrooms. Zhang et al. (2007) and Guan et al. (2008) used ISSR markers for strain identification in *Lentinula edodes* and isolation of different strains in *A. bisporus*, respectively. Malekzadeh et al. (2011) have given a vivid description of how ISSR markers can be utilized as a powerful tool for detection of polymorphism among closely related genotypes of *A. bisporus*.

9.4 Future Scope

Research for new improved varieties, with increase in yields and quality is required today to make mushroom production highly productive. Increasing the yield and quality of crops as well as resistance to diseases are the primary goals for mushroom breeders and mushroom research. Breeders now use DNA molecular markers to identify and select specific genes to locate superior traits. Utilizing molecular markers to explore traits from wild isolates can expand the genetic base of the cultivated mushrooms. Genetic mapping of the mushroom genome will help in understanding their complex traits such as yield, size, color, flavor and shelf life. Thus, the exploitation of genetic resources to broaden genetic variability is promising to develop new cultivars with higher yield, enhanced flavor, resistance to diseases or adaptation to climate changes with a greater shelf life. The use of genetic engineering in mushroom industry will be determined by economic factors related to necessity and resources. Due to funding constraints, mushrooms are very much lagging behind other crops in terms of advancement in molecular biotechnology. Public acceptance of genetically modified foods and greater consumption of mushrooms can increase research efforts.

With mapping of the mushroom genome and understanding of the functional genomics in mushrooms, complex traits such as yield, size, color, self-life and physical stress, which are controlled by more than one gene, can be undertaken in the future. The production of new mushroom cultivars with novel and improved traits will provide the industry with options for solving food problems and increase the production efficiency. Improvement of tools available to the breeder, decoding mushroom genome and commercial pressure facing the industry can propel efforts for new strain development in the future. Consequently, the results of the present study can be seen as a starting point for future research to determine the level of intra and inter specific/genetic diversity in mushroom. In addition, their evolutionary relationships could provide an important clue for further exploration of the active compounds. Furthermore, molecular characterization is an authentication of wild mushrooms.

References

Agarwal K, Prasad MP, Rindhe G. Genomic discrimination of eleven commercial mushrooms by DNA fingerprinting using RAPD marker. Int Res J Biol Sci. 2013;2(10):1–5.

Bao D, Gong M, Zheng H, Chen M, Zhang L, Wang H, Jiang J, Wu L, Zhu Y, Zhu G. Sequencing and comparative analysis of the straw mushroom (*Volvariella volvacea*) genome. PLoS One. 2013;8, e58294.

Barroso G, Sonnenberg AS, Van-Griensven LJ, Labarere J. Molecular cloning of widely distributed microsatellite core sequence from the cultivated mushroom *Agaricus bisporus*. Fungal Genet Biol. 2000;31:115–23.

Bos CJ. Fungal genetics principles and practice. New York: Marcel Dekkar; 1996.

Bruns TD, White TJ, Taylor JW. Fungal molecular systematics. Ann Rev Ecol Sys. 1991;22:525–64.

Çaglarirmak N. The nutrients of exotic mushrooms (*Lentinula edodes* and *Pleurotus* species) and an estimated approach to the volatile compounds. Food Chem. 2007;105:1188–94.

Capelari M, Fungaro MH. Determination of biological species and analysis of genetic variability by RAPD of isolates of *Pleurotus* subgenus *Coremiopleurotus*. Mycol Res. 2003;107(9):1050–4.

Castle AJ, Horgen PA, Anderson JB. Restriction fragment length polymorphisms in the mushrooms *Agaricus brunnescens* and *Agaricus bitorquis*. Appl Environ Microbiol. 1987;53:816–22.

Chakravarty B. Trends in mushroom cultivation and breeding. Aus J Agric Eng. 2011;2(4):102–9.

Chang ST. Volvariella volvacea. In: Chang ST, Hayes WA, editors. The biology and cultivation of edible mushroom. New York: Academic; 1978. p. 573–605.

Chang ST, Li SX. Genetical studies on the sexuality pattern of *Volvariella volvacea*. In: Van Griensven LJLD, editor. Science and cultivation of edible fungi. Boca Raton: CRC Press/ Balkema; 1991. p. 119–22.

Chang ST, Miles PG, Wai CC. A study of monosporous isolates of *Volvariella volvacea*. Mushroom Sci. 1981;11:603–21.

Chang ST, Mshigeni KE. Mushroom and their human health: their growing significance as potent dietary supplements. Windhoek, Namibia: The University of Namibia; 2004. p. 1–79.

Chen S, Xu J, Liu C, Zhu Y, Nelson DR, Zhou S, Li C, Wang L, Guo X, Sun Y. Genome sequence of the model medicinal mushroom *Ganoderma lucidum*. Nat Commun. 2012;3:913.

Chen Q, Liao DC, Zhang XP, Zhang CB, Li DY (2003). Preliminary analysis of genetic aspects of edible fungi Tricholoma matsutake isolated from Yajiang, Sichuan, by AFLP technique. Scientia Agric. Sin. 36: 1588–1594.

Cheng KT, Chang HC, Huang H, Lin CT. RAPD analysis of *Lycium barbatum* medicine in Taiwan market. Bot Bull Acad Sin. 2000;41(1):11–4.

Chyi S, Hoencke ME, Sernyk JL. A genetic linkage map of restriction fragment length polymorphism loci for *Brassica rapa* (syn. *campestries*). Genome. 1992;35:746–57.

Daniel OO. Molecular characterization of the wild edible mushrooms of the *Pleurotus* species in Kenya. A Thesis for Master of Science in Biotechnology, Jomo Kenyatta University, Kenya; 2012.

Dooner HK, He L. Maize genome structure variation: interplay between retrotransposon polymorphisms and genetic recombination. Plant Cell Online. 2008;20:249–58.

Elmastas M, Isildak O, Turkekul I, Temur N. Determination of antioxidant activity and antioxidant compounds in wild edible mushrooms. J Food Comp Anal. 2007;20:337–45.

Feuk L, Carson AR, Scherer SW. Structural variation in the human genome. Nat Rev Genet. 2006;7:85–97.

Fonseca GG, Gandra EA, Sclowitz LF, Correa APA, Costa JAV, Levy JA. Oyster mushrooms species differentiation through molecular markers RAPD. Int J Plant Br Gen. 2008;2:13–8.

Foulongne-Oriol M, Spatar C, Cathalot V, Monllor S, Savoie JM. An expanded genetic linkage map of an intervarietal *Agaricus bisporus* var. *bisporus* × *A. bisporus* var. *burnettii* hybrid based on AFLP, SSR and CAPS markers sheds light on the recombination behavior of the species. Fungal Genet Biol. 2010;47:226–36.

Froslev T, Matheny P, Hibbett D. Lower level relationships in the mushroom genus *Cortinarius* (Basidiomycota, Agaricales): a comparison of RPB1, RPB2, and ITS phylogenies. Mol Phylog Evol. 2005;37:602–18.

Gardes M, Bruns TD. ITS primers with enhanced specificity for basidiomycetes – application to the identification of mycorrhizae and rusts. Mol Ecol. 1993;2:113–8.

Garnica S, Weiss M, Oertel B, Oberwinkler F. Phylogenetic relationships of European *Phlegmacium* species (*cortinarius*, Agaricales). Mycologia. 2003;95:1155–70.

Gong WB, Liu W, Lu YY, Bian YB, Zhou Y, Kwan HS, Cheung MK, Xiao Y. Constructing a new integrated genetic linkage map and mapping quantitative trait loci for vegetative mycelium growth rate in *Lentinula edodes*. Fungal Biol UK. 2014;118:295–308.

Gu YF, Zhang XP, Chen Q, Liao DC, Liu J. Genetic variation of *Agaricus bisporus* analyzed by AFLP technique. Southwest China J Agric Sci. 2003;16:39–43.

Guan XJ, Xu L, Shao YC, Wang ZR, Chen FS, Luo XC. Differentiation of commercial strains of *Agaricus* species in China with inter-simple sequence repeat marker. World J Microbiol Biotechnol. 2008;24:1617–22.

Guimarães EP. Marker-assisted selection: current status and future perspectives in crops, livestock, forestry and fish. Rome: Food & Agriculture Organization of the United Nations; 2007.

Harushima Y, Yano M, Shomura A, Sato M, Shimano T, Kuboki Y, Yamamoto T, Lin SY, Antonio BA, Parco A. A high density rice genetic linkage map with 2275 markers using a single F2 population. Genetics. 1998;148:479–94.

Hawksworth DL. The fungal dimension of biodiversity : magnitude, significance, and conservation. Mycol Res. 1991;95:641–55.

Hofstetter V, Clémençon H, Vilgalys R, Moncalvo JM. Phylogenetic analyses of the Lyophylleae (Agaricales, Basidiomycota) based on nuclear and mitochondrial rDNA sequences. Mycol Res. 2002;106:1043–59.

Hussein JM, Tibuhwa DD, Mshandete AM, Kivaisi AK. Molecular phylogeny of saprophytic wild edible mushroom species from Tanzania based on ITS and nLSU rDNA sequences. Curr Res Environ Appl Mycol. 2014;4(2):250–60.

Jones N, Ougham H, Thomas H. Markers and mapping: we are all geneticists now. New Phytol. 1997;137:165–77.

Kalendar R, Grob T, Regina M, Suoniemi A, Schulman A. IRAP and REMAP: two new retrotransposon-based DNA fingerprinting techniques. TAG Theor Appl Genet. 1999;98:704–11.

Kalendar R, Schulman AH. IRAP and REMAP for retrotransposon-based fingerprinting and fingerprinting. Nat Protoc. 2006;1:2478–84.

Karp A. Reproducibility testing of RAPD, AFLP and SSR markers in plants by a network of European laboratories. Mol Breed. 1997;3:381–90.

Kerrigan RW, Royer JC, Baller LM, Kohli Y, Horgen PA, Anderson JB. Meiotic behavior and linkage relationships in the secondarily homothallic fungus *Agaricus bisporus*. Genetics. 1993;133:225–36.

Khatun S, Islam A, Cakilcioglu U, Chatterjee NC. Research on mushroom as a potential source of nutraceuticals: a review on Indian perspective. Am J Exp Agric. 2012;2(1):47–73.

Khatun S, Islam A, Guler P, Cakilcioglu U, Chatterjee NC. Hypoglycemic activity of a dietary mushroom *Pleurotus florida* on alloxan induced diabetic rats. Biol Divers Conserv. 2013;6(2):91–6.

Khatun S, Islam A, Guler P, Cakilcioglu U, Chatterjee NC. Nutritional qualities and antioxidant activity of three edible mushrooms of *Pleurotus* spp. Wagening J Life Sci. 2015;72–73:1–5.

Khush R, Morhan I, Becker E, Wach M, Griensven LJ. Use of the polymerase chain reaction (PCR) in *Agaricus bisporus*. In: Proceedings of 1st International Seminar on Mushroom. 1991. Pudoc Press, Wageningen, The Netherlands.

Kidd JM, Cooper GM, Donahue WF, Hayden HS, Sampas N, Graves T, Hansen N, Teague B, Alkan C, Antonacci F. Mapping and sequencing of structural variation from eight human genomes. Nature. 2008;453:56–64.

Kwan HS, Xu HL. Construction of a genetic linkage map of shiitake mushroom *Lentinula edodes* strain L-54. J Biochem Mol Biol. 2002;35:465–71.

Labbe J, Zhang X, Yin T, Schmutz J, Grimwood J, Martin F, Tuskan GA, Tacon FL. A genetic linkage map for the ectomycorrhizal fungus *Laccaria bicolor* and its alignment to the whole-genome sequence assemblies. New Phytol. 2008;180:316–28.

Lanner C, Bryngelsson T, Gustafson M. Relationship of wild *Brassica* species with chromosome number 2n = 18, based on RFLP studies. Genome. 1997;40:302–8.

Larraya LM, Pérez G, Ritter E, et al. Genetic linkage map of the edible basidiomycete *Pleurotus ostreatus*. Appl Environ Microb. 2000;66:5290–300.

Le QV, Hyo-Kyung W, Tae-Soo L, Chang-Yun L, Hyun-Sook L, Hyeon-Su R. Retrotransposon microsatellite amplified polymorphism strain fingerprinting markers applicable to various mushroom species. Mycobiology. 2008;36(3):161–6.

Lee JS, Lim MO, Cho KY, Cho JH, Chang SY, Nam DH. Identification of medicinal mushroom species based on nuclear large subunit rDNA sequences. J Microbiol. 2006;44:29–34.

Li SF, Chang ST. Variation in the homothallic basidiomycetes *Volvariella volvacea*. Mushroom Sci. 1979;10:171–84.

Li SX, Chang ST. Study on the spore pattern in basidium of *Volvareilla volvacea*. In: Van Griensven LJLD, editor. Science and cultivation of edible fungi. Boca Raton: CRC Press/Balkema; 1991. p. 115–8.

Li Y, Zheng H, Luo R, Wu H, Zhu H, Li R, Cao H, Wu B, Huang S, Shao H. Structural variation in two human genomes mapped at single-nucleotide resolution by whole genome de novo assembly. Nat Biotechnol. 2011;29:723–30.

Lian B, Zang JP, Hou WG, Yuan S, Smith DL. PCR-based sensitive detection of the edible fungus *Boletus edulis* from rDNA ITS sequences. Electro J Biotechnol. 2008;11:102–9.

Lindequist U, Niedermeyer THJ, Julich WD. The pharmacological potential of mushrooms. eCAM. 2005;2(3):285–99.

Liu D, Gong J, Dai W, Kang X, Huang Z, Zhang HM, Liu W, Liu L, Ma J, Xia Z. The genome of *Ganderma lucidum* provide insights into triterpense biosynthesis and wood degradation. PLoS One. 2012;7, e36146.

Loftus M, Bouchti-King L, Robles C. Use of SCAR markers for cap color in *Agaricus bisporus* breeding programs. In: van Griensven LJLD, editor. Science and cultivation of edible fungi. Rotterdam: Balkema; 2000. p. 201–5.

Lopandic K, Molnar O, Prillinger H. Application of ITS sequence analysis, RAPD and AFLP fingerprinting in characterizing the yeast genus *Fellomyces*. Microbiol Res. 2005;160:13–26.

Lucia R, Victor M, Mikel A, Alberto GB, Leopoldo A, Antonio GP. Use of molecular markers to differentiate between commercial strains of the button mushroom *Agaricus bisporus*. FEMS Microbio Lett. 2001;198:45–8.

Ma FY, Luo XC. PCR-based restriction analysis of internal transcribed spacers of nuclear ribosomal DNA in the genus *Pleurotus*. Mycosystema. 2002;21:356–62.

Malekzadeh K, Jalalzadeh B, Moghsddam S, Ehsan M. Use of ISSR markers for strain identification in the button mushroom, *Agaricus bisporus*. Proc Int Conf Mush Biol Mush Produc (ICMBMP7). 2011;30–34.

Manaf SRA, Mustafa M, Amin NM, Ali AM. Genetic relatedness among isolates of *Acantham oeba* based on RAPD analysis. J Appl Sci. 2006;6(1):15–9.

Martin P, Muruke M, Hosea K, Kivaisi AK, Zerwas N, Bauerle C. A rapid PCR-RFLP method for monitoring genetic variation among commercial mushroom species. Biochem Mol Biol Educ. 2004;32:390–4.

Meksem K, Kahl G. The handbook of plant genome mapping: genetic and physical mapping. Weinheim: Wiley-VCH Verlag GMCH & CO. KgaA; 2005.

Meng Y, Jiang CS, Liao WT, Zhang YZ. AFLP fingerprinting map analysis of *Pleurotus ostreatus*. Acta Gen Sin. 2003;30:1140–6.

Miyazaki K, Huang F, Zhang B, Shiraishi S, Sakai M, Shimaya C, Shishido K. Genetic map of a basidiomycete fungus, *Lentinula edodes* (shiitake mushroom), constructed by tetrad analysis. Breed Sci. 2008;58:23–30.

Molcalvo JM, Wang HH, Hseu RS. Phylogenetic relationships in *Ganoderma* inferred from the internal transcribed spacer and 25S ribosomal DNA sequences. Mycologia. 1995;87:223–38.

Moquet F, Desmerger C, Mamoun M, Ramos-Guedes-Lafargue M, Olivier JM. A quantitative trait locus of *Agaricus bisporus* resistance to *Pseudomonas tolaasii* is closely linked to natural cap color. Fungal Genet Biol. 1999;28:34–42.

Moreau PA, Peintner U, Gardes M. Phylogeny of the ectomycorrhizal mushroom genus *Alnicola* (Basidiomycota, Cortinariaceae) based on rDNA sequences with special emphasis on host specificity and morphological characters. Mol phylogenet Evol. 2006;38:794–807.

Mueller UG, Wolfenbarger LL. AFLP genotyping and fingerprinting. Trend Ecol Evol. 1999;14:389–94.

Muraguchi H, Ito Y, Kamada T, Yanagi SO. A linkage map of the basidiomycete *Coprinus cinereus* based on random amplified polymorphic DNAs and restriction fragment length polymorphisms. Fungal Genet Biol. 2003;40:93–102.

Murata H, Miyazaki Y. Expression of marY1, a gypsy-type LTR-retroelement from the ectomycorrhizal homobasidiomycete *Tricholoma matsutake*, in the budding yeast *Saccharomyces cerevisiae*. Biosci Biotechnol Biochem. 2001;65:993–5.

Murata H, Babasaki K. Intra- and inter-specific variations in the copy number of two types of retrotransposons from the ectomycorrhizal basidiomycete *Tricholoma matsutake*. Mycorrhiza. 2005;15:381–6.

Murata H, Babasaki K, Yamada A. Highly polymorphic DNA markers to specify strains of the ectomycorrhizal basidiomycete *Tricholoma matsutake* based on sigmamarY1, the long terminal repeat of gypsy-type retroelement marY1. Mycorrhiza. 2005;15:179–86.

Murata H, Miyazaki Y, Babasaki K. The long terminal repeat (LTR) sequence of marY1, a retroelement from the ectomycorrhizal homobasidiomycete *Tricholoma matsutake*, is highly conserved in various higher fungi. Biosci Biotechnol Biochem. 2001;65:2297–300.

Murata H, Yamada A. marY1, a member of the gypsy group of long terminal repeat retroelements from the ectomycorrhizal basidiomycete *Tricholoma matsutake*. Appl Environ Microbiol. 2000;66:3642–5.

Ohm RA, de Jong JF, Lugones LG, Aerts A, Kothe E, Stajich JE, de Vries RP, Record E, Levasseur A, Baker SE. Genome sequence of the model mushroom *Schizophyllum commune*. Nat Biotechnol. 2010;28:957–63.

Okuda Y, Murakami S, Matsumoto T. A genetic linkage map of *Pleurotus pulmonarius* based on AFLP markers, and localization of the gene region for the spore less mutation. Genome. 2009;52:438–46.

Okuda Y, Ueda J, Obatake Y, Murakami S, Fukumasa Y, Matsumoto T. Construction of a genetic linkage map based on amplified fragment length polymorphism markers and development of sequence-tagged site markers for marker-assisted selection of the spore less trait in the oyster mushroom (*Pleurotus eryngii*). Appl Environ Microbiol. 2012;78:1496–504.

Paran I, Michelmore RW. Development of reliable PCR-based markers linked to downy mildew resistance genes in lettuce. Theor Appl Genet. 1993;85:985–93.

Park YJ, Baek JH, Lee S, Kim C, Rhee H, Kim H, Seo JS, Park HR, Yoon DE, Nam JY. Whole genome and global gene expression analyses of the model mushroom *Flammulina velutipes* reveal a high capacity for lignocellulose degradation. PLoS One. 2014;9, e93560.

Pei-Sheng Y, Luo XC, Zhou Q. RFLP analysis of amplified nuclear ribosomal DNA in the genus *Auricularia*. Myco System. 1999;18:206–13.

Peintner U, Moncalvo JM, Vilgalys R. Toward a better understanding of the infrageneric relationships in *Cortinarius* (Agaricales, Basidiomycota). Mycologia. 2004;96:1042–58.

Pradhan A, Prakash S, Mukhopadhyay A, Pental D. Phylogeny of *Brassica* and allied genera based on variation in chloroplast and mitochondrial DNA patterns: molecular and taxonomic classifications are incongruous. Theor Appl Genet. 1992;85:331–40.

Pringle A, Moncalvo JM, Vilgalys R. High levels of variation in ribosomal DNA sequences within and among spores of a natural population of the arbuscular mycorrhizal fungus *Acaulospora collosica*. Mycologia. 2000;92:259–68.

Ravash R, Shiran B, Alavi A, Zarvagis J. Evaluation of genetic diversity in Oyster mushroom (*Pleurotus eryngii*) isolates using RAPD marker. J Sci Technol Agr Nat Res. 2009;13:739–41.

Reddy MP, Sarla N, Siddiq EA. Inter simple sequence repeat (ISSR) polymorphism and its application in plant breeding. Eyphytica. 2002;128:9–17.

Ren Y, Zhao H, Kou Q, Jiang J, Guo S, Zhang H, Hou W, Zou X, Sun H, Gong G. A high resolution genetic map anchoring scaffolds of the sequenced watermelon genome. PLoS One. 2012;7, e29453.

Sadler M. Nutritional properties of edible fungi. Br Nutr Found Nutr Bull. 2003;28:305–8.

Sanchez-Ballesteros J, Gonzalez V, Salazar O, Acero J, Portal MA, Julian M, Rubio V, Bills GF, Polishook JD, Platas G, Mochale S, Pelaez F. Phylogenetic study of Hypoxylon and related genera based on ribosomal ITS sequences. Mycologia. 2000;92(5):964–77.

Sasidhara R, Thirunalasundari T. Molecular characterization of wild mushroom. Eur J Exp Biol. 2012;2(2):369–73.

Schoch CL, Seifert KA, Huhndorf S, Robert V, Spouge JL, Levesque CA. Nuclear ribosomal internal transcribed spacer (ITS) region as a universal DNA barcode marker for Fungi. Proc Natl Acad Sci U S A. 2012;109:6241–6.

Sharma SR. Scope of specialty mushrooms in India. In: Rai RD, Dhar BL, Verma RN, editors. Advances of mushroom biology and mushroom production. Solan: Mushroom society of India; 1997. p. 193–203.

Shnyreva AV, Belokon IS, Belokon MM. Use of molecular markers for differentiation of cultivated strains of oyster and button mushrooms. Genetika. 2003;39(11):1461–9.

Singh SK, Yadav MC, Upadhyay RC, Kamal Shwet Rai RD, Tewari RP. Mush Res. Molecular characterization of specialty mushroom germplasm of the National Mushroom Repository. 2003;12:67–8.

Stajic M, Sikorski J, Wasser SP, Nevo E. Genetic similarity and taxonomic relationships within the genus *Pleurotus* (higher Basidiomycetes) determined by RAPD analysis. Mycotoxon. 2005;93:247–55.

Stajich JE, Wilke SK, Ahrén D, Au CH, Birren BW, Borodovsky M, Burns C, Canbäck B, Casselton LA, Cheng CK. Insights into evolution of multicellular fungi from the assembled chromosomes of the mushroom *Coprinopsis cinerea* (*Coprinus cinereus*). Proc Natl Acad Sci U S A. 2010;107:11889–94.

Staniaszek M, Marczewski W, Szudyga K, Maszkiewicz J, Czaplicki A, Qian G. Genetic relationship between Polish and Chinese samples of the mushroom *Agaricus bisporus* (Lange) Sing., determined by the RAPD method. J Appl Genet. 2002;43:43–7.

Staub JE, Serquen FC. Genetic markers, map construction, and their application in plant breeding. Hort Sci. 1996;31:729–40.

Tanesaka E, Honda R, Sasaki S, Yoshida M. Assignment of RAPD marker probes designed from 12 linkage groups of *Flammulina velutipes* to CHEF-separated chromosomal DNAs. Mycoscience. 2012;53:238–43.

Terashima K, Matsumoto T, Hayashi E, Fukumasa-Nakai Y. A genetic linkage map of *Lentinula edodes* (shiitake) based on AFLP markers. Mycol Res. 2002;106:911–7.

Terashima K, Matsumoto T, Hayashi E, Kawasaki S, Fukumasa-Nakai Y. Construction of a linkage map of *Lentinula edodes* (shiitake) with the HEGS (high-efficiency genome scanning) system: Use of versatile AFLP and PCR-based gene markers. Mycoscience. 2006;47:336–46.

Terefework Z, Kaijalainen S, Lindström K. AFLP fingerprinting as a tool to study the genetic diversity of *Rhizobium galegae* isolated from *Galega orientalis* and *Galega officinalis*. J Biotechnol. 2001;91:169–80.

Tibuhwa DD. Folk taxonomy and use of mushrooms in communities around Ngorongoro and Serengeti National Park, Tanzania. J Ethnobiol Ethnomed. 2012;8:1–9.

Tibuhwa DD. Wild mushroom underutilized healthy food resource and income generator: experience from Tanzania rural areas. J Ethnobiol Ethnomed. 2013;9:1–14.

Tibuhwa DD, Savi S, Tibell L, Kivaisi AK. *Afrocantharellus* gen. nov. is part of a rich diversity of African Cantharellaceae. IMA Fungus. 2012;3:25–39.

Trappe M, Evans F, Trappe JM. Field guide to North American truffles: hunting, identifying and enjoying the world's most prized fungi, Natural history series. Berkeley: Ten Speed Press; 2007.

Virk PS, Zhu J, Newbury HJ, Bryan GJ, Jackson MT, Ford-lloyd BV. Effectiveness of different classes of molecular markers for classifying and revealing variations in rice (*Oryza sativa*) germplasm. Euphytica. 2000;112:275–84.

Vos PR, Hogers M, Bleeker M, Van de Lee RT, Hornes M, Fritjers A, Pot J, Peleman J, Kuiper M, Zabeau M. AFLP: a new concept for DNA fingerprinting. Nucleic Acid Res. 1995;23:4407–14.

Wang W, Chen B, Zhang L, Yan J, Lu Y, Zhang X, Jiang Y, Wu T, Arend Frans van P, Li S, Xie B. Structural Variation (SV) markers in the Basidiomycete *Volvariella volvacea* and their application in the construction of a genetic map. Int J Mol Sci. 2015a;16:16669–82.

Wang W, Liu F, Jiang Y, Wu G, Guo L, Chen R, Chen B, Lu Y, Dai Y, Xie B. The multigene family of fungal laccases and their expression in the white rot basidiomycete *Flammulina velutipes*. Gene. 2015b;563:142–9.

Welsh J, Mclelland M. Fingerprinting genomes using PCR with arbitrary primers. Nucleic Acid Res. 1990;18(24):7213–8.

White T, Bruns T, Lee S, Taylor J. Amplification and direct sequencing of fungal ribosomal RNA genes for phylogenetics. In: Innis M, Gelfand D, Sninsky J, White T, editors. PCR protocols. San Diego: Academic; 1990. p. 315–22.

Williams JGK, Kubelik A, Livak K, Rafalski JA, Tingey S. DNA polymorphisms amplified by arbitrary primers are useful as genetic markers. Nucleic Acid Res. 1990;18(22):6531–5.

Wipf D, Fribourg A, Nunch JC, Botton B, Buscot F. Diversity of the internal transcribed spacer of rDNA in morels. Can J Microbiol. 1999;45:769–78.

Xu J, Kerrigan RW, Callac P, Horgen PA, Anderson JB. The genetic structure of natural populations of *Agaricus bisporus*, the commercial mushroom. J Heredity. 1997;88:482–94.

Xu J, Kerrigan RW, Sonnenberg AS, Callac P, Horgen PA, Anderson JB. Mitochondrial DNA variation in natural populations of the mushroom *Agaricus bisporus*. Mol Ecol. 1998;7:19–33.

Zervakis GI, Venturella G, Papadopoulou K. Genetic polymorphism and taxonomic infrastructure of the *Pleurotus eryngii* species-complex as determined by RAPD analysis, isozyme profiles and ecomorphological characters. Microbiology. 2001;147:3183–94.

Zhang R, Huang C, Zheng S, Zhang J, Ng TB, Jiang R, Zuo X, Wang H. Strain-typing of Lentinula edodes in China with inter simple sequence repeat markers. Appl Microbiol Biotechnol. 2007;74:140–145.

Zheng LY, Jia DH, Luo X, Yang ZR. AFLP analysis for genetic diversity of *Ganoderma*. Ch J Chines Mat Med. 2007;32:1733–6.

Zheng Y, Zhang G, Lin F, Wang Z, Jin G, Yang L, Wang Y, Chen X, Xu Z, Zhao X. Development of microsatellite markers and construction of genetic map in rice blast pathogen *Magnaporthe grisea*. Fungal Genet Biol. 2008;45:1340–7.

Zhuo Y, Tan Q, Chen MJ, Cao H, Jia YN, Pan YJ. AFLP analysis of genetic diversity in main cultivated strains of *Lentinula edodes*. Mycosystema. 2006;25:203–10.

Chapter 10
Detection and Management of Fungal Respiratory Infection by Using Molecular Markers

Mashook Ali

10.1 Introduction

Pier Antonio Micheli, a priest and botanist published a book in 1729 entitled 'Nova Plantarum Genera' and gave the description of a fungus, named *Aspergillus* for the first time due to its resemblance to the brush of an aspergillum, which is a liturgical implement to sprinkle holy water. Fungus is ubiquitous in nature and is commonly isolated from the soil, plant debris, and the indoor environment, including hospitals. Fungal respiratory diseases consist of fungal colonization, allergy and infections of respiratory tract and lungs. Clinical presentations of fungal infection of the respiratory tract are nonspecific and may mimic viral or bacterial infection. The fungi responsible for fungal respiratory infections vary with the population selected and the geographical region. Fungal infection may be of two kinds nosocomial (mycoses are always opportunistic) and community associated (consists of both endemic mycoses and opportunistic mycoses). Fungal infection spread very fast in immunocopromised patient and are important cause of morbidity and mortality, thus early intervention is often crucial in effectively treating fungal infection. Although dramatic strides in the diagnosis of respiratory fungal infections have been made in recent years, it remains imperative for the clinician to maintain a high index of suspicion for the possibility of pulmonary fungal infection when evaluating an ill, immunocompromised patient. This chapter will review the current approaches to the diagnosis of fungal infection of respiratory tract.

M. Ali
Dr. B. R. Ambedkar Centre for Biomedical Research, University of Delhi, Delhi, India
e-mail: alimashook@gmail.com

B.P. Singh, V.K. Gupta (eds.), *Molecular Markers in Mycology*, Fungal Biology,
DOI 10.1007/978-3-319-34106-4_10

10.2 Diagnosis of Fungal Disease

Early diagnosis of fungal infection and initiation of appropriate therapy is of paramount importance to improve clinical outcome by impeding disease progression and mortality (Meyers 1990; von Eiff et al. 1995; Caillot et al. 1997; Greene et al. 2007). The currently used traditional diagnostic methods are far from adequate (Baddley et al. 2010; Park et al. 2011; Nguyen et al. 2012). In addition to it, the early clinical and laboratory diagnosis of fungal infection remains a challenge. The European Organization for Research and Treatment of Cancer/Mycoses Study Group (EORTC/MSG) created Research oriented diagnostic definitions of invasive fungal infection (IFI) in 2002 and modified it in 2008 for the purpose of creating homogenous groups for clinical trials and their clinical application (Ascioglu et al. 2002; DePauw et al. 2008).

The diagnosis of invasive aspergillosis within this classification is made considering a defined criterion for proven, probable, or possible infection; depending on a combination of host factors, clinical manifestations, and mycological factors (Ascioglu et al. 2002). Although galactomannan1, 3-β-D glucan are now incorporated as serological antigen biomarkers as part of the laboratory diagnostic criteria of invasive aspergillosis (DePauw et al. 2008). The EORTC/MSG definitions still rely heavily on tissue biopsy and analysis by traditional histopathology and fungal culture laboratory methods.

Sometimes patients with suspected fungal infection are too ill to undergo invasive procedures and the reliance on traditional methods hampers early confirmation of infection. Traditionally used diagnostic methods and their dependence on tissue sample and slow turnaround times delay early confirmation of invasive aspergillosis. Use of novel biomarkers and molecular tests, in combination with clinical criteria, may enable more accurate and earlier diagnosis of invasive aspergillosis (Halliday et al. 2006; Maertens et al. 2002). This will improve the appropriate prescription of early effective anti-fungal therapy and clinical outcomes (Maertens et al. 2012).

10.3 Molecular Marker in Fungal Disease Diagnostics

10.3.1 Galactomannan

Galactomannan is an important component of the fungal Cell wall. Galactomannan (GM), a heat-stable heteropolysaccharide consisting of a non-immunogenic mannan core with immune reactive galactofuranosyl units, is released during hyphal growth (Latge et al. 1994). Galactomannan is most commonly detected in serum and BALF using a double-sandwich enzyme-linked immunosorbent assay (ELISA). GM-ELISA assays use a rat monoclonal antibody EB-A2 that targets the β- (1,5)-linked galactofuranoside side-chains of galactomannan and defines a positive result as an Optical Density Index (ODI) ratio ≥0.5 (Stynen et al. 1995; Maertens et al. 2004; Marr et al. 2004). Measurement of GM antigen in other specimens including sputum samples, bronchial washings urine, pericardial fluid, pleural fluid, peritoneal fluid, and tissue samples, has not yet been clinically validated (Salonen et al. 2000; Klont et al. 2004; Musher et al. 2004; Kimura et al. 2009; Dufresne et al. 2012; Kono et al. 2013). In the EORTC/MSG criteria GM detection is also included in CSF sample for diagnosis of invasive aspergillosis with blood and BALF (DePauw et al. 2008). GM-ELISA assay have been evaluated in numerous studies and meta-analyses (Mennink et al. 2004; Allan et al. 2005; Pfeiffer et al. 2006; Leeflang et al. 2008).

In a meta-analysis of 27 studies to diagnose invasive Aspergillosis in which GM ELISA was used found a 71 % and 89 % sensitivity and specificity respectively (Pfeiffer et al. 2006). The specificity of GM-ELISA usually remains high, though inconstancy is repeated in sensitivity (29–100 %), (Maertens et al. 2002, 2004; Marr et al. 2004). The high negative predictive value (NPV) (98 %) and low positive predictive value (PPV) (26 %) suggest that the assay is good for ruling out disease but is less useful for confirming the disease (Rex 2006). About two third of patients with invasive aspergillosis using the technique can be diagnose at a mean of 8 days before in comparison to diagnosis by any another methods (Mennink et al. 2004). But GM assay gives false positive results in patients treated with semisynthetic β-lactam antibiotics such as piperacillin-tazobactam, amoxicillin-clavulanate, ceftriaxone (Aubry et al. 2006; Boonsarngsuk et al. 2010), in addition those given enteral nutrition (Girmenia et al. 2011),as well as in BAL fluid containing plasmolyte (Hage et al. 2007).

There are several factors which contribute to lowered sensitivity, false-negative and false-positive GM assay results (Hachem et al. 2009). False positive results occur due to administration of certain beta-lactam drugs, especially piperacillin tazobactam because components of piperacillin-tazobactam are derived from Penicillium sp., which leads to enhanced presence of GM within the drug itself (Sulahian et al. 2003; Viscoli et al. 2004; Singh et al. 2004; Walsh et al. 2004; Park et al. 2010). So it may remain positive even after discontinuation of the antibiotic for as long as five days (Mennink et al. 2004; Walsh et al. 2004). Other β-lactam antibiotics such as amoxicillin-clavulanic acid, cefipime, ceftriaxone, carbapenem and ampicillin shows cross reactivity (Boonsarngsuk et al. 2010; Fortun et al. 2009).

In order to make an early diagnosis the European Conference for Infection in Leukaemia (ECIL-3) recommends a twice a week GM antigen screening in those at high risk of developing invasive aspergillosis (Verweij et al. 1995; Foy et al. 2007; Arendrup et al. 2012; Marchetti et al. 2012). Although the serum GM-ELISA assay is being widely used particularly among hematology/HSCT patients, low sensitivity restricts its similar use in other at-risk non-neutropenic populations (Cordonnier et al. 2009). Lower fungal burden in non-neutropenic patients are also responsible for the reduced sensitivity which may cause a missed diagnosis even when the patient is predisposed to an increased probability of IA infection.

10.3.2 1, 3-β-D Glucan

1,3-β-D Glucan (BDG) is a polysaccharide present in the cell wall of most fungi including *Aspergillus*, *Candida*, *Fusarium*, *Acremonium*, *Penicillium*, *Paecilomyces*, and *Pneumocystis jirovecii*, but not in bacterial or virales cell wall (Odabasi et al 2004). BDG is found in the bloodstream of fungal infected patients. BDG has been used as a diagnostic marker of fungal infection, though it is not immunogenic in nature. It is a calorimetric assay which determines the concentration of BDG in the serum. BDG assay activates a coagulation cascade within amebocytes from the hemolymph of horseshoe crabs (Hope et al. 2005).

The pooled sensitivity and specificity of BDG assay for diagnosis of fungal infection evaluated in a recent meta-analysis of 16 studies was found to be 76.8 % and 85.3 % respectively (Karageorgopoulos et al. 2011). Although there has been variation in the diagnostic cut-off values used and range from 20 to 140 pg/ml (Obayashi et al. 1995; Odabasi et al. 2004; Pazos et al. 2005; De Vlieger et al. 2011). A recent study suggests that the patients with hematological malignancies cut-off value (80 pg/ml) is recommended for the screening purpose of invasive pulmonary aspergillosis (Metan et al. 2012). The clinical validity of the assay is mostly limited to studies of patients with hematological conditions and data further shows reduced diagnostic accuracy. Serum BDG assays may serve as a preliminary screening tool for the diagnosis of IA and have been included in the EORTC criteria for the diagnosis of probable IFI (DePauw et al. 2008). BDG assays appear to shorten the time interval of diagnosis when used as a screening tool in patients with hematological malignancies (Pazos et al. 2005). This helps in early diagnosis and initiation of antifungal therapy in this high risk population. The assay sensitivity does not decrease due to the relatively slow clearance of BDG in humans, despite treatment with antifungal agents (Koo et al. 2009). However the specificity of BDG assay decreases by the following factors: hemodialysis with cellulose membranes (Kanda et al. 2001), intravenous immunoglobulin (IVIG) (Ikemura et al. 1989; Ogawa et al. 2004), albumin or other commercial blood components, cellulose filters for IV administration (Usami et al. 2002; Nagasawa et al; 2003; Ohata et al. 2003),IV amoxicillin-clavulanic acid (Mennink et al. 2006), exposure of serosal surfaces to glucan containing gauze (Marty and Koo 2009), cardiopulmonary bypass (Otto et al. 2013) and bacteremia with *Pseudomonas aeruginosa* (Mennink et al. 2008), Azithromycin and pentamidine solutions inhibit the BDG assay (Marty et al. 2006).

Overall, the BDG assay is characterized by both low sensitivity and specificity for the detection of invasive fungal infection. This assay is most useful for detection of candidiasis and fungal infections other than IA in patients with hematological malignancies. In addition, the evaluation of BDG assay as a diagnostic test for IA specifically remains limited. The assay is not specific to *Aspergillus* and can be positive in other fungal infections including *Candida* and *Pneumocystis jirovecii* infections (Odabasi et al. 2004; Ostrosky et al. 2005; Marty et al. 2006; Mennink et al. 2008; Otto et al. 2013) widening its ambit of detection.

10.3.3 Nucleic Acid Based Molecular Markers

Nucleic acid is an important molecular marker. Nucleic acid-based diagnostics may be used to detect a specific gene or the expression of a gene associated with disease by amplifying the presence of specific DNA or RNA. Nucleic acid-based diagnostics have an advantage over immunoassays in that other than detecting the presence of a pathogen, they can detect genetic markers, such as those for drug resistance in bacteria. However, nucleic acid-based diagnostics may fail to detect the infectious organisms that are able to hide in the body of host where obtaining an infected sample is difficult. There are Common nucleic-acid based diagnostic techniques used to diagnose infectious diseases.

10.3.3.1 PCR Based

The sensitivity of molecular methods boosts the prospectus of identifying an infection at a very early stage, when it is easier to treat the patient or even completely prevent its clinical manifestation. Molecular diagnostic method has primarily targeted nucleic acid amplification, either through isothermal technique (Loeffler et al. 2001; Marty and Koo 2009) or more frequently through polymerase chain reaction (PCR) (Hope et al. 2005; Ostrosky-Zeichner 2012). PCR is one of the most widely applied and oldest molecular methods used in fungal diagnostics. It has been used as a potential method to diagnose *Aspergillus* sp., *Candida* sp., *Pneumocystis jirovecii*, *Pneumonia*, *mucormycosis*, and even rarer fungal infections, such as coccidioidomycosis and scedosporidiosis (Morris and Norris 2012).

A major shortcoming of conventional PCR techniques developed as potential fungal diagnostic tests is that they do not quantify the amount of amplified DNA. Therefore, there is no decisive way to identify the microbial load within the human body. This problem was solved by the development of real-time PCR techniques. Real-time PCR is able to quantify the amount of amplified DNA in real time (Maaroufi et al. 2003).

Several issue associated with isolation of fungal DNA create significant discrepancies between different assays and impede efforts toward standardization curtailing the great potential of PCR based methods. Fungal organisms have strong cell walls that are particularly difficult to lyse, thus required cumbersome methods for DNA isolation, such as phenol-chloroform, mechanical disruption, glass beads, and sonication (Muller et al. 1998; Gonzalez et al. 2010). To overcome this barrier, automated extraction methods have been developed which decreases the time for sample processing and lessen the possibility of errors (Francesconi et al. 2008), and provide homogenous nucleic acid overcoming this shortcoming.

However, there is no clarity regarding whether these techniques alone are adequate to disrupt the fungal cell wall and can significantly improve the sensitivity of fungal PCR assays (Francesconi et al. 2008; Sun et al. 2011). Contamination is another problem associated with fungal PCR. Fungi are ubiquitous and can easily contaminate surfaces and materials used in fungal PCR, including reagents (Miyajima et al. 2009) and collection tubes (Harrison et al. 2010). Therefore, highly skilled personnel and precaution are necessary to avoid false-positive results due to contamination (Zmeili and Soubani 2007).

Due to the aforementioned issues and given the wide variety of fungal infections, with different characteristics and problems associated with each specific fungal infection no single test has yet provided enough evidence of its accuracy to be incorporated into guidelines, and thus PCR is not yet widely used in the diagnosis of IFIs (Marchetti et al. 2012; Kourkoumpetis et al. 2012). PCR assay developed and evaluated for detection of fungal species, such as Scedosporium spp. (Castelli et al. 2008), *Coccidioides* sp. (Thompson et al. 2013), *Fusarium* sp. (Sugawara et al. 2013) and *Cryptococcus* sp. (Rappelli et al. 1998) do not see clinical implementation both because serologic assays are already highly sensitive and specific or clinical evaluation has not yet been accomplished (Castelli et al. 2008; Marchetti et al. 2012).

10.3.3.2 Multiplex PCR

Multiplex PCR is based on the use of primers specifically designed to amplify a region that is conserved among different fungal genera and can detect either some (Paterson et al. 2006) or all (Badiee et al. 2007) fungal species at once in the same specimen. Multiplex PCR has also been tested to detect fungal species in whole-blood (Pryce et al. 2003; Ribeiro et al. 2006; Badiee et al. 2007, 2009; Sugawara et al. 2013), serum (Pham et al. 2003), or BAL fluid (Orsi et al. 2012) samples. The results are variable, but nonetheless reports superior sensitivities and specificities of 80 % (Badiee et al. 2007; Badiee et al. 2009; Mauro et al. 2012; Sugawara et al. 2013). A multiplex assay that can identify 25 pathogens and antibiotic resistance genes in positive blood cultures within 1 h, known as the Film- Array system, was recently developed and commercialized for use in the microbial laboratory (Blaschke et al. 2012).

However, studies evaluating these methods report small numbers of fungal cases due to their use in sepsis cases, where fungal infections are less. Therefore, further research is needed to probe whether these tests would be useful for individuals at high risk for IFIs. Finally, broad-range PCR methods have the potential to be used as methods of rapid identification of the pathogen in cases of outbreaks.

10.3.3.3 Fluorescence In Situ Hybridization (FISH)

FISH is a technique that uses fluorescent probes to identify target areas on the genomes of microbial pathogens in human samples and can be detected by fluorescence microscopy (Fig. 10.1). FISH has been used as an auxiliary technique to culture (Wilson et al. 2005) or PCR (Rickerts et al. 2011) and has been proven to have high accuracy for the identification of Candida sp. infections from blood

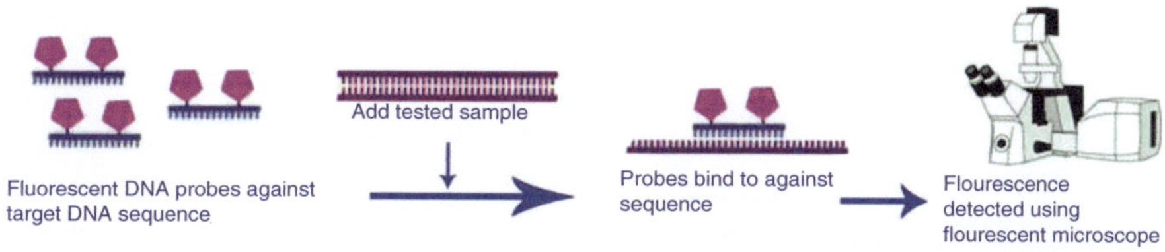

Fig. 10.1 Fluorescence *in situ* hybridization. Fluorescent probes against a specific target sequence and tested sample are mixed and allowed to bind to their complementary DNA sequence. The excess probes are then washed off and the bound probes are detected under a fluorescence microscope

culture bottles (Wilson et al. 2005). This method has a promising performance on frozen tissue sections of coccidioidomycosis (Montone et al. 2010) and invasive fungal rhinosinusitis (Montone et al. 2011), the test show positive results even in cases where cultures are not available or have not been performed.

10.3.3.4 Nucleic Acid Sequence-Based Amplification (NASBA)

This method is very similar to PCR but differs in the sense that it amplifies mRNA by using an RNA polymerase instead of DNA, and it is isothermal (Compton 1991). The methods ability to detect mRNA gives it the advantage of detecting infection of latent or previous infection (Yoo et al. 2005). Loeffler et al. (2001) evaluated the potential of this method as a diagnostic tool for invasive aspergillosis (Loeffler et al. 2001) with threshold for detection of 1 CFU per 100 μl of whole blood. In a later study, investigators developed and evaluated a real-time NASBA method on blood samples from 78 patients and reported a sensitivity and specificity of 100 % and 43 %, respectively (Yoo et al. 2007). The high sensitivities of the assay increase hopes for the future use of the method as a screening test for high-risk populations to rule out IA (Kim et al. 2012).

10.3.3.5 Surface Enhanced Resonance Raman Spectroscopy (SERRS)

This novel spectroscopic method employs specific sensors that can detect scattered light produced by DNA coupled with a specialized dye and placed against roughened surfaces consisting of metals such as gold or silver (Faulds et al. 2005). In this method, the tested sample is combined with multiple DNA probes, each targeting a sequence specific to a different fungal pathogen. After binding of a probe to its complementary fungal DNA within the tested sample, an exonuclease with activity on double-stranded DNA (dsDNA) digests the probe. This step is repeated several times, until most or all probes from the fungal species found in the sample are digested. Finally, the scattered light from the remaining probes is measured by the SERRS sensor, and the missing probe reveals the fungal pathogen (Fig. 10.2). The technique was tested on eight clinical blood culture samples that were positive for

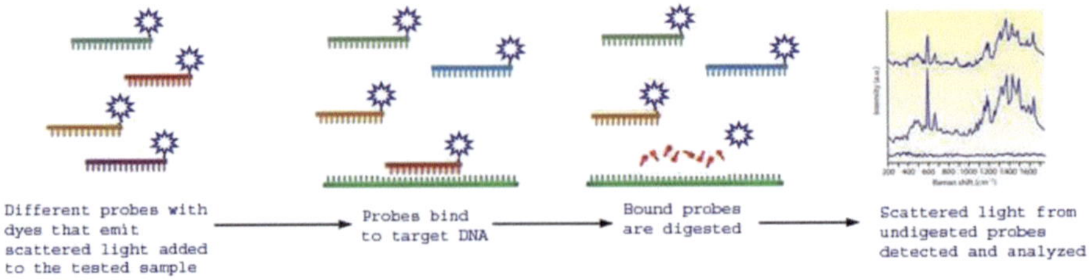

Fig. 10.2 Surface enhanced resonance Raman spectroscopy. The tested sample is placed on a rough surface and various DNA probes coupled with specialized dyes that emit light are added to the sample. The probes bind to the target DNA sample, and subsequently, a double-stranded DNA exonuclease is added to the mix and digests all bound probes, while the unbound probes are left indigested. Finally, the scattered light from the undigested probe is detected by a sensor and analyzed, thus identifying the DNA sequence of the digested probes

fungi and resulted in findings that were 100 % concordant with results from culture (Yoo et al. 2011). Although these results seem promising, prospective studies with larger patient sample populations are needed to establish the position of the method in the realm of clinical diagnosis.

10.3.4 Protein

Protein can be used as diagnostic molecular marker in the identification of fungal disease. Every microorganism has a characteristic pattern of protein expression that can be used as identification marker of microorganism.

10.3.4.1 MALDI-TOF MS

This method is based on mass spectrometry to identify the protein fingerprints of different microorganisms by direct comparison of the spectral pattern of the organism with databases of known patterns from different microorganisms. This method has created a revolution for microbiological laboratories to identify the detected microbe at the genus, species, and even strain levels (Hettick et al. 2008; Seng et al. 2009) (Fig. 10.3).

This system was recently cleared by the FDA for use in clinical microbiology laboratories. Studies shows the performance of this method is promising and is able to accurately and rapidly identify *Candida* sp. and *Aspergillus* sp. from positive cultures, with a high concordance (consistently 90 %) in comparison to conventional methods (Hettick et al. 2008; Marklein et al. 2009; Ferroni et al. 2010; Alanio et al. 2011; Bader et al. 2011; Bille et al. 2012; Ferreira et al. 2011; De Carolis et al. 2012; Iriart et al. 2012; Spanu et al. 2012; Yaman et al. 2012; Lohmann et al. 2013; Rosenvinge et al. 2013; Lacroix et al. 2014). Furthermore, in some reports, given the evidence of its superior and faster performance MALDI-TOF MS seems to

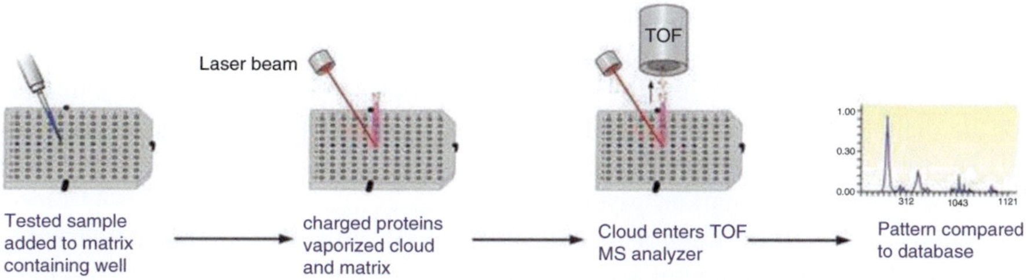

Fig. 10.3 Matrix-assisted laser desorption ionization-time of flight mass spectrometry. The tested sample is added to a matrix containing well which has the ability to absorb UV light and transform it into heat. A laser beam is targeted to the mix. The laser beam is absorbed by the matrix, and part of the analyte-matrix mix is vaporized and ionized, creating a cloud of ionized proteins and matrix. This cloud subsequently is subjected to an electric field, which leads the particles to accelerate toward a detector. The mass and charge of each particle determine the time needed to reach the detector. This allows the mass spectrometer to determine the characteristics of the particles within the tested sample. Comparison of the produced spectral pattern against a standard database allows for identification of the microorganism in the sample

outperform traditional methods of fungal identification techniques (Bader et al. 2011; Iriart et al. 2012).

10.3.4.2 Microscopic Resonating Cantilevers

These are microchips with surfaces coated with antibodies or other proteins that are able to bind microbial membranes from a fluid that is directed to pass over them. After binding of the pathogen, the mass of the cantilever increases, leading to a decrease in its resonance frequency, which can easily be detected (Ilic et al. 2000; Gupta et al. 2004). Cantilevers coated with concanavalin A, fibronectin, and IgG immunoglobulins are now frequently used to detect *Aspergillus niger* and *Saccharomyces cerevisiae* and have been reported to have detection limits of 103–106 CFU/ml (Nugaeva et al. 2005). Although the approach seems intriguing, refinement of the technique to achieve lower detection thresholds and evaluation with clinical samples would be pertinent before it can be considered clinically viable.

10.4 Other Novel Molecular Marker

Biochemical detection of circulating metabolites remains an attractive and relatively non-invasive means of making an early diagnosis of infection are other potential molecular diagnostic avenues that have not been thoroughly explored yet. Primary and secondary metabolites like mannitol and gliotoxin could be used as a potential biomarker for the detection of infection (O'Sullivan et al. 2003; Lewis et al. 2005). D-mannitol has been evaluated as a diagnostic marker in some animal models (Wong et al. 1989; Francis et al. 1994) but due to the complexity of measurements has been of limited clinical use. Gliotoxin produced by most *Aspergillus* strains have a potential to be used as a biochemical marker (Lewis et al. 2005). Gliotoxin precursor, cyclo (L-Phe-L-Ser) may serve as an alternative marker for the diagnosis of invasive aspergillosis caused by *A. fumigatus* (Sekonyela et al. 2013). Bismethylgliotoxin, an inactive derivative of gliotoxin has also been shown as a potential reliable marker, given its ability to be recovered from whole blood (Domingo et al. 2012). Lateral flow device (LFD), an immuno-chromatographic assay can detect an extracellular glycoprotein secreted during the active growth of *Aspergillus sp* using monoclonal antibody JF5 within 15 min and does not require complicated laboratory equipment (Thornton 2008; Held et al. 2013; White et al. 2013). With increasing numbers of fungal infection further research directed towards early and accurate diagnosis needs to be done to find the optimally specific biomarker.

10.5 Management of Fungal Infection

The rising incidence of immunocompromised patients during the past two decades has had profound effects on the management of patients with haematological malignancies. Molecular tests may be used to predict disease response to specific antimicrobial therapy and detection of specific mutation resulting in resistance has proven to be efficacious in managing disease. Molecular-based tests can be used to predict the response to specific therapies and to monitor the response of the disease to the agents administered. This will improve the disease management, reduce toxicity and facilitate cost-effective therapy, optimizing doses and timing regimen.

Clinicians should adapt themselves with an ever increasing number of previously rare fungal pathogens, utilization and interpretation of novel diagnostic modalities, selection and appropriate usage of systemic antifungal drugs and implementation of strategies to prevent infection. Published clinical practice guidelines summarize the best available evidence for managing patients with IFD (Walsh et al 2008; Girmenia et al. 2009; Herbrecht et al. 2011; Arendrup et al. 2012; Marchetti et al. 2012; National Comprehensive Cancer Network. Prevention and Treatment of Cancer Related Infections.). However; at the institutional level to translate such guidelines into effective management strategies one often encounters logistical barriers, chiefly the lack of communication and cooperation between the many specialists and services that need to be involved in the care of patients with IFD.

Patient needs with complex medical conditions today cannot be created by a single healthcare provider, leading to the fragmentation of medical care among multiple consultants and services and inflating the medical costs and expenditure. For that purpose, we need a clearly defined case manager so that responsibility for the patient may not become dispersed amongst numerous medical specialists and they have a clear view of the patient as a whole (Stavert and Lott 2013). Communication is of paramount importance with patients and amongst healthcare providers because interrupted continuity of care and lack of accountability should be considered as a threat to patient safety (Manser 2009). To counter this problem multidisciplinary team approach has to come up as a way of providing comprehensive medical care by bringing together professionals from a wide range of disciplines in a coordinated and effective manner (Manser 2009; Ezziane et al. 2012).

Multidisciplinary team members should be well aware of regional and national trends in fungal infection rates in high-risk groups and antifungal susceptibility patterns, as well as process and outcome measures, such as the quantity and appropriateness of antifungal drug use and survival rates among patients diagnosed with invasive fungal disease. This multidisciplinary team consist of healthcare professionals, from different disciplines directly involved in patient care, such as haemato-oncology, pulmonary medicine, infectious diseases, surgical specialists and nurses, as well as supporting services providers, such as medical microbiology, histopathology, radiology and clinical pharmacology.

For a comprehensive infection control program we need an active surveillance by both infection control practitioners and the clinical microbiology laboratory to identify clusters of infections with a common microbial phenotype. Molecular strain typing of microorganisms is an essential component of a comprehensive infection control program that also involves the infection control department, the infectious disease division, and pharmacy (Pfaller and Herwaldt 1997; Hacek et al. 1999). Molecular tests are effective in tracking the spread of nosocomial infections and streamlining the activities of the infection control program (Back et al. 1993; Pfaller and Herwaldt 1997).

References

Alanio A, Beretti JL, Dauphin B, Mellado E, Quesne G, Lacroix C, Amara A, Berche P, Nassif X, Bougnoux ME. Matrix-assisted laser desorption ionization time-of-flight mass spectrometry for fast and accurate identification of clinically relevant *Aspergillus* species. Clin Microbiol Infect. 2011;17:750–5.

Allan EK, Jordanides NE, McLintock LA, et al. Poor performance of galactomannan and mannan sandwich enzyme-linked immunosorbent assays in the diagnosis of invasive fungal infection. Br J Haematol. 2005;128:578–9.

Arendrup MC, Bille J, Dannaoui E, et al. ECIL-3 classical diagnostic procedures for the diagnosis of invasive fungal diseases in patients with leukaemia. Bone Marrow Transplant. 2012;47:1030–45.

Ascioglu S, Rex JH, de Pauw B, et al. Defining opportunistic invasive fungal infections in immunocompromised patients with cancer and hematopoietic stem cell transplants: an international consensus. Clin Infect Dis. 2002;34:7–14.

Aubry A, Porcher R, Bottero J, et al. Occurrence and kinetics of false-positive Aspergillus galactomannan test results following treatment with beta-lactam antibiotics in patients with hematological disorders. J Clin Microbiol. 2006;44:389–94.

Back NA, Linnemann CC, Pfaller MA, Staneck JL, Morthland V. Recurrent epidemics caused by a single strain of erythromycinresistant *Staphylococcus aureus*: The importance of molecular epidemiology. JAMA. 1993;270:1329–33.

Baddley JW, Andes DR, Marr KA, Kontoyiannis DP, Alexander BD, Kauffman CA, Oster RA, Anaissie EJ, Walsh TJ, Schuster MG, Wingard JR, Patterson TF, Ito JI, Williams OD, Chiller

T, Pappas PG. Factors associated with mortality in transplant patients with invasive aspergillosis. Clin Infect Dis. 2010;50:1559–67.

Bader O, Weig M, Taverne-Ghadwal L, Lugert R, Gross U, Kuhns M. Improved clinical laboratory identification of human pathogenic yeasts by matrix-assisted laser desorption ionization time-of-flight mass spectrometry. Clin Microbiol Infect. 2011;17:1359–65.

Badiee P, Kordbacheh P, Alborzi A, Malekhoseini S, Zeini F, Mirhendi H, Mahmoodi M. Prospective screening in liver transplant recipients by panfungal PCR-ELISA for early diagnosis of invasive fungal infections. Liver Transpl. 2007;13:1011–6.

Badiee P, Kordbacheh P, Alborzi A, Malekhoseini S, Ramzi M, MirhendiH MM, Shakiba E. Study on invasive fungal infections in immunocompromised patients to present a suitable early diagnostic procedure. Int J Infect Dis. 2009;13:97–102.

Bille E, Dauphin B, Leto J, Bougnoux ME, Beretti JL, Lotz A, Suarez S, Meyer J, Join-Lambert O, Descamps P, Grall N, Mory F, Dubreuil L, Berche P, Nassif X, Ferroni A. MALDI-TOF MS Andromas strategy for the routine identification of bacteria, mycobacteria, yeasts, Aspergillus spp. and positive blood cultures. Clin Microbiol Infect. 2012;18:1117–25.

Blaschke AJ, Heyrend C, Byington CL, Fisher MA, Barker E, Garrone NF, Thatcher SA, Pavia AT, Barney T, Alger GD, Daly JA, Ririe KM, Ota I, Poritz MA. Rapid identification of pathogens from positive blood cultures by multiplex polymerase chain reaction using the Film-Array system. Diagn Microbiol Infect Dis. 2012;74:349–55.

Boonsarngsuk V, Niyompattama A, Teosirimongkol C, Sriwanichrak K. False positive serum and bronchoalveolar lavage Aspergillus galactomannan assays caused by different antibiotics. Scand J Infect Dis. 2010;42:461–8.

Caillot D, Casasnovas O, Bernard A, et al. Improved management of invasive pulmonary aspergillosis in neutropenic patients using early thoracic computed tomographic scan and surgery. J Clin Oncol. 1997;15:139–47.

Castelli MV, Buitrago MJ, Bernal-Martinez L, Gomez-Lopez A, Rodriguez-Tudela JL, Cuenca-Estrella M. Development and validation of a quantitative PCR assay for diagnosis of scedosporiosis. J Clin Microbiol. 2008;46:3412–6.

Compton J. Nucleic acid sequence-based amplification. Nature. 1991;350:91–2.

Cordonnier C, Botterel F, Ben Amor R, et al. Correlation between galactomannan antigen levels in serum and neutrophil counts in haematological patients with invasive aspergillosis. Clin Microbiol Infect. 2009;15:81–6.

De Carolis E, Posteraro B, Lass-Flörl C, Vella A, Florio AR, Torelli R, Girmenia C, Colozza C, Tortorano AM, Sanguinetti M, Fadda G. Species identification of *Aspergillus*, *Fusarium* and *Mucorales* with direct surface analysis by matrix-assisted laser desorption ionization time-of-flight mass spectrometry. Clin Microbiol Infect. 2012;18:475–84.

DePauw B, Walsh TJ, Donnelly JP, et al. Revised definitions of invasive fungal disease from the European Organization for Research and Treatment of Cancer/Invasive Fungal Infections Cooperative Group and the National Institute of Allergy and Infectious Diseases Mycoses Study Group (EORTC/MSG) Consensus Group. Clin Infect Dis. 2008;46:1813–21.

De Vlieger G, Lagrou K, Maertens J, et al. Beta-D-glucan detection as a diagnostic test for invasive aspergillosis in immunocompromised critically ill patients with symptoms of respiratory infection: an autopsy-based study. J Clin Microbiol. 2011;49:3783–7.

Domingo MP, Colmenarejo C, Martinez-Lostao L, et al. Bis(methyl)gliotoxin proves to be a more stable and reliable marker for invasive aspergillosis than gliotoxin and suitable for use in diagnosis. Diagn Microbiol Infect Dis. 2012;73:57–64.

Dufresne SF, Datta K, Li X, et al. Detection of urinary excreted fungal galactomannan-like antigens for diagnosis of invasive aspergillosis. PLoS One. 2012;7, e42736.

Ezziane Z, Maruthappu M, Gawn L, et al. Building effective clinical teams in healthcare. J Health Organ Manag. 2012;26:428–36.

Faulds K, Smith WE, Graham D. DNA detection by surface enhanced resonance Raman scattering (SERRS). Analyst. 2005;130:1125–31.

Ferreira L, Sanchez-Juanes F, Porras-Guerra I, Garcia-Garcia MI, Garcia-Sanchez JE, Gonzalez-Buitrago JM, Munoz-Bellido JL. Microorganisms direct identification from blood culture by matrix assisted laser desorption/ionization time-of-flight mass spectrometry. Clin Microbiol Infect. 2011;17:546–51.

Ferroni A, Suarez S, Beretti JL, Dauphin B, Bille E, Meyer J, Bougnoux ME, Alanio A, Berche P, Nassif X. Real-time identification of bacteria and *Candida* species in positive blood culture broths by matrix-assisted laser desorption ionization-time of flight mass spectrometry. J Clin Microbiol. 2010;48:1542–8.

Fortun J, Martin-Davila P, Alvarez ME, et al. False-positive results of Aspergillus galactomannan antigenemia in liver transplant recipients. Transplantation. 2009;87:256–60.

Foy PC, van Burik JA, Weisdorf DJ. Galactomannan antigen enzyme-linked immunosorbent assay for diagnosis of invasive aspergillosis after hematopoietic stem cell transplantation. Biol Blood Marrow Transplant. 2007;13:440–3.

Francis P, Lee JW, Hoffman A, et al. Efficacy of unilamellar liposomal amphotericin B in treatment of pulmonary aspergillosis in persistently granulocytopenic rabbits: the potential role of bronchoalveolar D-mannitol and serum galactomannan as markers of infection. J Infect Dis. 1994;169:356–68.

Francesconi A, Kasai M, Harrington SM, Beveridge MG, Petraitiene R, Petraitis V, Schaufele RL, Walsh TJ. Automated and manual methods of DNA extraction for Aspergillus fumigatus and Rhizopus oryzae analyzed by quantitative real-time PCR. J Clin Microbiol. 2008;46:1978–84.

Girmenia C, Barosi G, Aversa F, et al. Prophylaxis and treatment of invasive fungal diseases in allogeneic stem cell transplantation: results of a consensus process by Gruppo Italiano Trapianto di Midollo Osseo (GITMO). Clin Infect Dis. 2009;49:1226–36.

Girmenia C, Santilli S, Ballarò D, et al. Enteral nutrition may cause false-positive results of Aspergillus galactomannan assay in absence of gastrointestinal diseases. Mycoses. 2011;54:e883–4.

Gonzalez-Mendoza D, Argumedo-Delira R, Morales-Trejo A, Pulido-Herrera A, Cervantes-Diaz L, Grimaldo-Juarez O, Alarcon A. A rapid method for isolation of total DNA from pathogenic filamentous plant fungi. Genet Mol Res. 2010;9:162–6.

Greene RE, Schlamm HT, Oestmann JW, et al. Imaging findings in acute invasive pulmonary aspergillosis: clinical significance of the halo sign. Clin Infect Dis. 2007;44:373–9.

Gupta A, Akin D, Bashir R. Single virus particle mass detection using microresonators with nanoscale thickness. Appl Phys Lett. 2004;84:1976–8.

Hacek DM, Suriano T, Noskin GA, Kruszynski J, Reisberg B, Peterson LR. Medical and economic benefit of a comprehensive infection control program that includes routine determination of microbial clonality. Am J Clin Pathol. 1999;111:647–54.

Hachem RY, Kontoyiannis DP, Chemaly RF, et al. Utility of galactomannan enzyme immunoassay and (1,3) beta-D-glucan in diagnosis of invasive fungal infections: low sensitivity for Aspergillus fumigatus infection in hematologic malignancy patients. J Clin Microbiol. 2009;47:129–33.

Hage CA, Reynolds JM, Durkin M, et al. Plasmalyte as a cause of false-positive results for Aspergillus galactomannan in bronchoalveolar lavage fluid. J Clin Microbiol. 2007;45:676–7.

Halliday C, Hoile R, Sorrell T, et al. Role of prospective screening of blood for invasive aspergillosis by polymerase chain reaction in febrile neutropenic recipients of haematopoietic stem cell transplants and patients with acute leukaemia. Br J Haematol. 2006;132:478–86.

Harrison E, Stalhberger T, Whelan R, Sugrue M, Wingard JR, Alexander BD, Follett SA, Bowyer P, Denning DW, Aspergillus Technology Consortium. Aspergillus DNA contamination in blood collection tubes. Diagn Microbiol Infect Dis. 2010;67:392–4.

Held J, Schmidt T, Thornton CR, et al. Comparison of a novel *Aspergillus* lateral-flow device and the platelia(R) galactomannan assay for the diagnosis of invasive aspergillosis following haematopoietic stem cell transplantation. Infection. 2013;41:1163–9.

Herbrecht R, Fluckiger U, Gachot B et al. Antifungal therapy in leukemia patients—update ECIL4. 2011. http://www.kobe.fr/ecil/telechargements/ECIL%204Update%202011%20Antifungal%20therapy.pdf.

Hettick JM, Green BJ, Buskirk AD, Kashon ML, Slaven JE, Janotka E, Blachere FM, Schmechel D, Beezhold DH. Discrimination of *Aspergillus* isolates at the species and strain level by matrix-assisted laser desorption/ionization time-of-flight mass spectrometry fingerprinting. Anal Biochem. 2008;380:276–81.

Hope WW, Walsh TJ, Denning DW. Laboratory diagnosis of invasive aspergillosis. Lancet Infect Dis. 2005;5:609–22.

Ikemura K, Ikegami K, Shimazu T, et al. False-positive result in Limulus test caused by Limulus amebocyte lysate-reactive material in immunoglobulin products. J Clin Microbiol. 1989;27: 1965–8.

Ilic B, Czaplewski D, Craighead HG, Neuzil P, Campagnolo C, Batt C. Mechanical resonant immunospecific biological detector. Appl Phys Lett. 2000;77:450–2.

Iriart X, Lavergne RA, Fillaux J, Valentin A, Magnaval JF, Berry A, Cassaing S. Routine identification of medical fungi by the new Vitek MS matrix-assisted laser desorption ionization-time of flight system with a new time-effective strategy. J Clin Microbiol. 2012;50:2107–10.

Kanda H, Kubo K, Hamasaki K, et al. Influence of various hemodialysis membranes on the plasma (1→3)-beta-D-glucan level. Kidney Int. 2001;60:319–23.

Karageorgopoulos DE, Vouloumanou EK, Ntziora F, et al. Beta-D-glucan assay for the diagnosis of invasive fungal infections: a meta-analysis. Clin Infect Dis. 2011;52:750–70.

Kim SH, Park C, Kwon EY, Shin NY, Kwon JC, Park SH, Choi SM, Lee DG, Choi JH, Yoo JH. Real-time nucleic acid sequence-based amplification to predict the clinical outcome of invasive aspergillosis. J Korean Med Sci. 2012;27:10–5.

Kimura S, Odawara J, Aoki T, et al. Detection of sputum Aspergillus galactomannan for diagnosis of invasive pulmonary aspergillosis in haematological patients. Int J Hematol. 2009;90:463–70.

Klont RR, Mennink-Kersten MA, Verweij PE. Utility of Aspergillus antigen detection in specimens other than serum specimens. Clin Infect Dis. 2004;39:1467–74.

Kono Y, Tsushima K, Yamaguchi K, et al. The utility of galactomannan antigen in the bronchial washing and serum for diagnosing pulmonary aspergillosis. Respir Med. 2013;107:1094–100.

Koo S, Bryar JM, Page JH, et al. Diagnostic performance of the (1→3)-beta-D-glucan assay for invasive fungal disease. Clin Infect Dis. 2009;49:1650–9.

Kourkoumpetis TK, Fuchs BB, Coleman JJ, Desalermos A, Mylonakis E. Polymerase chain reaction-based assays for the diagnosis of invasive fungal infections. Clin Infect Dis. 2012;54:1322–31.

Latge JP, Kobayashi H, Debeaupuis JP, et al. Chemical and immunological characterization of the extracellular galactomannan of Aspergillus fumigatus. Infect Immun. 1994;62:5424–33.

Lacroix C, Gicquel A, Sendid B, Meyer J, Accoceberry I, Francois N, Morio F, Desoubeaux G, Chandenier J, Kauffmann-Lacroix C, Hennequin C, Guitard J, Nassif X, Bougnoux ME. Evaluation of two matrix-assisted laser desorption ionization-time of flight mass spectrometry(MALDI-TOFMS)systems for the identification of *Candida* species. Clin Microbiol Infect. 2014;20:153–8.

Leeflang MM, Debets-Ossenkopp YJ, Visser CE et al. Galactomannan detection for invasive aspergillosis in immunocompromized patients. Cochrane Database Syst Rev. 2008;CD007394. doi: 10.1002/14651858.CD007394.

Lewis RE, Wiederhold NP, Chi J, et al. Detection of gliotoxin in experimental and human aspergillosis. Infect Immun. 2005;73:635–7.

Loeffler J, Hebart H, Cox P, et al. Nucleic acid sequence-based amplification of Aspergillus RNA in blood samples. J Clin Microbiol. 2001;39:1626–9.

Lohmann C, Sabou M, Moussaoui W, Prevost G, Delarbre JM, Candolfi E, Gravet A, Letscher-Bru V. Comparison between the Biflex III-Biotyper and the Axima-SARAMIS systems for yeast identification by matrix-assisted laser desorption ionization-time of flight mass spectrometry. J Clin Microbiol. 2013;51:1231–6.

Maaroufi Y, Heymans C, De Bruyne JM, Duchateau V, Rodriguez-Villalobos H, Aoun M, Crokaert F. Rapid detection of *Candida* albicans in clinical blood samples by using a TaqMan-based PCR assay. J Clin Microbiol. 2003;41:3293–8.

Maertens J, Van Eldere J, Verhaegen J, et al. Use of circulating galactomannan screening for early diagnosis of invasive aspergillosis in allogeneic stem cell transplant recipients. J Infect Dis. 2002;186:1297–306.

Maertens J, Theunissen K, Verbeken E, et al. Prospective clinical evaluation of lower cut-offs for galactomannan detection in adult neutropenic cancer patients and haematological stem cell transplant recipients. Br J Haematol. 2004;126:852–60.

Maertens JA, Nucci M, Donnelly JP. The role of antifungal treatment in hematology. Haematologica. 2012;97:325–7.

Manser T. Teamwork and patient safety in dynamic domains of healthcare: a review of the literature. Acta Anaesthesiol Scand. 2009;53:143–51.

Marchetti O, Lamoth F, Mikulska M, Viscoli C, Verweij P, Bretagne S, European Conference on Infections in Leukemia Laboratory Working Group. ECIL recommendations for the use of biological markers for the diagnosis of invasive fungal diseases in leukemic patients and hematopoietic SCT recipients. Bone Marrow Transplant. 2012;47:846–54.

Marklein G, Josten M, Klanke U, Muller E, Horre R, Maier T, Wenzel T, Kostrzewa M, Bierbaum G, Hoerauf A, Sahl HG. Matrixassisted laser desorption ionization-time of flight mass spectrometry for fast and reliable identification of clinical yeast isolates. J Clin Microbiol. 2009;47:2912–7.

Marty FM, Lowry CM, Lempitski SJ, et al. Reactivity of (1→3)-beta-d-glucan assay with commonly used intravenous antimicrobials. Antimicrob Agents Chemother. 2006;50:3450–3.

Marty FM, Koo S. Role of (1→3)-beta-D-glucan in the diagnosis of invasive aspergillosis. Med Mycol. 2009;47:S233–40.

Marr KA, Balajee SA, McLaughlin L, et al. Detection of galactomannan antigenemia by enzyme immunoassay for the diagnosis of invasive aspergillosis: variables that affect performance. J Infect Dis. 2004;190:641–9.

Mauro MV, Cavalcanti P, Perugini D, Noto A, Sperli D, Giraldi C. Diagnostic utility of LightCycler SeptiFast and procalcitonin assays in the diagnosis of bloodstream infection in immunocompromised patients. Diagn Microbiol Infect Dis. 2012;73:308–11.

Mennink-Kersten MA, Donnelly JP, Verweij PE. Detection of circulating galactomannan for the diagnosis and management of invasive aspergillosis. Lancet Infect Dis. 2004;4:349–57.

Mennink-Kersten MA, Warris A, Verweij PE. 1,3-beta-Dglucan in patients receiving intravenous amoxicillin-clavulanic acid. N Engl J Med. 2006;354:2834–5.

Mennink-Kersten MA, Ruegebrink D, Verweij PE. Pseudomonas aeruginosa as a cause of 1,3-beta-D-glucan assay reactivity. Clin Infect Dis. 2008;46:1930–1.

Metan G, Koc AN, Atalay A, et al. What should be the optimal cut-off of serum 1,3-beta-D-glucan for the detection of invasive pulmonary aspergillosis in patients with haematological malignancies? Scand J Infect Dis. 2012;44:330–6.

Meyers JD. Fungal infections in bone marrow transplant patients. Semin Oncol. 1990;17:10–3.

Miyajima Y, Satoh K, Umeda Y, Makimura K. Quantitation of fungal DNA contamination in commercial zymolyase and lyticase used in the preparation of fungi. Nippon Ishinkin Gakkai Zasshi. 2009;50:259–62.

Montone KT, Litzky LA, Feldman MD, Peterman H, Mathis B, Baliff J, Kaiser LR, Kucharczuk J, Nachamkin I. In situ hybridization for Coccidioides immitis 5.8S ribosomalRNAsequences in formalin-fixed, paraffin-embedded pulmonary specimens using a locked nucleic acid probe: a rapid means for identification in tissue sections. Diagn Mol Pathol. 2010;19:99–104.

Montone KT, LiVolsi VA, Lanza DC, Kennedy DW, Palmer J, Chiu AG, Feldman MD, Loevner LA, Nachamkin I. In situ hybridization for specific fungal organisms in acute invasive fungal rhinosinusitis. Am J Clin Pathol. 2011;135:190–9.

Morris A, Norris KA. Colonization by Pneumocystis jirovecii and its role in disease. Clin Microbiol Rev. 2012;25:297–317.

Muller FM, Werner KE, Kasai M, Francesconi A, Chanock SJ, Walsh TJ. Rapid extraction of genomic DNA from medically important yeasts and filamentous fungi by high-speed cell disruption. J Clin Microbiol. 1998;36:1625–9.

Musher B, Fredricks D, Leisenring W, et al. Aspergillus galactomannan enzyme immunoassay and quantitative PCR for diagnosis of invasive aspergillosis with bronchoalveolar lavage fluid. J Clin Microbiol. 2004;42:5517–22.

Nagasawa K, Yano T, Kitabayashi G, et al. Experimental proof of contamination of blood components by (1→3)-beta-D glucan caused by filtration with cellulose filters in the manufacturing process. J Artif Organs. 2003;6:49–54.

National Comprehensive Cancer Network. Prevention and treatment of cancer related infections. http://www.nccn.org/professionals/physician_gls/pdf/infections.pdf.

Nguyen MH, Wissel MC, Shields RK, Salomoni MA, Hao B, Press EG, Shields RM, Cheng S, Mitsani D, Vadnerkar A, Silveira FP, Kleiboeker SB, Clancy CJ. Performance of Candida real-time polymerase chain reaction, beta-D-glucan assay, and blood cultures in the diagnosis of invasive candidiasis. Clin Infect Dis. 2012;54:1240–8.

Nugaeva N, Gfeller KY, Backmann N, Lang HP, Duggelin M, Hegner M. Micromechanical cantilever array sensors for selective fungal immobilization and fast growth detection. Biosens Bioelectron. 2005;21:849–56.

Obayashi T, Yoshida M, Mori T, et al. Plasma (1→3)-beta-Dglucan measurement in diagnosis of invasive deep mycosis and fungal febrile episodes. Lancet. 1995;345:17–20.

Odabasi Z, Mattiuzzi G, Estey E, et al. Beta-D-glucan as a diagnostic adjunct for invasive fungal infections: validation, cutoff development, and performance in patients with acute myelogenous leukemia and myelodysplastic syndrome. Clin Infect Dis. 2004;39:199–205.

Ogawa M, Hori H, Niiguchi S, et al. False-positive plasma (1→3)-beta-D-glucan test following immunoglobulin product replacement in an adult bone marrow recipient. Int J Hematol. 2004;80:97–8.

Ohata A, Usami M, Horiuchi T, et al. Release of (1→3)-beta-D-glucan from depth-type membrane filters and their in vitro effects on proinflammatory cytokine production. Artif Organs. 2003; 27:728–35.

Orsi CF, Gennari W, Venturelli C, La Regina A, Pecorari M, Righi E, Machetti M, Blasi E. Performance of 2 commercial real-time polymerase chain reaction assays for the detection of *Aspergillus* and *Pneumocystis* DNA in bronchoalveolar lavage fluid samples from critical care patients. Diagn Microbiol Infect Dis. 2012;73:138–43.

Ostrosky-Zeichner L. Invasive mycoses: diagnostic challenges. Am J Med. 2012;125:S14–24.

Ostrosky-Zeichner L, Alexander BD, Kett DH, et al. Multicenter clinical evaluation of the (1→3) beta-D-glucan assay as an aid to diagnosis of fungal infections in humans. Clin Infect Dis. 2005;41:654–9.

O'Sullivan CE, Kasai M, Francesconi A, et al. Development and validation of a quantitative real-time PCR assay using fluorescence resonance energy transfer technology for detection of *Aspergillus fumigatus* in experimental invasive pulmonary aspergillosis. J Clin Microbiol. 2003;41:5676–82.

Otto GP, Ludewig K, Jacobsen ID, et al. Limitation of (1→3)-beta-D-glucan monitoring in major elective surgery involving cardiopulmonary bypass. Crit Care. 2013;17:437.

Park BJ, Pappas PG, Wannemuehler KA, Alexander BD, Anaissie EJ, Andes DR, Baddley JW, Brown JM, Brumble LM, Freifeld AG, Hadley S, Herwaldt L, Ito JI, Kauffman CA, Lyon GM, Marr KA, Morrison VA, Papanicolaou G, Patterson TF, Perl TM, Schuster MG, Walker R, Wingard JR, Walsh TJ, Kontoyiannis DP. Invasive non-Aspergillus mold infections in transplant recipients, United States, 2001–2006. Emerg Infect Dis. 2011;17:1855–64.

Park SY, Lee SO, Choi SH, et al. Aspergillus galactomannan antigen assay in bronchoalveolar lavage fluid for diagnosis of invasive pulmonary aspergillosis. J Infect. 2010;61:492–8.

Paterson PJ, Seaton S, McHugh TD, McLaughlin J, Potter M, Prentice HG, Kibbler CC. Validation and clinical application of molecular methods for the identification of molds in tissue. Clin Infect Dis. 2006;42:51–6.

Pazos C, Ponton J, Del Palacio A. Contribution of (1→3)-beta-D-glucan chromogenic assay to diagnosis and therapeutic monitoring of invasive aspergillosis in neutropenic adult patients: a comparison with serial screening for circulating galactomannan. J Clin Microbiol. 2005;43:299–305.

Pfaller MA, Herwaldt LA. The clinical microbiology laboratory and infection control: emerging pathogens, antimicrobial resistance, and new technology. Clin Infect Dis. 1997;25:858–70.

Pfeiffer CD, Fine JP, Safdar N. Diagnosis of invasive aspergillosis using a galactomannan assay: a meta-analysis. Clin Infect Dis. 2006;42:1417–27.

Pham AS, Tarrand JJ, May GS, Lee MS, Kontoyiannis DP, Han XY. Diagnosis of invasive mold infection by real-time quantitative PCR. Am J Clin Pathol. 2003;119:38–44.

Pryce TM, Kay ID, Palladino S, Heath CH. Real-time automated polymerase chain reaction (PCR) to detect *Candida albicans* and *Aspergillus fumigatus* DNA in whole blood from high-risk patients. Diagn Microbiol Infect Dis. 2003;47:487–96.

Rappelli P, Are R, Casu G, Fiori PL, Cappuccinelli P, Aceti A. Development of a nested PCR for detection of *Cryptococcus neoformans* in cerebrospinal fluid. J Clin Microbiol. 1998;36:3438–40.

Rex JH. Galactomannan and the diagnosis of invasive aspergillosis. Clin Infect Dis. 2006;42:1428.

Ribeiro P, Costa F, Monteiro A, Caldas J, Silva M, Ferreira G, Veiga J, Sousa MO, Viegas MP, Santos E, Goncalves AJ, Sousa AB. Polymerase chain reaction screening for fungemia and/or invasive fungal infections in patients with hematologic malignancies. Support Care Cancer. 2006;14:469–74.

Rickerts V, Khot PD, Myerson D, Ko DL, Lambrecht E, Fredricks DN. Comparison of quantitative real time PCR with sequencing and ribosomal RNA-FISH for the identification of fungi in formalin fixed, paraffin-embedded tissue specimens. BMC Infect Dis. 2011;11:202.

Rosenvinge FS, Dzajic E, Knudsen E, Malig S, Andersen LB, Lovig A, Arendrup MC, Jensen TG, Gahrn-Hansen B, Kemp M. Performance of matrix-assisted laser desorption-time of flight mass spectrometry for identification of clinical yeast isolates. Mycoses. 2013;56:229–35.

Salonen J, Lehtonen OP, Terasjarvi MR, et al. Aspergillus antigen in serum, urine and bronchoalveolar lavage specimens of neutropenic patients in relation to clinical outcome. Scand J Infect Dis. 2000;32:485–90.

Sekonyela R, Palmer JM, Bok JW, et al. RsmA regulates *Aspergillus fumigatus* gliotoxin cluster metabolites including cyclo(L-Phe-L-Ser), a potential new diagnostic marker for invasive aspergillosis. PLoS One. 2013;8, e62591.

Seng P, Drancourt M, Gouriet F, La Scola B, Fournier PE, Rolain JM, Raoult D. Ongoing revolution in bacteriology: routine identification of bacteria by matrix-assisted laser desorption ionization time-offlight mass spectrometry. Clin Infect Dis. 2009;49:543–51.

Singh N, Obman A, Husain S, et al. Reactivity of platelia Aspergillus galactomannan antigen with piperacillin-tazobactam: clinical implications based on achievable concentrations in serum. Antimicrob Agents Chemother. 2004;48:1989–92.

Spanu T, Posteraro B, Fiori B, D'Inzeo T, Campoli S, Ruggeri A, Tumbarello M, Canu G, Trecarichi EM, Parisi G, Tronci M, Sanguinetti M, Fadda G. Direct matrix-assisted laser desorption ionization-time of flight mass spectrometry assay of blood culture broths for rapid identification of *Candida* species causing bloodstream infections: an observational study in two large microbiology laboratories. J Clin Microbiol. 2012;50:176–9.

Stavert RR, Lott JP. The bystander effect in medical care. N Engl J Med. 2013;368:8–9.

Stynen D, Goris A, Sarfati J, et al. A new sensitive sandwich enzyme-linked immunosorbent assay to detect galactofuran in patients with invasive aspergillosis. J Clin Microbiol. 1995;33: 497–500.

Sugawara Y, Nakase K, Nakamura A, Ohishi K, Sugimoto Y, Fujieda A, Monma F, Suzuki K, Masuya M, Matsushima Y, Wada H, Nobori T, Katayama N. Clinical utility of a panfungal polymerase chain reaction assay for invasive fungal diseases in patients with haematologic disorders. Eur J Haematol. 2013;90:331–9.

Sulahian A, Touratier S, Ribaud P. False positive test for Aspergillus antigenemia related to concomitant administration of piperacillin and tazobactam. N Engl J Med. 2003;349:2366–7.

Sun W, Wang K, Gao W, Su X, Qian Q, Lu X, Song Y, Guo Y, Shi Y. Evaluation of PCR on bronchoalveolar lavage fluid for diagnosis of invasive aspergillosis: a bivariate metaanalysis and systematic review. PLoS One. 2011;6, e28467.

Thompson GR, Sharma S, Bays DJ, Pruitt R, Engelthaler DM, Bowers J, Driebe EM, Davis M, Libke R, Cohen SH, Pappagianis D. Coccidioidomycosis: adenosine deaminase levels, serologic parameters, culture results, and polymerase chain reaction testing in pleural fluid. Chest. 2013;143:776–81.

Thornton CR. Development of an immunochromatographic lateral-flow device for rapid serodiagnosis of invasive aspergillosis. Clin Vaccine Immunol. 2008;15:1095–105.

Usami M, Ohata A, Horiuchi T, et al. Positive (1→3)-beta-D-glucan in blood components and release of (1→3)-beta-D glucan from depth-type membrane filters for blood processing. Transfusion. 2002;42:1189–95.

Viscoli C, Machetti M, Cappellano P, et al. False-positive galactomannan platelia Aspergillus test results for patients receiving piperacillin-tazobactam. Clin Infect Dis. 2004;38:913–6.

Verweij PE, Latge JP, Rijs AJ, et al. Comparison of antigen detection and PCR assay using bronchoalveolar lavage fluid for diagnosing invasive pulmonary aspergillosis in patients receiving treatment for hematological malignancies. J Clin Microbiol. 1995;33:3150–3.

von Eiff M, Roos N, Schulten R, et al. Pulmonary aspergillosis: early diagnosis improves survival. Respiration. 1995;62:341–7.

Walsh TJ, Shoham S, Petraitiene R, et al. Detection of galactomannan antigenemia in patients receiving piperacillintazobactam and correlations between in vitro, in vivo, and clinical properties of the drug-antigen interaction. J Clin Microbiol. 2004;42:4744–8.

Walsh TJ, Anaissie EJ, Denning DW, et al. Treatment of aspergillosis: clinical practice guidelines of the Infectious Diseases Society of America. Clin Infect Dis. 2008;46:327–60.

Wilson DA, Joyce MJ, Hall LS, Reller LB, Roberts GD, Hall GS, Alexander BD, Procop GW. Multicenter evaluation of a Candida albicans peptide nucleic acid fluorescent in situ hybridization probe for characterization of yeast isolates from blood cultures. J Clin Microbiol. 2005;43:2909–12.

White PL, Parr C, Thornton C, et al. Evaluation of realtime PCR, galactomannan enzyme-linked immunosorbent assay (ELISA), and a novel lateral-flow device for diagnosis of invasive aspergillosis. J Clin Microbiol. 2013;51:1510–6.

Wong B, Brauer KL, Tsai RR, et al. Increased amounts of the *Aspergillus* metabolite D-mannitol in tissue and serum of rats with experimental aspergillosis. J Infect Dis. 1989;160:95–103.

Yaman G, Akyar I, Can S. Evaluation of the MALDI TOF-MS method for identification of *Candida* strains isolated from blood cultures. Diagn Microbiol Infect Dis. 2012;73:65–7.

Yoo JH, Choi JH, Choi SM, Lee DG, Shin WS, Min WS, Kim CC. Application of nucleic acid sequence-based amplification for diagnosis of and monitoring the clinical course of invasive aspergillosis in patients with hematologic diseases. Clin Infect Dis. 2005;40:392–8.

Yoo JH, Choi SM, Lee DG, Park SH, Choi JH, Kwon EY, Shin WS. Comparison of the real-time nucleic acid sequence-based amplification (RTi-NASBA) with conventional NASBA, and galactomannan assay for the diagnosis of invasive aspergillosis. J Korean Med Sci. 2007;22:672–6.

Yoo SM, Kang T, Kang H, Lee H, Kang M, Lee SY, Kim B. Combining a nanowire SERRS sensor and a target recycling reaction for ultrasensitive and multiplex identification of pathogenic fungi. Small. 2011;7:3371–6.

Zmeili OS, Soubani AO. Pulmonary aspergillosis: a clinical update. QJM. 2007;100:317–34.

Chapter 11
Mitochondrial DNA Based Molecular Markers in Arbuscular Mycorrhizal Fungi (AMF) Research

R.K. Sarma, Ratul Saikia, and N.C. Talukdar

11.1 Introduction

Arbuscular mychorrhizal (AM) fungi (Phylum Glomeromycota) are the plant root hindering obligate symbionts with more than 80 % of plant species (Schüßler et al. 2001). The symbiotic association provides nutrients and also resistance from soil borne pathogens to the host plants in exchange for plant assimilated carbon for the fungi (Smith and Read 2008). The AM fungi (AMF) in soil helps the plants to access many nutrients (e.g. phosphate) and simultaneously provides suitable food materials for multiple rhizosphere associated microorganisms. Thus the unseen fungi-plant assembly might have the capability to modulate plant and soil community interaction with respect to different ecological factors and thereby modulate a proper terrestrial ecosystem functioning (Finlay 2008). However, a proper investigation of their ecological significance is still in a premature stage because of the controversies in both their classification and phylogeny (Redecker et al. 2013). Uncertainty in the biological characters makes genetic investigations of AMF highly complicated. Earlier they have been recognized as asexual entities; however, recent finding stated them to be the sexually reproducing organisms (Riley et al. 2014). Additionally, the

R.K. Sarma (✉) • R. Saikia
Biotechnology Group, Biological Science & Technology Division, CSIR-North East Institute of Science and Technology, Jorhat, Assam 785006, India

Life Science Division, Institute of Advanced Study in Science and Technology (DST, Govt. India), Guwahati, Assam 781035, India
e-mail: rsarmaonline@gmail.com

N.C. Talukdar
Life Science Division, Institute of Advanced Study in Science and Technology (DST, Govt. India), Guwahati, Assam 781035, India

B.P. Singh, V.K. Gupta (eds.), *Molecular Markers in Mycology*, Fungal Biology,
DOI 10.1007/978-3-319-34106-4_11

origin of interspecific and intraspecific genetic diversity is still confusing. These complex diversity pattern within the AMF individual or populations was the main reason of delaying the use of molecular markers to figure out their community profiling, diversity and ecological importance (Beaudet et al. 2013a, b).

So far, near about 230 numbers of AMF species have been identified (www.amf-phylogeny.com). Most of the previous classification was based on spore and hyphae morphology. However in recent years, rapid development in molecular biology change this picture too many extent. Most of the molecular biology based classification depends on sequence driven information of ribosomal RNA (*rRNA*) genes. Nucleotide sequences of nuclear small subunit (SSU) *rRNA* gene (Clapp et al. 1999; Krüger et al. 2012; Montes-Borrego et al. 2014), internal transcribes spacer 1 (*ITS1*) (Melanie et al. 2002; Lindahl et al. 2013), *5.8S rRNA* gene and *ITS2* (Renker et al. 2003; Krüger et al. 2012) as well the nuclear large subunit (*LSU*) (Krüger et al. 2012; Porter and Golding 2012) are now getting the prime importance in this aspect. However, the main demerits of these markers were their inability to distinguish the AMF up to species level. This is because of their elusiveness in genome organization (Rosendahl 2012). Again multiple allelic variants of some nuclear genes within an isolate derived from a single spore is not an uncommon phenomena in AMF (Sanders et al. 1995; Helgason et al. 2003; Koch et al. 2004). In this case, AMF mitochondrial DNA (mtDNA) could be the more appropriate target for marker development due to their homogeneity characteristics within single isolates (Beaudet et al. 2015). Again, their maternal inheritance and almost complete absence of recombination make the organelle genomes a unique tool for population biology. Therefore, mtDNA based markers are now gaining much importance as a marker in AMF taxonomy.

11.2 Brief Glimpses on AMF Mitochondrial Genome

Mitochondria are one of the important membrane bound cell organelle that takes part in different cell processes like oxidative phosphorylation, respiration, RNA maturation, and protein synthesis etc. This cell organelle have their own genetic makeup (mt genome) with definite structure, size and function (Fukuhara et al. 1993; Nadimi et al. 2012). According to endosymbiotic theory (Margulis 1971), there was a parallel evolution of both nuclear and mitochondrial genome in a eukaryotic cell. Like the nuclear genome; gene content, intron size, mobile elements like open reading frames (orfs), intergenic regions, plasmid-related DNA polymerase sequence are the key elements for polymorphism in eukaryotic mt genomes (Nadimi et al. 2015).

The First sequenced AMF mt genome was from *Rhizophagus irregularis* isolate 494, (previously named *Glomus intraradices* and then *Glomus irregulare*) (Lee and Young 2009), followed by 11 different starins of *Rhizophagus*, *Glomus*, and *Gigaspora* (Formey et al. 2012; Nadimi et al. 2012; Pelin et al. 2012; Beaudet et al. 2013a, b; de la Providencia et al. 2013). Lee and Young (2009) had sequenced the 70606 bp of mt genome of *Rhizophagus irregularis* isolate 494 (Fig. 11.1). They

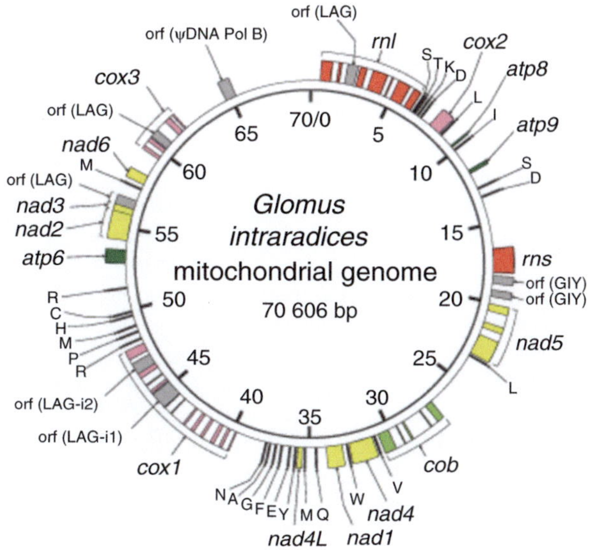

Fig. 11.1 Mitochondrial genome of *Glomus intraradices* FACE494. Location of genes (*shaded*) and introns (*white*) has been shown on the outer cycle with clockwise orientation. The scale is in kb (Adapted from Lee and Young 2009)

represented it by a circular map with a G+C content of 37.2 %. The mitochondrial gene assembly universally encodes the gene products either for (a) tRNAs, the small subunit (*rns*) and the large subunit (*rnl*) ribosomal RNAs or (b) for oxidative phosphorylation, i.e. NADH dehydrogenase complex (*nad*), apocytochrome b (*cob*), subunits 1–3 of cytochrome c oxidase (*cox*) and subunit 6, 8 and 9 of ATP synthase (*atp*). Lee and Young (2009) observed very low level of intra-isolate genetic variation in the whole mitochondrial genome. Unlike the nuclear genome, variants in mitochondrial sequences were obscure. The genome maintained the state of homoplasmy, i.e. all the mitochondrial genomes in a cell were essentially identical. Henceforth, they concluded the mitochondrial sequences to be the most unique for marker development rather than nuclear genes in the field of AMF research.

11.3 Mitochondrial Large Ribosomal Subunit (*mtLSU*) as Molecular Marker in AMF

Mitochondrial DNA has a long past of using as a molecular marker in AMF research (Table 11.1). Bruns et al. (1998) was the first to use an *mtLSU* region for identification of ectomycorrhizal fungal species from colonized roots. However, application of *mtLSU* in AMF identification was first introduced by Raab et al. (2005). The intraindividual as well as intraspecific sequence variability was been reported from the first AMF *mtLSU* sequence analysis at *Glomus intraradices* and *G. proliferum*. Similar variability was also reported in other AMF species (Formey et al. 2012; Nadimi et al. 2012; Pelin et al. 2012; Beaudet et al. 2013a, b, 2015). The *mtLSU* and its introns were found to be of highly accurate molecular markers for understanding

Table 11.1 The use of mitochondrial (*mt*) genes in taxonomy of different AMF

Mitochndrial (*mt*) gene	AMF individual	Reference
mtLSU and *mtLSU* introns	*Glomus* Spp	Raab et al. (2005)
mtLSU (homing endonucleases, DNA polymerase domain-containing open reading frames and small inverted repeats)	*Glomus irregulare*	Formey et al. (2012)
mtLSU introns	*Gigaspora rosea*	Nadimi et al. (2012)
mtLSU group I introns	*Gigaspora margarita*	Pelin et al. (2012)
mtLSU introns, mitochondrial plasmid-like DNA polymerase genes and mobile open reading frames	*Glomus sp*	Beaudet et al. (2013a)
Whole mtDNA	*Rhizophagus irregularis*	Beaudet et al. (2015)
mtLSU rRNA, *mtLSU* exon	*Glomus intraradices*	Börstler et al. (2008)
mtLSU Group I introns	*Glomus intraradices*	Lee and Young (2009)
mtLSU Group I introns	*Rhizophagus* sp.	Nadimi et al. (2015)
mtLSU	*Rhizophagus irregularis*	Peyret-Guzzon et al. (2015)

the diversity and community dynamics of field AMF populations (Börstler et al. 2008). Similarly, the *mtLSU* exon phylogeny also showed broader distinction within the strains of *G. intraradices*, while compared to nuclear rDNA based markers (Börstler et al. 2008). Lee and Young (2009) stated group I introns of *mtLSU* to be the wide spread genetic elements in *G. intraradices*. These introns are proved to be the most sensitive molecular markers in AMF population studies as they tend to change very fast in evolution due to low selection pressure (Nadimi et al. 2015). The group I introns with the characters of sequential ester-transfer splicing, encodes homing endonucleases i.e H.E. (selfish mobile elements). These enzymes have the capacity to cleave exons at specific recognition sites (15–35 bp) with the subsequent function of double-strand repair and insertion of the intron-containing allele at the respective site via recombination (Chevalier and Stoddard 2001). Haugen et al. (2005) reported the HE invasion in introns and simultaneous propagation and horizontal transfer of group I introns between the species of same or different kingdoms. Alternate way of intron transfer may be prompted by reverse splicing, i.e. pairing of few nucleotides with the internal guide sequence (IGS) of the intron (Woodson and Cech 1989). Thus the reverse splicing could turn the introns into new genes.

Raab et al. (2005) first stated the homogeneity of *mtLSU* in *G. intraradices* and *G. proliferum*, two glomeromycotan fungi species. Later, Lee and Young (2009) proved it through analysis of the complete mitochondrial genome of *G. intraradices*. Further, the genetic homogeneity have been proved to be a general feature of other AMF taxa (Thiery et al. 2010). Now a days, the *mtLSU* gene have been using as a sole marker gene in AMF taxonomy (Croll et al. 2008; Kiers et al. 2011; Thonar et al. 2012; Engelmoer et al. 2014). Recently, Peyret-Guzzon et al. (2015) have used *mtLSU* as a marker gene to understand the effect of physical disturbance and chemical fertilizers on intraspecific populations of *Rhizophagus irregularis* in a buffer strip surrounded by arable fields.

11.4 Cytochrome c Oxidase (*cox*) as Molecular Marker

Cytochrome oxidase I (*cox1*) is a conserved gene of mitochondrial genomes encoding the cytochrome c oxidase largest subunit, the constituent of the complex IV participates in the respiratory chain (Capaldi et al. 1983). The expression of *cox1* gene transcript are the indicative of cell respiration and thereby controls the cell physiological activity. A portion of *cox1* have been used in case of animal taxonomy. However, the gene gets little attention in fungal classification due to the frequent existence of either group I or II introns of varying sizes that makes problems in amplification of coding regions. A comprehensive investigation revealed length distinction of 1584 to 22006 bp in *cox1* gene of 56 fungal species (Seifert et al. 2007). In *G. intraradices* FACE494 (an AM fungi), 11 introns had been reported in *cox1* (Lee and Young 2009). Borriello (2010) investigated the *cox1* sequence diversity in different AMF, i.e. *Scutellospora*, *Gigaspora and Glomus* group A species, stated the low intraspecific sequence polymorphism (Fig. 11.1). Moreover, Borriello (2010) was not able to deign universal Glomeromycotan *cox1* primes of different AMF descents. The overall study of Borriello (2010) proved *cox1* as an unsuitable marker gene for AMF taxonomy. Damon et al. (2010) had argued in solving this problem through the application of reverse-transcribed RNA, including environmental RNA. However, there is a still a question mark of wide acceptability of *cox1*as universal molecular marker in AMF research.

11.5 Pitfalls of mtDNA Based Markers in AMF Research

The mitochondrial DNA based AMF diagnosis is one of the developing filed in AMF research. The mtDNA sequences are proved to be the virtually homogeneous, which might raise opportunities in answering many questions in AMF research. It will be useful in AMF identification up to strain level, even at single spore level. Secondly, the mtDNA based marker could be used in selection of new and more effective mycorrhizal strains. However, identification of most reliable differential mtDNA sequence; preferably insertions, deletions or genome re-arrangement are found to be most important criteria for distinctions between AMF isolates. Present investigations revealed many possible hindrances in restoring the homogeneity within AMF mtDNA of a single isolate. The work of Lee and Young (2009) on mitochondrial genome homogeneity was a hypothesis based on effective segregation and repair mechanisms. Till date knowledge on mitochondrial inheritance pattern in AMF is very scanty. Rabb et al. (2005) first reported the homogeneity of AMF mitochondrial marker and followed by different workers in plants and fungi.

The comprehensive investigation of de la Providencia et al. (2013) showed the occurrence of transient mitochondrial length heteroplasmy in spores nearer to anastomosis region in geographically distant *Rhizophagus irregularis* in vitro isolates. The length divergence between the two mtDNA haplotypes is due to the variation in selfish mobile elements (MGE), i.e. homing endonuclease, plasmid related DNA polymerase (*dpo*) and short inverted repeats (SIRs) (Formey et al. 2012; Beaudet et al. 2013a, b). However, till now their movement and recombination mechanisms are not clearly understood. These MGE are recognized as the DNA fragments that encodes enzymes or other proteins related to the movement of chromosomal segments through intracellular or intercellular mobility. Two similar SIR pairing may lead to intrachromosomal homologous recombination, i.e. genome rearrangements (Beaudet et al. 2013a, b).

11.6 Conclusion and Perspectives

The use of mitochondrial DNA sequence is one of the prime developing field in AMF genetics. Comparative mitochondrial genomics together with AMF whole genome sequencing may pave a new door towards development of species specific molecular marker with comparable evolutionary distances. A complete database of mtDNA sequencing within various AMF isolates may give rise to many information on its genome organization, segregation and also identification of mobile selfish elements. The identification of selfish mobile elements could be beneficial to estimate their predominance within the AMF mitochondrial genome and thereby assessing the consequences on mitochondrial genome organization.

Acknowledgement Authors are thankful to the Department of Biotechnology, for establishment of DBT-BIF centre and DBT-state Biotech Hub in the Biological Science & Technology Division, CSIR-NEIST Jorhat, which has been used for the present book chapter.

References

Beaudet D, de la Providencia IE, Labridy M, Roy-Bolduc A, Daubois L, Hijri M. Intraisolate mitochondrial genetic polymorphism and gene variants coexpression in arbuscular mycorrhizal fungi. Genome Biol Evol. 2015;7(1):218–27.

Beaudet D, Terrat Y, Halary S, de la Providencia IE, Hijri M. Mitochondrial genome rearrangements in *Glomus* species triggered by homologous recombination between distinct mtDNA haplotypes. Genome Biol Evol. 2013a;5(9):1628–43.

Beaudet D, Nadimi M, Iffis B, Hijri M. Rapid mitochondrial genome evolution through invasion of mobile elements in two closely related species of arbuscular mycorrhizal fungi. PLoS One. 2013b;8(4), e60768.

Borriello R. Arbuscular mycorrhizal fungal diversity in agricultural and natural ecosystem soils. In: Looking for new molecular tools and targets for AM fungal characterization, Università degli Studi di Torino, Torino (PhD Thesis); 2010.

Börstler B, Raab PA, Thiéry O, Morton JB, Redecker D. Genetic diversity of the arbuscular mycorrhizal fungus *Glomus intraradices* as determined by mitochondrial large subunit rRNA gene sequences is considerably higher than previously expected. New Phytol. 2008;180(2):452–65.

Bruns TD, Szaro TM, Gardes M, Cullings KW, Pan JJ, Taylor DL, Horton TR, Kretzer A, Garbelotto M, Li Y. A sequence database for the identification of ectomycorrhizal basidiomycetes by phylogenetic analysis. Mol Ecol. 1998;7:257–72.

Capaldi RA, Malatesta F, Darley-Usmar VM. Structure of cytochrome c oxidase. Biochim Biophys Acta. 1983;726:135–48.

Chevalier BS, Stoddard BL. Homing endonucleases: structural and functional insight into the catalysts of intron/intein mobility. Nucleic Acids Res. 2001;29:3757–74.

Clapp JP, Fitter AH, Young JPW. Ribosomal small subunit sequence variation within spores of an arbuscular mycorrhizal fungus, *Scutellospora* sp. Mol Ecol. 1999;8:915–21.

Croll D, Corradi N, Gamper HA, Sanders IR. Multilocus genotyping of arbuscular mycorrhizal fungi and marker suitability for population genetics. New Phytol. 2008;180(3):564–8.

Damon C, Barroso G, Férandon C, Ranger J, Fraissinet-Tachet L, Marmeisse R. Performance of the *cox1* gene as a marker for the study of metabolically active Pezizomycotina and Agaricomycetes fungal communities from the analysis of soil RNA. FEMS Microbiol Ecol. 2010;74(3):693–705.

de la Providencia IE, Nadimi M, Beaudet D, Morales GR, Hijri M. Detection of a transient mitochondrial DNA heteroplasmy in the progeny of crossed genetically divergent isolates of arbuscular mycorrhizal fungi. New Phytol. 2013;200:211–21.

Engelmoer DJ, Behm JE, Toby Kiers E. Intense competition between arbuscular mycorrhizal mutualists in an in vitro root microbiome negatively affects total fungal abundance. Mol Ecol. 2014;23(6):1584–93.

Finlay RD. Ecological aspects of mycorrhizal symbiosis: with special emphasis on the functional diversity of interactions involving the extraradical mycelium. J Exp Bot. 2008;59:1115–26.

Formey D, Molès M, Haouy A, Savelli B, Bouchez O, Bécard G, Roux C. Comparative analysis of mitochondrial genomes of *Rhizophagus irregularis* – syn. *Glomus irregulare* - reveals a polymorphism induced by variability generating elements. New Phytol. 2012;196(4):1217–27.

Fukuhara H, Sor F, Drissi R, Dinouël N, Miyakawa I, Rousset S, Viola AM. Linear mitochondrial DNAs of yeasts: frequency of occurrence and general features. Mol Cell Biol. 1993;13(4): 2309–14.

Haugen P, Wikmark OG, Vader A, Coucheron DH, Sjøttem E, Johansen SD. The recent transfer of a homing endonuclease gene. Nucleic Acids Res. 2005;33:2734–41.

Helgason T, Watson IJ, Young JPW. Phylogeny of the Glomerales and Diversisporales (Fungi: Glomeromycota) from actin and elongation factor 1-alpha sequences. FEMS Microbiol Lett. 2003;229:127–32.

Kiers ET, Duhamel M, Beesetty Y, Mensah JA, Franken O, Verbruggen E, Fellbaum CR, Kowalchuk GA, Hart MM, Bago A, Palmer TM, West SA, Vandenkoornhuyse P, Jansa J, Bücking H. Reciprocal rewards stabilize cooperation in the mycorrhizal symbiosis. Science. 2011;333(6044):880–2.

Koch AM, Kuhn G, Fontanillas P, Fumagalli L, Goudet J, Sanders IR. High genetic variability and low local diversity in a population of arbuscular mycorrhizal fungi. Proc Natl Acad Sci U S A. 2004;101:2369–74.

Krüger M, Krüger C, Walker C, Stockinger H, Schüssler A. Phylogenetic reference data for systematics and phylotaxonomy of arbuscular mycorrhizal fungi from phylum to species level. New Phytol. 2012;193(4):970–84.

Lee J, Young JPW. The mitochondrial genome sequence of the arbuscular mycorrhizal fungus *Glomus intraradices* isolate 494 and implications for the phylogenetic placement of *Glomus*. New Phytol. 2009;183:200–11.

Lindahl BD, Nilsson RH, Tedersoo L, Abarenkov K, Carlsen T, Kjøller R, Kõljalg U, Pennanen T, Rosendahl S, Stenlid J, Kauserud H. Fungal community analysis by high-throughput sequencing of amplified markers – a user's guide. New Phytol. 2013;199(1):288–99.

Margulis L. Origin of eukaryotic cells. New Haven: Yale University Press; 1971.

Melanie L, Wilde UHP, Nawrath K, Biró TTB, Bothe H. The arbuscular mycorrhizal fungus *Glomus geosporum* in European saline, sodic and gypsum soils. Mycorrhiza. 2002;12:199–211.

Montes-Borrego M, Metsis M, Landa BB. Arbuscular mycorhizal fungi associated with the olive crop across the Andalusian Landscape: factors driving community differentiation. PLoS One. 2014;9(5), e96397.

Nadimi M, Stefani FOP, Hijri M. The Mitochondrial genome of the Glomeromycete *Rhizophagus* sp. DAOM 213198 reveals an unusual organization consisting of two circular chromosomes. Genome Biol Evol. 2015;7(1):96–105.

Nadimi M, Beaudet D, Forget L, Hijri M, Lang BF. Group I intron-mediated trans-splicing in mitochondria of *Gigaspora rosea* and a robust phylogenetic affiliation of arbuscular mycorrhizal fungi with Mortierellales. Mol Biol Evol. 2012;29(9):2199–210.

Pelin A, Pombert JF, Salvioli A, Bonen L, Bonfante P, Corradi N. The mitochondrial genome of the arbuscular mycorrhizal fungus *Gigaspora margarita* reveals two unsuspected trans-splicing events of group I introns. New Phytol. 2012;194(3):836–45.

Peyret-Guzzon M, Stockinger H, Bouffaud ML, Farcy P, Wipf D, Redecker D. Arbuscular mycorrhizal fungal communities and *Rhizophagus irregularis* populations shift in response to short-term ploughing and fertilization in a buffer strip. Mycorrhiza. 2015. doi:10.1007/s00572-015-0644-5.

Porter TM, Golding GB. Factors that affect large subunit ribosomal DNA amplicon sequencing studies of fungal communities: classification method, primer choice, and error. PLoS One. 2012;7(4), e35749.

Raab PA, Brennwald A, Redecker D. Mitochondrial large ribosomal subunit sequences are homogeneous within isolates of Glomus (arbuscular mycorrhizal fungi, Glomeromycota). Mycol Res. 2005;109:1315–22

Redecker D, Schüssler A, Stockinger H, Stürmer SL, Morton JB, Walker C. An evidence-based consensus for the classification of arbuscular mycorrhizal fungi (Glomeromycota). Mycorrhiza. 2013;23(7):515–31.

Renker C, Heinrichs J, Kaldorf M, Buscot F. Combining nested PCR and restriction digest of the internal transcribed spacer region to characterize arbuscular mycorrhizal fungi on roots from the field. Mycorrhiza. 2003;13:191–8.

Riley R, Charron P, Idnurm A, Farinelli L, Dalpé Y, Martin F, Corradi N. Extreme diversification of the mating type–high-mobility group (*MATA-HMG*) gene family in a plant-associated arbuscular mycorrhizal fungus. New Phytol. 2014;201:254–68.

Rosendahl S. The first glance into the *Glomus* genome: an ancient asexual scandal with meiosis? New Phytol. 2012;193:546–8.

Sanders IR, Alt M, Groppe K, Boller T, Wiemken A. Identification of ribosomal DNA polymorphisms among and within spores of the Glomales -application to studies on the genetic diversity of arbuscular mycorrhizal fungal communities. New Phytol. 1995;130:419–27.

Schüβler A, Schwarzott D, Walker C. A new fungal phylum, the *Glomeromycota*: phylogeny and evolution. Mycol Res. 2001;105:1413–21.

Seifert KA, Samson RA, Dewaard JR, Houbraken J, Levesque CA, Moncalvo JM, Louis-Seize G, Hebert PD. Prospects for fungus identification using CO1 DNA barcodes, with *Penicillium* as a test case. Proc Natl Acad Sci U S A. 2007;104:3901–6.

Smith SE, Read DJ. Mineral nutrition, toxic element accumulation and water relations of arbuscular mycorrhizal plants. In: Smith SE, Read DJ, editors. Mycorrhizal symbiosis 3rd ed. London: Academic Press; 2008. p. 145–18.

Thiery O, Borstler B, Ineichen K, Redecker D. Evolutionary dynamics of introns and homing endonuclease ORFs in a region of the large subunit of the mitochondrial rRNA in *Glomus* species (arbuscular mycorrhizal fungi, Glomeromycota). Mol Phylogenet Evol. 2010;55:599–610.

Thonar C, Erb A, Jansa J. Real-time PCR to quantify composition of arbuscular mycorrhizal fungal communities – marker design, verification, calibration and field validation. Mol Ecol Resour. 2012;12(2):219–32.

Woodson SA, Cech TR. Reverse self-splicing of the tetrahymena group I intron: implication for the directionality of splicing and for intron transposition. Cell. 1989;57(2):335–45.

Chapter 12
Taxonomic Updates with Key Focus on Medically Important Aspergilli

Kunhiraman Chandanadukkam Rajeshkumar and Peralam Yegneswaran Prakash

12.1 Introduction

After the Melbourne Code (McNeill et al. 2012), taxonomy of fungi has been under significant debates over the last years due to the drastic changes in the fungal species concepts and species authentication. The article 59 of the ICBN was revised at IBC Nomenclature Section at Melbourne and the principle of "one fungus: one name" was established (Norvell 2011). In accordance to these developments in fungal systematics, innovative changes were made in *Penicillium* and *Aspergillus* taxonomy and nomenclature. The International Commission of *Penicillium* and *Aspergillus* was established by the IUMS to set protocol for the accurate identification of species belonging to *Aspergillus* and *Penicillium*. The classification of the genus *Aspergillus* is traditionally based on morphological characters. Micheli (1729) introduced the name *Aspergillus*, with Haller (1768) validating the genus and Fries (1832) sanctioning the generic name with *Aspergillus glaucus* (L.) link as the generic type.

Thom and Raper (1945) and Raper and Fennell (1965) published major monographic treatments on the genus *Aspergillus* and respectively accepted 89 and 150 species. Tamura et al. (2000) determined the relationships within *Aspergillus* using 18S rDNA for the first time. Peterson (2008) studied the relationship among Aspergilli using a multigene phylogeny and accepted 5 subgenera (*Aspergillus, Circumdati, Fumigati, Nidulantes,* and *Ornati*) and 16 sections. Houbraken and Samson (2011) and Houbraken et al. (2014) used a four-gene phylogeny (RPB1,

K.C. Rajeshkumar
National Fungal Culture Collection of India (NFCCI), Biodiversity and Palaeobiology Group, Agharkar Research Institute, Pune, Maharashtra 411 004, India

P.Y. Prakash (✉)
Medical Mycology Division, Department of Microbiology, Kasturba Medical College – Manipal, Manipal University, Madhav Nagar, Manipal, Karnataka 576104, India
e-mail: prakashpy123@yahoo.co.in

B.P. Singh, V.K. Gupta (eds.), *Molecular Markers in Mycology*, Fungal Biology,
DOI 10.1007/978-3-319-34106-4_12

RPB2, Tsr1 and Cct8) and established monophyly of *Aspergillus*. Presently, the species concept in *Aspergillus* is based on phenotypic (morphology and extrolite profiles) and molecular (ITS, calmodulin, β-tubulin and RPB2 gene sequences) characters in a polyphasic approach. The consequence of the single-name system is that teleomorph-based genera, such as *Neosartorya*, *Emericella*, *Eurotium*, and *Petromyces*, are synonymized with *Aspergillus*. Currently, the genus *Aspergillus* belongs to the recently established fungal family *Aspergillaceae* segregated from *Trichocomaceae* (Houbraken and Samson 2011).

The *Aspergillaceae* are characterized by the formation flask-shaped or cylindrical phialides, asci produced inside cleistothecia or surrounded by hülle cells and mainly ascospores with a furrow or slit. Houbraken and Samson (2011) placed *Aspergillus_sen. str.* (*Aspergillus* in strict sense) under Clade 2 close to *Penicillium* sensu stricto (Clade 1) in family *Aspergillaceae*. They have identified four (04) Subgenus; and a distinct Aspergillus *section Cremei* along with *Phialosimplex* and *Polypaecilum* under *Aspergillus_sen. str.*

1. ***Circumdati*** (sections: *Circumdati*, *Flavi*, *Candidi*, *Flavipedes*, *Nigri*, *Terrei*);
2. ***Nidulantes*** (sections: *Nidulantes*, *Usti*, *Ochraceorosei*, *Sparsi*, *Aeni*);
3. ***Fumigati*** (sections: *Fumigati*, *Clavati*, *Cervini*);
4. ***Aspergillus*** (sections: *Restricti*, *Aspergillus*)

Samson et al. (2014) proposed a recombination for accepted species presently lacking an *Aspergillus* name and provided an updated accepted *Aspergillus* species list containing 339 species. The major medically important Aspergillus distributed across the sections; *Circumdati, Nidulantes, Fumigati* is discussed further;

12.2 Sub-genus: *Circumdati*

Aflatoxins and the producing *Aspergillus_flavus* group (Raper and Fennell 1965) was distinguished by uni to biseriate phialides, with latter predominant, radiate conidial heads in shades of green-yellowish green and brown with coarsely verruculose conidiophores. Peterson (2008) observed 12 lineages in *Aspergillus* section *Flavi*. The maximum parsimony trees obtained based on phylogenetic analysis calmodulin, β-tubulinand ITS sequence data of *Aspergillus* section *Flavi* (Varga et al. 2011b) involves 22 species, which can be grouped into seven clades. *A. flavus* Clade, *A. tamari* Clade (*A. tamarii, A. pseudotamarii, A. caelatus, A. pseudocaelatus*) and *A. nominus*_Clade (*A. nomius, A. pseudonomius, A. zhaoqingensis, A. bombycis*) forms unique lineages with high bootstrap value. Varga et al. (2011b) also stated that multi-locussequence data indicated that several species assigned to section *Flavi* are synonyms of *A. flavus*, including *A. flavus var. columnaris, A. kambarensis A. fasciculatus, A. thomii and A. subolivaceus.*

New molecular approaches have shown high diversity in black aspergilli (*Aspergillus* section *Nigri*), however species level authentication is difficult solely based on their phenotypic characters (Samson et al. 2007). These species cannot be reliably separated from each other using extrolite data too. But, molecular sequence-based approaches using either β-tubulin, calmodulin or translation elongation factor α sequences and AFLP analysis were found to be useful for distinguishing these species (Perrone et al. 2011). For instance, *A. awamori* has recently been revalidated as a cryptic species within the *A. niger* species (Perrone et al. 2011). Varga et al. (2011a) revisited species in *Aspergillus* section *Nigri*; the maximum parsimony trees obtained based on phylogenetic analysis of calmodulin sequence data shown *A. acidus*, *A. tubingensis* cladded inside major *A. niger* Clade along with flagship species *A. niger* and *A. awamori*.

Aspergillus terreus group was conventionally characterized by strictly bi-seriate aspergilli with compact columnar conidial heads in shades of cinnamon colour (Raper and Fennell 1965). Samson et al. (2011) attempted to resolve taxonomy of *Aspergillus* section *Terrei* using polyphasic approach. Based on their phylogenetic analysis of calmodulin and β-tubulin sequences seven lineages were identified among isolates that have previously been treated as *A. terreus* and its subspecies by Raper and Fennell (1965) and others. *A. alabamensis, A. terreus* var. *floccosus, A. terreus* var._*africanus, A. terreus* var. *aureus, A. hortai and A. terreus* (isolate NRRL 4017) all represent distinct lineages from the *A. terreus* clade. The ITS sequence data are also found reliable to distinguish *A. terreus* var. *floccosus*, *A. terreus* (isolate NRRL 4017) and *A. terreus* var. *aureus* from *A. terreus*.

12.3 Sub-genus: *Nidulantes*

Aspergillus subgenus *Nidulantes* is one of the largest subgenera of the genus *Aspergillus* including about 80 species (Peterson 2008; Peterson et al. 2008). Several species of this subgenus were earlier assigned to a teleomorph *Emericella* (Pitt et al. 2000; Samson 2000). Morpho-taxonomy concepts segregate *A. nidulans* from *A. terreus* based on its distinguishable radiate greenish conidial head colour, flask shaped vescicle and presence of globose hulle cells. Varga et al. (2010) on establishing a new *Aspergillus* sect. *Aeni* nov; studies the phylogenetic affinities of *Aspergillus* section *Aenei* to section *Nidulantes* based on neighbor-joining analysis of calmodulin sequence and placed *A. nidulans* (*E. nidulans*) and *A. quadrilineata (E. quadrilineata)* under_section *Nidulantes* in *Aspergillus* subgenus Nidulantes.

Raper and Fennell (1965) classified *A. ustus* to the *Ustus* group together with four other species: *A. panamensis, A. puniceus, A. conjunctus* and *A. deflectus*. Based on phylogenetic analysis of sequence data by Samson et al. (2011) *Aspergillus* section *Usti* includes 21 species, including two teleomorphic species *A. heterothallicus* (*E. heterothallica*) and *Fennellia monodii*. Genotypic studies using partial sequences of the β-tubulin and calmodulin genes placed *A. calidoustus* forming a unique clade in *section Usti* close to *A. carlsbadensis*. Recent studies clarified that infections attributed to *A. ustus* are caused in most cases by another species, *A. calidoustus* (Houbraken et al. 2007; Varga et al. 2008; Balajee et al. 2009).

12.4 Sub-genus: *Fumigati*

Aspergillus section *Fumigati* includes species characterized by uni-seriate aspergilli, columnar conidial heads in shades of green and flask shaped vesicles (Raper and Fennell 1965). Polyphasic taxonomic concepts in *Aspergillus* section *Fumigati* based on analysis of macro and micromorphology, extrolite profiles and β-tubulin, calmodulin, ITS and actin gene sequences of the isolates. There are 37 *Aspergillus* species currently placed in the section_*Fumigati* (Hong et al. 2008; Samson et al. 2007).

Several species such *A. fumigatus, A. novofumigatus, fumigatiaffinis, A. fumisynnematus* and *A. lentulus* show strong morphological resemblance and in the light microscope these species can be difficult to be separated. *A. udagawae* and *A. fennelliae* also shows similar morphology. Therefore an accurate species delimitation and recognition should be carried out using calmodin and β-tubulin gene sequence analysis (Samson et al. 2007).

12.5 Conclusion

Understanding the evolving taxonomy of medically important *Aspergillus* is of utmost importance since they contribute towards accurate identification, enables diagnostic development and targeted therapy based on susceptibility to antifungal agents in an era where large number of immunocompromised individuals would benefit with appropriate medical intervention.

References

Balajee SA, Kano R, Baddley JW, Moser SA, Marr KA, Alexander BD, Andes D, Kontoyiannis DP, Perrone G, Peterson S, Brandt ME, Pappas PG, Chiller T. Molecular identification of *Aspergillus* species collected for the Transplant-Associated Infection Surveillance Network. J Clin Microbiol. 2009;47:3138–41.

Fries EM. Systema mycologicum III. 1832;2:383.

Haller A von. Historia stirpium indigenarum Helvetiae inchoata. Bernae. 1768;3:113.

Hong S, Shin H, Hong J, Frisvad JC, Nielsen PV, Varga J. Samson RA. New taxa of neosartorya and aspergillus in aspergillus section fumigati. Antonie van Leeuwenhoek. 2008;93:87–98.

Houbraken J, Samson RA. Phylogeny of *Penicillium* and the segregation of *Trichocomaceae* into three families. Stud Mycol. 2011;70:1–51.

Houbraken J, Due M, Varga J, Meijer M, Frisvad JC, Samson RA. Polyphasic taxonomy of *Aspergillus* section *Usti*. Stud Mycol. 2007;59:107–28.

Houbraken J, de Vries RP, Samson RA. Modern taxonomy of biotechnologically important *Aspergillus* and *Penicillium* species. Adv Appl Microbiol. 2014;86:199–249.

McNeill J, et al. editors. International Code of Nomenclature for algae, fungi, and plants (Melbourne Code), Adopted by the Eighteenth International Botanical Congress Melbourne,

Australia, July 2011 (electronic ed.). Bratislava: International Association for Plant Taxonomy; 2012. Retrieved 20 Dec 2012.

Micheli PA. Nova plantarum genera ivxtaTovrnefortiimethodvmdisposita. Florentiae Typis Bernardi Paperinii; 1729.

Norvell LL. Melbourne approves a new code. Mycotaxon. 2011;116:481–90. http://dx.doi.org/10.5248/116.481.

Perrone G, Stea G, Epifani F, Varga J, Frisvad JC, Samson RA. *Aspergillusniger* contains a cryptic phylogenetic species A. awamorii. Fungal Biol. 2011;115(11):1138–50.

Peterson SW. Phylogenetic analyses of *Aspergillus* species using DNA sequences from four loci. Mycologia. 2008;100:205–26.

Peterson SW, Varga J, Frisvad JC, Samson RA. Phylogeny and subgeneric taxonomy of *Aspergillus*. In: Varga J, Samson RA, editors. Aspergillus in the genomic era. Wageningen Academic Publishers; 2008. p. 33–56.

Pitt JI, Samson RA, Frisvad JC. List of accepted species and their synonyms in the family *Trichocomaceae*. In: Samson RA, Pitt JI, editors. Integration of modern taxonomic methods for penicillium and aspergillus classification. Amsterdam: Harwood Academic Publishers; 2000. p. 9–49.

Raper KB, Fennell DI. The genus *Aspergillus*. Baltimore: Williams & Wilkins; 1965.

Samson RA. List of names of *Trichocomaceae* published between 1992 and 1999. In: Samson RA, Pitt JI, editors. Integration of modern taxonomic methods for *Penicillium* and *Aspergillus* classification. Amsterdam: Harwood Academic Publishers; 2000. p. 73–9.

Samson RA, Hong S, Peterson SW, Frisvad JC, Varga J. Polyphasic taxonomy of *Aspergillus* section *Fumigati* and its teleomorph *Neosartorya*. Stud Mycol. 2007;59:147–203.

Samson RA, Varga J, Meijer M, Frisvad JC. New taxa in *Aspergillus* section Usti. Stud Mycol. 2011;69:81–97.

Samson RA, Noonim P, Meijer M, Houbraken CM, Frisvad JC, Varga J. Diagnostic tools to identify black *Aspergilli*. Stud Mycol. 2007;59:129–45.

Samson RA, Visagie CM, Houbraken J, Hubka V, Perrone G, Seifert KA, Tanney JB, Varga J, Kocsub S, Szigeti G, Yaguchi T, Frisvad JC. Phylogeny, identification and nomenclature of the genus Aspergillus. Stud Mycol. 2014;78:141–73.

Tamura M, Kawahara K, Sugiyama J. Molecular phylogeny of *Aspergillus* and associated teleomorphs in the *Trichocomaceae* (*Eurotiales*). In: Samson RA, Pitt JI, editors. Integration of modern taxonomic methods for *Penicillium* and *Aspergillus* classification. Amsterdam: Harwood Academic Publishers; 2000. p. 357–72.

Thom C, Raper KB. A manual of the Aspergilli. Baltimore: The Williams & Wilkins Company; 1945.

Varga J, Houbraken J, Van Der Lee HA, Verweij PE, Samson RA. *Aspergillus calidoustus* sp. nov., causative agent of human infections previously assigned to *Aspergillusustus*. Eukaryot Cell. 2008;7:630–8.

Varga J, Frisvad JC, Samson RA. Aspergillus sect. Aeni sect. nov., a new section of the genus for A. karnatakaensis sp. nov. and some allied fungi. IMA Fungus. 2010;1(2):197–205.

Varga J, Frisvad JC, Kocsubé S, Brankovics B, Tóth B, Szigeti G, Samson RA. New and revisited species in *Aspergillus* section *Nigri*. Stud Mycol. 2011a;69:1–17.

Varga J, Frisvad JC, Samson RA. Two new aflatoxin producing species, and an overview of *Aspergillus* section *Flavi*. Stud Mycol. 2011b;69:57–80.

Chapter 13
Current Approaches Towards Development of Molecular Markers in Diagnostics of Invasive Aspergillosis

Preetida J. Bhetariya, Taruna Madan, P. Usha Sarma, and Peralam Yegneswaran Prakash

13.1 Introduction

Aspergillus genus causes spectrum of diseases with clinical manifestations that ranges from colonization of fungi in organs to allergic bronchopulmonary aspergillosis and disseminated forms of infection. The prevalence of invasive aspergillosis (IA) is on rise because of the growing population of transplant recipients and other immunosuppressed populations. The diagnosis of invasive aspergillosis remains a challenge due to lack of specific, sensitive and rapiddiagnostic tests. Detection of galactomannan in the clinical practice has improved the detection of IA. However cross reactivity of the test with other fungi and sensitivity delimits its use and successful treatment is still a challenging endeavor. Established infection is difficult to eradicate, resulting in associated mortality rates ranging from 40 to 90 % (Kumar et al. 2011). The rate of mortality and the poor prognosis in cases of invasive aspergillosis is mainly due to delay in diagnosis, ineffective therapy due to misidentification of etiological agent and lastly due to the resistance of antifungals (Frias-de et al. 2014).

P.J. Bhetariya
Department of Infectious Disease, University of Utah School of Medicine, Salt Lake City, UT 84132, USA

T. Madan
Innate Immunity Lab, National Institute for Research in Reproductive Health, J.M.Street, Parel, Mumbai-12, India

P.U. Sarma
Division of Plant Pathology, Indian Agricultural Research Institute, Pusa Campus, New Delhi 110012, India

P.Y. Prakash (✉)
Medical Mycology Division, Department of Microbiology, Kasturba Medical College – Manipal, Manipal University, Madhav Nagar, Manipal, Karnataka 576104, India
e-mail: prakashpy123@yahoo.co.in

B.P. Singh, V.K. Gupta (eds.), *Molecular Markers in Mycology*, Fungal Biology,
DOI 10.1007/978-3-319-34106-4_13

Major concern with respect to diagnosis is the extensive use of empiric antifungal therapy leading to development of resistance in the pathogen, making current diagnostic platform unsuitable for efficient therapeutic decisions. Azole-resistant strains of *Aspergillus fumigatus* (Domingo et al. 2012) are well documented. Such drug resistant pathogen infection stresses the need for personalized diagnostic and risk prediction approaches based on individual traits beforehand (Jegorov et al. 2006). Molecular markers have proven to be useful tools for the diagnosis, treatment and epidemiology of mycoses, such as in the case of aspergillosis. Molecular markers such as presence of fungal DNA signatures in bodily fluids have been developed which can differentiate between species within the same genus. Such diagnostic markers provide precise genetic profile of organism in infection and their response towards antifungal therapies (Zarrinfar et al. 2013). Our limited understanding of the critical immune defects that predispose to these opportunistic fungal infections and the lack of validation of genetic diagnostics platform to do so only adds to diagnostic challenges (Oliveira-Coelho et al. 2015). In this review we provide an update on biomarkers for the diagnosis of invasive aspergillosis. Further, this chapter reflects the current developments on diagnosing invasive fungal infections. Molecular techniques under development and their potential for routine diagnostic applications will be discussed.

13.2 Spectrum of Human Disease in Aspergillosis

Among *Aspergillus*, *Aspergillus fumigatus* is the common etiological agenthuman aspergillosisresponsible for >90 % of the cases in humans followed by *A. flavus*, *A. niger*, and *A. terreus* (Latge 2001). Other species that have been associated with IA include *A. nidulans* in those with chronic granulomatous disease (CGD) (Liss et al. 2015), as well as *A. ustus*, *A. lentulus*, and *A. versicolor* (Greene et al. 2007). Lung is the major site of *Aspergillus* infection, but other infected sites include: the tracheobronchialtree, sinuses, skin and soft tissue, central nervous system (CNS), bone, heart, and eyes (Ellis et al. 2009). *A. fumigatus* is a filamentous, thermophilic saprophyte widely distributed in the environment. It has high sporulating capacity, thus, conidia (1–100 conidia m^{-3}) are present ubiquitously in the air, indoors and outdoors (Latge 2001). These airborne conidia have a diameter small enough (2–3 μm) to reach the alveoli in the lung. Conidia are continuously inhaled by humans and are eliminated efficiently by the innate immune response; however in immunocompromised individuals it can often cause fatal invasive infections.

Aspergillosis encompasses a spectrum of diseases related to host factors. Acute invasive aspergillosis is rapidly progressive and frequently fatal in highly immunocompromised persons. IA is seen primarily in patients with haematological malignancy and in solid organ and stem cell transplant recipients (Fig. 13.1). Aspergilloma is a fungal mass that develops in a preexisting lung cavity. In immunocompetent individuals, *Aspergillus* species can induce allergic responses, as sinusitis and asthma. Allergic bronchopulmonary aspergillosis is an allergic disease mainly

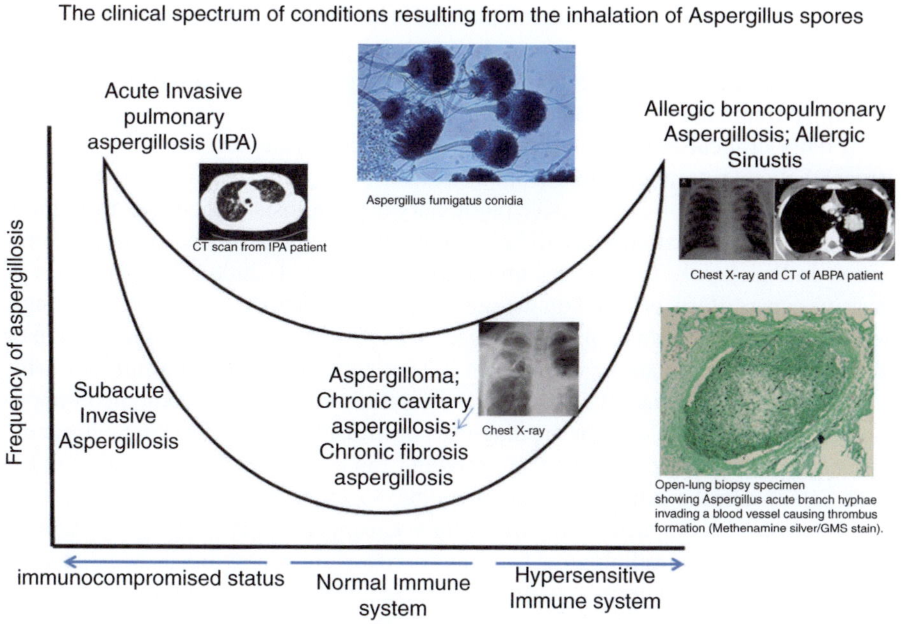

Fig. 13.1 Spectrum of clinical disease by *Aspergillus fumgiatus* (Chest x-ray, CT scan and Hyphal staining adopted from Chris and Denning (2015); Soubani and Chandrasekar (2002); Agarwal et al. (2011); Zmeili and Soubani (2007))

mediated by Th2 CD4+ cell response in 1–2 % of patients with asthma globally while in India it reaches as high as 5–20 % and in 1–15 % of patients with cystic fibrosis (Neustadt et al. 2009; Agarwal et al. 2014).

13.3 Criteria for Definition of Proven, Probable and Possible IA

In 2002, a consensus group working on the field of Invasive fungal infections worldwide; EORTC/MSG (European Organization for Research and Treatment of Cancer/Invasive Fungal Infections Cooperative Group and the National Institute of Allergy and Infectious Diseases Mycoses Study Group) has established guidelines for classifying invasive fungal infections for clinical and epidemiological research as "proven," "probable," and "possible" invasive fungal infection in immunocompromised patients (Badiee and Hashemizadeh 2014). The proven aspergillosis cases require histopathological examination of infection and a positive result of culture of a specimen from a normally sterile site while probable aspergillosis requires to prove clinical manifestations (symptoms and radiological features), presence of host factors and microbiological evidence (Badiee and Hashemizadeh 2014). Current criteria for the diagnosis of invasive aspergillosis are restricted to specific groups, such as patients with cancer, patients with COPD, and critically ill patients (De Pauw et al. 2008). This approach is helpful to clinicians to detect high-risk patients with clinical suspicion of invasive aspergillosis who could benefit from antifungal treatment (Arvanitis et al. 2014).

In daily clinical practice, culture-based diagnostic tests are the gold standard. Broncho alveolar lavage, tap from spinal fluid or biopsy are standard procedures for establishing a diagnosis of invasive pulmonary aspergillosis. Specimens from these procedures may reveal presence of branching septate hyphae on microscopic examination and/or *Aspergillus* species on culture (Walsh et al. 2008). The diagnosis of invasive aspergillosis is based on the presence of host factors, radiological findings, and mycological criteria (isolation of *Aspergillus* from a non-sterile compartment or detection of galactomannan or β-glucan). This combined approach fails for critically ill patients and patients with COPD (Arvanitis et al. 2014). Computed tomography is a useful technique for the diagnosis of invasive aspergillosis in neutropenic patients, with up to 66 % of patients showing the characteristic halo sign (Godoy et al. 2012). On the other end, Infections due to other angioinvasive filamentous fungi, such as Zygomycetes, *Fusarium* and *Scedosporium* species may also cause a halo sign and similar radiological features described for aspergillosis. More characteristic radiological patterns specific to invasive pulmonary aspergillosis have been well described in neutropenic hosts, but still less is known about the features of these lesions in other immunocompromised patients (Caillot et al. 2001).

13.4 Current Methods Used in Diagnosis of IA and Their Limitations

Diagnosis of IA is often considered complicated due to negative results in blood culture for *A. fumigatus*. Sampling of the Broncho alveolar lavage to culture the fungus is not always the option in patients with severe conditions. Lack of specific diagnostic tools has led to development and evaluation of nonculture diagnostic approaches including imaging, PCR-based detection of *Aspergillus* DNA and the detection of galactomannan in serum and broncho-alveolar lavage (Posteraro and Sanguinetti 2014). Galactomannan and 1,3-β-D Glucan are the two fungal biomarkers that have been extensively studied and used in diagnostic platform (Orsi et al. 2015).

13.5 Galactomannan Antigen Detection

Galactomannan (GM) is a component of the *Aspergillus* fungal wall that is released during fungal growth. It is a heat-stable heteropolysaccharide. The molecule is comprised of a nonimmunogenicmannan core with side chains of varying lengths containing galactofuranosyl units, and is released during hyphal growth (Lunel et al. 2009). GM is anchored to the membrane via a glycosylphosphatidylinositolor is covalently linked to the cell wall and isreadily released as a lipopeptide GM of about 20 kDa (Morelle et al. 2005). The composition of GM varies between strains, as well as the conditions used for its production, extraction, and purification (Pfaller and Diekema 2007). GM is most commonly detected in serum and BALF using a double-sandwich enzyme-linked immunosorbent assay (GM-ELISA). Measurement of GM antigen in other specimens including sputum samples, bronchial washings, urine, peritoneal fluid, pericardial fluid, pleural fluid, and tissue samples, has not yet been clinically validated (Greene et al. 2007). However, GM detection in CSF is also included with blood and BALF in the EORTC/MSG criteria for diagnosis of IA (De Pauw et al. 2008).

There are two commercial assays for the detection of GM—the Pastorex kit (Sanofi Diagnostics Pasteur, France) and Platelia ELISA (BioRad, Marnes-La-Coquette, France). GM assays use a monoclonal antibody, EB-A2 derived from rats directed against β-(1,5)-linked galactofuranoside side-chain residues of theGM molecules (Stynen et al. 1992; Chong et al. 2015). The current assay defines a positive result as an Optical Density Index (ODI) ratio ≥0.5 (Kuper et al. 2012). Several factors contribute to lowered sensitivity and both false-negative and false-positive results. The binding of EB-A2 requires more than four galactofuranoside epitopes and thus sensitivity can be compromised by the inability to detect secreted antigens that bear fewer residues. In assay, a pretreatment step is required to remove complexing antibody which can block EB-A2 binding. In this step, acid-sensitive galactofuranoside residues may be degraded by the edetic acid and thus hamper the results. Cross-reactivity of Fusarium spp. with *Aspergillus* GM also presents a drawback with respect to the specificity of the GM-ELISA (Groll and Tragiannidis 2009). Cross-reactivity of the test has been noted with other organisms containing GM, like Histoplasma capsulatum (Gupta et al. 2015), and Penicillium species (Jeong et al. 2015). False positive results also occur initially with the concomitant administration of certain beta-lactic drugs, like piperacillin tazobactam. Cross reactivity also occurs with other β-lactam antibiotics including amoxicillin-clavulanic acid, cefipime, ceftriaxone, carbapenem (De Pascale and Tumbarello 2015), and ampicillin (Blyth et al. 2014). False positive results occur within the first 100 days after HSCT and in patients with GI tract mucositis. This is due to the translocation of food-borne GM or bacteria with cross reactive isotopes into the bloodstream from the GI tract (Thursky et al. 2008). Cross reactivity is also seen with *Bifidobacterial lipoglycan*, a bacteria present in human gut, especially in neonates (Van Peer et al. 1989). The clinical sensitivity of GM ELISA is somewhat variable, with a range of 29–100 % (Pfaller and Diekema 2004).

GM detection success is limited by the turn-around time (from one day to several days), and the facilities needed to perform the assay. *Aspergillus* Lateral-Flow Device (LFD), a novel point-of-care (POC) test addresses these issues as it is simple, rapid (15 min), and single-use test can be performed for BAL samples. The LFD test has been evaluated in 39 BAL samples from hematologic malignancy patients and solid organ transplant recipients and showed sensitivity and specificity of 100 % and 81 %, respectively (Pan et al. 2015). Multicenter studies with larger sample sizes are currently evaluating this point of care test.

13.6 β-(1,3)-D Glucan Assay

1,3-β-D Glucan (BDG) is a polysaccharide present in the cell wall of most fungi including *Aspergillus*, *Candida*, *Fusarium*, *Acremonium*, *Penicillium*, *Paecilomyces*, and *Pneumocystis jirovecii*, but not bacteria or viruses (Chen et al. 2010). Numerous assays to detect BDG in the serum have been developed and studied as a marker of fungal infections. The common feature of all of the glucan assays is the ability of β-(1,3)-D glucan to activate a coagulation cascade within amoebocytes derived from the haemolymph of horseshoe crabs. BDG glucan induces clot formation (independently of factor C) via second serine protease zymogen, factor G, providing the basis for the development of the current assays. The FungitellR (commercial) and GlucatellR (research) versions of the same assay marketed by Associates of Cape Cod, Inc. have been the most extensively studied. Fungitell is approved test by the FDA in the USA for the diagnosis of invasive fungal infections and its analytical sensitivity to detect BDG is in the order of 1 pg/mL. However, the cut-off of 60 pg/mL for BDG was used in a recent clinical study, making it unsuitable (Odabasi et al. 2004). Human plasma contains a number of inhibitors of serine proteases that need to be removed in a pretreatment step in assay. Removal of endotoxin which also activates the horseshoe crab coagulation pathway is required during pretreatment. This enhances analytical specificity via the removal of non-specific activators of serine proteases present in human serum but affects the sensitivity of the test. A positive BDG result in serum along with the presence of appropriate host factors, clinical features and othermycological evidence allows possible IA to be upgraded to probable IA in the revised EORTC/MSG criteria (De Pauw et al. 2008). However the presence of BDG in other fungal species such as Candida spp, Fusarium spp, Acremonium spp, and *Pneumocystis jiroveci*, means that it cannot be used for specific diagnosis of invasive aspergillosis (Chen et al. 2010).

13.7 PCR Detection for IA

Use of nucleic acid based dectection in fungal diagnostics assays is promising as these techniques are ultrasensitive, allowing amplifying and detecting of extremely low concentrations (from 1 to 10 fg) of *Aspergillus* DNA or RNA corresponding to 10 to 100 conidia per milliliter of sample. Techniques can be used on nucleic acid extracts of various clinical specimens such as BAL, blood and other tissues. *Aspergillus* DNA in nucleus and mitochondria has been targeted for detection. PCR methods reported for *Aspergillus* detection are based on detection of ribosomal DNA (18S rDNA, 28S rDNA, and ITS2 and ITS1 regions) as the presence of high copy number of genes; but is not genus specific as these regions are conserved across a wide range of fungi with minimum nucleotide changes. These PCR studies are done on spiked volunteer blood samples or a murine model of IA to test the utility of qPCR assays and still lack the standardization of BAL and blood samples

(Morton et al. 2012). Thus, PCR-based test results are not yet accepted as a mycological evidence of infection by EORTC/MSG.

qPCRtest based on unique ITS1 regions for genus- and species-specific detection from four most common medically important *Aspergillus*species (*Aspergillus fumigatus*, *A. flavus*, *A. niger*, and *A. terreus*) infections using TaqMan technology is reported from BAL specimens (Walsh et al. 2011). After the availability of genome sequence of various *Aspergillus* species, array of virulence factors, secondary metabolite pathways and uniques genes has been proposed as drug target as well as detection target based on their prevalence *in vivo*. A virulence factor, the aspHS geneencoding a haemolysin product, which is overexpressed *in vivo* during infection has been tested as diagnostic target for specific detection of *A. fumigatus* by Qpcr from human bronchoalveolar lavage (BAL) samples (Abad-Diaz-De-Cerio et al. 2013). Transcriptome studies show over expression of several genes in drug resistant *Aspergillus*isolates and provide clues to understand pathogenicity and improve the selection of targets for clinical diagnosis. Several virulence-related genes have been proposed for specific identification of *A. fumigatus* by conventional PCR which are *abr1*, *alb1*, *arp1*, *aspHS*, *catA*, *dppIV*, *mep20*, *mitF*, *pabaA*, *pep*, *pyrG*, *rhbA* and *rodB* (Abad-Diaz-De-Cerio et al. 2013; Sueiro-Olivares et al. 2015). This information in combination of tested dignostic chemistry may be of great relevance to develop the senstive and specific diagnostic assays which not only detects the presence of *Aspergillus* DNA but specify the species level detection for drug resistance species. AsperGenius, a new multiplex real-timePCR assay (PathoNostics, Netherlands) which detects resistant *A. fumigatus* isolatesspecifically from BALfrom hematology patients has been reported (Chong et al. 2015). This assay targets the single-copy CYP51A gene of *A. fumigatus* and detects the mutation at specific sites (TR34, L98H, Y121F, and T289A) to differentiate wild-type from mutant *A. fumigatus* strains via melting curve analysis in real-time PCR.

The conserved region of the *Aspergillus* collagen-like (acl) genes has been studied and exploredfor the species level detection of *Aspergillus* using using Microfluidic PCR applications from sputum and bronchoalveolar fluid from clinical samples (Tuntevski et al. 2013). DNA topoisomerase II genes were also tested for their specificities in PCR amplifications from major pathogenic fungal species such as Candida spp and *Aspergillus fumigatus* (Kanbe et al. 2003). A multiplex qPCR assay with TaqMan probes based on internal transcribed spacer regions of benA gene discriminating *Aspergillus* species, mainly *A. fumigatus* in BAL samples has been also reported (Fernandez-Molina et al. 2014).

Apart from qPCR, amplification of RNA is also a powerful technique when combined with molecular beacon chemistry (Yoo et al. 2007; Zhao et al. 2009; Zhao and Perlin 2013).

Another widley popular molecular technique, Microarray analysis also has great potential to detect many target nucleic acids from multiple fungal organisms for simultaneous species identification (Spiess et al. 2007). DNA microarray with the simultaneous use of internal transcribed spacer region (ITS) of ribosomal RNA gene and β-tubulin gene has been exploited to detect fungal species of genus *Aspergillus*, the Eurotium and Penicillium (Isshiki et al. 2014). A multifungal

DNA-microarray detecting 15 different fungi mainly *Aspergillus*, Candida, Fusarium, Rhizopus, and Trichosporon species together with nested *Aspergillus* specific PCR assay from biopsies, bronchoalveolar lavage and peripheral blood samples from immunocompromised patients were investigated to improve identification pathogens in invasive fungal disease (Boch et al. 2015). Such platforms can detectthe virulence factors and antifungal resistance determinants as well as gene expression profiling in the context of invasive infections (Simitsopoulou et al. 2011). However, itsuse in clinical practise for diagnosis is hindered by its variable hybridization efficiencies, resulting in unreliable quantification and false-positive/negative results (Bustin and Dorudi 2002).

Detection based on mycotoxin pathway is known in agriculturally relevant *Aspergillus* speceis using multiplex PCR and HPLC techniques (Kim et al. 2014). *Aspergillus fumigatus*shows the presence of several virulence factors which are secreted as secondary metabolites such as gliotoxin, fumigilin and also polyketides such as melanin pigments. Some cyclic peptides and depsipeptides are synthesized in microorganisms by large multienzymes called nonribosomal peptide synthetases. These proteins are specific tofungus and can serve as dependable molecular targets for diagnosing fungal infections at an early stage (Jegorov et al. 2006). The virulence factor gliotoxin (GT) and its inactive derivative, bis(methylthio) gliotoxin (bmGT), are produced by pathogens of the genus *Aspergillus*. Use of these products as diagnostic markers has been demonstrated experimentally by spiked human serum or plasmawith *A. fumigatus*. (Domingo et al. 2012). Detection of genes involved in these secondary metabolite pathways in clinical specimens has potential to be developed as diagnostic assays. However, these genes may not be multicopy but senstive detection may deliver the presence of fungus *invivo* and can help in early detection with combination of already availablediagnostics. Seveal phylogenetic studies have been used for genus level identificationand can be used for diagnosis (Houbraken and Samson 2011).

There are no validated PCR-based diagnositic assays as they lack the 'standardization using various chemistries and specific target combinations (Donnelly 2006). Working Group of different country ISHAM (International Society for Human and Animal Mycology), and EAPCRI (European *Aspergillus* PCR Initiative) are involved in the concerns regarding the standardization of the *Aspergillus* PCR methodology for clinical application (White et al. 2011), but yet PCR methods, have not been accepted as standard diagnositic criteria. Table 13.1 shows molecular markers tested for diagnosis of *Aspergillus* spp from tissue and in culture by various methods.

13.8 Proteomics Studies for Biomarkers

Various proteomics studies have been performed using mice models to explore drug and diagnostic targets for *Aspergillus* infections. Proteomic profiles of bronchoalveolar lavage fluid (BALF) from a neutropenic rabbit model of IPA; infected and treated with ravuconazole/amphotericin B, were studied and analyzed by two-dimensional (2D) gel electrophoresis, followed by mass spectrometry (matrix-assisted laser

Table 13.1 Summary of molecular markers for the diagnosis of *Aspergillus* spp

Molecular Marker	Locus targeted	Use in diagnostic platform	Target species	References
rRNA	ITS1-5·8S rRNA-ITS2	Sequencing and PCR	*A. fumigatus*, *A. flavus*, *A. terreus*, *A. niger*, *A. ustus*	Hinrikson et al. (2005)
rRNA	5·8S rRNA-ITS2 region	PCR using Automated fluorescent capillary electrophoresis (detection of different length of amplicon)	*A. fumigatus*, *A. flavus*, *A.terreus*, *A. niger*	Turenne et al. (1999)
rRNA and GM	18S rRNA gene	Light Cycler PCR-ELISA	*A. fumigatus*	Scotter and Chambers (2005)
rRNA and CYP51A [encoding cytochrome p450 sterol 14α-demethylase]	28S rRNA and TR34-L98H and TR46-Y121F-T289A mutation in CYP51A gene	Multiplex real-time PCR assay	*A. fumigatus*, *Aspergillus lentulus*, *A. udagawae*, *A. viridinutans* and azole resistant *A.fumigatus* isolates	Chong et al. (2015)
β-tubulin	Beta tubulin gene (Bt2a & Bt2b primers)	PCR-RFLP analysis	*A. fumigatus*, *A. flavus*, *A. niger*, *A. terreus*, *A. clavatus*, *A. nidulans*	Glass and Donaldson (1995)
Calmodulin	Primer sets for black Aspergilli using conserved sequence differences in the calmodulin gene	PCR	*A. niger*, *A. welwitschiae*, *A. carbonarius*, *A. tubingensis*	Palumbo and O'Keeffe (2015)
Minichromosome maintenance protein	ITS 1–5.8S -ITS2, minichromosome maintenance protein, subunit of RNA polymerase II	PCR and sequencing	*A. fumigatus and Aspergillus section Circumdati*	Sugui et al. (2012)
mitochondrial cytochrome *b* genes	Specific sequence of the mitochondrial cytochrome *b* gene.	Light Cycler PCR	*A. fumigatus*	Spiess et al. (2003); Bu et al. (2005)

desorption ionization-time of flight/time of flight-MALDI-TOF/TOF) and quantified by enzyme-linked immunosorbent assay (ELISA). This study showed host-derived proteins haptoglobin (Hp), C-reactive protein (CRP), and annexin A1 (Anx A1) expressed prominently in BALF during the IPA infection and showed significant changes in response toantifungal therapy. These host protein biomarkers can be of value in monitoring therapeutic response to antifungal agents in IPA patients with confirmed disease (Krel et al. 2014). Another study showing secretome analysis of proteins isolated from the culture filtrate of *Aspergillus fumigatus* has been done by 2D gel electrophoresis coupled with MS/MS. The immunosecretome analysis probed with the sera of patients, immunized rabbit and identified 15 proteins as immunogenic in human. These proteins can be useful as diagnostic markers (Kumar et al. 2011). *Aspergillus* species are known to secrete various pathologically relevant proteases. Monitoring of their protease activity in serum may serve as a direct diagnostic approach. Several proteases were identified that facilitate the optimization of protease profiling as a diagnostic tool for invasive aspergillosis (Neustadt et al. 2009). Proteomics-targeted MALDI-TOF MS is now being applied to the identification of filamentous fungi for fungal diagnostic procedures. The flexibility that allows MS to analyze all kinds of molecules under optimal conditions makes it reliable for detection of *Aspergillus* – specific metabolites and nucleic acids.

13.9 Antibodies Directed Toward *Aspergillus* spp

Antibody detection could be useful as a means of establishing a retrospective diagnosis of invasive aspergillosis mainly in immunocompromised hosts who have undergone chemotherapy. To establish the diagnosis of chronic pulmonary aspergillosisdemonstration of specific antibody presence in serum or body fluid is essential criteria. Many assay formats have been reported in literature to detect antibodies as best non-invasive means of establishing the diagnosis of invasive aspergillosis in non-neutropenic patients (Schlageter and Kozel 1990). Several assay formats using specific and combination of antigens have been used to detect antibodies to *Aspergillus* spp, including immunodiffusion, particle haemagglutination, indirect-immunofluoresence, radioimmunoassay, and ELISA (Schlageter and Kozel 1990; Rachini et al. 2007).

The *A. fumigatus* antigen Crf2; a novel variant of a group of surface proteins (Crf1, Asp f9, Asp f16) which belong to the glycosylhydrolase family was isolated from a patient with proven IA. Crf2 and the corresponding recombinant antibodies were suggested for the early diagnosis of IA caused by *A. fumigatus* (Schutte et al. 2009). *Aspergillus* antigen-capture ELISA based on monoclonal antibodies, which recognize epitopes present on the cell walls of the hyphae and the conidia of *Aspergillus* species, and these are also circulating or excreted as immunodominant antigens during the acute phase of IA (established in the animal models); is shown as a promising tool for the diagnosis of IA (Hao et al. 2008). The use of recombinant antigens; eg, dipeptidylpeptidases, superoxide dismutase, catalase, metalloprotease, mitogillin and galactomannoprotein (Speed and Dunt 1995; Chen et al. 2000;

Dromer et al. 2007; Ananda-Rajah et al. 2012) can be tested in appropriate combination and platform. This work requires strategic standardization. This approach can facilitate the generation of assays that may confer prognostic information using single and combination epitope.

Aspergillus fumigatus is a potent source of allergens with immune reactivity shown to more than 20 allergens derived from this organism, with known proteases (Nierman et al. 2005; Singh et al. 2014). A number of studies haveevaluated the use of *Aspergillus* allergens, and recombinant allergens in the diagnosis of *Aspergillus*-caused diseases. These will be useful for ABPA detection.

13.10 Update on Molecular Markers

Sequencing of fungal ribosomal targets is an attractive and widely exploited diagnostic method. Fungal ribosomal genes are multi-copy and thus increase detection sensitivity during PCR amplification. Also the multiple conserved ribosomal (18 s, 5.8 s, and 28 s) subunit genes are in close proximity in fungi, providing better binding sites for conserved PCR primer to yield PCR products. The conserved nature of the subunits makes sequence identification possiblefor any unknown fungus to be amplified with universal primers targeted to these regions (Kozel and Wickes 2014). Use of Panfungal primers based diagnostics are promising as it is directed toward conserved regions amplification and also detects species specific polymorphisms.

The recent genome sequencing of *A. fumigatus*, using strain Af293, revealed mitochondrial genes encoding some of the tRNA genes and (apo) cytochrome b which have also been used as primer targetsfor *Aspergillus* detection (Bu et al. 2005; Nierman et al. 2005). There are 12 copies of the mitochondrial genome present for every copy of the nuclear genome, thus mitochondrial targets can be considered "multicopy" and reliable target for detection. Suitability of using mitochondrial single nucleotide polymorphisms (SNP) as markers was examined recently and were found suitable for detection and identification of *Aspergillus fumigatus* in clinical specimens. The assay requires further optimization and clinical validation to allow its application in routine diagnostics (Oliveira et al. 2014).

The ability of *Aspergillus* spp to produce D-mannitol has been known for many years and its diagnostic potential examined in several experimental models of invasive aspergillosis (Wong et al. 1989). A comprehensive summary of the various secondary metabolites (mycotoxins) produced by *Aspergillus* spp can be found at http://www.*Aspergillus*.man.ac.uk. The detection of metabolites represents an under-researched area in terms of their possible application as diagnostic modalities for invasive aspergillosis.

Bioinformatic analysis of the sequenced products from clinical specimens using various genomic regions can also shed light on genus diversity and can help in developing diagnostics. Enzymes/proteins involved in secondary metabolites can be good targets to study the diversity and may have diagnosticpotential. This diversity has been exploited in development of the diagnostic assays for species specific *Aspergillus* detection (Bhetariya et al. 2009); (PCT application : US 20130266941 A1). Domain

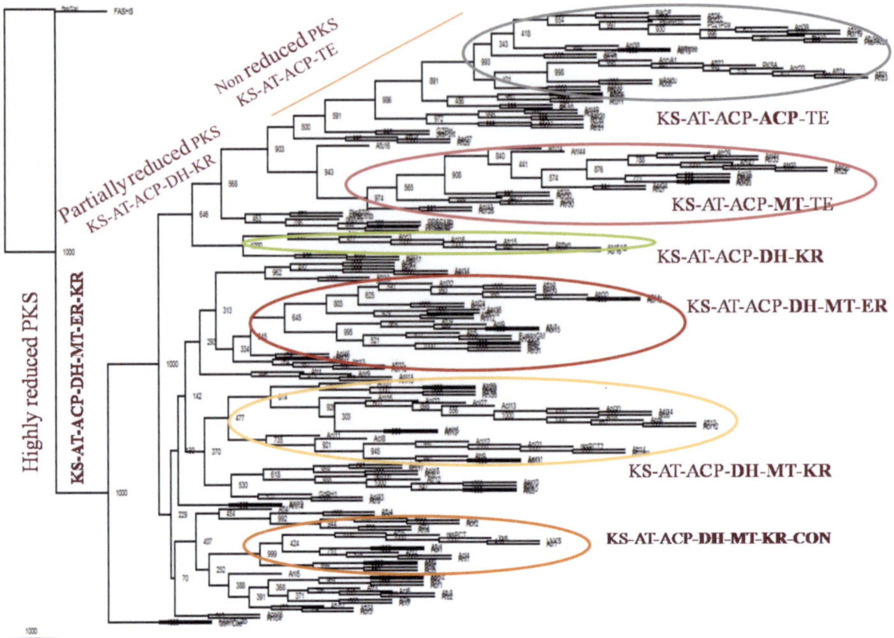

Fig. 13.2 Phylogram of *Aspergillus* Polyketide Synthase protein inferred by maximum parsimony analysis of putative AA sequences of full length polyketide synthases (KS-AT-ACP-DH-MT--ER-KR). **KS* ketosynthase, *ACP* Acyl carrier protein, *AT* Acyl transferase, *TE* thioesterase, *DH* dehydrogenase, *MT* methyltransferase, *ER* enylreductase. Major clades are marked by horizontal lines to the right of the tree. *Aspergillus* PKSs were classified in Non Reduced and Highly Reduced, Partially Reduced PKSs based on presence and absence of domains

diversity in the megaenzyme such as Polyketide synthase has been mapped out by research groups (Varga et al. 2003). Large scale bioinformatic study was also carried out using bioinformatics platforms. The phylogenetic tree based on the amino acid sequences of PKSs proteins of *Aspergillus* showed greater diversity of their domains. Aspergilli PKSs were categorised into non reducing (NR), partially reducing (PR), and highly reducing (HR) PKS enzymes (Fig. 13.2; unpublished data). Non-Conserved regions from PKS gene was selected for primer target and is being used for detection of *Aspergillus* spp. (unpublished data).

13.11 Conclusion

In view of the complexity involved in invasive aspergillosis with respect to host immune status and pathogen's virulence and also the differential drug responses,there is an urgent need to explore more sensitive and specific detection methods. Multiple approaches based on genomic and proteomics along with advanced sensitive diagnostic technologies can definitely help in achieving this unmet need.

References

Abad-Diaz-De-Cerio A, Fernandez-Molina JV, Ramirez-Garcia A, Sendino J, Hernando FL, Peman J, Garaizar J, Rementeria A. The aspHS gene as a new target for detecting *Aspergillus* fumigatus during infections by quantitative real-time PCR. Med Mycol. 2013;51(5):545–54.

Agarwal R, Khan A, Garg M, Aggarwal AN, Gupta D. Pictorial essay: allergic bronchopulmonary aspergillosis. Indian J Radiol Imaging. 2011;21(4):242–52.

Agarwal R, Denning DW, Chakrabarti A. Estimation of the burden of chronic and allergic pulmonary aspergillosis in India. PLoS One. 2014;9(12):e114745.

Ananda-Rajah MR, Slavin MA, Thursky KT. The case for antifungal stewardship. Curr Opin Infect Dis. 2012;25(1):107–15.

Arvanitis M, Anagnostou T, Fuchs BB, Caliendo AM, Mylonakis E. Molecular and non molecular diagnostic methods for invasive fungal infections. Clin Microbiol Rev. 2014;27(3):490–526.

Badiee P, Hashemizadeh Z. Opportunistic invasive fungal infections: diagnosis & clinical management. Indian J Med Res. 2014;139(2):195–204.

Bhetariya P, Shankar J, Singh Y, Madan T, Basir SF, Sarma PU. Multiplex PCR for detection of *Aspergillus* fumigatus, *Aspergillus* flavus and *Aspergillus* niger. J Allergy Clin Immunol. 2009; 123(2):160.

Blyth CC, Gilroy NM, Guy SD, Chambers ST, Cheong EY, Gottlieb T, McGuinness SL, Thursky KA. Consensus guidelines for the treatment of invasive mould infections in haematological malignancy and haemopoietic stem cell transplantation. Intern Med. 2014;J44(12b):1333–49.

Boch T, Reinwald M, Postina P, Cornely OA, Vehreschild JJ, Heussel CP, Heinz WJ, Hoenigl M, Eigl S, Lehrnbecher T, Hahn J, Claus B, Lauten M, Egerer G, Muller MC, Will S, Merker N, Hofmann WK, Buchheidt D, Spiess B. Identification of invasive fungal diseases in immunocompromised patients by combining an *Aspergillus* specific PCR with a multifungal DNA-microarray from primary clinical samples. Mycoses. 2015;58(12):735–45.

Bu R, Sathiapalan RK, Ibrahim MM, Al-Mohsen I, Almodavar E, Gutierrez MI, Bhatia K. Monochrome Light Cycler PCR assay for detection and quantification of five common species of Candida and *Aspergillus*. J Med Microbiol. 2005;54(3):243–8.

Bustin SA, Dorudi S. The value of microarray techniques for quantitative gene profiling in molecular diagnostics. Trends Mol Med. 2002;8(6):269–72.

Caillot D, Couaillier JF, Bernard A, Casasnovas O, Denning DW, Mannone L, Lopez J, Couillault G, Piard F, Vagner O, Guy H. Increasing volume and changing characteristics of invasive pulmonary aspergillosis on sequential thoracic computed tomography scans in patients with neutropenia. J Clin Oncol. 2001;19(1):253–9.

Chen ST, Sorrell T, Nimmo G, Speed B, Currie B, Ellis D, Marriott D, Pfeiffer T, Parr D, Byth K. Epidemiology and host- and variety-dependent characteristics of infection due to Cryptococcus neoformans in Australia and New Zealand. Australasian Cryptococcal Study Group. Clin Infect Dis. 2000;31(2):499–508.

Chen SC, Playford EG, Sorrell TC. Antifungal therapy in invasive fungal infections. Curr Opin Pharmacol. 2010;10(5):522–30.

Chong GL, Van de Sande WW, Dingemans DJ, Gaajetaan GR, Vonk AG, Hayette MP, Van Tegelen DW, Simons GF, Rijnders BJ. Validation of a new *Aspergillus* real-time PCR assay for direct detection of *Aspergillus* and azole resistance of *Aspergillus* fumigatus on bronchoalveolar lavage fluid. J Clin Microbiol. 2015;53(3):868–74.

Chris K, Denning DW. The clinical spectrum of pulmonary aspergillosis. Postgrad Med J. 2015;91:403–10.

De Pascale G, Tumbarello M. Fungal infections in the ICU: advances in treatment and diagnosis. Curr Opin Crit Care. 2015;21(5):421–9.

De Pauw B, Walsh TJ, Donnelly JP, Stevens DA, Edwards JE, Calandra T, Pappas PG, Maertens J, Lortholary O, Kauffman CA, Denning DW, Patterson TF, Maschmeyer G, Bille J, Dismukes WE, Herbrecht R, Hope WW, Kibbler CC, Kullberg BJ, Marr KA, Munoz P, Odds FC, Perfect JR, Restrepo A, Ruhnke M, Segal BH, Sobel JD, Sorrell TC, Viscoli C, Wingard JR, Zaoutis T,

Bennett JE. Revised definitions of invasive fungal disease from the European Organization for Research and Treatment of Cancer/Invasive Fungal Infections Cooperative Group and the National Institute of Allergy and Infectious Diseases Mycoses Study Group (EORTC/MSG) Consensus Group. Clin Infect Dis. 2008;46(12):1813–21.

Domingo MP, Colmenarejo C, Martinez-Lostao L, Mullbacher A, Jarne C, Revillo KJ, Delgado P, Roc L, Meis JF, Rezusta A, Pardo J, Galvez EM. Bis(methyl)gliotoxin proves to be a more stable and reliable marker for invasive aspergillosis than gliotoxin and suitable for use in diagnosis. Diagn Microbiol Infect Dis. 2012;73(1):57–64.

Donnelly JP. Polymerase chain reaction for diagnosing invasive aspergillosis: getting closer but still a ways to go. Clin Infect Dis. 2006;42(4):487–9.

Dromer F, Mathoulin-Pelissier S, Launay O, Lortholary O. Determinants of disease presentation and outcome during cryptococcosis: the CryptoA/D study. PLoS Med. 2007;4(2):e21.

Ellis M, Al-Ramadi B, Bernsen R, Kristensen J, Alizadeh H, Hedstrom U. Prospective evaluation of mannan and anti-mannan antibodies for diagnosis of invasive Candida infections in patients with neutropenic fever. J Med Microbiol. 2009;58(5):606–15.

Fernandez-Molina JV, Abad-Diaz-de-Cerio A, Sueiro-Olivares M, Pellon A, Ramirez-Garcia A, Garaizar J, Peman J, Hernando FL, Rementeria A. Rapid and specific detection of section Fumigati and *Aspergillus* fumigatus in human samples using a new multiplex real-time PCR. Diagn Microbiol Infect Dis. 2014;80(2):111–8.

Frias-de Leon MG, Acosta-Altamirano G, Duarte-Escalante E, Martinez-Hernandez JE, Martinez-Rivera Mde L, Mdel R-M. Molecular markers: an important tool in the diagnosis, treatment and epidemiology of invasive aspergillosis. Cir Cir. 2014;82(1):109–18.

Glass NL, Donaldson GC. Development of primer sets designed for use with the PCR to amplify conserved genes from filamentous ascomycetes. Appl Environ Microbiol. 1995;61(4):1323–30.

Godoy MC, Viswanathan C, Marchiori E, Truong MT, Benveniste MF, Rossi S, Marom EM. The reversed halo sign: update and differential diagnosis. Br J Radiol. 2012;85(1017):1226–35.

Greene RE, Schlamm HT, Oestmann JW, Stark P, Durand C, Lortholary O, Wingard JR, Herbrecht R, Ribaud P, Patterson TF, Troke PF, Denning DW, Bennett JE, De Pauw B, Rubin RH. Imaging findings in acute invasive pulmonary aspergillosis: clinical significance of the halo sign. Clin Infect Dis. 2007;44(3):373–9.

Groll AH, Tragiannidis A. Recent advances in antifungal prevention and treatment. Semin Hematol. 2009;46(3):212–29.

Gupta P, Khare V, Kumar D, Ahmad A, Banerjee G, Singh M. Comparative evaluation of disc diffusion and E-test with Broth Micro-dilution in susceptibility testing of Amphotericin B, Voriconazole and Caspofungin against clinical *Aspergillus* isolates. J Clin Diagn Res. 2015;9(1):04–7.

Hao W, Pan YX, Ding YQ, Xiao S, Yin K, Wang YD, Qiu LW, Zhang QL, Woo PC, Lau SK, Yuen KY, Che XY. Well-characterized monoclonal antibodies against cell wall antigen of *Aspergillus* species improve immunoassay specificity and sensitivity. Clin Vaccine Immunol. 2008;15(2):194–202.

Hinrikson HP, Hurst SF, Lott TJ, Warnock DW, Morrison CJ. Assessment of ribosomal large-subunit D1-D2, internal transcribed spacer 1, and internal transcribed spacer 2 regions as targets for molecular identification of medically important *Aspergillus* species. J Clin Microbiol. 2005;43(5):2092–103.

Houbraken J, Samson RA. Phylogeny of Penicillium and the segregation of Trichocomaceae into three families. Stud Mycol. 2011;70(1):1–51.

Isshiki A, Takeharu H, Aoki S, Kokaji M, Tanabe S, Kasetani T, Yoshida M. Development of a multiple detection technique for fungi by DNA microarray with the simultaneous use of internal transcribed spacer region of ribosomal RNA gene and beta-tubulin gene probes. Biocontrol Sci. 2014;19(3):139–45.

Jegorov A, Hajduch M, Sulc M, Havlicek V. Nonribosomal cyclic peptides: specific markers of fungal infections. J Mass Spectrom. 2006;41(5):563–76.

Jeong SJ, Lee JU, Song YG, Lee KH, Lee MJ. Delaying diagnostic procedure significantly increases mortality in patients with invasive mucormycosis. Mycoses. 2015;58(12):746–52.

Kanbe T, Arishima T, Horii T, Kikuchi A. Improvements of PCR-based identification targeting the DNA topoisomerase II gene to determine major species of the opportunistic fungi Candida and *Aspergillus* fumigatus. Microbiol Immunol. 2003;47(9):631–8.

Kim NY, Lee I, Ji GE. Reliable and simple detection of ochratoxin and fumonisin production in black *Aspergillus*. J Food Prot. 2014;77(4):653–8.

Kozel TR, Wickes B. Fungal diagnostics. Cold Spring Harb Perspect Med. 2014;4(4):a019299.

Krel M, Petraitis V, Petraitiene R, Jain MR, Zhao Y, Li H, Walsh TJ, Perlin DS. Host biomarkers of invasive pulmonary aspergillosis to monitor therapeutic response. Antimicrob Agents Chemother. 2014;58(6):3373–8.

Kumar A, Ahmed R, Singh PK, Shukla PK. Identification of virulence factors and diagnostic markers using immunosecretome of *Aspergillus* fumigatus. J Proteomics. 2011;74(7): 1104–12.

Kuper KM, Coyle EA, Wanger A. Antifungal susceptibility testing: a primer for clinicians. Pharmacotherapy. 2012;32(12):1112–22.

Latge JP. The pathobiology of *Aspergillus* fumigatus. Trends Microbiol. 2001;9(8):382–9.

Liss B, Cornely OA, Hoffmann D, Dimitriou V, Wisplinghoff H (2015) 1,3- ß-D-glucanconcentrations in blood products predict false positive post-transfusion results. Mycoses 59(1):39–42. doi:10.1111/myc.12432

Lunel FM, Mennink-Kersten MA, Ruegebrink D, Van der Lee HA, Donnelly JP, Blijlevens NM, Verweij PE. Value of Candida serum markers in patients with invasive candidiasis after myeloablative chemotherapy. Diagn Microbiol Infect Dis. 2009;64(4):408–15.

Morelle W, Bernard M, Debeaupuis JP, Buitrago M, Tabouret M, Latge JP. Galactomannoproteins of *Aspergillus* fumigatus. Eukaryot Cell. 2005;4(7):1308–16.

Morton CO, De Luca A, Romani L, Rogers TR. RT-qPCR detection of *Aspergillus* fumigatus RNA in vitro and in a murine model of invasive aspergillosis utilizing the PAXgene(R) and Tempus RNA stabilization systems. Med Mycol. 2012;50(6):661–6.

Neustadt M, Costina V, Kupfahl C, Buchheidt D, Eckerskorn C, Neumaier M, Findeisen P. Characterization and identification of proteases secreted by *Aspergillus* fumigatus using free flow electrophoresis and MS. Electrophoresis. 2009;30(12):2142–50.

Nierman WC, Pain A, Anderson MJ, Wortman JR, Kim HS, Arroyo J, Berriman M, Abe K, Archer DB, Bermejo C, Bennett J, Bowyer P, Chen D, Collins M, Coulsen R, Davies R, Dyer PS, Farman M, Fedorova N, Feldblyum TV, Fischer R, Fosker N, Fraser A, Garcia JL, Garcia MJ, Goble A, Goldman GH, Gomi K, Griffith-Jones S, Gwilliam R, Haas B, Haas H, Harris D, Horiuchi H, Huang J, Humphray S, Jimenez J, Keller N, Khouri H, Kitamoto K, Kobayashi T, Konzack S, Kulkarni R, Kumagai T, Lafon A, Latge JP, Li W, Lord A, Lu C, Majoros WH, May GS, Miller BL, Mohamoud Y, Molina M, Monod M, Mouyna I, Mulligan S, Murphy L, O'Neil S, Paulsen I, Penalva MA, Pertea M, Price C, Pritchard BL, Quail MA, Rabbinowitsch E, Rawlins N, Rajandream MA, Reichard U, Renauld H, Robson GD, Rodriguez de Cordoba S, Rodriguez-Pena JM, Ronning CM, Rutter S, Salzberg SL, Sanchez M, Sanchez-Ferrero JC, Saunders D, Seeger K, Squares R, Squares S, Takeuchi M, Tekaia F, Turner G, Vazquez de Aldana CR, Weidman J, White O, Woodward J, Yu JH, Fraser C, Galagan JE, Asai K, Machida M, Hall N, Barrell B, Denning DW. Genomic sequence of the pathogenic and allergenic filamentous fungus *Aspergillus* fumigatus. Nature. 2005;438(7071):1151–6.

Odabasi Z, Mattiuzzi G, Estey E, Kantarjian H, Saeki F, Ridge RJ, Ketchum PA, Finkelman MA, Rex JH, Ostrosky-Zeichner L. Beta-D-glucan as a diagnostic adjunct for invasive fungal infections: validation, cutoff development, and performance in patients with acute myelogenous leukemia and myelodysplastic syndrome. Clin Infect Dis. 2004;39(2):199–205.

Oliveira-Coelho A, Rodrigues F, Campos JA, Lacerda JF, Carvalho A, Cunha C. Paving the way for predictive diagnostics and personalized treatment of invasive aspergillosis. Front Microbiol. 2015;6:411.

Oliveira M, Lackner M, Amorim A, Araujo R. Feasibility of mitochondrial single nucleotide polymorphisms to detect and identify *Aspergillus* fumigatus in clinical samples. Diagn Microbiol Infect Dis. 2014;80(1):53–8.

Orsi CF, Bettua C, Pini P, Venturelli C, La Regina A, Morace G, Luppi M, Forghieri F, Bigliardi S, Luppi F, Codeluppi M, Girardis M, Blasi E. Detection of Pneumocystis jirovecii and *Aspergillus* spp. DNa in bronchoalveolar lavage fluids by commercial real-time PCr assays: comparison with conventional diagnostic tests. New Microbiol. 2015;38(1):75–84.

Palumbo JD, O'Keeffe TL. Detection and discrimination of four *Aspergillus* section Nigri species by PCR. Lett Appl Microbiol. 2015;60(2):188–95.

Pan Z, Fu M, Zhang J, Zhou H, Fu Y, Zhou J. Diagnostic accuracy of a novel lateral-flow device in invasive aspergillosis: a meta-analysis. J Med Microbiol. 2015;64(7):702–7.

Pfaller MA, Diekema DJ. Rare and emerging opportunistic fungal pathogens: concern for resistance beyond Candida albicans and *Aspergillus* fumigatus. J Clin Microbiol. 2004;42(10):4419–31.

Pfaller MA, Diekema DJ. Epidemiology of invasive candidiasis: a persistent public health problem. Clin Microbiol Rev. 2007;20(1):133–63.

Posteraro B, Sanguinetti M. The future of fungal susceptibility testing. Future Microbiol. 2014;9(8):947–67.

Rachini A, Pietrella D, Lupo P, Torosantucci A, Chiani P, Bromuro C, Proietti C, Bistoni F, Cassone A, Vecchiarelli A. An anti-beta-glucan monoclonal antibody inhibits growth and capsule formation of Cryptococcus neoformans in vitro and exerts therapeutic, anticryptococcal activity in vivo. Infect Immun. 2007;75(11):5085–94.

Schlageter AM, Kozel TR. Opsonization of Cryptococcus neoformans by a family of isotype-switch variant antibodies specific for the capsular polysaccharide. Infect Immun. 1990;58(6):1914–8.

Schutte M, Thullier P, Pelat T, Wezler X, Rosenstock P, Hinz D, Kirsch MI, Hasenberg M, Frank R, Schirrmann T, Gunzer M, Hust M, Dubel S. Identification of a putative Crf splice variant and generation of recombinant antibodies for the specific detection of *Aspergillus* fumigatus. PLoS One. 2009;4(8):e6625.

Scotter JM, Chambers ST. Comparison of galactomannan detection, PCR-enzyme-linked immunosorbent assay, and real-time PCR for diagnosis of invasive aspergillosis in a neutropenic rat model and effect of caspofungin acetate. Clin Diagn Lab Immunol. 2005;12(11):1322–7.

Simitsopoulou M, Roilides E, Georgiadou E, Paliogianni F, Walsh TJ. Differential transcriptional profiles induced by amphotericin B formulations on human monocytes during response to hyphae of *Aspergillus* fumigatus. Med Mycol. 2011;49(2):176–85.

Singh B, Singh S, Asif AR, Oellerich M, Sharma GL. Allergic aspergillosis and the antigens of *Aspergillus* fumigatus. Curr Protein Pept Sci. 2014;15(5):403–23.

Soubani AO, Chandrasekar PH. The clinical spectrum of pulmonary aspergillosis. Chest. 2002;121(6):1988–99.

Speed B, Dunt D. Clinical and host differences between infections with the two varieties of Cryptococcus neoformans. Clin Infect Dis. 1995;21(1):28–34.

Spiess B, Buchheidt D, Baust C, Skladny H, Seifarth W, Zeilfelder U, Leib-Mosch C, Morz H, Hehlmann R. Development of a LightCycler PCR assay for detection and quantification of *Aspergillus* fumigatus DNA in clinical samples from neutropenic patients. J Clin Microbiol. 2003;41(5):1811–8.

Spiess B, Seifarth W, Hummel M, Frank O, Fabarius A, Zheng C, Morz H, Hehlmann R, Buchheidt D. DNA microarray-based detection and identification of fungal pathogens in clinical samples from neutropenic patients. J Clin Microbiol. 2007;45(11):3743–53.

Stynen D, Sarfati J, Goris A, Prevost MC, Lesourd M, Kamphuis H, Darras V, Latge JP. Rat monoclonal antibodies against *Aspergillus* galactomannan. Infect Immun. 1992;60(6):2237–45.

Sueiro-Olivares M, Fernandez-Molina JV, Abad-Diaz-de-Cerio A, Gorospe E, Pascual E, Guruceaga X, Ramirez-Garcia A, Garaizar J, Hernando FL, Margareto J, Rementeria A. *Aspergillus* fumigatus transcriptome response to a higher temperature during the earliest steps of germination monitored using a new customized expression microarray. Microbiology. 2015;161(Pt 3):490–502.

Sugui JA, Peterson SW, Clark LP, Nardone G, Folio L, Riedlinger G, Zerbe CS, Shea Y, Henderson CM, Zelazny AM, Holland SM, Kwon-Chung KJ. *Aspergillus* tanneri sp. nov., a new pathogen that causes invasive disease refractory to antifungal therapy. J Clin Microbiol. 2012;50(10):3309–17.

Thursky KA, Playford EG, Seymour JF, Sorrell TC, Ellis DH, Guy SD, Gilroy N, Chu J, Shaw DR. Recommendations for the treatment of established fungal infections. Intern Med J. 2008;38(6b):496–520.

Tuntevski K, Durney BC, Snyder AK, Lasala PR, Nayak AP, Green BJ, Beezhold DH, Rio RV, Holland LA, Lukomski S. *Aspergillus* collagen-like genes (acl): identification, sequence polymorphism, and assessment for PCR-based pathogen detection. Appl Environ Microbiol. 2013;79(24):7882–95.

Turenne CY, Sanche SE, Hoban DJ, Karlowsky JA, Kabani AM. Rapid identification of fungi by using the ITS2 genetic region and an automated fluorescent capillary electrophoresis system. J Clin Microbiol. 1999;37(6):1846–51.

Van Peer A, Woestenborghs R, Heykants J, Gasparini R, Gauwenbergh G. The effects of food and dose on the oral systemic availability of itraconazole in healthy subjects. Eur J Clin Pharmacol. 1989;36(4):423–6.

Varga J, Rigo K, Kocsube S, Farkas B, Pal K. Diversity of polyketide synthase gene sequences in *Aspergillus* species. Res Microbiol. 2003;154(8):593–600.

Walsh TJ, Anaissie EJ, Denning DW, Herbrecht R, Kontoyiannis DP, Marr KA, Morrison VA, Segal BH, Steinbach WJ, Stevens DA, Van Burik JA, Wingard JR, Patterson TF. Treatment of aspergillosis: clinical practice guidelines of the Infectious Diseases Society of America. Clin Infect Dis. 2008;46(3):327–60.

Walsh TJ, Wissel MC, Grantham KJ, Petraitiene R, Petraitis V, Kasai M, Francesconi A, CottonMP HJE, Greene L, Bacher JD, Manna P, Salomoni M, Kleiboeker SB, Reddy SK. Molecular detection and species-specific identification of medically important *Aspergillus* species by real-time PCR in experimental invasive pulmonary aspergillosis. J Clin Microbiol. 2011;49(12):4150–7.

White PL, Mengoli C, Bretagne S, Cuenca-Estrella M, Finnstrom N, Klingspor L, Melchers WJ, McCulloch E, Barnes RA, Donnelly JP, Loeffler J. Evaluation of *Aspergillus* PCR protocols for testing serum specimens. J Clin Microbiol. 2011;49(11):3842–8.

Wong B, Brauer KL, Tsai RR, Jayasimhulu K. Increased amounts of the *Aspergillus* metabolite D-mannitol in tissue and serum of rats with experimental aspergillosis. J Infect Dis. 1989;160(1): 95–103.

Yoo JH, Choi SM, Lee DG, Park SH, Choi JH, Kwon EY, Shin WS. Comparison of the real-time nucleic acid sequence-based amplification (RTi-NASBA) with conventional NASBA, and galactomannan assay for the diagnosis of invasive aspergillosis. J Korean Med Sci. 2007;22(4):672–6.

Zarrinfar H, Mirhendi H, Makimura K, Satoh K, Khodadadi H, Paknejad O. Use of mycological, nested PCR, and real-time PCR methods on BAL fluids for detection of *Aspergillus* fumigatus and A. flavus in solid organ transplant recipients. Mycopathologia. 2013;176(5–6):377–85.

Zhao Y, Park S, Kreiswirth BN, Ginocchio CC, Veyret R, Laayoun A, Troesch A, Perlin DS. Rapid real-time nucleic Acid sequence-based amplification-molecular beacon platform to detect fungal and bacterial bloodstream infections. J Clin Microbiol. 2009;47(7):2067–78.

Zhao Y, Perlin DS. Quantitative detection of *Aspergillus* spp. by real-time nucleic acid sequence-based amplification. Methods Mol Biol. 2013;968:83–92.

Zmeili OS, Soubani AO. Pulmonary aspergillosis: a clinical update. QJM. 2007;100(6):317–34.

Chapter 14
An Overview of Major Fungal Diseases of Sugarcane in India: Detection and Management Strategies

Ajay Kumar Tiwari, Atul Singh, Surendra Pratap Singh, Awadesh Dagar, Kavita Kumari, Dipendra Kumar, Naveen Pandey, and Pradeep Kumar

14.1 Introduction

Sugarcane is one of the most important commercially crop grown in more than 115 tropical and sub tropical countries of the world. India is a major producer as well as consumer of sugar in the world and produced about 25 metric tons (MT) of sugar from 360 MT of sugarcane in year 2011–2013, contributing about 15 % of the totals sugar production of the world. A quantum of sugar is produced from sugarcane; however, this crop faces a number of problems such as low cane productivity, biotic and abiotic stresses, high cost of cultivation, unavailability of seed cane of newly released varieties, post harvest losses and low sugar recovery (Tiwari et al. 2015a). Sugarcane is long duration vegetative propagated crop and is being attacked by a number of pathogens (Fungi, Bacteria, Phytoplasma and Viruses) according to the prevailing climatic conditions from germination till maturity. About 60 diseases

A.K. Tiwari • A. Dagar • K. Kumari
Central Lab, UP Council of Sugarcane Research, Shahjahanpur, Uttar Pradesh 242001, India
e-mail: ajju1985@gmail.com

A. Singh • S.P. Singh
Division of Plant Pathology, UP Council of Sugarcane Research, Shahjahanpur, Uttar Pradesh 242001, India

D. Kumar
Division of Food Science and Post Harvest Technology, Indian Agricultural Research Institute, Pusa, New Delhi, India

N. Pandey
Agriculture Sciences, Arunanchal University of Studies, Namsai, Arunachal Pradesh 792103, India

P. Kumar (✉)
School of Biotechnology, Yeungnam University, Gyeongsan 712–749, Korea
e-mail: pkbiotech@gmail.com

B.P. Singh, V.K. Gupta (eds.), *Molecular Markers in Mycology*, Fungal Biology,
DOI 10.1007/978-3-319-34106-4_14

have been reported in India to attack this crop. Out of these about dozen diseases are of major and economically important. Most of the major diseases are seed piece transmissible and inter state transportation of seed material without any quarantine system results in the spread of these diseases. On an average the loss caused in sugarcane by fungal diseases alone ranges from 15 to 30 % (Tiwari et al. 2010). But in case of epiphytotic conditions the losses in yield and sugar goes up to 100 % (Viswanathan and Rao 2011). The fungal diseases are comparatively more damaging than others. In this chapter, fungal disease and their impact on sugarcane has been described.

14.2 Major Fungal Diseases of Sugarcane

14.2.1 Red Rot of Sugarcane

The term sugarcane and red rot diseases of sugarcane go hand in hand. Even in early eighth to seventh century BC the disease was mentioned in the Vedic text of Aitereya Aryanaka. Where it is stated that in Indian epic Mahabaratha there was a Pundra kingdom where the people suffered from a skin disease called Pandu and it is said the name Pandu originated from a sugarcane variety then found in that region which was affected by a diseases of red patches. One also finds the mention of red rot in the Buddhist literature. The disease was so prevalent in UP and Bihar where Gautama Buddha started preaching after enlightenment. Buddha referred to red rot of sugarcane as Manjitthika after a red dye. Gautama Buddha quoted "just as the disease known as Manjitthika falls on a field of ripened sugarcane that field does not last long". In India the first red rot epidemics occurred in the Godavari delta in 1895. The disease devastated the entire sugarcane cultivation in the Godavari delta to the extend that area under sugarcane reduced to 500 hectare (ha) from more than 2500 ha by 1899. Hence, the government deputed Mr C.A Barber to study the crop and the disease. In his report he described the causal agent a fungal pathogen named as *Colletotrichum falcatum* Went (Panicker and Velazhahan 2015).

Red rot is most serious sett borne disease and also known as cancer of sugarcane. It has appeared many times in the form of epiphytotic in U.P., Bihar, Haryana, Punjab and Kerala. In Uttar Pradesh and Bihar (two major cultivars i.e. Co 213 and Co 312) were badly affected by this disease causing epiphytotic conditions in the year 1938–1940 and 1946–1947. A very recently, important and popular variety CoS 07250 was rejected and gone out of cultivation because of red rot disease found at various places of eastern Uttar Pradesh, India. It is one of the most widespread sugarcane diseases in the country and has been a constraint for sugarcane productivity for the past 100 years in India. The disease is responsible for the elimination of many commercial varieties in India in the earlier decades. Epidemics of the disease have been very common ever since its occurrence in India. Currently the disease occurs in all the sugarcane growing states in India except Karnataka and Maharashtra states. Recently reports of red rot occurrence in CoC 671 in parts of Kolhapur and Solapur Districts in Maharashtra were reported. In the recent years we have witnessed break down of cvs CoS 8436, CoSe 95422, 92423 and BO138, the important commercial varieties of subtropical region to red rot. Severe damage to crop stand is found in these varieties due to disease epidemics in the states of Haryana, Uttar Pradesh and Bihar.

14.2.1.1 Economic Importance of Red Rot Disease of Sugarcane

The pathogen affects the economically valuable stalk tissues and even the limited infection can bring about drastic changes in the juice quality. The disease affected cane gives poor sugar recovery because of impaired sucrose metabolism. The red rot infection reduced total carbohydrates in the diseased canes and the reduction was more in the highly susceptible varieties (Agnihotri 1996). Moreover, the pathogen produces abundant quantities of acid invertases which break the sucrose into glucose and fructose which are consumed by the pathogen. Higher production of acid invertases in the highly susceptible varieties was recorded upon pathogen infection as compared to resistant varieties (De Silva et al. 1977). Pathogen infection also resulted in increased level of total soluble salts, acid, reducing sugars, gum and simultaneously decrease in pH, sucrose and purity of cane juice (Singh and Waraitch 1977). Increased activity of the enzyme invertase or inhibition of normal synthesis of carbohydrates is indicated by degradation of sucrose and increase in the levels of reducing sugars. Similar studies conducted at SBI, Coimbatore revealed that pathogen infection has drastically reduced brix, sucrose percentage, purity and CCS per cent in the diseased canes. The affected canes recorded 25–75 % reduction in sucrose content than the healthy canes (Viswanathan and Samiyappan 1999a, b). Studies of Satyavir et al. (2002) in Haryana revealed that red rot infection reduces 7.1–32.5 % in extraction, 7.4–38.7 % in polarity, 0.5–8.3 % in purity co-efficient, 7.8–39 % in commercial cane sugar and an increase of 19.2–40.95 % in reducing sugars. During the milling process, mixing of juice from healthy and diseased canes result in spoilage of entire juice due to inversion of sucrose. Similarly 'jaggery' setting will also be affected if red rot affected canes are crushed with healthy ones. Resistance to red rot in sugarcane varieties is not static; hence, practically all varieties under cultivation are susceptible. Once a resistant variety occupying large areas succumbs to the disease, extensive damage to cane cultivation occurs.

14.2.1.2 Epidemiology and Symptom of Red Rot Disease

Sett borne inoculum is considered to be the major mode of inoculum transmission. Planting the infected material is the prime source of inoculum for recurrent disease incidence. Post germination, infection on young emerging shoots is caused by the dormant mycelium present in the bud scales. In the field, the various sources of pathogen transmission are midrib lesions, disease stalks, infected setts, crop debris and infested soil. The means of inoculum transmission depend on the season and the cane growing circumstances. These sources of inoculum initiate infections in young developing canes in the nearby fields. The secondary spread of the disease is affected by conidial dissemination through irrigation water, rain splashing, dew washings from mid rib lesions and dispersal by wind. Bore holes could also act as portals of entry. Singh and Lal (1996) trapped propagules or structures of *C. falcatum* other than ascospores from air and confirmed development of nodal infection in healthy plants in the vicinity of the diseased ones. The resting structures of the fungus such as appressoria, chlamydospores, thick-walled hyphae and setae play a vital role in the soil borne transmission. Crop debris comprised of diseased stalk or stubble pieces left after harvest provide a bulk of inoculum to the subsequent infection under optimal high humid conditions. Though the fungus is not a true soil borne pathogen, enough evidence suggests that debris borne inoculum have a role in perpetuating the fungal propagules (Viswanathan and Rao 2011).

Expression of the disease may vary depending upon nature of infection and prevailing environmental conditions (Satyavir 2003). In the early stages of infection, it is difficult to recognize the presence of the disease in the field as reddening of the internal tissues with interrupted red and white patches, the characteristic symptoms of the disease, develops on the stem only at later stages. Furthermore, latent infection occurs frequently, making visual diagnosis impossible (Nithya et al. 2012). The symptoms of this disease start from the month of July/August and found till the end of the crop. Third to fourth leaf of the top starts withering along the margins. Later on, the whole crown withers away. At initial stages purple colour develops on the nodes of cane, which turns to reddish-brown later. Midrib lesions are the major source of inoculum during the cane growing season. Typical symptoms are observed in the internodes of the stalk after splitting the cane longitudinally. Dull reddening of tissues with white spots are seen which omit alcoholic smell. Red rot affected canes are very easily broken from the nodes. Infected standing stalks serve as a greater inoculum source at later stages of the crop (Fig. 14.1).

The first symptoms of the disease are seen when the vegetative growth of the plant is stopped and sucrose formation begins, i.e., after rainy season. The pathogen, *C. falcatum* Went, can attack any part of the sugarcane plant; be it stalk, leaf, buds or roots. *C. falcatum* completes its life cycle on the sugarcane leaf and usually damage to leaf does not pose a serious threat to cane or cause much harm to the plant (Duttamajumder 2008). Discolorations of the leaves are the first symptom in the field. The spindle leaves (third and fourth leaf) display drying which withers away at the tips along the margins. This discoloration from tip to the base is continued till all the leaves of the crown wilt (Agnihotri 1996). Tiny reddish lesions occur on the upper surface of the lamina with minute red spots in both the directions of the upper surface of the midrib. Infection also resulted in change in the colour of the leaves that become straw colored in the center and dark reddish brown at the margins with the development of black acervuli. The infected leaves may break at the lesions and hang down (Sharma and Tamta 2015).

Fig. 14.1 Sugarcane stalk (LS) showing the symptoms of red dot disease

14.2.1.3 Molecular Markers for Red Rot Disease Diagnosis

Correct diagnosis of pathogens is the primary requirement in any sound disease management practice. Disease diagnosis and pathogen identification by conventional methods involves isolating the pathogen and characterizing it by inoculation tests (Nithya et al. 2012). As vegetative propagation in sugarcane favors harboring of the pathogen in the planting setts, adequate care should be taken while selecting seed canes. Since, it is difficult to detect incipient infections of *C. falcatum* in seed-pieces, it is recommended to take the planting material from a disease free crop. Three-tier seed production system is being recommended to raise disease free seed cane in the sugar factory area. PCR results confirmed that the primers are able to detect the *C. falcatum* specifically in mixed state of infection in sugarcane which helps to detect the fungal infections more accurately (Viswanathan et al. 2010). The characterization of pathogen has primarily been based upon variation in colour, conidial size and shape, appressoria, colony characters, host association and molecular methods based on polymerase chain reaction (PCR) are frequently being used now a day for detecting fungal pathogens in plant tissues (Bonants et al. 1997; Madan et al. 2000; Wang et al. 2006), due to their increased specificity and sensitivity in comparison to the traditional techniques. Variability in *C. falcatum* isolates have been characterized by using arbitrary marker system such as RAPD (Madan et al. 2000; Mohan Raj et al. 2002; Suman et al. 2005; Saksena et al. 2013). Kumar et al. (2010) used inter simple sequence repeat (ISSR), universal rice primers (URPs) and RAPD markers to characterize variability among *C. falcatum* isolates. ITS primers were also used to study the variability among pathogen (Malathi et al. 2010). Nithya et al. (2012) developed SCAR markers specific to races and isolates of *C. falcatum* sporulation. But this phenotypic identification is time consuming, expertise specific and not always fully discriminative (Kumar et al. 2010; Abbas et al. 2010).

14.2.1.4 Management of Red Rot

Management of red rot has been a challenging area of work for the pathologists and sugarcane breeders. The epiphytotic of the disease depends upon weather conditions, genotypes, presence of virulent pathogen and time for disease development. These factors must be studied in depth so as to achieve effective control of the disease. It has been observed that once the disease has appeared in the field it is impossible to control. Most of the recommended management practices hence are aimed at prophylactic measures to reduce pathogen build up in the field (Viswanathan et al. 2011). Kumar et al (2012) developed red rot resistance through soma clones, but developed soma clones are loosed originality. Role of endochitinase (ech42) gene in *Trichoderma* was worked out by Singh et al (2013) to prove the efficacy against most dreaded red rot disease of sugarcane caused by *C. falcatum* Went. *Trichoderma* spp. along with *Trichoderma harzianum* strain and *Aspergillus awamori* were tested against the virulent pathotypes of *C. falcatum* (Cf08, Cf09 and Cf401) under *in vitro* and *in vivo* conditions. *T. harzianum* was significantly able to control the incidence of red rot disease in highly susceptible sugarcane cultivar. A single method would not be useful to mitigate the losses from red rot (Agnihotri 1996), hence Integrated Disease Management (IDM) should be practiced. The combinations of different control methods could be adopted for controlling the red rot disease.

14.2.1.5 Resistant Varieties Against Red Rot

Effective control of red rot has been mainly through the use of resistant varieties. Even though genetics of inheritance of red rot resistance is not well established, considerable progress has been made in the production of red rot resistant varieties (Viswanathan et al. 2011). In India, the breeding work is primarily focused on developing red rot resistant varieties (Agnihotri 1996). But as the pathogen is highly variable in nature, therefore, even if a disease resistant variety is released for cultivation, it gets susceptible to red rot disease within 8–10 years because of the development of new more virulent races of the pathogen (Yadav 2006).

14.2.1.6 Chemotherapy and Thermotherapy for Management of Red Rot

A number of fungitoxicants have been tried against the red rot but a little success is found in controlling this devastating disease. This may be because of impervious nature of rind, presence of fibrous nodes at the cut ends, low solubility of fungicides, and lack of broad spectrum fungicides and presence of abundant nutrients in the sett (Agnihotri 1996). Some reduction in red rot incidence has been reported when infected setts were treated with carbendazim and benomyl, which is not commercially produced now for 1–2 h (Anzalone 1970; Chand et al. 1974; Waraitch 1983). Soaking of sugarcane setts in 0.25 % suspension of thiophanate methyl and its metabolite carbendazim for 24 h before planting was found to be effective in controlling debris-borne infection (Malathi et al. 2004). In an experiment conducted by Bharadwaj and Sahu (Bharadwaj and Sahu 2014), bavistin showed complete inhibition of mycelial growth of the *C. falcatum*.

Heat therapy has been used for controlling sett borne infection of red rot by various workers. Joshi (Joshi 1954) and Singh (1973) reported complete elimination of sett-borne infection by hot air treatment (54 °C for 8 h). Some workers have used heat and chemotherapy in combination by incorporating the chemical into the hot water tank for adequate control of red rot (Agnihotri 1983). Findings of various workers have proved that moist hot air therapy (MHAT) of seed cane at 54 °C for 4 h (R.H. 95–100) was most effective against red rot (Singh et al. 1980; Dhillon et al. 1983). Anonymous (1980) reported that moist hot air treatment (54 °C for 2 h) was more effective than hot water treatment (50 °C–2 h) in reducing red rot. Aerated steam treatment at 52 °C or the soaking of setts in cold running water for 48 h followed by hot-water treatment (50 °C for 150–180 min) also helps in eliminating the pathogen from infected setts (Singh and Singh 1989).

14.2.1.7 Biological Control of Red Rot

Red rot disease of sugarcane was observed to be biologically controlled through *Trichoderma harzianum* and *T. viride* (Singh et al. 2008). *T. harzianum* and *Pseudomonas spp.* possess the ability to protect the crop from soil borne inoculum of red rot and the efficacy is because of the chitinase enzyme produced by them (Malathi and Viswanathan 2013). Singh et al (2013) clearly supported that *ech42* gene of Trichoderma spp. is responsible for controlling the red rot incidence in sugarcane. In an experiment conducted by Singh et al. (2010), addition of SA (salicylic acid) has boosted the protection level significantly against red rot disease and helped *T. harzianum* for inducing systemic resistance in sugarcane. *Ocimum*, *Ginger*, *Onion* and *Garlic* were found to inhibit the mycelial growth. It was also examined that the essential oils, *viz.* Peppermint oil, Mentha oil, Geranium oil, Patchouli oil and Palmaroza oil were effective in inhibiting the growth of mycelia of *C. falcatum* (Bharadwaj and Sahu 2014). In sugarcane, Viswanathan and Samiyappan (1999a, b) established PGPR (Plant growth promoting rhizobacteria) mediated ISR (Induced systemic resistance) against *C. falcatum* causing red rot disease. The study conducted by Viswanathan et al. (2005) indicates a possible role of PR (pathogenesis related) proteins in conferring red rot resistance in sugarcane. The peptides MUC1 60 mer and Purothionins mixture significantly inhibited the mycelial growth and spore germination of *C. falcatum* (Edward et al. 2013). It was showed in an experiment that leaf extracts of *Curcuma domestica* and *Datura metel* inhibited the conidial as well as mycelial growth. Smoke of dhup (incense) and tobacco also showed inhibition against conidial germination (Imtiaj et al. 2007).

14.2.2 Smut Disease of Sugarcane

One of the easily recognized diseases of sugarcane with worldwide distribution. Smut disease is generally observed in Maharashtra, Andhra Pradesh, Tamil Nadu, Kerala and Gujrat. In U.P. its occurrence is noticed in the months of April to June, November and February. A number of varieties were knocked down of cultivation only due to this disease in the past. Smut disease of sugarcane, caused by the fungus *Ustilago scitaminea* (Sydow 1924) leads to considerable yield losses and reduction in cane quality (Ferreira and Comstock 1989) which ultimately affects the sugar industries throughout the world. Smut was first reported from Natal in South Africa in 1877 (McMartin 1945), and thereafter out brokc all over the African continent and nearby islands, India, the other south-east Asian countries, the Philippines, Taiwan, and South American countries (Antoine 1961).

Smut was recorded in 1971 on Hawaii's island of Oahu (Ferreira and Comstock 1989) and in the Caribbean state of Guyana in 1974 (Bates 1975). It then spread to all the Caribbean islands, Central America and mainland USA. In Australia, the disease was first observed in 1998 in a marginal production area on the west coast (Riley et al. 1999) and was detected in the main producing area (east coast) in 2006. Only Papua New Guinea which is considered as the center of origin of *Saccharum officinarum*, the main species involved in modern cultivars, and Fiji Islands are still free of this disease.

14.2.2.1 Symptomology and Epidemiology

This is also a sett borne disease. The leaves of the crown become shorter and pointed making a fan like structure of the top. The affected plants become thin with elongated internode. A whip like structure emerges out from the top which ranges from 6 in. to 1 m. The black colored whip is initially covered with a silvery membrane which ruptures later to disseminate the spores in air for secondary infection on other plants (Viswanathan et al. 2000) (Fig. 14.2). The first appearance of the apical whips is seen around 120 days of planting. The second flush of whip emergence produces an enormous quantity of teliospores which infect the terminal and lateral buds in the rapidly growing crop. The infected buds may remain dormant and may germinate to produce lateral whips in the third flush of whip production. The

Fig. 14.2 Smut symptoms in field crop of sugarcane

infection producing the third level of whips is believed to be critical in the epidemiology of the disease (Ramesh Sundar et al. 2012). Smut is generally favored by hot dry weather conditions. Plant stress increases frequency of whip development. Under high stress conditions the cultivars may show symptoms, otherwise they normally do not produce whips. Teliospore survival is decreased rapidly by soil moisture. Similarly, high rainfall reduces the severity of smut development. In subtropical areas, after severe winter the level of smut decreases probably due to death of smut infected plants. Disease severity usually increases in the ratoon crops (Viswanathan and Rao 2011).

14.2.2.2 Molecular Markers for Smut Disease Diagnosis

Smut is morphologically diagnosed by the presence of different types of whip such as long, closed, twisted, short and multiple whips. Smut infected sugarcane clump also shows profuse tillering and poor cane formation. Some unusual symptoms of smut infection are apical deformatity, floral infection, malformed spindle, and bud proliferation (Ramesh Sundar et al. 2012).

The polymerase chain reaction (PCR) techniques, including PCR and nested-PCR are widely used as rapid, specific and sensitive tools for the detection and evaluation of plant diseases (Kamel et al. 2003; Farid et al. 2006; Kawther 2008; Li et al. 2009; Liu et al. 2009). For detection of *U. scitaminea*, so far only one PCR-based method was reported, which was based on the b loci (bE4/bE8E) mating-type gene sequence of *Ustilago maydis* (Albert and Schenck 1996). Nested-PCR detection offers higher sensitivity in detection of plant diseases as compared to PCR-based methods (Li et al. 2009; Liu et al. 2009). The life cycle of the fungus is regulated by *a* and *b* mating-type loci within the sporidia. Mating type *a* has two alleles which encode a pheromone and a receptor whilst *b* is multi allelic and appears to control pathogenicity and sexual development (Bakkeren et al. 1992). By using primers based on the *U. maydis b*E mating-type gene, Albert and Schenck (Albert and Schenck 1996) sequenced the corresponding gene in *U. scitaminea*. Since then molecular detection of the smut pathogen has become possible using PCR to amplify the *b* E mating-type gene of *U. scitaminea*. Based on nucleotide differences in the internal transcribed spacer (ITS) sequences of *U. scitaminea*, a pair of species-specific primers, SL1 (5′-CAGTGCACGAAAGTACCTGTGG-3′) and SR2 (5′-CTAGGGCGGTGTTCAGAAGCAC-3′) was designed by using a panel of fungal and bacterial species as controls. The primers SL1/SR2 specifically amplified a unique PCR product about 530 bp in length from *U. scitaminea* strains with a detecting sensitivity at 200 fg of the fungal genomic DNA in a 25 μl reaction solution. To increase sensitivity, a nested-PCR protocol was further established, which used ITS4/ITS5 as the first-round primers followed by the primer pair SL1/SR2. This protocol increased the detection sensitivity by 10,000-fold compared to the PCR method and could detect the fungal DNA as low as 20 ag (Shen et al. 2012).

Amplified Fragment Length Polymorphism (AFLP) fingerprinting technique was used to estimate the degree of genetic variation among 38 isolates of sugarcane smut collected from 16 sugar industries in 13 countries by Braithwaite et al. (2004). The AFLPs revealed a low level of genetic variation among 31 isolates, however, seven isolates collected from Taiwan, the Philippines and Thailand, were genetically more divergent. There was no significant difference between isolates collected in 1984 and those collected 16 years later from the same sugar industry, suggesting that genetic change occurs slowly in this fungus and very limited changes have occurred in the existing populations. The induced resistance mechanisms in sugarcane in response to challenge by the pathogen *U. Scitaminea* was investigated using the cDNA-AFLP differential display technique by Thokoane and Rutherford (2001).

Smut disease on sugarcane though first recorded in South Africa in 1877 on the *Saccharum sinense* clone known as 'China cane', it was most likely present in Asia for much longer. There are circumstantial reports that smut has always been present in India (Antoine 1961). Indian sugarcane (*Saccharum barberi*) is very susceptible to sugarcane smut and the wild grass *Saccharum spontaneum* is considered to be a collateral host of sugarcane smut and a major reservoir of inoculum in India (Chona 1956). An Asian origin for *U. scitaminea* may explain the higher level of genetic variation among the Asian smut isolates. Further sampling from Asia, particularly India, and also from relatives of sugarcane (e.g. *S. spontaneum*, *S. barberi*, *S. sinense*) may reveal more variation within the Asian region. Interestingly, the centre of origin of *Saccharum officinarum* is believed to be New Guinea (Daniels and Roach 1987), and this region remains free of smut even to this date. Alternative fingerprinting techniques, such as simple sequence repeats (SSRs or microsatellites), may provide higher sensitivity and generate more polymorphisms to allow to verify the significance of the weak clusters being observed, or to reveal the existence of yet other clusters.

14.2.2.3 Role of Different Compounds in Resistant Varieties Developments

The resistance to a disease occurs by accumulation of different compounds such as: phytoalexins (i.e. low molecular mass antimicrobial compounds that accumulate at sites of infection); systemic enzymes that degrade pathogens (e.g. chitinases, β-1,3-glucanases and proteases); systemic enzymes that generate antimicrobial compounds and protective biopolymers (e.g. peroxidases and phenoloxidases); biopolymers that restrict the spread of pathogens (e.g. hydroxyproline-rich glycoproteins, lignin, callose); and regulators of the induction and/or activity of defensive compounds (e.g. elicitors of plant and microbial origin, immune signals from primed plants and compounds, which release immune signals) (Kuc 1990). Three varieties of sugarcane have been studied for the production of glycoproteins after smut infection viz., Barbados which is extremely susceptible; Jaronu shows moderate resistance, while Mayari is extremely resistant to smut. When vegetative buds of sugarcane was inoculated with teliospores and cultured under field conditions for 12 months, it was observed that soluble polysaccharides and glycoproteins of juices increased after infection in Jaronu and Mayari plants, but the same decreased in Barbados specimens (Martinez et al. 2000). Several glycoproteins, defined as mid-molecular mass (MMMG) or high molecular mass (HMMG) macromolecules are produced as a primary response to smut infection (Legaz et al. 1998a, b). Teliospore germination in the presence of both MMMG and HMMG decreased about 50 % following 5 h of teliospore contact with glycoproteins which may be related to the ability of glycoproteins to produce cyto-agglutination. Binding of fluorescein-labelled glycoproteins was studied by fluorescence microscopy, which reveals that staining of cells was not even, but mainly restricted to the contact zone between two individual teliospores when aggregated (Fontaniella et al. 2002). Ungerminated fungal spores generally lack specific localization of their organelles and most are apparently able to germinate from any point. Hence, one of the major changes that must occur during germination is the establishment of cytoplasmic polarity (Fontaniella et al. 2002). Thus disease resistance seems to be a multifactorial process.

14.2.2.4 Management of Smut Disease

Breeding and Selection

Breeding for resistance to sugarcane smut, caused by *U. scitaminea* H. & *P. Sydow*, is a feature of sugarcane breeding programs in all countries where the disease is present. The disease can be successfully managed with resistant varieties (Comstock 2000). Breeding a new variety starts with the crossing of parent clones to produce sexual seed. In 2004, the smut breeding strategy was reviewed and a new strategy was developed that targeted 25 % of crosses with mid-parent ratings in the resistant range and at least 25 % in the intermediate range. This new strategy was almost achieved in 2005 and was exceeded in 2006 in Australia. Overseas research has shown that, even in crosses between two highly susceptible parents, between 6 and 20 % of progeny will be smut-resistant (Wu et al. 1983; Chao et al. 1990). These resistant progeny from susceptible crosses will be assessed as potential varieties and will be used as parents. The methods for screening for smut resistance have been reviewed by Ferreira et al. (1980). Screening is conducted in most countries by dipping setts in a suspension of smut spores and rating clones on the percentage of infected plants in a plant and ratoon crop. Most researchers recommend that the spore suspension should contain approximately 1×10^6 viable spores per mL. Collecting sufficient spores to conduct large screening trials can be difficult when heavily infested crops are not common. When spores are in short supply, painting a spore paste or spraying spores onto buds are alternative methods. Maintaining the inoculated setts in a warm (31 °C), humid environment provides the ideal conditions for infection (Bock 1964). There are no commonly accepted methods for screening true seedlings for smut resistance. A number of methods have been published but none have been widely adopted.

Integrated Pest Management

IPM is a tactic strategy for minimizing the damage caused by the diseases which vary according to the variety of modifying factors (Kendrick 1988). IPM emphasizes on the combination of methods in contrast to single strategy control for minimizing side effects (Allsopp and Manners 1997). Pathogens are regarded as indicators of unmanaged design system for control of disease rather than an enemy in an IPM system.

Application of Fungicides

The use of resistant varieties has proved to be an efficient and economic method for the control of sugarcane smut (Villalon 1982; Pruett and Waller 1989; Phelps and Donelan 1991). However, the development of resistant cane varieties takes time to achieve, so fungicides are applied for short term. Infection due to *U. scitaminea* is through air-borne spores which gain entry into buds of standing canes and from spores in the soil or irrigation water that enter buds of planted setts (Bebbee 1988). As the bud sprouts, the fungus grows within the resulting seedling and arrests its growth by turning the terminal axes into its own reproductive structure—the sorus—which appears when the cane is 2.5–5 months old (Antoine 1961; Peros 1984). Systemic fungicides such as benomyl and carbendazim (Gul and Hassan 1989; Olufolaju 1993) as well as several protectant fungicides such as mancozeb, chlorothalonil and captan have also been applied by soaking the setts in their solutions or suspensions (Kidd and James 1991). The plant sets are dip-inoculated in smut spore suspension and then in solutions or a suspension of six fungicides for 10 min and planted out in glasshouse and field trials. Mancozeb, carbendazim + maneb, metalaxyl + carboxin + furathiocarb, pyroquilon, benomyl and chlorothalonil are all effective against *U. scitaminea*. However, the best disease control was obtained with pyroquilon at 4.0, carbendazim + maneb at 4.57 and metalaxyl + carboxin + furathiocarb at 9.9 g a.i.l^{-1}, respectively (Wada 2003).

Hot Water Treatment (HWT)

The hot water treatment of 50–55 °C for 2 h has been reported for the control of ratoon stunting disease (RSD), and other combinations of time and temperature, also control smut disease in sugarcane (Srinivasan 1971; Steiner and Byther 1971) in Rhodesia. Thompson HA (1973, personal communication) in Mocambique observed the development of smut in canes that were previously hot water treated (HWT), and revealed higher incidences of disease in first and subsequent ratoon crops of smut susceptible varieties compared to non-HWT canes of the same varieties. So, hot water treatment of seed cane for the establishment of RSD-free nurseries was discontinued in Rhodesia, where smut was endemic. The major commercial variety in Rhodesia is NCo376, which- is very susceptible to smut. James (1974) also confirmed a greater incidence of smut in the ratoon crops-of both heat-treated compared to RSD-inoculated cane and healthy compared to RSD-infected cane in Rhodesia and suggested an interaction between the RSD bacterium and *Ustilago scitaminea*, in cane infected by both pathogens, whereby smut is suppressed. In Zambia also demonstrated a greater incidence of smut in the variety NCo310 after HWT.

14.2.3 *Wilt Disease of Sugarcane*

Among the disease of the sugarcane, wilt disease has also received the attention recently by the researchers working on sugarcane. Serological characterization has been made by the several workers across the world but the information at molecular level is scanty. Limited reported has been made by the workers especially in India (Viswanathan et al. 2010). This pathogen has been reported throughout the world in various research publications. The disease is still spreading in various sugarcane producing areas of the world and now a days it is being judged and counted as the second major disease of sugarcane in India.

14.2.3.1 Epidemiology and Symptoms of Disease

Wilt of sugarcane is a serious stalk disease in India and other South Asian countries. Although wilt has been associated with red rot in causing severe damage in different countries, wilt alone causes severe damage to certain sugarcane varieties in India (Viswanathan and Padmanaban 2008). However, conflicting claims have been made regarding the true causal organism of sugarcane wilt as species of *Fusarium*,

Cephalosporium and Acremonium by different authors (Agnihotri and Rao 2002) and information on the pathogen(s) involved and its variation, are not clearly brought out. The pathogens associated with the disease were recorded to be different from area to area and between investigators. Butler and Khan (1913) were the first to describe the disease in India and they identified the pathogen as *C. sacchari*. A similar disease named as "*Fusarium* sett" or "Stem rot" was recorded by Bourne (1922) in Barbados. Abbott (1932) also observed the disease in Louisiana. Gams (1971) who reviewed the taxonomy of *C. sacchari* also considered it as *F. moniliforme* var. subglutinans and coined a new combination as *F. Sacchari* (Butler) W. Gams to which both *C. sacchari* and *F. moniliforme* var subglutinans were made synonyms. Later, Nirenberg (1976) distinguished two varieties of *F. Sacchari* i.e., *F. sacchari* var sacchari and *F. sacchari* var subglutinans, the former having mostly unseptate conidia in the aerial mycelium, no sporodochia, while the latter with one to three septate conidia, macroconidia more commonly formed in sporodochia. Wilt incidence is always higher in ratoon crops compared with the plant crop. Besides yield reduction, wilt also causes 15–30 % reduction in juice extraction and up to 20 % in sugar recovery. The pathogen is primarily transmitted through infected seed canes in the field. Also the fungus surviving in sugarcane debris in the soil serves as a source of inoculum to infect sugarcane from soil. Symptoms of wilt appear in monsoon with the yellowing of the crown. Later on, the weight of cane starts to become lighter due to increase in pith area. From October onwards, the whole crown withers away and the cane becomes hollow. After splitting longitudinally, light pink to dark purple coloured tissues are seen at nodal area along with the presence of grayish mycelium in the hollow internodes. The dried internodes remain intact at the nodes and do not detach easily from the nodes (Agnihotri and Rao 2002) (Fig. 14.3). The wilt fungus can survive in soil for 2.5–3 years. Secondary spread from field to field occurs through rain and irrigation water. Wilt is very common in certain locations where conducive environment and susceptible hosts are available. However, the disease expression and its severity are being influenced by various biotic and abiotic factors in the field. Although abiotic factors play a decisive role on disease severity, biotic agents facilitates entry of the pathogen inside the root or stalk. Hence, role of biotic factors are of paramount importance in wilt initiation, development and severity. Wilt is more dangerous and causes enormous damage to crop in association with red rot, pineapple disease, borers or scale insects. Wherever root borer, *Emmalocera depressella*, infestation is severe, more wilt infection is noticed. However such borer-wilt complex was rarely noticed in the tropical region except, Gujarat state. In this case nearly 80–90 % borer infestation resulted in about 60 % wilt infection (Viswanathan and Rao 2011).

The disease symptoms appear during the monsoon and post monsoon periods, affected plants, either singly or in small groups, displays the disease symptoms. This is followed by yellowing and/or withering of upper leaves. The midrib of all leaves in a crown generally turns yellow, while the leaf lamina may remain green. Towards the end of the crop growth, the cane becomes light and hollow and dry. On splitting up on the affected canes, at the early stage of the infection, the diffused purple or muddy red colour is seen as conical patches on each node just above the

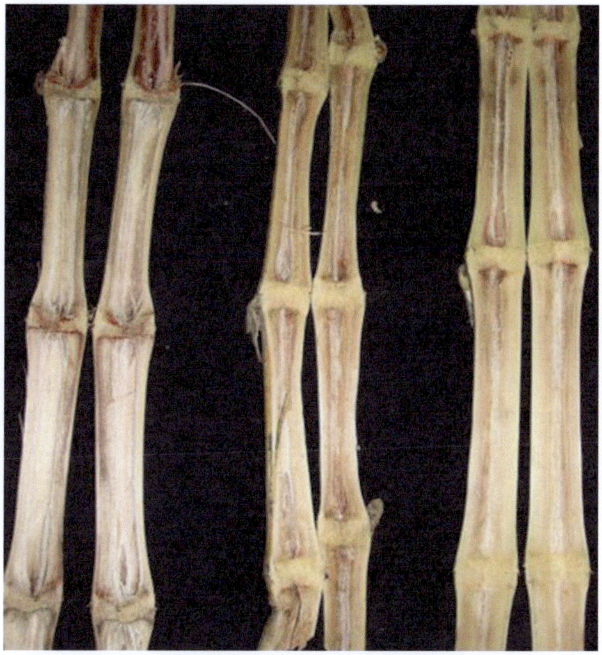

Fig. 14.3 Sugarcane stalk (LS) showing the symptoms of Wilt disease

growth rind. Occasionally, above the badly affected internodes are seen one or two red stripes, (vascular strands) that pass from one internodes to another. In severe cases, the spindle shaped cavities, tapering towards the nodes; develop in each internode because of general recession and rapid desiccation of tissues. Sometimes cavities also develops in the nodal tissues, this makes the canes tubular. In severe cases, affected stools withered and dry away.

14.2.3.2 Managements of Wilt Disease

Whenever severe wilt epidemics occurs and devastated the cane cultivation during the past, replacements of varieties with disease resistance were done and sugar industry was saved from this threats in India. Hence, growing resistant varieties are the cost effective and it will assure sustained cane cultivation in a region. So far, there is no systematic screening programme for screening wilt resistance, as it is difficult to reproduce the symptoms by artificial inoculation unlike red rot and smut which have different methods for screening. Viswanathan et al. (2009) have identified effective biocontrol isolates of *T. harzianum*, *T. viride* and *T. pseudokoningii* from sugarcane rhizosphere. Detailed field trials to manage the disease revealed that soil application of *Trichoderma* multiplied on press mud is effective in managing wilt in endemic locations of Gujarat and Andhra Pradesh. However, more trials needs to be conducted to validate the results in different endemic locations in the country (Viswanathan and Rao 2011).

14.2.4 Pineapple Disease

This is a soil borne disease. Due to smell like pineapple from the diseased setts, the name of disease was given as pineapple disease. The odor comes out from the diseased setts due to the formation of ethyl acetate because of metabolic changes. In India, this disease has been reported from Punjab, Maharashtra, Karnataka, Tamil Nadu, Kerala and U.P. Sett rot or pineapple disease, caused by ascomycetous fungus, *Ceratocystis paradoxa* (de Seynes) Moreau, is one of the sett borne and soil borne disease of the sugarcane. The symptoms associated with this organism include a smell of matured pineapple fruit, attributed to the production of ethyl acetate by the metabolic activity of the fungus (Coale 1989). The disease has been reported from many of the sugar cane growing regions with heavy losses under severe condition. The crop is propagated by means of stem cuttings. The disease mainly affects the germination of the setts at early stages of planting and the most serious losses are through the failure of infected cuttings to germinate (Comstock et al. 1984), although standing cane may also become infected (Manzo 1975; Natarajan and Subba Raja 1976). In India the disease was reported from different varieties of the crop (Singh et al. 1990).

14.2.4.1 Symptoms and Development of the Fungus

The symptoms of disease appear 2–3 weak after planting. The pathogen gets entry through the cut ends of the setts. The affected setts become red and there after the colour changes to blackish brown. The internodes of the sett become hollow. The pathogen grows and multiplies rapidly. The setts ultimately are rotted away and the buds do not germinate (Fig. 14.4). The disease is also noticed with association of white leaf disease (phytoplasma) in India at few places (Tiwari et al 2015b). Major factors that affect the growth and proliferation of the fungus *Ceratocystis paradoxa*, are soil moisture level, pH, relative humidity (RH), carbon and nitrogen sources as well as existing interactive microflora in the soil (Adiver 1996; Yadahalli et al. 2006, 2007). The soil moisture level of 60 % favoured significantly highest sett germination (65.60 %) followed by 80 and 40 % soil levels (55.30 and 46.70 % germination, respectively). The 100 and 20 % soil moisture levels recorded significantly low set germination (28.60 and 38.50 % germination, respectively). Soil moisture level of 60 % led to least development of sett rot (5.5 %). Increase or decrease in soil moisture above 60 % led to the increase in sett rot development. The interactions between days after planting and soil moisture status were also significant (Yadahalli et al. 2006). Soil moisture is growth determinant for most of the soil dwelling organisms. It is a medium through which the soil fungi absorb their nutrients and carryout most of their metabolic and physiological functions (Bolkan et al. 1982). Disease incidence was more in sterile soil than the unsterile soils, as the pathogen has to compete with other soil microbes for nutrition and space. The results are in conformity with the findings of Hooda and Grover (1988). The carbon is essential ingredient of all the living organisms and almost half of the dry weight of the fungal cell

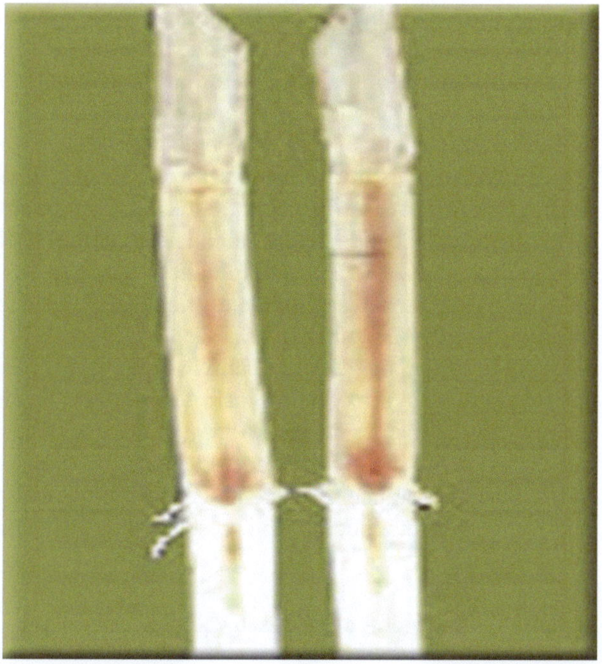

Fig. 14.4 Sugarcane stalk (LS) showing the pineapple disease disease symptoms in sugarcane

consists of carbon. Carbon compounds are equally important in fungus nutrition (Cochrance 1958). Starch induces highest growth of *C. paradoxa* with maximum biomass of 1039.99 mg as well as sporulation (79.67 spores/mL) followed by glucose (72.30 spores/mL). The lowest sporulation (7.67 × 106 spores/mL) is recorded in treatments with inositol as carbon source. However, treatments with the maltose and xylose as carbon sources were on par and non-significant (Yadahalli et al. 2006). Among nitrogen sources, nitrate forms have been reported to be excellent sources for imperfect fungi (Suryanarayanan 1958). *C. paradoxa* produce significantly highest biomass (1590.70 mg) in potassium nitrate, ammonium nitrate (1496.97 mg), ammonium oxalate (1405.00 mg) and magnesium nitrate (1234.30 mg) were next best preferred sources. Colony growth and conidial formation are sharply affected in different levels of pH. The growth of *C. paradoxa* is maximum in pH 6.5 (82.67 mm) irrespective of media. Similarly, the highest mean growth is recorded in PDA (62.75 mm) irrespective of pH levels tested. Lowest sett rot incidence is recorded at 60 % RH at all the intervals (Yadahalli et al. 2007). These results concur with the findings of Laxminarayana (1981).

14.2.4.2 Mechanism of Inhibition of Bud Germination

Pineapple disease infected sugarcane tissue may contain ethyl acetate up to 1 % that is sufficient to inhibit germination of buds (Kuo et al. 1969). The disease can reduce the sett germination up to 47 % (Anonymous 1999) which subsequently hamper the yield of cane and the yield reduction may be as high as 31–35 % (Anonymous 2000). Invading pathogens often produce certain compounds inside host that alter the metabolism, paving a way for further proliferation and growth of the pathogen. Rotting of sugarcane setts after planting often is a result of certain toxins produced by the invading pathogen, causing poor stand of the crop. The major cause of rotting of sugarcane setts is invasion by *Ceratocystis paradoxa*, an ascomycetous fungus, causing sett rot disease. The volatile toxic substance was identified as ethyl acetate by the experiment conducted by Yadahalli et al. (2007). In the inoculated setts, there was no germination and radical initiation and in treatment involving ethyl acetate, the bud germination was completely inhabited but only few roots (15 roots/node were observed). The studies conducted by Flori Bai (1985) also revealed similar results. To confirm the release of ethyl acetate (the phytotoxic volatile substance) by the pathogen, a hydroxylamine acid test was conducted in which 1 ml of 0.5 M hydroxylamine hydrochloride in 95 % ethanol and 0.1 ml of 6 M aqueous NaOH was heated to boiling. Slightly cooled and to this 2 ml of 1 M hydrochloric acid was added. The appearance of magenta colour by the addition of one drop of five per cent ferric chloride indicated the presence of ethyl acetate (Flori Bai 1985).

14.2.4.3 Management of Pineapple Disease

Pineapple disease can cause significant reductions and the most serious losses are through the failure of infected seed pieces to germinate (Wismer and Bailey 1989), although standing cane may also become infected (Manzo 1975; Natarajan and Subba Raja 1976). In Florida, two cultivars of commercial significance, CP 74–2005 and CP 72–2086, are highly susceptible to this disease (Coale 1989), these two cultivars together accounted for 10.3 % of the Florida sugarcane hectarage in 1989 (Coale 1989). On the organic soils of the Everglades Agricultural Area (EAA), disease development appears to be most favoured by cool, wet soil conditions (Coale 1989). Although Florida's dry season generally coincides with the cool winter months (December–February), heavy localized rainfalls during this period are not uncommon. Stand reductions due to pineapple disease of severities sufficient to require selective replanting have been observed in poorly drained fields in the EAA. A number of control measures have been recommended for pineapple disease, including the use of resistant cultivars, site selection, increased seed piece length, and avoidance of factors which down the seed piece germination (Wismer and Bailey 1989). In many cane-growing regions of the world, seed pieces are routinely treated with a fungicide as part of the planting operation (Steindl 1970; Comstock et al. 1984; Taylor and Ryan 1984; Sivanesan and Waller 1986). In Florida, where seed piece treatment is not practiced, seed pieces are typically overlapped and at least doubled in the furrow at planting. This practice is to insure establishment of an adequate number of primary shoots. When a pineapple disease susceptible cultivar is being planted under conditions favorable for disease development, the seeding density may be further augmented. Although pineapple disease is widespread in Hawaii (Rashid and Trujillo 1974) and Australia (Aberdeen 1969), the standard planting procedure involves only a single line of fungicide-treated seed pieces per row. This seeding rate, in combination with seed piece treatment, is capable of producing stands sufficient for maximizing yields.

Biocontrol of plant pathogen involves the use of biological processes to reduce the inoculum density of pathogen and to maintain their soil population below the disease threshold level. The pathogen suppression in the soil is considered as an important step in the control of disease as it involves the direct disinfestations of the soil. The pathogen in the absence of their hosts survives either as dormant propagules or actively as saprophytes on dead organic substrates of the host in the soil (Mahalingam et al. 2011). The antagonistic activity of some soil fungi against *C. paradoxa* has been investigated in vitro dual culture and with cell free culture filtrates of fungi amended in medium. The toxic metabolite produced by the initial fungal colonies of natural substrate may act to slow or present invasion by other species (Ambikapathy et al. 1994). *Trichoderma* sp. is most common fungal biological control agents that have been comprehensively researched and deployed throughout the world.

14.2.5 Pokkah Boeng Disease

Pokkah boeng is now playing a very important role and spreading rapidly in UP. Vishwakarma et al. (2013) showed increasing trend of disease incidence and most of the commercial cultivars affected by the disease ranged from 1 to 90 %. Although pokkah boeng comes under minor concern but these days it is going to be major on basis of their rapid epidemiology during last few years. Nowadays, the incidence and severity of pokkah boeng disease has been reported from major sugarcane growing states like Uttar Pradesh, Maharashtra, Punjab, Haryana, Assam, Tamil Nadu and Bihar in India and other sugarcane growing countries. This disease appears with the onset of the monsoon. Broad leaf cultivars are more vulnerable to get infection by this disease. The incidence of this disease reduces with the offset of the monsoon itself. The crown leaves of the affected plant become pale yellow or white which later on turn to dark reddish brown due to rotting and are detached from the top. Thus the affected plant looks without top and becomes thin (Sharma and Kumar 2015).

14.2.5.1 Epidemiology

Pokkah boeng of sugarcane caused by *Fusarium moniliforme* and the pathogen was first described by Sheldon (1904) and the perfect stage of pathogen is *Gibberella fujikuroi* (Sawada). *Fusarium* now confirm by several worker as a causal agent of pokkah boeng in sugarcane in Asia as well established pathogen by many worker (Singh et al. 2006; Govender et al. 2010; Mohammadi et al. 2012). Pokkah boeng is the serious and devastating disease not only in central Uttar Pradesh but also in the whole of the Southern and Northern sugarcane growing zone of India. Pokkah boeng is a Javanese term denoting a malformation or distorted top was originally in Ja1va but in that time no causal agent was established and its incidence was recorded by Padwick (1940). Pokkah boeng disease recorded in all over the countries where sugarcane grown and pathogen spreads in wind-blown rain, infected cane cuttings, pupae and adults of sugarcane stem borers (Patil and Hapase 1987; Whittle and Irawan 2000). This disease was well-known in sugarcane for long time but severity of disease reported in two commercial varieties Co7219 and Co C671 in Maharashtra 1983–1984 (Patil and Hapase 1987). *Fusarium moniliforme* var. *subglutinans* reduce the quality of the harvested crop mainly among varieties with high sugar yields the sugar production depending upon the variety up to 40.8–64.5 % (Duttamajumder 2004).

The *pokkah boeng* pathogen also attacks sorghum and had been reported that the disease was caused by *F. moniliforme* (*Gibberella fujikuroi*) (Mohammadi et al. 2012), now known as *F. verticoilides*. Fungus infects a wide range of species including monocotyledons and dicotyledons causing various diseases such as seedling blight, scorch, stalk and root rot, abnormal stunting or hypertrophy. Pathogen of *Pokkah boeng* disease has been associated with several diseases of sugarcane such as sett rot, root rot (Verma et al. 1984), wilt (Saponaro 1980; Waraitch and Kumar 1982) and knife cut (Zummo 1970; Kamal and Singh 1979).

14.2.5.2 Symptomology

The characteristics symptoms of *pokkah boeng* disease are the appearance of chlorotic patches towards the base of the young leaves, in acute cases disease shows distortion of stalk with external and internal cut like lesions and rotting of apical part of stalk. Under field conditions, the disease may develop many variations from the general symptoms, but the final result is usually a malformed or damaged top and stalk. The base of affected leaves is often narrower as compare to normal leaves. Knife cut symptom of the disease reported in varieties CoS767, CoC671, CoC8014, Co1158, CoS8315 and CoS 8436 (Patil 1995). Development of disease symptom in four phases namely chlorotic phases I & II, top rot and knife cut phase. The apical leaves may also show pronounced wrinkling and twisting depending upon the susceptibility of varieties and existing climatic conditions also malformed or damaged top and stalk due to this disease (Martin et al. 1961). The symptoms of *pokkah boeng* disease were mainly two types i.e. chlorotic phase and acute phase of top rot (Patil and Hapase 1987) and Knife cut (fourth phase) of this disease in Maharashtra (Patil 1995). Similar findings of about the disease symptom were also reported by workers (Eira et al. 1974; Higgy et al. 1977).

Major and minor (N, P, K, S. Zn, Fe, Cu and Mg) study of the sugarcane plant affected by the *pokkah boeng* showed the decreasing pattern in stalks and leaves of diseased plant as compare to healthy ones and its incidence have been noticed on broad leaves varieties of sugarcane during monsoon. This may be leads to the reduction in weight of cane, length of internodes, juice percent, girth, pol per cent and total sugars in juice in infected canes of varieties CoS 8436 and CoS 88230 (Singh et al. 2006).

14.2.5.3 Pathogenesis

The pathogen enters into the host tissues through any injury by insects/borers or natural growth cracks etc. After the entry of pathogen, the infection thread develops normal hypha which grows within the host tissues for some time and then emerges out through the cells to the outer surface and develops acervuli. Rains and heavy dews usually wash the acervuli developed on nodes and internodes and the spores get lodged around the nodes behind the leaf sheath. The spores germinate and the mycelium gets established in bud scales, root primordial or leaf scars and later within the plant tissues. Electron microscopic of infected leaves revealed that after lodging of conidia and incubation of minimum 1 month at the time of germination, the thin walled bulliform cells of the epidermis are attacked and soon collapse then the older cells of the epidermis are attacked (Patil 1995). From the epidermal cells, the hyphae enter the underlying tissues. Changes in the structure of stomata were also observed in the infected leaves but there was no evidence found for the entry of pathogen through stomata.

14.2.5.4 Molecular Detection of Pokkah Boeng Pathogen

PCR-based methods offer rapid and cost-effective tools of molecular diagnosis. Real-time PCR coupled with conventional PCR techniques can reliably detect and quantify pathogens in the fungal isolates or from infected leaf samples (Sayler et al. 2012). Several PCR assays have been developed to identify species of *Fusarium* based on the translation elongation factor gene (*tef1*) (Sampietro et al. 2010; Wang et al. 2011; Amatulli et al. 2012), polyketide synthase gene (pks) (González-Jaén et al. 2004), and trichodiene synthase gene (*tri5*) (Zhu et al. 2010). The sequence of the ITS-rDNA is highly conserved but sufficiently variable among species of Fusarium species complex. The ability of the ITS region to differentiate and provide accurate and rapid detection of fungus has been reported at the species-level (Zeng et al. 2007). Specific PCR primers based on ITS sequences for Fusarium species complex have been designed based on two single nucleotide polymorphisms (SNPs) present in the ITS region (Dita et al. 2010). Lin et al. (2014a, b) developed rapid, efficient, and specific PCR assay to detect pathogens of sugarcane pokkah boeng based on the differences in the ITS-rDNA sequences between two species of *F. verticolides* and *F. proliferatum*. Species-specific primers and probes were used for their specificity and sensitivity. These primers were evaluated to detect *F. verticolides* and *F. proliferatum* using purified DNA from 84 isolates or 28 infected sugarcane leaves. TaqMan PCR using FAM-gx1 and FAM-gx2 primer–probe combinations was very sensitive, with a detection limit of 10 pg/μl fungal DNA. Conventional PCR using Pgx1 and Pgx2 primers, which also targeted the ITS-rDNA region, resulted in fragments at the size of 439 bp with Pgx1 and 400 bp with Pgx2. However, the detection limit for the conventional PCR was approximately 10 ng of fungal DNA input.

Lin et al. (2014a, b) used the sequences of the internal transcribed spacer from 103 isolates were amplified by PCR using fungus-conserved ITS1 and ITS4 primers and sequence alignment to studies the diversity and differences among the isolates of *Fusarium* species complex. Two pathogens (*F. verticolides* and *F. proliferatum*) to cause sugarcane pokahh boeng in China were identified by morphological observation, pathogenicity test, and phylogenetic analysis. Both conventional PCR and TaqMan PCR were developed to reliably detect the species of Fusarium using purified fungal DNA as input based on their ITS sequence. The TaqMan PCR was a more efficient tool for the early diagnosis of disease, and it was capable of detecting pokkah boeng in symptomatic young tissues of infected plants. They also found similar symptoms to sugarcane pokkah boeng, such as the twisted and curling symptoms of crown leaves, and was found to be infected by *Phoma* sp. based on the morphological characteristics and the ITS sequence of rDNA (Lin et al. 2014a, b). TaqMan PCR is more sensitive method to diagnose pokkah boeng pathogen which could not be easily detected in most of leaf tissues by conventional PCR.

14.2.5.5 Control Management

Basically it is an air-borne disease and primarily transmitted through the air-circulation and secondary infection takes place through the infected setts, irrigation water, splashed rains and soil. The pathogen (*F. moniliforme*) can survive for 12 months in the plant debris under natural conditions and can remain viable for more than 10 months under laboratory conditions. Spraying of different fungicides like Bavistin (1 g/l of water) or Blitox (0.2 %) or Copper oxychloride or 0.3 % Dithane M-45 (3 g/l of water) are the effective for reducing the *pokkah boeng* disease (Kamal and Singh 1979). Two to three sprayings with an interval of 15 days interval reduces the multiplication of a pathogen and losses in yield and quality of cane and therefore, paired row or wider spacing planting of sugarcane is necessary to facilitate the plant protection operations. Canes showing 'top rot' or 'knife cut' should be rouged out from the fields as they are shown. Planting of healthy seed material/use of resistant verities and follow of Integrated Disease Management practices are the best way to prevent disease incidence (Patil et al. 2007). *F. moniliforme* can be disseminated horizontally by airborne spores or crop debris and vertically through the seed pieces. So it is important to uses of resistant varieties and fungicide applications. Both the processes for controlling, is limited and there is increasing need for novel and environmentally sound strategies to control this and other diseases of sugarcane. *Burkholderia* isolates from sugarcane plants is a crucial step toward further development of these isolates for biological control of *Pokkah boeng* and other sugarcane diseases. The endophytic bacteria community associated with sugarcane harbours multiple genera with potential for plant growth promotion and disease control (Mendes et al. 2007).

14.2.6 Leaf Binding Disease

Leaf binding disease of sugarcane was first observed in 1918 in Argentina (Abbott 1964) caused by *Myrigenospora aciculispore* V & *Fusarium moniliforme* (*F. verticolides*) and it was called *Myrigenospora* leaf binding disease or Tangle top disease.

Fig. 14.5 Symptoms of *Pokkah Boeng* of sugarcane (**a**) Leaf (**b**) Shoot

In India, leaf binding disease was identified by biological, microscopic and molecular characteristics and also attempted the control prospective of this disease by utilizing fungicides. The disease falls in the minor category and is not prevalent in Asian countries. The occurrence of the disease is in monsoon period (June last week to September) which is the grand growth period of the host plants (Tiwari et al 2016).

14.2.6.1 Symptomatology

Mechanical binding with necrosis of leaves is major symptoms of the disease. Early death of infected shoots and the dwarfing and fascination is also found. The fascination occurred due to mechanical binding by fungus tissues of the adjacent leaves and clumps during the growth of the shoots. No characteristic symptoms found on the stalk of the sugarcane plants (Fig. 14.5). The growth of the affected plants completely checked due to the disease and thus reduction in yield was observed around 2 % in all the affected varieties (Tiwari et al 2016).

14.3 Conclusion

The fungal diseases of sugarcane are responsible for the quality and stability of the crop yield. Being as long durational crop and its vegetative proppogation, monoculturing and rationing practices makes it easily prone to build up of the diseases.

Hence, an integrated approach involving cultural, chemical/physical methods, host resistance and legislative measures is suggested for the sustainable management of sugarcane diseases. Periodical monitoring and scouting of the disease population estimates are very essential and basic pre-requisites for deciding the need based control measures and to develop IDM. This is the key for successful implementation of IDM. Such survey and surveillance should be regular so as to know the changing disease situations due to agro- climatic factors. Molecular breeding approaches help in developing resistance verieties. Early disease diagnosis and marker development helps in detection and developing management strategies for the diseases. Genomics, transcriptomics and proteomics approaches need to further explore to understand the exact mechanisms of host-pathogen interaction and involvements of different up-regulating and down regulating gene involvements.

References

Abbas H, Anwar SA, Javed N, Iqbal MA, Abid N. Morphological variability among isolates of *Colletotrichum falcatum* Went. Infecting four cultivars of sugarcane. Pak J Phytopathol. 2010;22:101–4.

Abbott EV. Seed rots of sugarcane in Louisiana. Proceedings of Interna- tional Society of Sugar Cane Technologists, 4 Congress Bulletin 48. 1932, 1-2

Aberdeen JEC. The measurement of the concentration and distribution of Ceratocystis paradoxa (de Seynes) Moreau in soil. Aust J Agr Res. 1969;20:843–85.

Adiver SS. Studies on pod rot of groundnut (*Arachishypogaea* L.) caused by species of *Fusarium*. PhD. thesis, University of Agricultural Sciences, Dharwad; 1996.

Agnihotri VP. Diseases of sugarcane. New Delhi: Oxford and Ibh Publishing Company Private Limited; 1983. p. 373.

Agnihotri VP. Current sugarcane disease scenario and management strategies. Indian Phytopathol. 1996;49:109–26.

Agnihotri VP, Rao GP. A century status of sugarcane wilt in India. In: Singh SB, Rao GP, Eswaramoorthy S, editors. Sugarcane crop management. Houston: SciTech Publishing LLC; 2002. p. 145–60.

Albert HH, Schenck S. PCR amplification from a homolog of the *b*E mating-type gene as a sensitive assay for the presence of *Ustilago scitaminea* DNA. Plant Dis. 1996;10:1189–92.

Allsopp PG and Manners JM (1997) Novel approaches for managing pests and diseases in sugarcane. In: Keating BA, JR Wilson, editors. Intensive sugarcane production: meeting the challenges beyond 2000. Proceedings of the sugar 2000 symposium. Brisbane, Australia, 20-23 August 1996. International, Willingford. p. 173–88.

Amatulli MT, Spadaro D, Gullino ML, Garibaldi A. Conventional and real-time PCR for the identification of *Fusarium fujikuroi* and *Fusarium proliferatum* from diseased rice tissues and seeds. Eur J Plant Pathol. 2012;134:401–8.

Ambikapathy V, Panneerselvam A, Chandrasekaran R. Population Dynamics of Mycoflora in the paddy field of Mannargudi, Nagai Quaid-e-Milleth District, Tamil Nadu. Intl J Geobios News Report. 1994;13:171–4.

Ambikapathy V, Panneerselvam A, Chandrasekaran R, Geobios J. New Rep. 1994;13:171–4.

Anonymous. Proceedings and Recommendations. All India Coordinated Project On Sugarcane, I.I.S.R., Lucknow, 1980.

Anonymous. Smut disease and hot water treatment. S Afr Sugar J. 1970;54(5):309.

Anonymous. BSRI annual report (1997). Ishurdi, Pabna: Bangladesh Sugarcane Research Institute; 1999. p. 89.

Anonymous. Investigation of varietal resistance/susceptibility against pineapple disease. BSRI annual report (1998-99). Ishurdi Pabna: Bangladesh Sugarcane Research Institute; 2000. p. 86.

Antoine R. Smut. In: Martin JP, Abbott EV, Hughes CG, editors. Sugar cane diseases of the world, vol. 1. Amsterdam: Elsevier Pub Co; 1961. p. 542.

Anzalone Jr L. Air pressure incorporation of fungicides into short stalk sections of sugarcane for the control of red rot. Phytopathology. 1970;60:741–2.

Bakkeren G, Gibbard B, Yee A, Froeliger E, Leong S, Kronstad J. The a and b loci of Ustilago maydis hybridize with DNA sequences from other smut fungi. Mol Plant Microbe Interact. 1992;5:347–55.

Bates JF. Smut disease in Guiana. Sugarcane Pathol Newslett. 1975;13/14:8.

Bebbee CN. Smut. In: Chattes N, Bebbee CN, editors. The protection of sugarcane and sugarbeets. New York: Longman Pub. Inc; 1988. p. 230–4.

Bharadwaj N, Sahu RK. Evaluation of some fungicides, botanicals and essential oils against the fungus *Colletotrichum falcatum* causing red rot of sugarcane. Bioscan. 2014;9:175–8.

Bock KR. Studies on sugar-cane smut (*Ustilago scitaminea*) in Kenya. Trans Br Mycol Soc. 1964;47:403–7.

Bolkan HA, Dianese JC, Cupertha FP. Survival and colonization potential of *Fusarium moniliforme var. sub-glutinans* in soil. Phytopathology. 1982;72:1298–301.

Bonants P, Hagenaar-De Weerdt M, Van Gent-Pelzer M, Cooke D, Duncan J. Detection and identification of Phytophthora fragariae Hickman by the polymerase chain reaction. Eur J Plant Pathol. 1997;103:345–55.

Bourne BA. Researches on the root diseases of sugarcane. Barbados: Department of Agriculture; 1922. p. 1–17.

Braithwaite KS, Bakkeren G, Croft BJ, Brumbley SM. Genetic variation in a worldwide collection of the sugarcane smut fungus *Ustilago scitaminea*. Proc Aust Soc Sugar Cane Technol. 2004;48–56.

Butler EJ, Khan AH. Some new sugarcane diseases. Part I, Wilt. Memoirs of the Department of Agriculture, India, Botany series. 6th ed. 1913. p. 180–90.

Chand JN, Dang JK, Kapoor JR. Systemic chemicals as sett protects ants. Sci Cult. 1974;40: 69–70.

Chao CP, Hoy JW, Saxton AM, Martin FA. Heritability of resistance and repeatability of clone reactions to sugarcane smut in Louisiana. Phytopathology. 1990;80:622–6.

Chona BL. Chairman's opening address. Proc Int Soc Sugar Cane Technol. 1956;9(1):975–86.

Coale FJ. An integrated approach for control of Pineapple disease of sugarcane. Florida cooperative Extension Service. Sugarcane Growers Newslett. 1989;3:1–5.

Comstock JC. Smut. In: Rott P, Bailey RA, Comstock JC, Croft BJ, Salem, Saumtally A, editors. A guide to sugarcane diseases. CIRAD and ISSCT; 2000. p. 181–5.

Comstock JC, Ferreira SA, Ching SA, Hilton HW. Control of pineapple disease of sugarcane with propiconazole. Plant Dis. 1984;68:1072–5.

Daniels J, Roach BT. Taxonomy and evolution. In: Heinz DJ, editor Sugarcane, Elsevier 1987.

De Silva GA, Silava MAAA, De Carvalho PCT. Preliminary studies on invertases of sugarcane stalk inoculated with Colletotrichum falcatum Went. Proc Int Soc Sugarcane Technol. 1977;16:407–14.

Dhillon PS, Kanwar RS, Bedi JS. Effect of moist hot air treatment on the control of Gsd and Rsd and productivity of sugarcane. Indian Sugar Crops J. 1983;9:3–4.

Dita M, Waalwijk C, Buddenhagen I, Souza Jr M, Kema G. A molecular diagnostic for tropical race 4 of the banana *Fusarium* wilt pathogen. Plant Pathol. 2010;59:348–57.

Duttamajumder SK. Bacterial diseases of sugarcane in India: a bird's eye view. In: Rao GP, Saumtally AS, Rott P, Rao GP, Saumtally AS, Rott P, editors. Sugarcane pathology: bacterial and nematodes diseases. Enfield: Science Publishers; 2004. p. 15–50.

Duttamajumder SK. Red rot of sugarcane. Lucknow: Indian Institute of Sugarcane Research; 2008.

Edward A, Dusane DH, Melchias G, Daninigo V, Silambarasan S, et al. Evaluation of antifungal activity of peptides and enzymes against the sugarcane red rot pathogen, *Colletotrichum Falcatum*. Middle East J Sci Res. 2013;15:748–56.

Eira AFDA, Carvalto PCT, Sangiono DE. Studies on aggressiveness of *Fusarium moniiforme* sheld. causal agent of *Pokkah boeng* in sugarcane. Proc ISSCT. 1974;15:374–83.

Farid B, Maria JF, Aziz M. A rapid PCR based method to distinguish between *Enterococcus* species by using degenerate and species-specific *sod*A gene primers. Afr J Biotechnol. 2006;5(9):697–702.

Ferreira SA, Comstock JC. Smut. In: Ricaud C, Egan BT, Gillaspie Jr AG, Hughes CG, editors. Disease of sugarcane. Amsterdam: Elsevier Press; 1989. p. 211–29.

Ferreira SA, Comstock JC, Wu KK. Evaluating sugarcane smut resistance. Proc Int Soc Sugar Cane Technol. 1980;17:1463–76.

Flori Bai A. Studies on pineapple disease of sugarcane. M. Sc. (Agri) thesis, Tamilnadu Agricultural University (TNAU), Coimbatore; 1985. p. 85.

Fontaniella B, Marquez A, Rodriguez CW, Pinon D, Solas MT, Vicente C, Legaz ME. A role for sugarcane glycoproteins in the resistance of sugarcane to *Ustilago scitaminea*. Plant Physiol Biochem. 2002;40:881–9.

Gams W. Cephalosporium – artige schimmelpize (hyphomycetes). Stuttgart: G. Fischer; 1971. p. 262.

González-Jaén MT, Mirete S, Patiño B, López-Errasquín E, Vazquez C. Genetic markers for the analysis of variability and for production of specific diagnostic sequences in fumonisin-producing strains of *Fusarium verticillioides*. Eur J Plant Pathol. 2004;110:525–32.

Govender P, MCFarlane SA, Rutherford RS. *Fusarium* species causing *pokkah boeng* and their effect on *Eldana saccharina* walker (lepidoptera: pyralidae). Proc S Afr Sug Technol Ass. 2010;83:267–70.

Gul F, Hassan S. Efficacy of different fungicides to control whip smut of sugarcane. Path Sec Sugar Crops Res Insa Mardan Pak Sarhad J Agric. 1989;5(1):87–9.

Higgy AH, Abd-Elrazic EA, Rushdi MH. Occurrence of *Pokkah boeng* disease in Egypt. Proceeding of ISSCT, XVI. 1977. p. 473–81.

Hooda I, Grover RK. Studies on different isolates, age and quantity of inoculum of *Rhizoctonia bataticola* in relation to disease development in mungbean. Indian Phytopathol. 1988; 78:619–23.

Imtiaj A, Alam SM, Islam AKMR, Alam S, Lee TS. In vitro studies on *Colletotrichum falcatum* the causal of red rot disease of sugarcane. Am Eurasian J Agric Environ Sci. 2007;2:511–7.

James GL. Culmicolous smut of sugarcane, and the effects of its control on yield. ISSCT Proc. 1974;15:292.

James GL. The effect of ratoon stunting disease on the expression of smut symptoms. Proc S Afr Sugar Technol Assoc. 1976;50:69–72.

Joshi NC. Hot water treatment of setts for the control of red rots and smut diseases of sugarcane. Indian sugar NS. 1954;4:2.

Kamal, Singh RP. Knife cut disease of sugarcane in India. Sugarcane Pathol News Lett. 1979;23:34.

Kamel AA, Ibrahim NA, Mohmed AA, Mohmed SK, Joseph AV. PCR identification of Fusarium genus based on nuclear ribosomal DNA sequence data. Afr J Biotechnol. 2003;2(4):82–5.

Kawther S. Genotyping of bovine viral diarrhea virus using multiples PCR with and without RNA extraction. Afr J Microbiol Res. 2008;2:316–8.

Kendrick Jr JB. A view point on integrated pest management. Plant Dis. 1988;72(80):647.

Kidd H, James DR, editors. The agronomical handbook. 3rd ed. Cambridge: The Royal Society of Chemistry; CRC press, USA 1991.

Kuc J. Compounds from plants that regulate or participate in disease resistance. Ciba Found Symp. 1990;154:213–24.

Kumar N, Jhang T, Satyavir, Sharma TR. Molecular and pathological characterization of *Colletotrichum falcatum* infecting subtropical Indian sugarcane. J Phytopathol. 2010;154:260–7.

Kumar P, Agrawal A, Tiwari AK, Jabari ML, Lal M. Possibilities of development of red rot resistance in sugarcane through somaclonal variation. Sugar Tech. 2012;14(2):192–4.

Kuo TT, Chien MM, Li HW. Ethyl acetate produced by *Ceratocystis paradoxa* and *C. adiposum* and its role in the germination of sugarcane buds. Can J Bot. 1969;47:1459–63.

Laxminarayana HK. Studies on *Cercospora solanimelongenae* Chupp causing leaf spot of brinjal. M.Sc. (Agric) thesis, University of Agricultural Sciences, Dharwad; 1981. p. 85.

Legaz ME, de Armas R, Pinon D, Vicente C. Relationships between phenolics-conjugated polyamines and sensitivity of sugarcane to smut (*Ustilago scitaminea*). J Exp Bot. 1998a;49:1723–8.

Legaz ME, Pedrosa MM, de Armas R, Rodriguez CW, de los Rios V, Vicente C. Separation of soluble glycoproteins from sugarcane juice by capillary electrophoresis. Anal Chim Acta. 1998b;372:201–8.
Li H, Song GT, He MJ, Hao Y, Zhou GY. Detection of Camellia oleifera root rot pathogen with nested-PCR. J Zhejiang For Coll. 2009;26(6):849–53.
Lin Z, Que Y, Deng Z, Xu S, Rao GP, et al. First report of *Phoma* sp. causing twisting and curling of crown leaves of sugarcane in mainland of China. Plant Dis. 2014a;98:850.
Lin Z, Xu S, Que Y, Wang J, Comstock JC, et al. Species-specific detection and identification of *Fusarium* species complex, the causal agent of sugarcane *pokkah boeng* in China. PLoS One. 2014b;9(8):e104195.
Liu JN, Li H, Zhou GY. Specific and rapid detection of *Camellia oleifera* anthracnose pathogen by nested-PCR. Afr J Biotechol. 2009;8(6):1056–61.
Madan VK, Bikas M, Ansari M, Srivastava A, Soni N, et al. Rapd-Pcr analysis of molecular variability in the red rot pathogen (*Colletotrichum Falcatum*) of sugarcane. Sugarcane Int. 2000;3:5–8.
Mahalingam R, Ambikapathy V, Panneerselvam A. Biocontrol measures of pineapple disease in sugarcane. Pelagia research library. Eur J Exp Biol. 2011;1(2):64–7.
Malathi P, Viswanathan R. Role of microbial chitinase in the biocontrol of sugarcane red rot caused by *Colletotrichum falcatum* Went. Sciences. 2013;6:17–23. 117.
Malathi P, Padmanaban P, Viswanathan R, Mohanraj D, Sundar AR. Efficacy of thiophanate methyl against red rot of sugarcane. Acta Phytopathol Entomol Hung. 2004;39:39–47.
Malathi P, Viswanathan R, Sundar AR, Prakasam N, Padmanaban P, et al. Variability among *Colletotrichum falcatum* pathotypes used for screening red rot resistance in sugarcane. Sugarcane Int. 2010;28:47–52.
Manzo SK. Pineapple disease on standing sugarcane in Nigeria. Sugarcane Pathol Newslett. 1975;14:3–4.
Martin JP, Hong HL, Wismar CA. Pokkah boeng. In: Sugar-cane diseases of the world. Vol. 1. New York: Elsevier Publ. Co.;1961. p. 247–257.
Martinez M, Medina I, Naranjo S, Rodriguez CW, de Armas R, Pinon D, Vicente C, Legaz ME. Changes of some chemical parameters, involved in sucrose recovery from sugarcane juices, related to the susceptibility or resistance of sugarcane plants to smut (*Ustilago scitaminea*). Int Sugar J. 2000;102:445–8.
McMartin A. Sugarcane smut: reappearance in Natal. South Afr Sugar J. 1945;29:55–7.
Mendes R, Pizziarani-Kleiner AA, Araujo WL, Raaijmakers JM. Diversity of cultivated endophytic bacteria from sugarcane: genetic and biochemical characterization of *Burkholderia cepacia* complex isolates. Appl Environ Microbiol. 2007;73(22):7259–67.
Mohammadi A, Nejad RF, Mofrad NN. *Fusarium verticillioides* from sugarcane, vegetative compatibility groups and pathogenicity. Plant Protect Sci. 2012;48:80–4.
Mohan Raj D, Padmanabhan P, Karunakaran M. Association of phytotoxin produced by *Colletotrichum falcatum* Went in the red rot disease of sugarcane. Sugarcane. 2002;5:21–3.
Natarajan S, Subba Raja KT. Infection by Ceratocystis paradoxa Moreau on standing canes of some sugarcane clones. Sugarcane Pathol Newslett. 1976;17:25–8.
Nirenberg H. Untersuchungen Uber die morphologische und biologische Differenzierung in der Fusararium-Sektion Liseola (Investigations on the morphological and biological differentiation in the Fusarium section Liseola). Mitt aus der Biologischen Bundesanstalt fur Land und Forstwirt- schaft Berling-Dahlem. 1976;169:1–117.
Nithya K, Bukhari KAIM, Valluvaparidasan V, Paranidharan V, Velazhahan R. Molecular detection of *Colletotrichum falcatum* causing red rot disease of sugarcane (Saccharum officinarum) using a scar marker. Ann Appl Biol. 2012;160:168–73.
Olufolaju DB. Evaluation of some relatively new fungicides for smut control in. Niger Crop Prot. 1993;12(4):293–5.
Padwick CW. Soc Rept Ind Agril Res Inst. New Delhi. 1940. 198–39: 103–115.
Panicker S, Velazhahan R. A century of sugarcane red rot research in India. In: Tiwari AK, Lal M, Singh AK, editors. Current status of sugarcane research in India. New York: Nova Publisher; 2015. p. 160–71.

Patil AS. Studies on *Pokkah boeng* and pine apple disease of sugarcane in Maharashtra with their economic losses in yield and quality of sugarcane. Final project report, ICAR, VSI, Pune, 1995. p. 441.

Patil AS, Hapase DG. Studies on *Pokkah boeng* disease in Maharashtra. Ind Phytopath. 1987;40:290.

Patil AS, Singh H, Sharma SR, Rao GP. Morphology and pathogenecity of isolates of *Fusarium moniliformae* causing *Pokkah boeng* of sugarcane in Maharastra. In: Ram RC, Singh A, editors. Microbial diversity: modern trends. New Delhi: Daya Publishers; 2007. p. 234–63.

Peros JJ. Effects of Ustilago scitaminea Syd., on the development of sugar content and invertase activity of sugarcane. Agron Trop. 1984;39:81–8.

Phelps RH, Donelan AF. Disease and other problems affecting sugarcane variety change in Trinidad and Tobago 1975 to 1989. In: Proceedings of the 24th West Indies sugar Technologist' Conference, Kington, Bridgetown, Barbacaos, April 8–12 (Sugar Association of the Caribbean (1989) Caroni Research Station, Carapichaina, Tridinad). 1991. p. 88–95.

Pruett CJH, Waller JM. A report on sugarcane diseases in the Santa Cruz area of Bolivia. Sugar Azucar. 1989;84(11):35–7.

Ramesh Sundar A, Leonard Barnabas E, Malathi P and Viswanathan R. A mini-review on smut disease of sugarcane caused by Sporisorium scitamineum, Botany, Dr. John Mworia, editor. InTech publisher; 2012. ISBN: 978-953-51-0355-4.

Rashid AR, Trujillo EE. Ecology of *C. paradoxa* in field soils. Hawaii Sugar Planters' Assoc Ann Rep. 1974:50.

Riley IT, Jubb TF, Egan BT, Croft BJ. First outbreak of sugarcane smut in Australia. Proc Int Soc Sugarcane Technol. 1999;23(2):333–6.

Saksena P, Vishwakarma SK, Tiwari AK, Singh A, Kumar A. Pathological and molecular variation in *Colletotrichum falcatum* went isolates causing red rot of sugarcane in the Northwest Zone of India. J Plant Protection Res. 2013;53:37–41.

Sampietro D, Marín P, Iglesias J, Presello D, Vattuone M, et al. A molecular based strategy for rapid diagnosis of toxigenic *Fusarium* species associated to cereal grains from Argentina. Fungal Biol. 2010;114:74–81.

Saponaro A. Observation of some cases of maize stalk and root rot due to *Fusarium*. Ann Exp Inst Plant Pathol. 1980;6:5–14.

Satyavir. Red rot of sugarcane – current scenario. Indian Phytopathol. 2003;56:245–54.

Satyavir KA, Raj K, Virk KS. Red rot of sugarcane: the research scene in Haryana. In: Singh SB, Rao GP, Eswaramoorthy S, editors. Sugarcane crop management. Houston: SCI TECH Publishing LLC; 2002. p. 109–26.

Sayler RJ, Walker C, Goggin F, Agudelo P, Kirkpatrick T. Conventional PCR detection and real-time PCR quantification of reniform nematodes. Plant Dis. 2012;96:1757–62.

Sharma DDK, Kumar A. Morphological, physiological and pathological variations among the isolates of *Fusarium moniliforme* Sheldon causing pokkah boeng of sugarcane. Agrica. 2015;4(2):119–29.

Sharma R, Tamta S. A review on red rot: the "cancer" of sugarcane. J Plant Pathol Microbiol. 2015;S1:003. doi:10.4172/2157-7471.S1-003.

Sheldon JL. A corn mold (Fusarium moniliforme sp). New Agric Exp Sta Ann Report. 1904; 17:23–32.

Shen W, Xi P, Li M, Sun L, Zhang L, Jiang Z. Development of a sensitive nested-polymerase chain reaction (PCR) assay for the detection of *Ustilago scitaminea*. Afr J Biotechnol. 2012; 11(46):10541–7.

Singh K. Hot air therapy against red rot of sugarcane. Plant Dis Rep. 1973;57:220–2.

Singh RP, Lal S. Air borne propagules of *Colletotrichum falcatum* and their role in the epidemiology of sugarcane red rot. Ind Phytopathol. 1996;49:89–91.

Singh K, Singh RP. Red rot. In: Red R, Ricaud C, Egan BT, Gillaspie AG, Hughes CG, editors. Sugarcane diseases – major diseases. Amsterdam: Elsevier; 1989. p. 169–88.

Singh ON, Waraitch KS. Metabolic changes induced by *Colletotrichum falcatum* Went in sugarcane. Sugarcane Pathologists Newslett. 1977;19:7–9.

Singh K, Misra SR, Shukla US, Singh RP. Moist hot air therapy of sugarcane: control of sett-borne infections of Gsd, smut and red rot. Sugar J. 1980;43:26–8. 112.

Singh H, Patil AS, Hapase DG, Singh H. Pineapple disease of sugarcane – a review. Bharatiya Sugar. 1990;16:187–92.
Singh A, Chauhan SS, Singh A, Singh SB. Deterioration in sugarcane due to *Pokkah boeng* disease. Sugar Tech. 2006;8(2&3):187–90.
Singh V, Srivastava RL, Awasthi SK, Joshi BB. Biological control of red rot disease of sugarcane through Trichoderma harzianum and Trichoderma viride. Ind Phytopathol. 2008;61: 486–91. 116.
Singh V, Singh PN, Yadav RL, Awasthi SK, Joshi BB. Increasing the efficacy of Trichoderma harzianum for nutrient uptake and control of red rot in sugarcane. J Hortic For. 2010;2:66–71.
Singh RK, Kumar P, Tiwari NN, Singh SP, Tiwari AK, Kumar A. Role of endochitinase gene and efficacy of *Trichoderma* against *Colletotrichum falcatum* Went. Causing red rot disease in sugarcane. Sugar Tech. 2013;16:180–8. 118.
Srinivasan KV. Hot water treatment for disease control. Sugarcane Pathol Newslett. 1971;6:46.
Sivanesan A, Waller JM. Sugarcane diseases. CMI Phytopathol Paper No. 29. 1986.
Steindl DRL. The control of pineapple disease and the stimulation of germination in cane setts in Queensland. Sugarcane Pathol Newslett. 1970;5:53–4.
Steiner GW, Byther RS. Effect of moist soil, fungicides and hot water on smut teliospores. Expt Sta Hawaii Sugar Planters Assoc. 1971:39.
Suman A, Lal S, Shasany AK, Gaur A, Singh P. Molecular assessment of diversity among pathotypes of *Colletotrichum falcatum* prevalent in subtropical Indian sugarcane. World J Microbiol Biotechnol. 2005;21:1135–40.
Suryanarayanan S. Sugarcane disease management. Proc Indian Acad Sci. 1958;48:154–88.
Sydow H. Notizen Uber Ushlagineen. Ann Mycol. 1924;22:277.
Taylor PWJ, Ryan CC. Propiconazole fungicide as a sett treatment for the control of pineapple disease. Sugar Cane. 1984;5:5–8.
Thokoane LN, Rutherford RS. cDNA-AFLP differential display of sugarcane (*Saccharum* Spp. Hybrids) genes induced by challenge with the fungal Pathogen *Ustilago scitaminea* (Sugarcane Smut). Proc S Afr Sug Technol Assoc. 2001;75:104–7.
Tiwari AK, Bharti YP, Tripathi S, Mishra N, Lal M, Rao GP, Sharma PK, Sharma ML. Biotechnological approaches to improve sugarcane crop with special reference to disease resistance. Acta Phytopathol Entomol Hungric. 2010;45(2):235–49.
Tiwari AK, Lal M, Singh AK. Current status of sugarcane research in India. New York: Nova Publisher; 2015a. p. 1–170.
Tiwari AK, Singh SP, Dagar A, Mishra N, Sharma BL, Rao GP. Mixed infection of pineapple and grassy shoot diseases on commercial sugarcane varieties in Uttar Pradesh, India. Phytopathogenic Mollecutes. 2015b;5(2):91–4.
Tiwari AK, Singh SP, Kumari K, Singh A, Sharma BL. *Fusarium moniliforme* associated with Sugarcane leaf binding disease in India and its possible management. In: Kumar P, Gupta VK, Tiwari AK, Kamle M, editors. Recent trends in plant disease diagnostic and management. Springer; 2016. p. 187–91.
Verma KP, Singh MP, Upadhyay UC, Singh HN. Knife cut, the first report of the incidence of this disease in India. Sugarcane. 1984;40(11):817–20.
Villalon B. Sugarcane smut in lower Rico Grande Valley of south Texas. Plant Dis. 1982;66(7): 605–6.
Vishwakarma SK, Kumar P, Nigam A, Singh A, Kumar A. Pokkah Boeng: an emerging disease of sugarcane. J Plant Pathol Microb. 2013;4:170. doi:10.4172/2157-7471.1000170.
Viswanathan R, Padmanaban P. Hand book on sugarcane diseases and their management. Coimbatore: Sugarcane Breeding Institute; 2008.
Viswanathan R, Rao GP. Disease scenario and management of major sugarcane diseases in India. Sugar Tech. 2011;13(4):336–53.
Viswanathan R, Samiyappan R. Red rot disease of sugarcane: major constraint for Indian sugar industry. Sugar Cane. 1999a;5:9–15.
Viswanathan R, Samiyappan R. Induction of systemic resistance by plant growth promoting rhizobacteria against red rot disease caused by *Collectotrichum falcatum* Went in sugarcane. Proc Sugar Technol Assoc India. 1999b;61:24–39. 120.

Viswanathan R, Padmanaban P, Mohanraj D, Jothi R. Indirect-ELISA technique for the detection of the red rot pathogen in sugarcane (Saccharum spp. hybrid) and resistance screening. Indian J Agric Sci. 2000;70(5):308–11.

Viswanathan R, Malathi P, Sundar AR, Aarthi S, Premkumari SM, et al. Differential induction of chitinases and thaumatin-like proteins in sugarcane in response to infection by *Colletotrichum falcatum* causing red rot disease. J Plant Dis Protection. 2005;112:417–25.

Viswanathan R, Karuppaiah R, Malathi P, Ganesh Kumar V, Chinnaraja C. Diagnosis of sugarcane yellow leaf virus in asymptomatic sugarcane by RT-PCR. Sugar Tech. 2009;11:368–72.

Viswanathan R, Karuppaiah R, Balamuralikrishnan M. Detection of three major RNA viruses infecting sugarcane by multiplex reverse transcription-polymerase chain reaction (multiplex-RT-PCR). Australas Plant Pathol. 2010;39:1–6.

Viswanathan R, Sundar AR, Malathi P, Padmanaban P. Red rot of sugarcane (Shanthy TR, editor). Sugarcane Breeding Institute: Coimbatore; 2011.

Wada AC. Control of sugarcane smut disease in Nigeria with fungicides. Crop Prot. 2003;22: 45–9.

Wang Y, Zhang W, Wang Y, Zheng X. Rapid and sensitive detection of Phytophthora sojae in soil and infected soybeans by species-specific polymerase chain reaction assays. Phytopathology. 2006;96:1315–21.

Wang H, Xiao M, Kong F, Chen S, Dou HT, et al. Accurate and practical identification of 20 *Fusarium* species by seven-locus sequence analysis and reverse line blot hybridization, and an in vitro antifungal susceptibility study. J Clin Microbiol. 2011;49:1890–8.

Waraitch KS. Control of red rot of sugarcane with systemic fungicides. Indian J Mycol Plant Pathol. 1983;13:343–5.

Waraitch KS, Kumar B. Pathogenic behaviour and varietal performance of *Fusarium* causing sugarcane wilt. Indian Sugar. 1982;32(5):317–20.

Whittle PJL. Pokkah boeng: A Guide to Sugarcane Diseases. In: Rott P, Bailey RA, Comstock JC, Croft BJ, Saumtally AS, editors. A guide to sugarcane diseases. Montpellier: CIRAD/ISSCT; 2000. p. 136–40.

Whittle PJL, Irawan (2000) *Pokkah boeng*: a guide to sugarcane diseases. Rott P, Bailey RA, Comstock JC, Croft BJ, Saumtally AS, editors. CIRAD/ISSCT: Montpellier. pp136–40.

Wismer CA, Bailey RA. Pineapple disease. In: Ricaud C, Egan BT, Gillaspie Jr AG, Hughes CG, editors. Diseases of sugarcane. Amsterdam: Elsevier; 1989. p. 399.

Wu KK, Heinz DJ, Meyer HK. Heritability of sugarcane smut resistance and correlation between smut grade and yield components. Crop Sci. 1983;23:54–6.

Yadahalli KB, Adiver SS, Kulkarni S. Influence of 'c' and 'n' sources on growth and development of *Ceratocystis paradoxa*. Int J Plant Sci. 2006;1(2):322–4.

Yadahalli KB, Adiver SS, Kulkarni S. Effect of pH, temperature and relative humidity on growth and development of *Ceratocystis paradoxa* – a causal organism of pineapple disease of sugarcane. J Agric Sci. 2007;20(1):159–61.

Yadav RL. Research vision to manage red-rot disease of sugarcane in India. Sugar Tech. 2006;8:99–100.

Zeng X, Kong F, Halliday C, Chen S, Lau A, et al. Reverse line blot hybridization assay for identification of medically important fungi from culture and clinical specimens. J Clin Microbiol. 2007;45:2872–80.

Zhu YJ, Lim ST, Schenck S, Arcinas A, Komor E. RT-PCR and quantitative real-time RT-PCR detection of sugarcane yellow leaf virus (SCYLV) in symptomatic and asymptomatic plants of Hawaiian sugarcane cultivars and the correlation of SCYLV titre to yield. Eur J Plant Pathol. 2010;127:263–73.

Zummo N. Unusual banding of sugarcane stalks associated with Knife cut. Sugarcane Pathol News Letter. 1970;5:10–1.

Chapter 15
Assessing Fungal Biodiversity Using Molecular Markers

Md. Shamim, Pankaj Kumar, Ravi Ranjan Kumar, Mahesh Kumar, Ranjeet Ranjan Kumar, and K.N. Singh

15.1 Introduction

The kingdom Fungi covered marvelous genetic diversity, with their spanning members, wide range of lifestyles, forms, habitats, and their sizes. It is well known that fungi are sister to animals and contain thousands of ancestry, ranging from the mushroom fungi to yeasts, molds, smuts, rusts, and other conspicuous critters with interesting morphologies. Fungi inclusive requisite in the ecological roles most notably decomposition processes of different organic matters, they also involved in very important symbiotic associations and are well known to comprise significant parasites (Alexopoulos et al. 1996). Traditionally, fungi and other microbes were classified on the basis of their morphological, chemical, and anatomical characters which are mainly associated with spore-bearing structures and other outer surface structures (McLaughlin et al. 2009). The documentation of fungi has been poorly investigated as organisms and the phylogenetic relationships within the kingdom are not thus far fully understood, although modern efforts have been flaking light on the evolutionary

M. Shamim
Department of Plant Molecular Biology and Genetic Engineering,
N.D. University of Agriculture and Technology, Kumarganj, Faizabad, U.P. 224229, India

Department of Molecular Biology and Genetic Engineering, Bihar Agricultural University, Sabour, Bhagalpur, Bihar 813210, India

P. Kumar • R.R. Kumar • M. Kumar
Department of Molecular Biology and Genetic Engineering, Bihar Agricultural University, Sabour, Bhagalpur, Bihar 813210, India

R.R. Kumar
Division of Biochemistry, Indian Agricultural Research Institute, Pusa, New Delhi 110012, India

K.N. Singh (✉)
Department of Plant Molecular Biology and Genetic Engineering,
N.D. University of Agriculture and Technology, Kumarganj, Faizabad, U.P. 224229, India
e-mail: kapildeos@hotmail.com

B.P. Singh, V.K. Gupta (eds.), *Molecular Markers in Mycology*, Fungal Biology,
DOI 10.1007/978-3-319-34106-4_15

record of the fungi (James et al. 2006; Hibbett et al. 2007). Arrangement of taxa into diverse morphotypes, though, does not reflect species phylogeny, since morphotypes are not existent taxonomic entities. Rapid and accurate characterization of fungal pathogens to species or strain stage is frequently necessary for disease examination and implementing a disease management for a specific are. Molecular data sets like DNA/RNA based molecular markers have allowed the advancement of a more expected classification and improved appreciative fungal diversity. However, molecular approaches exposed the subsistence of repeated peculiarity evolution and therefore the occurrence of artificial groupings in the traditional classifications used before the molecular characterization. There are numerous plant fungal species that are morphologically very similar, and their correct species description can be complicated. With the recent advancement of biotechnological and molecular tools, the rapid identification and characterization of different kinds of fungi has become quite easier (Tanabe et al. 1990; White et al. 1990; Morris et al. 2000; Leickfeldt and Seifert 2000; Peever et al. 2002; Njambere et al. 2008; Attanayake et al. 2009; Dhar et al. 2011).

Several molecular markers like RAPD, RFLP, SSR and AF LP have been utilized to assess large number of plant fungal population (Walker et al. 2001; Guo et al. 2004; Datta et al. 2011a). The rapid expansion of gene-specific primers for accurate amplification (White et al. 1990) has facilitated systematic learning, and the recognition & identification of fungal pathogens. The internal transcribed spacer (ITS) region of nuclear ribosomal DNA has generally been considered a convenient marker for molecular identification and phylogeny analysis of fungi at species level because of its conserved feature within species and multi-copy number per genome (Sanchez-Ballesteros et al. 2000; Datta et al. 2011a).

Advances in molecular tools and techniques have shaped the support for a improve in studies concerning fungal diversity, and the quick progress of next generation sequencing technologies pledges extra progress towards a more thorough understanding of fungal diversity and function. The development of molecular tools allowed a much more clear-cut, sensible and quick approach to the study fungal biodiversity. Recent development in DNA sequencing platforms like next-generation sequencing (NGS) permitted for the democratization of genomic approaches due to cost reduction and wide availability (Shendure and Ji 2008). Such technological development has been welcomed by the fungal research community, as it permits rapid studies of deeper scope than have been possible to date. Fungal communities can now be described based on millions of sequences in a very short time frame and at relatively minimum expenditure. Moreover, NGS is also enabling and enhance in the number of sequenced fungal genomes, provided that important information, fundamental for a better understanding of fungal biology and evolution (Brockhurst et al. 2011). Undoubtedly, the technological improvements on high-throughput sequencing coupled with refinement of analytic tools will significantly increase the quality of biodiversity results in the near future, construction of NGS level more powerful and informative approach. With all the recent technological advancements for the detection of biodiversity use of genomic tools will facilitate mycology to prosper in the close to future, making this a very exciting time to be a mycologist. Usually, genetic diversity of plant fungi has been reviewed by computing their morphological and physiological distinction among personages within specific populations. The above feature depends on the specific gene and their expression, thus they are persuaded by environmental situations. Finally, morphological and physiological differences, which do not imitate genetic distinction at the genetic (DNA) level, may come about among entities. Similarly, merely part of the inherited variation among individuals may be articulated as differences in morphological and physiological traits. Presently, huge number of studies on fungal diversity in the plant systems has been executed using molecular techniques.

15.2 Diversity Assessments of Plant Fungi

15.2.1 Conventional Methods

Traditional identification, characterization and diversity assessment of several plant fungi were simply identified conventionally according to their macroscopic and microscopic features such as colony colour, size and shape of conidia and appressoria, optimal temperature for growth, growth rate, presence or absence of setae (Smith and Black 1990; Gunnell and Gubler 1992). Unfortunately, throughout the infections to the plants, most of the pathogenic fungi show only their vegetative phase (absence of sporulation); in the host tissue, either hyphal elements or other nonspecific structures. Though, the pigmentation and shape of these hyphae of the fungi and the occurrence or absence of septa can provide the idea of their uniqueness. There are several fungal culture is needed for precise identification and characterization of plant fungi. Consequently, the growth of fungal isolates in suitable culture media, enabling their main characteristic aspects to be known, is tranquil the most widespread practices used.

However, due to environmental influences on the stability of morphological traits and the existence of intermediate forms of several important fungi, these criteria are not always adequate for reliable differentiation and characterization. The overlap of morphological characters and phenotypes among species makes characterization very difficult. Molecular techniques provide alternative methods for the characterization and diversity studies and are important tools in solving the problems of species delimitation (MacLean et al. 1993).

15.2.2 Biochemical Marker

15.2.2.1 Isozyme Electrophoresis

The electrophoretic movement of proteins such as enzymes is among the most cost efficient methods to examine genetic variation at the molecular level. There are presently five widespread methods of protein electrophoresis which are routinely used and differ in the nature of the supporting medium of separation either they are run horizontally or vertically (*i.e.,* starch gel is horizontal and vertical; polyacrylamide gel is vertical; agarose gel horizontal). Isozymes are widely reported as biochemical genetic markers and widely used to study genetic diversity in fungus and plants. Isozymes are enzymes with different molecular forms due to their differences in amino acid residues and others exhibits same catalytic activity. Isozyme analysis is a powerful biochemical technique with numerous applications in plant pathology. Mycologists and plant pathologists more recently adopted the procedure, and it is now being used routinely to settle taxonomic disputes, identify different cultures and fingerprinting of patentable fungal lines for diversity porpuse (Bonde et al. 1993). Amino acid sequences are most useful in isozymes analysis for the fungal and other species of plants. Differences in amino acid sequences can add to both the charge of the molecule and its 3-dimensional structure to their respective amino acid sequence. The basic isozyme techniques depend on the net charges among the different amino acids can found if a basic amino acids (*i.e.* lysine will substituted by an acidic amino acid aspartic amino acid). If there is 28.7 % of all amino acid substitutions occur, the net charge of a protein changed (Shaw 1970). Some amino acid substitutions that do not involve charge differences can also affect the electrophoretic mobility of a protein, presumably by altering the tertiary structure of the enzyme. Thus, about one-third of all single amino acid substitutions will be electrophoretically detectable, and several simultaneous substitutions can cancel out the effect. Isozymes analysis therefore provides a very conservative estimate of the extent of genetic variability within a population (Shaw 1970). Detectable isozymes can arise from three different genetic and biochemical conditions: (i) multiple alleles at a single locus, (ii) single or multiple alleles at multiple loci, and (iii) secondary isozymes, usually arising from post-translational processing.

Isozyme analysis is a powerful technique and has been successfully applied for the identification, taxonomic and genetic diversity in different fungal species and genera like *Trichoderma harzianum* (Siddiquee et al. 2010), *Ganoderma* (Smith and Sivasithamparam 2000), *Fusarium* (Laday and Szecsi 2001), and *Phytophthora* (Goodwin et al. 1995). In other different *Trichoderma*, first characterization was reported by Zamir and Chet (1985), who successfully examined 23 different geographically diverse isolates of *T. harzianum*. Finally *T. harzianum* isolates were grouped into 5 different types according to their isozyme pattern and helpful for distinguishing *Trichoderma* at the intra species level. In a similar study by Stasz et al. (1989), 16 enzyme loci showed 109 alleles in a study of 71 strains that were distributed between 5 morphological species. Cellulose-acetate electrophoresis (CAE) technique has been also used and successfully detects several species of *T. pseudokoningii*, *T. koningii T. citrinoviride T. longibrachiatum T. virens*, and *T. harzianum*. (Szekeres et al. 2006; Bissett 1992). The major disadvantage to isozyme analysis method is the comparatively bulky quantities of an organism that are often required for extracting sufficient amount of enzyme. However, this is usually not a problem with facultative fungi that can be cultured on artificial media. Obligate pathogens may also require considerable effort to obtain the necessary quantities of fungal tissue. For example, with maize downy mildew fungi, conidia can be washed directly from the plant surface and concentrated by centrifugation (Bonde et al. 1984; Michales et al. 1988). Such type of difficulties arises when urediniospores of rust fungi experiments carried out. In general, large amounts of the samples are required for the isozyme characterization like, 50–100 mg (wet weight) of mycelium, 50 mg (wet weight) of downy mildew condia, or 30–50 mg (pregermination dry weight) of germinating rust urediniospores are needed for isozyme analysis. Other disadvantage of isozyme analysis is the time which needs more than the other biochemical characterization. However isozyme electrophoresis can be conducted rapidly, several days or even weeks are often needed to isolate and grow the organisms. Such time requirements are unacceptable in situations where identification is required within hours. In such cases, alternative procedures, such as gene probes, may be more satisfactory.

15.2.3 DNA Based Molecular Method

15.2.3.1 Restriction Fragment Length Polymorphism (RFLP) Marker

Restriction fragment length polymorphism (RFLP) marker has been very valuable tools in the identification and characterization of plant fungal pathogens. The RFLP procedure is based on the natural differences in the sequences of the genomes of different groups or strains of organisms. The variations (polymorphisms) in DNA fragment sizes may be produced by the losing or expanding or by other proceedings that influence fragment sizes such as deletions or insertions in the DNA sequences of particular strains. The DNA of the target organism is digested with restriction enzymes and the fragments are separated by electrophoresis in agarose or polyacrylamide gel to detect the differences in the size of the DNA fragments.

The number and size of the fragments formed after digestion are determined by the distribution of restriction sites in the DNA. Hence, depending on the combination of each restriction enzyme and target DNA, a specific set of fragments that can be considered as the 'fingerprint' for a given strain is formed. The specific sites of fragments are usually identified by Southern blot analysis (Hamer et al. 1989; Leach and White 1991). They can also be directly observed by staining the gels with ethidium bromide and observing under ultraviolet light (Klich and Mullaney 1987). In the RFLP method high quality genomic DNA digested with restriction enzymes which slice the genomic DNA at specific nucleotide sequences. Further the digested DNA samples electrophoresed and blotted on a membrane and probed with labeled probes. The DNA fragments are then transferred to nitrocellulose or nylon membrane and hybridized with an appropriate probe (Southern 1975) on to a nitrocellulose membrane. Labeling of the probe may be performed with a radioactive isotope or with alternative non-radioactive stains, such as digoxigenin or fluorescein. Polymorphism in the hybridization pattern is revealed and attributed to sequence difference between individual pathogen. Botstein et al (1980) firstly used RFLP markers for construction of genetic map.

RFLPs have found wide application in the detection and characterization of fungi (Goodwin et al. 1990; Whitehead et al. 1992; Poupard et al. 1995; Woo et al. 1996). In the *Phytophthora* isolates, RFLP marker acquiesce a genetic fingerprint of and has proved a precious implement in observing *P. infestans* genetic diversity (Goodwin et al. 1992b; Forbes et al. 1998). Several of isolates of *Phytophthora* worldwide have been fingerprinted and an international database of the outcomes assembled (Forbes et al. 1998). The dataset has been significant in defining and monitoring (Goodwin and Drenth 1997) lineages of *P. infestans* and tracking inoculum sources (Zwankhuizen et al. 2000). The method does have disadvantages, however; large amounts of pure DNA are required, it is time consuming, the banding patterns can be difficult to interpret and the resultant data are dominant. The convenience of RFLPs has been vulnerable due to the bulk quantities (1–10 µg) of purified DNA are required for DNA digestion and southern blotting procedures.

There is also some precise method of diversity analysis have been investigated of fungi by using the polymerase chain reaction (PCR) and restriction fragment length polymorphisms (RFLP) of the internal transcribed spacer (ITS) of the fungal nuclear ribosomal DNA repeat (Egger 1995; Tejesvi et al. 2009). The PCR product amplified by specific oligonucleotide primers ITS 1 & ITS 4 (White et al. 1990) and there after restricting with different endonucleases, has been successfully used to analyze variations in the regions of ribosomal DNA of various groups of fungi (Gomes et al. 1999; Glen et al. 2001; Hyde and Soytong 2008). The ITS region and PCR-RFLP analysis were successfully used for identification of ectomycorrhizal morphotypes of *Scots pine* by Timonen et al. (1997) and *Dendroctonus ponderosae* and *Ips pini* (Lim et al. 2005). Similarly, Diguta et al. (2011) isolated different isolates like species *C. cladiosporioides*, *B. cinerea*, *E. nigrum*, *A. alternata*, *T. koningiopsis*, *P. diplodiella*, *C. herbarum*, *A. alternatum*, *T. cucumeris* and *F. oxysporum* from the grapes and characterized by PCR ITS-RFLP marker. This technique is a rapid, reliable and appropriate method for the characterization and identification of several fungi.

15.2.3.2 Random Amplified Polymorphic DNA (RAPD) Marker

Random amplified polymorphic DNA (RAPD) is a quick and cost-effective method to study the genetic diversity among fungal pathogens (Williams et al. 1990). The RAPD method is based on the amplification of DNA with single short (typically 10 bp) primers of arbitrary sequence. The single primer used in this method bind at random to the target DNA in the respective genome, resulting in the amplification of pieces of unknown sequence. The amplification process of RAPD reaction is conceded out under conditions of little stringency (typically 35–40 °C). The amplified products are separated on an agarose gel visualized the difference among the organism/fungi studied. The understanding of RAPD patterns is based on the number and the size of the amplified fragments from the different organisms. In general, the RAPD evaluation generates comparatively multifarious blueprints, which very much diverge among dissimilar fungal isolates. This technique are rapidly used in the diversity analysis of different fungal pathogens including *F. oxysporum* formae speciales, and races (Grajal-Martin et al. 1993; Bentley et al. 1994; Kelly et al. 1994; Datta et al. 2011b; Vishwa et al. 2011). Pastor-Corrales and Jara (1995) and Abadio et al. (2012) estimated the genetic diversity of *P. griseola* and its co-evolution with the common bean in Latin America by developing differential cultivars and using RAPD (random amplification polymorphic DNA) molecular markers. Studies related to the classification of isolates in races using differential varieties and characterization by RAPD have shown that *P. griseola* is extremely diverse (Mahuku et al. 2002; Sartorato 2004). Stenglein and Balatti (2006) estimated the genetic diversity of 45 *P. griseola* isolates using differential varieties as well as RAPD and ISSR (inter-simple sequence repeat) molecular techniques. The RAPD technique has been used to differentiate a collection of isolates into races corresponding to pathogenicity tests in cotton (Assigbetse et al. 1994) and basil (Chiocchetti et al. 1999; Chiocchetti 2001). Lievens et al. (2007) developed a robust RAPD marker-based assay to specifically detect and identify the pathogens *F. oxysporum* f. sp. *cucumerinum* and *F. oxysporum* f.sp. *radicis-cucumerinum* and *Gliricidia sepium* and *G. maculate* (Chalmers et al. 1992).

The main disadvantage of using RAPD technique is the reproducibility between the different runs which is due to the short primer length and low annealing temperature (Schierwater and Ender 1993), and therefore highly standardized experimental procedures are required because of their sensitivity to the reaction conditions.

The inherent problems of reproducibility create RAPDs incongruous markers for the devolution or comparison of results amongst the different research groups. RAPD markers are not locus-specific because banding patterns cannot be interpreted in terms of loci and alleles thus rated as dominance of markers. However, RAPD markers were found to be very easy to carry out by different laboratories throughout the word, still reproducibility was not achieved to a satisfactory level (Jones et al. 1997).

15.2.3.3 Amplified Fragment Length Polymorphism (AFLP) Marker

Amplified Fragment Length Polymorphisms (AFLP) based genomic DNA fingerprinting is a technique used to detect DNA polymorphism, is essentially intermediate between RFLPs and PCR (Vos et al. 1995). AFLP markers are usually dominant and do not need previous knowledge of the genomic composition of the particular organism. AFLPs are produced in great numbers and are reproducible (Bragard et al. 1997) band after the gel electrophoresis in the PAGE (poly acryl amide gel electrophoresis). Amplified fragment length polymorphism (AFLP) has been used in many studies for the analysis of fungal population structure (Majer et al. 1996; Gonzalez et al. 1998; DeScenzo et al. 1999; Purwantara et al. 2000; Zeller et al. 2000; Silva et al. 2013). Genetic variation among pathogenic isolates of *F. oxysporum* was estimated using AFLP markers by several workers (Baayen et al. 2000; Bao et al. 2002; Sivaramakrishnan et al. 2001; Groenewald et al. 2006; Stewart et al. 2006). AFLP marker have higher reproducibility and efficiency hat led to its broader application in the analysis of population diversity and identification of pathogens (Baayen et al. 2000; Abd-Elsalam et al. 2002a, b; Kiprop et al. 2002; Sivaramakrishnan et al. 2002; Abdel-Satar et al. 2003; Zeller et al. 2003; Leslie et al. 2005; Gurjar et al. 2009). This technique was utilized to scrutinize genetic relationships among the different isolates of *F. oxysporum* f. sp. *vasinfectum* (Abd-Elsalam et al. 2004; Wang et al. 2006). Gurjar et al. (2009) also characterized two *F. oxysporum* f.sp. *ciceri* races (1 and 2) based on unique AFLP patterns from chickpea growing areas. Sequence characterization of these race-specific AFLP products revealed significant homologies with metabolically important fungal genes. AFLPs marker also have been used to distinguish *Fusarium* species by several researchers (Baayen et al. 2000a; Zeller et al. 2003; Leslie et al. 2004; Belabid et al. 2004). However, as AFLP is relatively costly and has a rather complicated technical procedure, it is being increasingly replaced by simpler PCR-based methods. There are many benefit of AFPL marker: (i) there is no need for prior knowledge of any sequence information of particular fungi, (ii) each experiments produced large no. of reproducible bands, (iii) the amplified bands are amplified from all over the genome of fungi (Blears et al. 1998; Vos and Kuiper 1998), (vi) the AFLP data can be stored in database like AmpliBASE MT (Majeed et al. 2004) for comparison purposes of particular organisms. There are disadvantages of AFLP technique are that the amplified alleles are not easily recognized in the gel. This technique has also medium reproducibility, labor exhaustive and has high equipped and expansion charges (Karp et al. 1997). Furthermore, AFLP necessitate knowledge of the genomic sequence to design primers with specific selective bases for the amplification purposes.

15.2.3.4 Simple Sequence Repeats (SSR)/Microsatellite Marker

Simple sequence repeats (SSRs), also known as microsatellites, provide a powerful tool for taxonomic and diversity analysis in the plants and several fungal population. This marker has also been frequently used in fungal diversity studies. Originally term, a microsatellite was introduced by Litt and Lutty (1989). SSR or microsatellites are parts of DNA, consisting of tenderly repeating mono-, di-, tri-, tetra- or penta-nucleotide components that are assembled throughout the genomes (Rafalski et al. 1996; Powell et al. 1996). SSR marker provides high resolution and co-dominant in nature among the several markers used presently (Bogale et al. 2005, 2006; Bayraktar et al. 2008). It is proposed that the differences in SSR markers are a outcome of polymerase slippage at some stage in DNA replication or imbalanced crossing over (Levinson and Gutman 1987). SSRs are not only very common, also are hypervariable for numbers of repetitive DNA motifs in the genomes of eukaryotes (Vosman and Arens 1997; Rallo et al. 2000; van der Schoot et al. 2000). The microsatellite markers, developed from genomic libraries, may be come from the transcribed region (exons) or the non transcribed region (introns) of the genome. These microsatellite sequence variations are particularly appropriate to distinguish closely related genotypes; since of their elevated quantity of unevenness in their respective genomes. Datta et al. (2011a) assessed molecular diversity in the different *Fusarium oxysporum* f. sp. *lentis*, causing agent of vascular wilt in lentil were collected from different agro-climatic regions of India by SSR marker. Berbegal et al. (2011) analyzed genetic diversity in the fungal soil borne plant pathogen *Verticillium dahlia* by SSR markers. Leyva-Madrigal et al. (2014) identified a total of 470 perfect microsatellites in the genome of *F. verticillioides* and found that these markers are highly polymorphic and thus useful for *F. verticillioides* population studies. One of the main drawbacks of microsatellites is that high development costs are involved if adequate primer sequences for the species of interest are unavailable, making them difficult to apply to unstudied groups. One advantage of microsatellite markers is that some are transferable across species, particularly between closely related species. Several studies have reported that SSR markers are very valuable for evaluating genetic diversity and relationship (Eujayl et al. 2001; Russell et al. 2004; Kumar et al. 2009).

15.2.3.5 Inter Simple Sequence Repeat (ISSR) Marker

ISSRs are DNA fragments of about 100–3000 bp lengths situated between neighboring, oppositely oriented microsatellite regions of genomic regions. ISSRs are amplified by PCR using single microsatellite core sequences as primers with a few selective nucleotides as anchors into the non-repeat adjacent regions (16–18 bp). About 10–60 amplified fragments from multiple loci are produced concurrently. ISSR procedure originally developed by Zietkiewicz et al. (1994) and this tool were successfully used in several fungal populations, plants, insects and vertebrates (Wolfe 2005). Tymon and Pell (2005) estimated genetic diversity of 30 isolates of the entomopathogenic fungus *P. neoaphidis* of different geographic origins to be calculated using ISSR, ERIC and RAPD with a larger number of polymorphic bands obtained from ISSR. Abadio et al. (2012) found that ISSR-PCR technique is suitable for assessing intraspecific variability of *P. griseola*. van der Nest et al. (2000) used inter-simple sequence repeat (ISSR) and SSR primers (random amplified microsatellites, RAMS) in PCR to expand SSR markers for *F. oxysporum*. Barve et al. (2001) assessed the genetic variability *in F. oxysporum* f.sp. *ciceri* (Foc) populations prevalent in India. The procedure of the ISSR technique has numerous advantages over other molecular techniques such as RAPD. The ISSR technique is more specific than the RAPD technique due to the use of longer oligonucleotide sequences, allowing more stringent annealing conditions in PCR amplification. This technique is more reproducible than RAPD and has higher levels of polymorphism (Reddy et al. 2002). However, ISSRs, like RAPDs, can have reproducibility problems. Further, changeable character of microsatellite provinces, minimum requirements and simple function as well as the logical cost, has made ISSR marker as a very useful tool for most systematic and ecological evaluations (Reddy et al. 2002).

15.2.3.6 Sequence Characterized Amplified Region (SCAR) Marker

SCAR marker developed by the sequencing of RAPD marker termini (22–24 nucleotide bases long) when a longer primers are designed for specific amplification of a particular locus (Michelmore et al. 1991; Martin et al. 1991). A specific 15–30 bp length designed primers from nucleotide sequences established from cloned RAPD fragments linked to a specific trait of interest were amplified by the PCR. Sequence characterized amplified region (SCAR) approach has proven to be effective for the identification of several formae speciales and races of *F. oxysporum*. Kelly et al. (1998) developed an in planta PCR method to detect isolates of race 5 of Foc in chickpea. The assay using RAPD-derived SCAR markers specifically identified race 5 of the pathogen from infected chickpea plants. Similarly, Jimenez-Gasco and Jimenez-Diaz (2003) sequenced previously identified Foc specific RAPD markers and designed SCAR markers to identify Foc and its four pathogenic races 0, 1A, 5, and 6. The assays were sensitive enough to detect as low as 100 pg of fungal genomic DNA. Based on RAPD analysis, Shimazu et al. (2005) widened STS markers for specific identification of three races of *F. oxysporum* f.sp. *lactucae*. These markers were specific to *F. oxysporum* f.sp. lactucae and did not amplify DNA from isolates of five other *F. oxysporum* formae speciales as well as other plant pathogenic fungi, and bacteria. Ganeshamoorthi and Dubey (2013) developed SCAR marker for the detection of *R. solani* isolates that are useful for early detection and quantification of wet root rot pathogen in chickpea. Liu et al. (2012a, b) developed 29 sequence characterized amplified region (SCAR) markers were developed for the screening of genetic diversity of *Lentinula edodes* strains. By using longer PCR primers, SCARs do not face the problem of low reproducibility generally encountered with RAPDs. Further, to investigate specific genomic regions of different fungi by PCR sequencing, SSCP, CAPS or SCAR, sequence data of the sites of interest (structural genes mainly) are required for primer construction.

15.2.3.7 Single Nucleotide Polymorphisms (SNPs) Marker

DNA sequence-based markers have been widely used for the different plant fungus characterization and diversity analysis purpose over recent years due to the reduced cost of DNA sequencing. Among these markers, single nucleotide polymorphisms (SNPs) are among the fastest-developing categories in biomedical and biological research since, SNPs are the most regularly pragmatic differences between DNA sequences obtained from different individuals or between alleles from within an individual in diploid or higher ploidy organisms. SNPs refer to polymorphisms generated by nucleotide substitutions or the insertion and deletion of single nucleotides between homologous DNA sequences. SNPs have become popular molecular markers for studying a variety of biological issues. So far, most such studies have focused on model organisms, and relatively little is known about the SNPs in non model organisms including many fungi (Xu et al. 2007). In addition, SNPs have several properties, such as a relatively low mutation rate and the ease of scoring and data sharing, that make them highly desirable for a variety of biological analyses (Brumfield et al. 2003; Xu 2006b). SNPs will provide a very promising type of molecular marker, if there is sufficient genomic data have been established in the particular group of individuals (Schlotterer 2004). In plant pathogens, SNPs have sparsely been used so far (Morin et al. 2004), and this is true in particular for Oomycota. SNP marker based population studies were carried out in the fungus *Phytophthora ramorum* (Martin 2008) and *Hyaloperonospora parasitica* (Clewes et al. 2007). In other study on *P. halstedii* field accessions, independent introduction events of the pathogen in French sunflower cultivation were traced and identified on the basis of SNP data (Delmotte et al. 2008). Closely related pathogens showing different host ranges or pathogenecity often differ in only a single to a few base pairs in target genes commonly used for identification. Therefore, the ability to discriminate single nucleotide polymorphisms (SNPs) should be pursued in any diagnostic assay. Based on the DNA nucleotide sequence difference in the mitochondrial cox I gene, a SNP method has been developed to detect and differentiate isolates of *Phytophthora ramorum* from Europe and those originating in the United States (Kroon et al. 2004). In other experiment, polymorphisms detected in the microsatellite flanking regions of *Phytophthora infestans* allowed the development of a SNP genetic marker system for typing this pathogen (Abbott et al. 2010). Similarly, a survey of seven genomic loci (3013 nt in total) for 84 natural strains of the model yeast *Saccharomyces cerevisiae* from Asia has identified a total of 62 SNPs, yielding an SNP frequency of 2.05 % per nucleotide (Ayoub et al. 2006). The SNP frequencies are higher in populations of several opportunistic pathogenic yeasts, such as *Candida parapsilosis* (~3.4 %; Fundyga et al. 2004), and the species complexes of *Candida guilliermondii* (~6.3 %; Lan and Xu 2006) and *Cryptococcus neoformans* (~20 % per nucleotide; Xu et al. 2000b). However, for *C. parapsilosis, C. guilliermondii* and *C. neoformans*, the population samples contain strains from distinct cryptic species and/or species (Fundyga et al. 2004; Lan and Xu, 2006; Xu et al. 2000b). The observed SNP frequency in *T. matsutake* was likely an underestimate of the true SNP frequency in this species. This is because sequences from only two strains were analyzed for SNP discovery and both strains were from the same general geographical area in southwestern China. Analyzing more strains should yield additional SNPs within the sequenced DNA fragments. The SNP frequencies in fungi are reported much higher than that in the human genome, where SNPs are observed approximately once every 250 bp (Miller et al. 2005). Xu et al. (2006) characterized *Tricholoma matsutake* a gourmet mycorrhizal mushroom primarily associated with pine forests by the SNP markers. Since an individual gene contains limited information, multi-locus sequence typing (MLST) has become a mainstream method for both molecular systematic and population genetics analyses as SNPs markers have very vast no. of differences in the respective genes (Taylor and Fisher, 2003; Xu 2006). MLST has been demonstrated to be highly discriminatory when analyzing the genetic diversities of many fungal species that are pathogenic to humans, such as: *Histoplasma capsulatum* (Kasuga et al. 2003), *Candida albicans* (Li et al. 2008), *Aspergillus fumigatus* (Bain et al. 2007) and Cryptococcus neoformans (Meyer et al. 2009). Most of the DNA sequences used for MLST in these studies were limited to housekeeping genes.

However, in case of plant pathogenic fungi, the genetic diversity based on MLST, only found in *Magnaporthe grisea* (Choi et al. 2013).

15.3 PCR- Denaturing Gradient Gel Electrophoresis (DGGE)

DGGE (Denaturing Gradient Gel Electrophoresis) simply uses chemical gradient such as urea to denature and separate DNA samples when they are separating across an acrylamide gel during the electrophoresis. DDGE is applied for the analysis of the genetic diversity of several microbial communities including important plant fungal population without the need of any prior knowledge of the species (Muyzer 1999; Gothwal et al. 2007; Portillo et al. 2011). A very sensitive detection of polymorphisms in DNA sequences were found by this methods as sequence variants of particular fragments migrate at different positions in the denaturing gradient gel. In addition, PCR-DGGE primers contain a GC rich tail in their 5' end to detect small variations (Myers et al. 1985). The bands obtained in the gel can be extracted, cloned or re amplified and sequenced for identification, being even possible to identify constituents that represent only 1 % of the total microbial community. These techniques are much suitable for the identification of novel or unknown organisms and the most abundant species can be readily detected. This method is, however, time-consuming, poorly reproducible and provides relative information about the abundance of detected species. Interpretation of the results may be difficult since micro heterogeneity present in some target genes may appear as multiple bands in the gel for a single species, leading to an overestimation of the community diversity. Furthermore, fragments with different sequences but similar melting behavior are not always correctly separated. In other cases, the analysis of complex communities of microorganisms may result in blurred gels due to the large number of bands obtained. A PCR-DGGE detection tool based in the amplification of the ITS region has been recently applied to detect multiple species of *Phytophthora* from plant material and environmental samples (Rytkonen et al. 2012). Other researchers have used this technique to compare the structure of fungal communities growing in different conditions or environments, e.g. to study the impact of culture management such as biofumigation, chemifumigation or fertilization on the relative abundance of soil fungal species (Omirou et al. 2011; Wakelin et al. 2008).

15.4 Internal Transcribed Spacer (ITS) Marker

Eukaryotic ribosomal RNA coding genes (also known as ribosomal DNA or rDNA) are elements of repeat entities that are assembled in tandem arrays. rDNA regions are located at the chromosomal sites and known as nucleolar organizing regions (NORs). The particular nuclear ribosomal DNA has two internal transcribed spacers: ITS-1 that one is located in between the small subunit (16s–18s) and 5.8S rRNA cistronic regions, and the second one is ITS-2 which located between the 5.8S and large subunit (23S–28S) rRNA cistronic regions. The two spacers combined with the 5.8S subunit are jointly known as the internal transcribed spacer (ITS) region (Anderson et al. 2003; Lord et al. 2002). The ITS regions of ribosomal DNA (600–700 bp) repeats are supposed to be quick evolving and therefore may differ in their length and sequences. The flanking regions of ITS region are conserved, thus used for the universal PCR primers to easy amplification of ITS region of the eukaryotes including fungi. The number of copies of rDNA repeats is up to-30,000 per cell (Dubouzet and Shinoda 1999). This makes the ITS region an interesting subject for evolutionary and phylogenetic investigations (Baldwin et al. 1995) as well as biogeographic investigations (Baldwin 1993). Internal transcribed spacer (ITS) region of nuclear ribosomal DNA has generally been considered a convenient and widely utilized marker for molecular identification of fungi at species level. The ITS regions have conserved feature within species and multi-copy number per genome that would help to characterize the different species including the plant fungus. Analysis of ITS sequence is usually applied to determine species identity (or sometimes higher taxonomic categories) and to identify and discriminate populations within a species (Sanchez-Ballesteros et al. 2000). *Aspergillus* fungus has been characterized at species level and differentiated it from other true pathogenic and opportunistic molds using ITS 1 and ITS 2 (Henry et al. 2000). Schneider et al. (1997) have expanded a precise a technique for revealing and characterization of major wilt causing fungus *Rhizoctonia solani* isolates. The developed techniques also have the capacity to detect pathogenic and nonpathogenic to tulips, using ITS rDNA sequences, and they could further identify various anastomosis groups. With the help of ITS based sequences and markers, *S. trifoliorum* was characterized and found that it contains group I introns in the nuclear small subunit ribosomal DNA, while *S. minor* and *S. sclerotiorum* do not contain any introns in the respective same DNA region. The nucleotide sequence encoding for the internal transcribed spacer (ITS) of the nuclear ribosomal RNA has recently been proposed as a standard marker for molecular identification of fungi and evaluation of fungal diversity. However, the analysis of large sets of ITS sequences involves many programs and steps, which makes this task intensive and laborious. The Ribosomal Database Project (RDP) was established to generate a credible bank for 28S rRNA gene information (Cole et al. 2011). Vishwa et al. (2011) characterized different *Fusarium udum* from different pigeonpea growing areas in India by ITS-RFLP marker. Similarly Datta et al. (2011b) also studied molecular diversity in Indian isolates of *Fusarium oxysporum* f.sp. *lentis* inciting wilt disease in lentil *(Lens culinaris* Medik), by RAPD, SSR and ITS-RFLP markers. Ferro et al. (2014) developed a database, ITScan which is an online and very user-friendly automated pipeline for fungal diversity examination and identification supported on ITS sequences. This database speeds up a process which would otherwise be repetitive and time-consuming for users. The ITS can tool and documentation are accessible at http://evol.rc.unesp.br:8083/itscan.

15.5 Intergenic Spacer Sequences (IGS) Marker

Higher fungi, have non-coding region of IGS1 (between the 25S and 5S coding units) and IGS2 (between the 5S and 17S coding units), which recognized to present length and sequence variations within the genome of particular species (Molina et al. 1993; Iracabal and Labarree 1994). The intergenic spacer sequence (IGS) situated in between the 28S and 18S rRNA genes is the region with the most amount of sequence difference in rDNA. Further, it is frequently used in PCR-based methods when ITS is fail to generate enough amount of variation. Bulky range and restriction site polymorphisms in IGS between rDNA repeats have been revealed in other fungal species (Bruns et al. 1991; Anderson and Stasovski 1992; Gryta et al. 1997). Primers in this region have been designed to detect and identify *Verticillium dahliae* and *V. alboatrum* (Schena et al. 2004) and to distinguish pathogenic and non-pathogenic *Fusarium oxysporum* in tomato (Validov et al. 2011). As another example, Inami et al. (2010) differentiated *Fusarium oxysporum* f. sp. *lycopersici*, and its races using primers and TaqMan-MGB probes based on IGS and avirulent SIX genes. IGS-RFLP markers have been used to differentiate varieties of *M. nivale* (Parry et al. 1995), analyse variation in *F. oxysporum* (Appel and Gordon 1995; Woudt et al. 1995), and detect length heterogeneity in *Pythium ultimum* (Buchko and Klassen 1990). However, few studies have used both RAPD and IGS-RFLP analysis to characterize variability within fungal populations (Fabre et al. 1995; Appel and Gordon 1996; Mahuku et al. 1998). IGS region is considered the most rapidly evolving region of the rDNA repeat unit (Hillis and Dixon 1991), it is constrained by secondary structure conservation (Baldridge et al. 1992). Also, there may be variation between IGS repeats within a genotype (Crease 1995) that could allow greater discrimination of different isolates, but such variation may not have been detected by the IGS-RFLP method.

15.6 PCR-Enzyme-Linked Immunosorbent Assay (ELISA)

This serological-based PCR method uses forward and reverse primers carrying at their 5′ end biotin and an antigenic group (e.g. fluorescein), respectively (Landgraf et al. 1991). PCR amplified DNA can be immobilized on avidin or streptavidin-coated microtiter plates via thebiotin moiety of the forward primer and then can be quantified by an ELISA specific for theantigenic group of the reverse primer (e.g. anti-fluorescein antibody detected by colorimetric reactions). PCR-ELISA method is as sensitive as nested PCR. In addition, it does not require electrophoretic separation and/or hybridisation, and can be easily automated. All reactions can be performed in 96-well microtiter plates for mass screening of PCR products making them very suitable for routine diagnostic purposes. This procedure has been used for detection and differentiation of *Didymella bryoniae* from related *Phoma* species in cucurbits (Somai et al. 2002) and for detection of several species of *Phytophtora* and *Pythium* (Bailey et al. 2002).

15.7 Real-Time PCR/qPCR

Real-time PCR is presently considered as one of the methods for characterization and detection of different plant pathogens. Real time- PCR allows the observing of the reaction throughout the amplification process by the use of fluorescent signal that increases proportionally to the number of amplicons (products) generated and to the number of targets present in the sample (Wittwer et al. 1997). Real time PCR has several advantages over the conventional PCR, as this system does not require the use of post PCR processing like electrophoresis, colorimetric reaction or hybridization with specific probes. Real time PCR also avoids the risk of carryover contamination and reducing assay labour and material costs.

The improved method of Real-time PCR has the sensitivity and specificity. This new technique also allowed the accurate quantification of the target pathogen genes, by interpolating the quantity measured to a standard curve with known amounts of target copies. This quantification characteristic of real time-PCR have very useful tools in phytopathology in order to correlate the amount of fungus gene in a biological sample with the disease condition, or to observe the development of the disease in an infected plant (Garrido et al. 2009). Real-time PCR have also the capability to perform multiplex detection of two or more pathogens in the single reaction.

Quantitative real-time PCR (qPCR) has been shown to be more sensitive than culture-based approaches against a wide range of fungal species (Meklin et al. 2004). Much progress has been made in developing qPCR assays that can detect diverse fungal species (Einsele et al. 1997; Haugland et al. 2004; Landlinger et al. 2010; Chemidlin Prevost-Boure et al. 2011), but an efficient method is required to develop a qPCR assay that should approach universal fungal coverage (Liu et al. 2012a, b).

15.8 Mass Spectrometry

Presence of considerable variability in morphology in the different fungal species, microscopy technique is frequently used as a tool for identification of individual species. Several microscopic techniques, commonly used for this purpose, which include the observation and evaluation of different infectious structures and reproductive organs (sexual and asexual spores) as regards to the color, shape and surface of the different fungus. In the present fungal molecular biology, matrix assisted laser desorption/ionization time of fight mass spectrometry (MALDI-TOF MS) systems are frequently used as biological application, their molecular characterization tools as well as for diagnostic applications in clinical medicine, biotechnology, and industry (Bizzini and Greub 2010). Mass spectrometry is already emerged as a very reliable molecular tool for fast identification and classification of several microorganisms (Chalupova et al. 2014). The high resolution and discriminative power of MALDI-TOF tool facilitated differentiation of closely related species or even different strains of the same fungal species. MALDI-TOF based identification and characterization of different fungal microorganisms employ peptide/protein profiles containing characteristic biomarker peaks in the m/z region of 1000–20,000.

Nowadays, commercial MALDI-TOF MS and MS/MS systems are accessible for several fungal research works as well as for diagnostic applications in clinical medicine, biotechnology, agriculture and industry. The development of intact cell/intact spore matrix-assisted laser desorption/ionization time-of-flight mass spectrometry (IC/IS MALDITOF MS) of fungal microorganisms began around the year 2000 by first experiments with *Penicillium* spp., *Scytalidium dimidiatum*, *Trichophyton rubrumand* and *Aspergillus niger* (Li et al. 2000; Welham et al. 2000). The experimental approaches used at that time were adopted from initial protocols for MALDI-TOF MS measurements with whole bacterial cells (Holland et al. 1996; Jarman et al. 1999; Welham et al. 1998). IC/IS MALDI-TOF MS of Hypocrea/Trichoderma has also been performed with a special focus on the detection of peptaibol profiles forstudying phylogenetic relationships (Neuhof et al. 2007c). There are several important fugus *i.e.*, *Aspergillus*, *Fusarium*, *Penicillium*, *Trichoderma* and also various yeasts have been successfully identified and characterized by MALDI-TOF MS (Bizzini and Greub 2010; Neuhof et al. 2007a).

There are different cell organelles molecules were undertaken for the analysis of diversity and characterization purpose like outer surface proteins named hydrophobins have been recognized as the cause of predominant ions during MALDI-TOF MS profiling of *Hypocrea* and *Trichoderma* strains (Neuhof et al. 2007b). Interestingly, spores of downy and powdery mildews release ribosomal proteins when treated with acidic matrix solutions, which is similar to the behavior of bacteria and yeasts (Fenselau and Demirev 2001). However there are needs to future development of MALDI-TOF/MS methodology for the different plant fungal analyses will absolutely involve a incessant updating of current profitable databases provided by vendors together with the instrumentation as well as building up new databases for specialized research purposes. There is still a big gap in linking MALDI-TOF MS peptide/protein profiles with proteomic identification of individual bio marker molecules, which needs to be expanded.

15.9 DNA Barcoding

DNA barcoding is a taxonomic method that uses a short genetic marker in the DNA to identify an organism as belonging to a particular species. It facilitates the explanation of several new species and characterisation of species complexes. Presently, fungal species recognition platforms are available: *Fusarium*-ID which is a easy, web-accessible BLAST server that consisted of sequences of the *TEF 1*α gene from representative species of *Fusarium* (Geiser et al. 2004). Sequences of numerous marker loci from almost all known *Fusarium* species have been progressively included for supporting strain identification and phylogenetic analyses. Two extra platforms have also been constructed: the Fusarium Comparative Genomics Platform (FCGP), which keeps five genomes from four species, supports genome browsing and analysis, and shows computed features of multiple gene families and functional groups; and the Fusarium Community Platform (FCP), an online research and learning forum. All together, these platforms form the Cyber infrastructure for *Fusarium* (CiF; http://www.fusariumdb.org/) For *Phytophthora* identification two web-based databases have been created: (i) *Phytophthora* Database (http://www.phytophthoradb.org/) based in nine loci sequences including the ITS region and the 5′ portion of the large subunit of rRNA genes; nuclear genes encoding 60S ribosomal protein L10, α -tubulin, enolase, heat shock protein 90, TigA fusion protein, and TEF 1 α; the mitochondrial gene *cox* II and spacer region between *cox* I and *cox* II genes (Park et al. 2008, 2011); and (ii) Phytophthora-ID (http://phytophthora-id.org/) based on ITS sequences and the *cox* I and *cox* II spacer regions (Grünwald et al. 2011). Additional web-based databases are available including UNITE (http://unite.ut.ee/index.php), an ITS database supporting the identification of ectomycorrhizal fungi (Koljalg et al. 2005); TrichOKey (http://www.isth.info/tools/molkey/index.php), a database supporting the identification of *Hypocrea* and *Trichoderma* species (Druzhinina et al. 2005); and BOLD (http://www.boldsystems.org/) containing ITS and *cox* I databases from oomycetes (Ratnasingham and Hebert 2007; Robideau et al. 2011). Consortium for the Barcode of Life (CBOL) is an international collaborative effort which aims to use DNA barcoding to generate a unique genetic barcode for every species of life on earth. The *cox* I mitochondrial gene is emerging as the usual barcode region for eukaryotes.

15.10 Next Generation Sequencing (NGS) Technology

Sequencing DNA is an attempt to decipher the permutation of its nucleotides. Early attempts to establish a sequencing system included the Maxam-Gilbert sequencing and chain termination methods. The Maxam-Gilbert method or the chemical sequencing method was developed by Walter-Gilbert and Alan Maxam in 1977. Purified DNA could be used directly in the sequencing reaction. End-labelled DNA is cleaved at specific bases using specific reagents. The labelled fragments cleave using dimethyl sulphate which selectively attacks purines and hydrazine which selectively attacks pyrimidines. The fragments are electrophoresed in a high resolution polyacrylamide gel to deduce the sequence of the DNA molecule. The chain termination method requires a single strand of the DNA molecule whose sequence must be determined, a DNA primer, a DNA polymerase, labelled nucleotides, nucleotides that serve as terminators because of their lack of 3′-OH group which are required to form phosphodiester bonds between two nucleotides. Based on the sequences of the template DNA, DNA strands of various lengths are synthesized. The newly synthesized strands are denatured by heat and run in a polyacrylamide gel. From the various size fragments on the gel, sequence of the DNA molecule can be deciphered. Now, Sanger sequencing method has been partially replaced by several "next-generation" sequencing (NGS) technologies. The NGS technology is able to produce a high number of short sequences from multiple organisms in short time. The next-generation technologies commercially available today include the 454 GS20 pyrosequencing-based instruments (Roche Applied Science), the Solexa 1G analyzer (Illumina, Inc.), and the SOLiD instrument (Applied Biosystems).

Prominently, NGST technologies capitulate not only qualitative information regarding the sequence of each DNA portion analyzed, but also give the quantitative information about the relative abundance of each DNA fragment in the library sequenced (Rokas and Abbot 2009). RNA-seq with the help of NGST proved a powerful tool for the study of genetic differentiation in *Aspergillus fumigatus*. NGST sequencing should provide in depth sampling of a few hundred loci, and thousands of SNPs, simply by sequencing transcripts in proportion to their representation in the library were found for the diversity purpose (Hittinger et al. 2010). Thus, RNA-seq is a powerful alternative to the standard multilocus sequence typing currently used for the study of isolate identification and population structure in *A. fumigatus*, and in filamentous fungi in general (Klaassen 2009).

15.11 Conclusion

Precise identification and characterization of plant fungal pathogens of important cereal, horticultural plants and other economically important crops can help to the societies a lot in better management of the diseases. Generally, fungal variability of the several plant systems were considered by using differential host reactions, culture characteristics, morphological markers and biochemical tests. These markers differentiate fungal pathogens on the basis of their physiological characters *i.e.* pathogenicity and growth performances and are highly persuaded by the host age, inoculums quality and environmental conditions. However, such type of the cultural approach alone or biochemical analyses are not sufficient for determining fungal diversity. Presently, several molecular markers have been advocated for characterization and identification for the genetic variability in phyto pathogenic fungal microbes. Molecular markers like RAPD, SSS, ITS-RFLP etc. have greatly contributed to explain the phenotypic and genetic diversity within and between the different fungal populations. Currently more sensitive methods like Real Time PCR and Microarrays are routinely being used for the identification and characterization of phyto fungal pathogen. Real Time PCR has emerged as one of the capable and very sensitive method of plant pathogen detection and diversity purposes. In the near future, RNA-seq based NGST applications will not only lead to far greater understanding of the parts, structure, and function of the several fungus genome, but will also identify the key differences between in vitro and in vivo models of the disease, as well as define the molecular interactions between the plant host and the fungal pathogen during the infection.

References

Abadio AKR, Lima SS, Santana MF, Salomao TMF, Sartorato A, Mizubuti ESG, Araujo EF, de Queiroz MV. Genetic diversity analysis of isolates of the fungal bean pathogen *Pseudocercospora griseola* from central and southern Brazilian. J Genet Mol Res. 2012;11(2):1272–9.

Abbott NJ, Patabendige AA, Dolman DE, Yusof SR, Begley DJ. Structure and function of the blood-brain barrier. Neurobiol Dis. 2010;37:13–25.

Abdel-Satar MA, Khalil MS, Mohmed IN, Abd-Elsalam KA, Verreet JA. Molecular phylogeny of Fusarium species by AFLP fingerprint. Afr J Biotechnol. 2003;2:51–6.

Abd-Elsalam KA, Khalil MS, Aly AA, Asran-Amal A. Genetic diversity among *Fusarium oxysporum* f.sp. *vasinfectum* isolates revealed by UP- PCR and AFLP markers. Phytopathol Mediterr. 2002a;41:1–7.

Abd-Elsalam KA, Khalil MS, Aly AA, Asran-Amal A. Population analysis of *Fusarium* spp. Phytomedizin. 2002b;3:18–9.

Abd-Elsalam KA, Omar MR, Migheli Q, Nirenberg HI. Genetic characterization of *Fusarium oxysporum* f. sp. *vasinfectum* isolates by random amplification of polymorphic DNA (RAPD) and amplified fragment length polymorphism (AFLP). J Plant Dis Prot. 2004;111:534–44.

Alexopoulos CJ, Mims CW, Blackwell M. Introductory mycology. 4th ed. New York: Wiley; 1996. p. 868.

Anderson JB, Stasovski E. Molecular physiology of Northern Hemisphere species of Armillaria. Mycologia. 1992;84:505–16.

Anderson IC, Campbell CD, Prosser JI. Potential bias of fungal 18S rDNA and internal transcribed spacer polymerase chain reaction primers for estimating fungal biodiversity in soil. Appl Environ Microbiol. 2003;5:36–47.

Appel DJ, Gordon TR. Intraspecific variation within populations of *Fusarium oxysporum* based on RFLP analysis of the intergenic spacer region of the rDNA. Exp Mycol. 1995;19:120–8.

Appel DJ, Gordon TR. Relationships among pathogenic and nonpathogenic isolates of *Fusarium oxysporum* based on the partial sequence of the intergenic spacer region of the ribosomal DNA. Mol Plant Microbe Interact. 1996;9:125–38.

Assigbetse KB, Fernandez D, Dubois MP, Geiger JP. Differentiation of *Fusarium oxysporum* f. sp. *vasinfectum* races on cotton by random amplified polymorphic DNA (RAPD) analysis. Phytopathology. 1994;84:622–6.

Attanayake V, Mckay R, Joffres M, Singh S, Burkle F, Mills E. Prevalence of mental disorders among children exposed to war: a systematic review of 7,920 children. Med Confl Surviv. 2009;25:4–19.

Ayoub MJ. Application of multi locus sequence typing to the analysis of the biodiversity of indigenous Saccharomyces cerevisiae wine yeasts from Lebanon. J Appl Microbiol. 2006;100:699–711.

Baayen RP, O'Donnell K, Bonants PJM, Cigelnik E, Kroon LPNM, Roebroeck EJA, Waalwijk C. Gene genealogies and AFLP analysis in the *Fusarium oxysporum* complex identify monophyletic and non-monophyletic *formae specials* causing wilt and rot diseases. Phytopathology. 2000;90:891–900.

Bailey AM, Mitchell DJ, Manjunath KL, Nolasco G, Niblett CL. Identification to the species level of the plant pathogens Phytophthora and Pythium by using unique sequences of the ITS1 region of ribosomal DNA as capture probes for PCR Elisa. FEMS Microbiol Lett. 2002;207:153–8.

Bain J, Plater L, Elliott M, Shpiro N, Hastie CJ, McLauchlan H, Klevernic I, Arthur JS, Alessi DR, Cohen P. The selectivity of protein kinase inhibitors: a further update. Biochem J. 2007;408:297–315.

Baldridge GD, Dalton MW, Fallon AM. Is high-order structure conserved in eukaryotic ribosomal DNA intergenic spacers? J Mol Evol. 1992;35:514–23.

Baldwin BG. Molecular phylogenetics of Calcydenia (Compositae) based on ITS sequences of nuclear ribosomal DNA: chromosomal and morphological evolution reexamined. Am J Bot. 1993;80:222–38.

Baldwin BG, Sanderson MJ, Porter JM, Wojciechowski MF, Campbell JS, Donoghue MJ. The ITS region of nuclear ribosomal DNA: a valuable source of evidence on angiosperm phylogeny. Ann Mo Bot Gard. 1995;82:247–77.

Bao JR, Fravel DR, O'Neill NR, Lazarovits G, van Berkum P. Genetic analysis of pathogenic and non-pathogenic *Fusarium oxysporum* from tomato plants. Can J Bot. 2002;80:271–9.

Barve MP, Haware MP, Sainani MN, Ranjekar PK, Gupta VS. Potential of micros atellites to distinguish four races of *Fusarium oxysporum* f. sp.ciceri prevalent in India. Theor Appl Genet. 2001;102:138–47.

Bayraktar H, Dolar FS, Maden S. Use of RAPD and ISSR markers in detection of genetic variation and population structure among *Fusarium oxysporum* f.sp. *ciceris* isolates on chickpea in Turkey. J Phytopathol. 2008;156:146–54.

Belabid L, Baum M, Forts Z, Bouznad Z, Eujayl I. Pathogenic and genetic characterization of Algerian isolates of *Fusarium oxysporum* f. sp. *lentis* by RAPD and AFLP analyses. Afr J Biotechnol. 2004;3:25–31.

Bentley NJ, Eisen T, Goding CR. Melanocyte-specific expression of the human tyrosinase promoter: activation by the microphthalmia gene product and role of the initiator. Mol Cell Biol. 1994;14:7996–8006.

Berbegal M, Garzon CD, Ortega A, Armengol J, Jimenez-Dıaz RM, Jimenez-Gasco MM. Development and application of new molecular markers for analysis of genetic diversity in *Verticillium dahlia* populations. Plant Pathol. 2011;60:866–77.

Bissett J. Trichoderma atroviride. Can J Bot. 1992;70:639–41.

Bizzini A, Greub G. Matrix-assisted laser desorption ionization time-of-flight mass spectrometry, a revolution in clinical microbial identification. Clin Microbiol Infect. 2010;16:1614–9.

Blears MJ, De Grandis SA, Lee H, Trevors JT. Amplified fragment length polymorphism (AFLP): review of the procedure and its applications. J Ind Microbiol Biotechnol. 1998;21:99–114.

Bogale M, Wingfield BD, Wingfield MJ, Steenkamp ET. Simple sequence repeats markers for species in the Fusarium oxysporum complex. Mol Ecol Notes. 2005;5:622–4.

Bogale M, Wingfield BD, Wingfield MJ, Steenkamp ET. Characterization of Fusarium oxysporum isolates from Ethiopia using AFLP, SSR and DNA sequence analyses. Fungal Divers. 2006;23:51–66.

Bonde MR, Micales JA, Peterson GL. The use of isozyme analysis for the identification of plant pathogenic fungi. Plant Dis. 1993;77:961–8.

Bonde MR, Peterson GL, Dowler WM, May B. Isozyme analysis to differentiate species of Peronosclerospora causing downy mildew of maize. Phytopathology. 1984;74:1278–83.

Botstein D, White RL, Skolnick M, Davis RW. Construction of a genetic map in man using restriction fragment length polymorphisms. Am J Hum Genet. 1980;32:314–31.

Bragard C, Singer E, Alizadeh A, Vauterin L, Maraite H, Swings J. *Xanthomonas translucens* from small grains: diversity and phylopathological relevance. Phytopathology. 1997;87:1111–7.

Brockhurst MA, Colegrave N, Rozen DE. Next-generation sequencing as a tool to study microbial evolution. Mol Ecol. 2011;20:972–80.

Brumfield RT, Beerli P, Nickerson DA, Edwards SV. The utility of single nucleotide polymorphisms in inferences of population history. Trends Ecol Evol. 2003;18:249–56.

Bruns T, White TJ, Taylor JW. Fungal molecular systematic. Annu Rev Ecol Syst. 1991;22:525–64.

Buchko J, Klassen GR. Detection of length heterogeneity in the ribosomal DNA of *Pythium ultimum* by PCR amplification of the intergenic region. Curr Genet. 1990;18:203–5.

Chalmers KJ, Waugh R, Sprent JI. Detection of genetic variation between and within populations of *Gliricidia sepium* and *G. maculata* using RAPD makers. Heredity. 1992;69:465–72.

Chalupova J, Raus M, Sedlarova M, Sebela M. Identification of fungal microorganisms by MALDI-TOF mass spectrometry. Biotechnol Adv. 2014;32:230–41.

Chemidlin Prevost-Boure N, Christen R, Dequiedt S, Mougel C, Lelievre M, Jolivet C, Shahbazkia HR, Guillou L, Arrouays D, Ranjard L. Validation and application of a PCR primer set to quantify fungal communities in the soil environment by real-time quantitative PCR. PLoS One. 2011;6(9):e24166.

Chiocchetti A. PCR detection of *Fusarium oxysporum* f sp *basilici* on Basil. Plant Dis. 2001;85:607–11.

Chiocchetti A, Bernardo I, Daboussi MJ, Garibaldi A, Gullino ML, Langin T, Migheli Q. Detection of *Fusarium oxysporum* f. sp. *dianthi* in carnation tissue by PCR amplification of transposon insertions. Phytopathology. 1999;89:1169–75.

Choi J, Park SY, Kim BR, Roh JH, Oh IS, Han SS, Lee YH. Comparative analysis of pathogenicity and phylogenetic relationship in *Magnaporthe grisea* species complex. PLoS One. 2013;8:e57196.

Clewes E, Barbara DJ, Kennedy R. Progress in the development of tools for molecular epidemiology of downy mildew on vegetable in Brassica. In: Lebeda A, Spencer-Phillips PTN, editors. Advances in downy mildew research, Vol. 3. Olomouc; 2007. p. 113–8.

Cole JR, Wang Q, Chai B, Tiedje J. The ribosomal database project: sequences and software for high-throughput rRNA analysis. In: Frans J. de Bruijn, Editor. Handbook of molecular microbial ecology I: Metagenomics and Complementary Approaches: John Wiley & Sons, Inc.; 2011. p. 313–324. DOI: 10.1002/9781118010518.ch36.

Crease TJ. Ribosomal DNA evolution at the population level: nucleotide variation in intergenic spacer arrays of *Daphnia pulex*. Genetics. 1995;141:1327–37.

Datta S, Choudhry RG, Shamim M, Vishwa D. Polymorphism in the internal transcribed spacer (ITS) of the ribosomal DNA among different *Fusarium* species. Arch Phytopathol Plant Protect. 2011a;44:558–66.

Datta S, Choudhry RG, Shamim M, Singh RK, Vishawa D. Molecular diversity in Indian isolates of *Fusarium oxysporum* f. sp. *lentis* inciting wilt disease in lentil (*Lens culinaris* Medik). Afr J Biotechnol. 2011b;10:7314–23.

Delmotte F, Giresse X, Richard-Cervera S, M'Baya J, Vear F, Tourvieille J, Walser P, Tourvieille de Labrouhe D. Single nucleotide polymorphisms reveal multiple introductions into France of Plasmopara halstedii, the plant pathogen causing sunflower downy mildew. Infect Genet Evol. 2008;8:534–40.

DeScenzo RA, Engel SR, Gomez G, Jackson EL, Munkvold GP, Weller J, Irelan NA. Genetic analysis of Eu-ypa strains from California supports the presence of two pathogenic species. Phytopathology. 1999;89:884–93.

Dhar R, Sagesser R, Weikert C, Yuan J, Wagner A. Adaptation of *Saccharomyces cerevisiae* to saline stress through laboratory evolution. J Evol Biol. 2011;24:1135–53.

Diguta CF, Vincent B, Guilloux-Benatier M, Alexandre H, Rousseaux S. PCR ITS-RFLP: a useful method for identifying filamentous fungi isolates on grapes. Food Microbiol. 2011;28:1145–54.

Druzhinina IS, Kopchinskiya AG, Komoja M, Bissettb J, Szakacs G, Kubiceka CP. An oligonucleotide barcode for species identification in *Trichoderma* and *Hypocrea*. Fungal Genet Biol. 2005;42:813–28.

Dubouzet JG, Shinoda K. Relationships among old and new world Alliums according to ITS DNA sequence analysis. Theor Appl Genet. 1999;98:422–33.

Egger KN. Molecular analysis of ectomycorrhizal fungal communities. Can J Bot. 1995;73:1415–22.

Einsele H, Hebart H, Roller G, Loffler J, Rothenhofer I, Muller CA, Bowden RA, van Burik J, Engelhard D, Kanz L. Detection and identification of fungal pathogens in blood by using molecular probes. J Clin Microbiol. 1997;35:1353–60.

Eujayl I, Sorrells M, Baum M, Woltersand P, Powell W. Assessment of genotypic variation among cultivated durum wheat based on EST-SSRs and genomic SSRs. Euphytica. 2001;119:39–43.

Fabre JV, Julien J, Parisot D, Dron M. Analysis of diverse isolates of *Colletotrichum lindemuthianum* infecting common bean using molecular markers. Mycol Res. 1995;99:429–35.

Fenselau C, Demirev PA. Characterization of intact microorganisms by MALDI mass spectrometry. Mass Spectrom Rev. 2001;20:157–71.

Ferro M, Antonio EA, Souza W, Jr MB. ITScan: a web-based analysis tool for Internal Transcribed Spacer (ITS) sequences. BMC Res Notes. 2014;7:857.

Forbes GA, Goodwin SB, Drenth A, Oyarzun P, Ordoñez ME, Fry WE. A global marker database for *Phytophthora infestans*. Plant Dis. 1998;82:811–8.

Fundyga RE, Kuykendall RJ, Lee-Yang W, Lott TJ. Evidence for aneuploidy and recombination in the human commensal yeast Candida parapsilosis. Infect Genet Evol. 2004;4:37–43.

Ganeshamoorthi P, Dubey SC. Phylogeny analysis of Indian strains of *Rhizoctonia solani* isolated from chickpea and development of sequence characterized amplified region (SCAR) marker for detection of the pathogen. Afr J Microbiol Res. 2013;7:5516–25.

Garrido C, Carbu M, Fernandez-Acero FJ, Boonham N, Colyer A, Cantoral JM, Budge G. Development of protocols for detection of *Colletotrichum acutatum* and monitoring of strawberry anthracnose using real-time PCR. Plant Pathol. 2009;58:43–51.

Geiser DM, Jimenez-Gasco M, Kang S, Makalowska I, Veeraraghavan N, Ward TJ, Zhang N, Kuldau GA, O'Donnell K. Fusarium-ID v. 1.0: a DNA sequence database for identifying Fusarium. Eur J Plant Pathol. 2004;110:473–9.

Glen M, Tommerup IC, Bougher NL, O'brien PA. Interspecific and intraspecific variation of ectomycorrhizal fungi associated with *Eucalyptus* ecosystems as revealed by ribosomal DNA PCR–RFLP. Mycol Res. 2001;105:843–58.

Gomes EA, Barros EG, Kasuya MCM, Araujo EF. Molecular characterization of *Pisolithus* spp. isolates by rDNA PCR-RFLP. Mycorrhiza. 1999;8:197–202.

Gonzalez JM, Masuchi Y, Robb FT, Ammerman JW, Maeder DL, Yanagibayashi M, Tamaoka J, Kato C. *Pyrococcus horikoshii* sp. *nov*. a hyperthermophilic archaeon isolated from a hydrothermal vent at the Okinawa Trough. Extremophiles. 1998;2:123–30.

Goodwin SB, Drenth A. Origin of the A2 mating type of *Phytophthora infestans* outside Mexico. Phytopathology. 1997;87:992–9.

Goodwin PH, English JT, Neher DA, Diniway JM. Detection of *Phytophthora parasitica* from soil and host tissue with a species-specific DNA probe. Phytopathology. 1990;80:277–81.

Goodwin SB, Schneider RE, Fry WE. Use of cellulose-acetate electrophoresis for rapid identification of allozyme genotypes of *Phytophthora infestans*. Plant Dis. 1995;79:1181–5.

Goodwin SB, Spielman LJ, Matuszak JM, Bergeron SN, Fry WE. Clonal diversity and genetic differentiation of *Phytophthora infestans* populations in northern and central Mexico. Phytopathology. 1992;82:955–61.

Gothwal RK, Nigam VK, Mohan MK, Sasmal D, Ghosh P. Extraction of bulk DNA from Thar Desert soils for optimization of PCRDGGE based microbial community analysis. Electron J Biotechnol. 2007;10(3):400–8.

Grajal-Martin MJ, Simon CJ, Muehleauer FJ. Use of random amplified polymorphic DNA (RAPD) to characterize race 2 of *Fusarium oxysporum* f. sp. *pisi*. Phytopathology. 1993;83:612–4.

Groenewald S, Berga NVD, Marasasb WFO, Viljoen A. The application of high-throughput AFLP's in assessing genetic diversity in *Fusarium oxysporum* f. sp. *cubense*. Mycol Res. 2006;110:297–305.

Grünwald NJ, Martin FN, Larsen MM, Sullivan CM, Press CM, Coffey MD, Hansen EM, Parke JL. *Phytophthora*-ID.org: a sequence-based *Phytophthora* identification tool. Plant Dis. 2011; 95:337–42.

Gryta H, Debaud JC, Effosse A, Gay G, Marmeisse R. Fine scale structure of populations of the ectomycorrhizal fungus Hebeloma cylindros Porvn, in coastal sand dune forest ecosystems. Mol Ecol. 1997;6:353–64.

Gunnell PS, Gubler WD. Taxonomy and morphology of *Colletotrichum* species pathogenic on strawberry. Mycologia. 1992;84:157–65.

Guo LD, Xu L, Zheng WH, Hyde KD. Genetic variation of *Alternaria alternata,* an endophytic fungus isolated from *Pinus tabulaeformis* as determined by random amplified microsatellites (RAMS). Fungal Divers. 2004;16:53–65.

Gurjar G, Barve M, Giri A, Gupta V. Identification of Indian pathogenic races of *Fusarium oxysporum* f.sp. *ciceris* with gene specific, ITS and random markers. Mycologia. 2009;101:480–91.

Hamer J, Farrel L, Orbach M, Valent A, Chumley F. Host species specific conservation of a family of repeated DNA sequences in the genome of a fungal pathogen. Proc Natl Acad Sci U S A. 1989;86:9981–5.

Haugland RA, Varma M, Wymer LJ, Vesper SJ. Quantitative PCR analysis of selected Aspergillus, Penicillium and Paecilomyces species. Syst Appl Microbiol. 2004;27:198–210.

Henry PC, Smith C, Charlton B, Boulianne G, Rotin D. Characterization of a *Drosophila melanogaster* homologue of Nedd4. Mol Biol Cell. 2000;11:222–3.

Hibbett DS, Binder M, Bischoff JF, Blackwell M, Cannon PF, Eriksson OE, Huhndorf S, James T, Kirk PM, Lücking R, Thorsten Lumbsch H, Lutzoni F, Matheny PB, McLaughlin DJ, Powell MJ, Redhead S, Schoch CL, Spatafora JW, Stalpers JA, Vilgalys R, Aime MC, Aptroot A, Bauer R, Begerow D, Benny GL, Castlebury LA, Crous PW, Dai YC, Gams W, Geiser DM, Griffith GW, Gueidan C, Hawksworth DL, Hestmark G, Hosaka K, Humber RA, Hyde KD, Ironside JE, Kõljalg U, Kurtzman CP, Larsson KH, Lichtwardt R, Longcore J, Miadlikowska J, Miller A, Moncalvo JM, Mozley-Standridge S, Oberwinkler F, Parmasto E, Reeb V, Rogers JD, Roux C, Ryvarden L, Sampaio JP, Schüssler A, Sugiyama J, Thorn RG, Tibell L, Untereiner WA, Walker C, Wang Z, Weir A, Weiss M, White MM, Winka K, Yao YJ, Zhang N. A higher-level phylogenetic classification of the fungi. Mycol Res. 2007;111:509–47.

Hills DM, Dixon MT. Ribosomal DNA: molecular evolution and phylogenetic interference. Q Rev Biol. 1991;66:411–53.

Hittinger CT, Johnston M, Tossberg JT, Rokas A. Leveraging skewed transcript abundance by RNA-seq to increase the genomic depth of the tree of life. Proc Natl Acad Sci U S A. 2010;107:1476–81.

Holland RD, Wilkes JG, Rafii F, Sutherland JB, Persons CC, Voorhees KJ, Ley JO. Rapid identification of intact whole bacteria based on spectral patterns using matrix-assisted laser desorption/ionization with time-of-flight mass spectrometry. Rapid Commun Mass Spectrom. 1996;10:1227–32.

Hyde KD, Soytong K. The fungal endophyte dilemma. Fungal Divers. 2008;33:163–73.

Inami K, Yoshioka C, Hirano Y, Kawabe M, Tsushima S, Teraoka T, Arie T. Real-time PCR for differential determination of the tomato wilt fungus, *Fusarium oxysporum f. sp lycopersici*, and its races. J Gen Plant Pathol. 2010;76:116–21.

Iracabal B, Labarree J. Restriction site and length polymorphism of the rDNA unit in the cultivated basidiomycete Pleurotv, s cornucopiae. Theor Appl Genet. 1994;88:824–30.

James TY, Kauff F, Schoch CL, Matheny PB, Hofstetter V, Cox CJ, Celio G, Gueidan C, Fraker E, Miadlikowska J, Lumbsch HT, Rauhut A, Reeb V, Arnold AE, Amtoft A, Stajich JE, Hosaka K, Sung GH, Johnson D, O'Rourke B, Crockett M, Binder M, Curtis JM, Slot JC, Wang Z, Wilson AW, Schüssler A, Longcore JE, O'Donnell K, Mozley-Standridge S, Porter D, Letcher PM, Powell MJ, Taylor JW, White MM, Griffith GW, Davies DR, Humber RA, Morton JB, Sugiyama J, Rossman AY, Rogers JD, Pfister DH, Hewitt D, Hansen K, Hambleton S, Shoemaker RA, Kohlmeyer J, Volkmann-Kohlmeyer B, Spotts RA, Serdani M, Crous PW, Hughes KW, Matsuura K, Langer E, Langer G, Untereiner WA, Lücking R, Büdel B, Geiser DM, Aptroot A, Diederich P, Schmitt I, Schultz M, Yahr R, Hibbett DS, Lutzoni F, McLaughlin DJ, Spatafora JW, Vilgalys R. Reconstructing the early evolution of fungi using a six-gene phylogeny. Nature. 2006;19:818–22.

Jarman KH, Daly DS, Petersen CE, Saenz AJ, Valentine NB, Wahl KL. Extracting and visualizing matrix-assisted laser desorption/ionization time-of-flight mass spectral fingerprints. Rapid Commun Mass Spectrom. 1999;13:1586–94.

Jimenez-Gasco M, Jimenez-Diaz RM. Development of a specific polymerase chain reaction-based assay for the identification of *Fusarium oxysporum* f. sp. *ciceris* and its pathogenic races 0, 1A, 5, and 6. Phytopathology. 2003;93:200–9.

Jones CJ, Edwards KJ, Castaglione S, Winfield MO, Sala F, van de Wiel C, Bredemeijer G, Vosman B, Matthes M, Daly A, Brettschneider R, Bettini P, Buiatti M, Maestri E, Malcevschi A, Marmiroli N, Aert R, Volckaert G, Rudea J, Linacero R, Vazquez A, Karp A. Reproducibility testing of RAPD, AFLP and SSR markers in plants by a network of European laboratories. Mol Breed. 1997;3:381–90.

Karp A, Kresovich S, Bhat KV, Ayad WG, Hodgkin T. Molecular tools in plant genetic resources conservation: a guide to the technologies. In: IPGRI Technical Bulletin No. 2. Rome: International Plant genetic Resources Institute; 1997.

Kasuga T, White TJ, Koenig G, McEwen J, Restrepo A, Castaneda E, Da Silva LC, Heins-Vaccari EM, De Freitas RS, Zancope-Oliveira RM, Qin Z, Negroni R, Carter DA, Mikami Y, Tamura M, Taylor ML, Miller GF, Poonwan N, Taylor JW. Phylogeography of the fungal pathogen Histoplasma capsulatum. Mol Ecol. 2003;12:3383–401.

Kelly DE, Rose ME, Kelly SL. Investigation of the role of sterol delta 8→7-isomerase in the sensitivity of Saccharomyces cerevisiae to fenpropimorph. FEMS Microbiol Lett. 1994;122: 223–26.

Kelly MG, Cazaubon A, Coring E. Recommendations for the routine sampling of diatoms for water quality assessments in Europe. J Appl Phycol. 1998;10:215–22.

Kiprop EK, Baudoin JP, Mwangombe AW, Kimani PM, Mergeai G, Maquet A. Characterization of Kenyan isolates of *Fusarium udum* from pigeonpea [*Cajanus cajan* (L.) Millsp.] by cultural characteristics, aggressiveness and AFLP Analysis. J Phytopathol. 2002;150:517–25.

Klaassen CH. MLST versus microsatellites for typing *Aspergillus fumigatus* isolates. Med Mycol. 2009;47:S27–33.

Klich MA, Mullaney EJ. DNA restriction enzyme fragment polymorphism as a tool for rapid differentiation of *Aspergillus flavus* from *Aspergillus oryzae*. Exp Mycol. 1987;11:170–5.

Koljalg U, Larsson KH, Abarenkov K, Nilsson RH, Alexander IJ, Eberhardt U, Erland S, Hoiland K, Kjoller R, Larsson E, Pennanen T, Sen R, Taylor AFS, Tedersoo L, Vralstad T, Ursing BM. UNITE: a database providing web-based methods for the molecular identification of ectomycorrhizal fungi. New Phytol. 2005;166:1063–8.

Kroon LPNM, Verstappen ECP, Kox LFF, Bonants PJM. A rapid diagnostic test to distinguish between American and European populations of Phytophthora ramorum. Phytopathology. 2004;94:613–20.

Kumar P, Gupta VK, Misra AK, Modi DR, Pandey BK. Potential of molecular markers in plant biotechnology. Plant Omics J. 2009;2:141–62.

Laday M, Szecsi A. Distinct electrophoretic isoenzyme profiles of *Fusarium graminearum* and closely related species. Syst Appl Microbiol. 2001;24:67–75.

Lan L, Xu J. Multiple gene genealogical analyses suggest divergence and recent clonal dispersal in the opportunistic human pathogen *Candida guilliermondii*. Microbiology. 2006;152:1539–49.

Landgraf A, Reckmann B, Pingoud A. Direct analysis of polymerase chainreaction products using enzyme-linked-immunosorbent-assay techniques. Anal Biochem. 1991;198:86–91.

Landlinger C, Preuner S, Baskova L, van Grotel M, Hartwig NG, Dworzak M, Mann G, Attarbaschi A, Kager L, Peters C. Diagnosis of invasive fungal infections by a real-time panfungal PCR assay in immunocompromised pediatric patients. Leukemia. 2010;24:2032–8.

Leach JE, White FE. Molecular probes for disease diagnosis and monitoring. In: Khush GS, Toenniessen GH, editors. Rice biotechnology. Philippines: CAB International UK and International Rice Resarch Institute; 1991. p. 281–307.

Leickfeldt E, Seifert KA. An evaluation of use of the use of its sequence in the taxonomy of hypocreales. Stud Mycol. 2000;45:35–44.

Leslie JF, Zeller KA, Logrieco A, Mule G, Moretti A, Ritieni A. Species diversity of and toxin production by *Gibberella fujikuroi* species complex strains isolated from native prairie grasses in Kansas. Appl Environ Microbiol. 2004;70:2254–62.

Leslie JF, Zeller KA, Lamprecht SC, Rheeder JP, Marasas WFO. Toxicity, pathogenicity, and genetic differentiation of five species of *Fusarium* from sorghum and millet. Phytopathology. 2005;95:275–83.

Levinson G, Gutman GA. Slipped-strand mispairing: a major mechanism for DNA sequence evolution. Mol Biol Evol. 1987;4:203–21.

Leyva-Madrigal KY, Larralde-Corona CP, Calderon-Vazquez CL, Maldonado-Mendoza IC. Genome distribution and validation of novel microsatellite markers of *Fusarium verticillioides* and their transferability to other Fusarium species. J Microbiol Methods. 2014;101:18–23.

Li J, Fan SR, Liu XP, Li DM, Nie ZH, Lin H, Huang WM, Zong LL, Jin JG, Lei H, Bai F. Biased genotype distributions of *Candida albicans*strains associated with vulvovaginal candidosis and candidal balanoposthitis in China. Clin Infect Dis. 2008;47:1119–25.

Li TY, Liu BH, Chen YC. Characterization of *Aspergillus* spores by matrix-assisted laser desorption/ionization time-of-flight mass spectrometry. Rapid Commun Mass Spectrom. 2000;14:2393–400.

Lievens B, Claes L, Vakalounakis DJ, Vanachter ACRC, Thomma BPHJ. A robust identification and detection assay to discriminate the cucumber pathogens *Fusarium oxysporum* f.sp. *cucumerinumand* f.sp. radicis-cucumerinum. Environ Microbiol. 2007;9:2145–61.

Lim YW, Kim JJ, Lu M, Breuil C. Determining fungal diversity on *Dendroctonus ponderosae* and *Ips pini* affecting lodgepole pine using cultural and molecular methods. Fungal Divers. 2005;19:79–94.

Litt M, Luty JA. A hypervariable microsatellite revealed by in vitro amplification of a dinucleotide repeat within the cardiac muscle actin gene. Am J Hum Genet. 1989;44:397–401.

Liu CM, Kachur S, Dwan MG, Abraham AG, Aziz M, Hsueh PR, Huang YT, Busch JD, Lamit LJ, Gehring CA, Keim P, Price LB. FungiQuant: a broad-coverage fungal quantitative real-time PCR assay. BMC Microbiol. 2012a;12:255.

Liu JY, Ying ZH, Liu F, Liu XR, Xie BG. Evaluation of the use of SCAR markers for screening genetic diversity of *Lentinula edodes* strains. Curr Microbiol. 2012b;64:317–25.

Lord NS, Kaplan CW, Shank P, Kitts CL, Elrod SL. Assessment of fungal diversity using terminal restriction fragment pattern analysis: comparison of 18S and ITS ribosomal regions. FEMS Microbiol Ecol. 2002;42:327–37.

MacLean DJ, Braithwaite KS, Manners JM, Irwing JAG. How do we identify and classify fungal pathogens in the era of DNA analysis? Adv Plant Pathol. 1993;10:207–44.

Mahuku GS, Hsiang T, Yang L. Genetic diversity of *Microdochium nivale* isolates from turfgrass. Mycol Res. 1998;102:559–67.

Mahuku GS, Jara C, Cuasquer JB, Castellanos G. Genetic variability within *Phaeoisariopsis griseola* from Central America and its implications for resistance breeding of common bean. Plant Pathol. 2002;51:594–604.

Majeed AA, Ahmed N, Rao KR, Ghousunnissa S, Kauser F, Bose B, Nagarajaram HA, Katoch VM, Cousins DV, Sechi LA, Golman RH, Hasnain SE. AmpliBASE MT: a Mycobacterium tuberculosis diversity knowledgebase. Bioinformatics. 2004;20:989–92.

Majer D, Mithen R, Lewis B, Vos P, Oliver RP. The use of AFLP fingerprinting for the detection of genetic variation in fungi. Mycol Res. 1996;100:1107–11.

Martin FN. Mitochondrial haplotype determination in the oomycete plant pathogen Phytophthora ramorum. Curr Genet. 2008;54:23–34.

Martin GB, Williams JGK, Tanksley SD. Rapid identification of markers linked to a Pseudomonas resistance gene in tomato by using random primers and near-isogenic lines. Proc Natl Acad Sci U S A. 1991;88:2336–40.

McLaughlin VV, Archer SL, Badesch DB, Barst RJ, Farber HW, Lindner JR, Mathier MA, McGoon MD, Park MH, Rosenson RS, Rubin LJ, Tapson VF, Varga J. ACCF/AHA 2009 expert consensus document on pulmonary hypertension a report of the American College of Cardiology Foundation Task Force on Expert Consensus Documents and the American Heart Association developed in collaboration with the American College of Chest Physicians; American Thoracic Society, Inc.; and the Pulmonary Hypertension Association. J Am Coll Cardiol. 2009;28:1573–619.

Meklin T, Haugland RA, Reponen T, Varma M, Lummus Z, Bernstein D, Wymer LJ, Vesper SJ. Quantitative PCR analysis of house dust can reveal abnormal mold conditions. J Environ Monit. 2004;6:615–20.

Meyer KE, Estrin S, Bhaumik SK, Peng MW. Institutions, resources and entry strategies in emerging economies. Strateg Manag J. 2009;30:61–80.

Michales JA, Bonde MR, Peterson GL. Isozyme analysis and aminopeptidase assays of the genus Peronosclerospora. Phytopathology. 1988;78:1402–8.

Michelmore RW, Paran I, Kesseli RV. Identification of markers linked to disease resistance genes by bulked segregant analysis: a rapid method to detect markers in specific genomic regions using segregating populations. Proc Natl Acad Sci U S A. 1991;88:9828–32.

Miller MR, Hankinson J, Brusasco V, Burgos F, Casaburi R, Coates A, Crapo R, Enright P, van der Grinten CPM, Gustafsson P, Jensen R, Johnson DC, MacIntyre N, McKay R, Navajas D, Pedersen OF, Pellegrino R, Viegi G, Wanger J, ATS/ERS Task Force. Standardisation of spirometry. Eur Respir J. 2005;26(2):319–38.

Molina FI, Jong SC, Huffman JL. PCR amplification of the 3′ external transcribed and intergenic spacers of the ribosomal DNA repeat unit in three species of Saccharomyces. FEMS Microbiol Lett. 1993;108:259–64.

Morin PA, Luikart G, Wayne RK, SNP-Workshop_Group. SNPs in ecology, evolution and conservation. Trends Ecolo Evol. 2004;19:208–16.

Morris PF, Connolly MS, St Clair DA. Genetic diversity of *Alternaria alternata* isolated from tomato in California assessed using RAPDs. Mycol Res. 2000;104:286–92.

Muyzer G. DGGE/TGGE a method for identifying genes from natural ecosystems. Curr Opin Microbiol. 1999;2:317–22.

Myers RM, Fischer SG, Lerman LS, Maniatis T. Nearly all single base substitutions in DNA fragments joined to a GC- clamp can be detected by denaturing gradient gel electrophoresis. Nucleic Acids Res. 1985;13:3131–45.

Neuhof T, Dieckmann R, Druzhinina IS, Kubicek CP, Von DH. Intact-cell MALDI-TOF mass spectrometry analysis of peptaibol formation by the genus Trichoderma/Hypocrea: can molecular phylogeny of species predict peptaibol structures? Microbiology. 2007a;153:3417–37.

Neuhof T, Dieckmann R, Druzhinina IS, Kubicek CP, Nakari-Setala T, Penttila M, von Dohren H. Direct identification of hydrophobins and their processing in Trichoderma using intact-cell MALDI-TOF MS. FEBS J. 2007b;274:841–52.

Neuhof T, Dieckmann R, Druzhinina IS, Kubicek CP, von Döhren H. Intact-cell MALDI-TOF mass spectrometry analysis of peptaibol formation by the genus Trichoderma/Hypocrea: can molecular phylogeny of species predict peptaibol structures? Microbiology. 2007c;153: 3417–37.

Njambere EN, Chen W, Frate C, Wu BM, Temple S, Muehlbauer F. Stem and crown rot of chickpea in California caused by Sclerotinia trifoliorum. Plant Dis. 2008;92:917–22.

Omirou M, Rousidou C, Bekris F, Papadopoulou KK, Menkissoglou-Spiroudi U, Ehaliotis C. The impact of biofumigation and chemical fumigation methods on the structure and function of the soil microbial community. Microb Ecol. 2011;61:201–13.

Park B, Park J, Cheon K-C, Choi J, Jung K, Kim D, Lee YH, Ward TJ, O'Donnell K, Geiser DM, Kang S. Cyber infrastructure for Fusarium: three integrated platforms supporting strain identification, phylogenetics, comparative genomics and knowledge sharing. Nucleic Acids Res. 2011;39:640–6.

Park J, Park B, Veeraraghavan N, Blair JE, Geiser DM, Isard S, Mansfield MA, Nikolaeva E, Park SY, Russo J, Kim SH, Greene M, Ivors KL, Balci Y, Peiman M, Erwin DC, Coffey MD, Jung K, Lee YH, Rossman A, Farr D, Cline E, Grünwald NJ, Luster DG, Schrandt J, Martin F, Ribeiro OK, Makalowska I, Kang S. *Phytophthora* database: forensic database supporting the identification and monitoring of *Phytophthora*. Plant Dis. 2008;92:966–72.

Parry DW, Rezanoor HN, Pettitt TR, Hare MC, Nicholson P. Analysis of *Microdochium nivale* isolates from wheat in the UK during 1993. Ann Appl Biol. 1995;126:449–55.

Pastor-Corrales MA, Jara CE. La evolucion de *Phaeoisariopsis griseola* con el frijol comun en America Latina. Fitopatol Colomb. 1995;19:15–22.

Peever TL, Ibanez A, Akimitsu K, Timmer LW. Worldwide phylogeography of the citrus brown spot pathogen, *Alternaria alternata*. Phytopathology. 2002;92:794–802.

Portillo MC, Villahermosa D, Corzo A, Gonzalez JM. Microbial community fingerprinting by differential display-denaturing gradient gel electrophoresis. Appl Environ Microbiol. 2011;77:351–4.

Poupard P, Frei U, Cavelier N, Lind V. Genetic diversity in W- and R-type populations of *Pseudocercosporella herpotrichoides* based on DNA restriction fragment length polymorphisms. J Phytopathol. 1995;143:99–104.

Powell W, Machray GC, Provan J. Polymorphism revealed by simple sequence repeats. Trends Plant Sci. 1996;1:215–22.

Purwantara A, Barrins JM, Cozijnsen AJ, Ades PK, Howlett BJ. Genetic diversity of isolates of the *Leptosphaeria maculans* species complex from Australia, Europe and North America using amplified fragment length polymorphism analysis. Mycol Res. 2000;104:772–81.

Rafalski J, Morgante M, Powell J, Tinggey S. Generating and using DNA markers in plants. In: Birren B, Lai E, editors. Analysis of non-mammalian genomes: a practical guide. Boca Raton: Academic; 1996. p. 75–134.

Rallo P, Dorado G, Martin A. Development of simple sequence repeats (SSRs) in olive tree (*Olea europaea* L.). Theor Appl Genet. 2000;101:984–9.

Ratnasingham S, Hebert PD. BOLD: the barcode of life data system (http://www.barcodinglife.org). Mol Ecol Notes. 2007;7:355–64.

Reddy MP, Sarla N, Siddiq EA. Inter simple sequence repeat (ISSR) polymorphism and its application in plant breeding. Euphytica. 2002;128:9–17.

Robideau GP, de Cock AW, Coffey MD, Voglmayr H, Brouwer H, Bala K, Chitty DW, Desaulniers N, Eggertson QA, Gachon CM, Hu CH, Kupper FC, Rintoul TL, Sarhan E, Verstappen EC, Zhang Y, Bonants PJ, Ristaino JB, Levesque CA. DNA barcoding of oomycetes with cytochrome c oxidase subunit I and internal transcribed spacer. Mol Ecol Resour. 2011;11:1002–11.

Rokas A, Abbot P. Harnessing genomics for evolutionary insights. Trends Ecol Evol. 2009;24:192–200.

Russell J, Booth A, Fuller J, Harrower B, Hedley P, Machray G, Powell W. A comparison of sequence-based polymorphism and haplotype content in transcribed and anonymous regions of the barley. Genome. 2004;47:389–98.

Rytkonen A, Lilja A, Hantula J. PCR–DGGE method for *in planta* detection and identification of *Phytophthora* species. For Pathol. 2012;42:22–7.

Sanchez-Ballesteros J, Gonzalez V, Salazar O, Acero J, Portal MA, Julian M, Rubio V, Bills GF, Polishook J, Platas G, Mochales S, Pelaez F. Phylogenetic study of Hypoxylon and related genera based on ribosomal ITS sequences. Mycologia. 2000;92:964–77.

Sartorato A. Pathogenic variability and genetic diversity of *Phaeoisariopsis griseola* isolates from two counties in the State of Goias. Braz J Phytopathol. 2004;152:385–90.

Schena L, Nigro F, Ippolito A. Real-time PCR detection and quantification of soilborne fungal pathogens: the case of *Rosellinia necatrix*, *Phytophthora nicotianae*, *P. citrophthora*, and *Verticillium dahlia*. Phytopathol Mediterr. 2004;43:273–80.

Schierwater B, Ender A. Different thermostable DNA polymerases may apply to different RAPD products. Nucleic Acids Res. 1993;21:4647–8.

Schlotterer C. The evolution of molecular markers – just a matter of fashion? Nat Rev Genet. 2004;5:63–9.

Schneider B, Gibb KS, Seemüller E. Sequence and RFLP analysis of the elongation factor Tu gene used in differentiation and classification of phytoplasmas. Microbiology. 1997;143:3381–9.

Shaw CR. How many gene evolve? Biochem Genet. 1970;4:275.

Shendure J, Ji H. Next-generation DNA sequencing. Nat Biotechnol. 2008;26:1135–45.

Shimazu T, Tsubono Y, Kuriyama S, Ohmori K, Koizumi Y, Nishino Y, Shibuya D, Tsuji I. Coffee consumption and the risk of primary liver cancer: pooled analysis of two prospective studies in Japan. Int J Cancer. 2005;116:150–4.

Siddiquee S, Soon GT, Umi KY. Isozyme analysis and relationships among three species in Malaysian *Trichoderma* isolates. J Microbiol Biotechnol. 2010;20:1266–75.

Silva AS, Oliveira EJ, Haddad F, de Jesus ON, de Oliveira SAS, de Carvalho CMAP. Molecular fingerprinting of *Fusarium oxysporum* f. sp. *passiflorae* isolates using AFLP markers. Sci Agric. 2013;70:108–15.

Sivaramakrishnan S, Kannan S, Singh SD. Genetic variability of Fusarium wilt pathogen isolates of chickpea (*Cicer arietinum* L.) assessed by molecular markers. Mycopathologia. 2001;155:171–8.

Sivaramakrishnan S, Kannan SS, Singh SD. Genetic variability of *Fusarium* wilt pathogen isolates of chickpea (*Cicer arietinum* L.) assessed by molecular markers. Mycopathologia. 2002; 155(3):171–8.

Smith BJ, Black LL. Morphological, cultural and pathogenic variation among *Colletotrichum* species isolated from strawberry. Plant Dis. 1990;74:69–76.

Smith BJ, Sivasithamparam K. Isozymes of *Ganoderma* species from Australia. Mycol Res. 2000;104:952–61.

Somai BM, Keinath AP, Dean RA. Development of PCR-ELISA detection and differentiation of *Didymella bryoniae* from related Phoma species. Plant Dis. 2002;7:710–6.

Southern EM. Detection of specific sequences among DNA fragments separated by gel electrophoresis. J Mol Biol. 1975;98:503–17.

Stasz TE, Nixon K, Harman GE, Weeden NF, Kuter A. Evaluation of phonetic species and phylogenetic relationships in the genus *Trichoderma* by cladistic analysis of isozyme polymorphism. Mycologia. 1989;81:391–403.

Stenglein SA, Balatti PA. Genetic diversity of *Phaeoisariopsis griseola* in Argentina as revealed by pathogenic and molecular markers. Physiol Mol Plant Pathol. 2006;68:158–67.

Stewart LE, Kim M, James RL, Dumroese RK, Klopfenstein NB. Molecular characterization of *Fusarium oxysporum* and *Fusarium commune* isolates from a conifer nursery. Phythopathology. 2006;96:1124–33.

Szekeres A, Laday M, Kredics L, Varga J, Antal Z, Hatvani L, Manczinger L, Vagvolgyi C, Nagy E. Rapid identification of clinical *Trichoderma longibrachiatum* isolates by cellulose-acetate electrophoresis-mediated isoenzyme analysis. Eur Soc Clin Microbiol Infect Dis. 2006;12: 369–75.

Tanabe K, Tsuge T, Nishimura S. Potential application of DNA restriction fragment length polymorphisms to the ecological studies of *Alternaria alternate* Japanese pear pathotype. Ann Phytopathol Soc Jpn. 1990;55:361–5.

Taylor JW, Fisher MC. Fungal multilocus sequence typing — it's not just for bacteria. Curr Opin Microbiol. 2003;6:351–6.

Tejesvi MV, Tamhankar SA, Kini KR, Rao VS, Prakash HS. Phylogenetic analysis of endophytic *Pestalotiopsis* species from ethnopharmaceutically important medicinal trees. Fungal Divers. 2009;38:167–83.

Timonen S, Tammi H, Sen R. Characterization of the host genotype and fungal diversity in Scots pine ectomycorrhiza from natural humus microcosms using isozyme and PCR-RELP analyses. New Phytol. 1997;135:313–23.

Tymon AM, Pell JK. ISSR, ERIC and RAPD techniques to detect genetic diversity in the aphid pathogen *Pandora neoaphidis*. Mycol Res. 2005;109:285–93.

Validov SZ, Kamilova FD, Lugtenberg BJJ. Monitoring of pathogenic and nonpathogenic *Fusarium oxysporum* strains during tomato plant infection. J Microbial Biotechnol. 2011;4:82–8.

van der Nest MA, Steenkamp ET, Wingfield BD, Wingfield MJ. Development of simple sequence repeat (SSR) markers in Eucalyptus from amplified inter- simple sequence repeats (ISSR). Plant Breed. 2000;119:433–6.

van der Schoot J, Pospiskova M, Vosman B. Development and characterization of microsatellite markers in black poplar (*Populus nigra* L.). Theor Appl Genet. 2000;101:317–22.

Vishwa D, Datta S, Chaudhary RG, Upadhyay JP, Saifulla M, Mishra S, Prajapti RK, Shamim M. Pathogenic and molecular characterisations of pigeonpea wilt pathogen *Fusarium udum*. Arch Phytopathol Plant Protect. 2011;44:558–66.

Vos P, Kuiper M. AFLP analysis. In: Caetano-Arnolles G, Gresshoff PH, editors. DNA markers, protocols, applications and overviews. New York: Wiley; 1998. p. 115–31.

Vos P, Hogers R, Bleeker M, Reijans M, Van der Lee T, Hornes M, Frijters A, Pot J, Peleman J, Kuiper M, Zabeau M. AFLP: a new technique for DNA fingerprinting. Nucleic Acids Res. 1995;23:4407–14.

Vosman B, Arens P. Molecular characterization of GATA/GACA microsatellite repeats in tomato. Genome. 1997;40:25–33.

Wakelin SA, Macdonald LM, Rogers SL, Gregg AL, Bolger TP, Baldock JA. Habitat selective factors influencing the structural composition and functional capacity of microbial communities in agricultural soils. Soil Biol Biochem. 2008;40:803–13.

Walker SL, Leath S, Hagler WM, Murphy JP. Variation among isolates of *Fusarium graminearum* associated with Fusarium head blight in North Carolina. Plant Dis. 2001;85:404–10.

Wang XF, Auler AS, Edwards RL, Cheng H, Ito E, Solheid M. Interhemispheric antiphaing of rainfall during last glacial period. Quat Sci Rev. 2006;25:3391–403.

Welham KJ, Domin MA, Johnson K, Jones L, Ashton DS. Characterization of fungal spores by laser desorption/ionization time of flight mass spectrometry. Rapid Commun Mass Spectrom. 2000;14:307–14.

Welham KJ, Domin MA, Scanell DE, Cohen E, Ashton DS. The characterization of microorganisms by matrix-assisted laser desorption/ionization time-of-flight mass spectrometry. Rapid Commun Mass Spectrom. 1998;12:176–80.

White TJ, Bruns T, Lee S, Taylor J. Amplification and direct sequencing of fungal ribosomal RNA genes for phylogenetics. In: Innis MA, Gelfand DH, Snisky JJ, White JJ, editors. PCR protocols: a guide to, methods and applications. San Diego: Academic; 1990. p. 315–22.

Whitehead DS, Coddington A, Lewis BG. Classification of races by DNA polymorphism analysis and vegetative compatibility grouping in *Fusarium oxysporum* f. sp. *pisi*. Physiol Mol Plant Pathol. 1992;41:295–305.

Williams JGK, Kubelik AR, Livak KJ, Rafalski JA, Tingey SV. DNA polymorphisms amplified by arbitrary primers are useful as genetic markers. Nucleic Acids Res. 1990;18:6231–5.

Wittwer CT, Herrmann MG, Moss AA, Rasmussen RP. Continuous fluorescence monitoring of rapid cycle DNA amplification. Biotechniques. 1997;22:130–8.

Wolfe AD. ISSR techniques for evolutionary biology. Methods Enzymol. 2005;395:134–44.

Woo SL, Zoina A, Sorbo GD, Lorito M, Nanni B, Scala F, Noviello C. Characterization of *Fusarium oxysporum* f. sp. *phaseoli* by pathogenic races, VCGs, RFLPs and RAPD. Phytopathology. 1996;86:966–73.

Woudt LP, Neuvel A, Sikkema A, van Grinsven MQJM, de Milliano WAJ, Campbell CL, Leslie JF. Genetic variation in *Fusarium oxysporum* from cyclamen. Phytopathology. 1995;85: 1348–55.

Xu F, Uszkoreit H, Li H. A seed-driven bottom-up machine learning framework for extracting relations of various complexity. Proceedings of the 45th annual meeting of the Association of Computational Linguistics. 2007. p. 584–91.

Xu J, Guo H, Yang ZL. Single nucleotide polymorphisms in the ectomycorrhizal mushroom *Tricholoma matsutake*. Microbiology. 2007b;153(7):2002–12.

Xu L, Porteous JE, Phillips MR, Zheng S. Development and validation of a calcium intake questionnaire for postmenopausal women in China. Ann Epidemiol. 2000b;10:169–75.

Xu J. Microbial ecology in the age of genomics and metagenomics: concepts, tools, and recent advances. Mol. Eco. 2006;15:1713–31.

Xu ZS. On generalized induced linguistic aggregation operators. Int J Gen Syst. 2006b;35:17–28.

Zamir D, Chet I. Application of enzyme electrophoresis for the identification of isolates in *Trichoderma harzianum*. Can J Microbiol. 1985;31:578–80.

Zeller B, Colin-Belgrand M, Dambrine E, Martin F. Decomposition of 15 N-labeled beech litter and fate of nitrogen derived from litter in a beech forest. Oecologia. 2000;123:550–9.

Zeller KA, Summerell BA, Bullock S, Leslie JF. *Gibberella konza* (*Fusarium konzum*) sp. nov. from prairie grasses, a new species in the *Gibberella fujikuroi* species complex. Mycologia. 2003;95:943–54.

Zietkiewicz E, Rafalski A, Labuda D. Genome fingerprinting by simple sequence repeat (SSR)-anchored polymerase chain reaction amplification. Genomics. 1994;20:176–83.

Zwankhuizen MJ, Govers F, Zadoks JC. Inoculum sources and genotypic diversity of *Phytophthora infestans* in Southern Flevoland, the Netherlands. Eur J Plant Pathol. 2000;106:667–80.

Chapter 16
Detection of *Multidrug-Resistant* Fungal Infections in Cancer Patients

K. Awasthi, Asmita Das, and T. Prasad

16.1 Introduction

Fungal infections especially the systemic mycoses are major factors responsible for morbidity and mortality in mainly immunosuppressed and hospitalized patients which include HIV/AIDS, cancer patients undergoing chemotherapy, patients with tuberculosis, organ transplantation, major surgery, high degree burns and diabetes. People who suffer from cancer are more prone to infectious diseases as compared to healthy individuals. Symptoms of such infections in cancer patients include high temperature 100. 5°F, chills or sweat, nasal congestion, burning or pain when urinating, redness swelling, drainage, warmth at the site of infection (Moriyama et al. 2011)

Usually *Candida albicans* is a harmless commensal found in approximately 70 % of the population. It is transmitted from mother to foetus during child birth and lives as a commensal, it is neither transmitted during sexual intercourse nor by any vector but in some cases immunocompromised individuals acquire infection by contact with the health care workers (Badiee et al. 2011). Change in the internal microenvironment like difference in pH due to decrease in *Lactobacillus* flora, long term uptake of antibiotics in cancer patients result in increase in occurrence of candidiasis. *Candida* infections are accompanied by morphogenesis triggered by various regulatory pathways (Brown and Gow 1997).

K. Awasthi • A. Das (✉)
Department of Biotechnology, Delhi Technological University, Delhi 110042, India
e-mail: asmita1710@gmail.com

T. Prasad
Special Centre for Nano Sciences and AIRF, Jawaharlal Nehru University, New Delhi 110067, India

B.P. Singh, V.K. Gupta (eds.), *Molecular Markers in Mycology*, Fungal Biology,
DOI 10.1007/978-3-319-34106-4_16

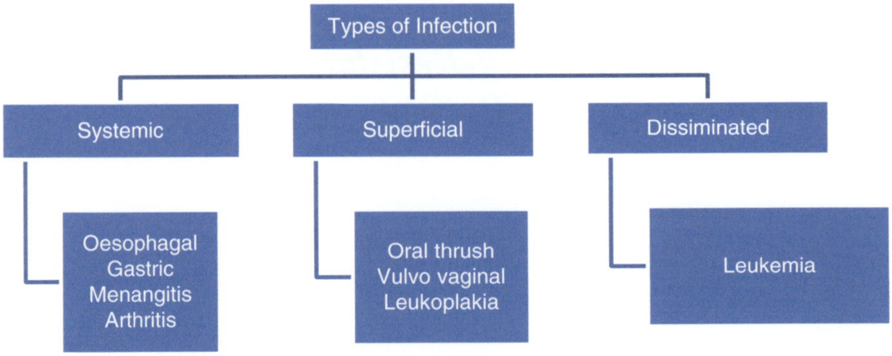

Fig. 16.1 Types of infection caused by *Candida* can be classified as (i) systemic in which *Candida* infects gastrointestinal tract and other organs like brain, joints etc. and cause serious problems like meningitis, ulcers, ulcer in gastric glands and salivary glands (ii) superficial infection, in this infection *Candida* do not penetrate deep into the body but remains in the body openings like vagina and cause vulvovaginal candidiasis (iii) dissiminated, in which *Candida* causes infection in multiple organs (Dean and Annilo 2005)

16.2 Resistance Mechanism of the Host Which Fails During Cancer Infection

16.2.1 Impaired Neutrophil Function

Neutrophils are the primary phagocytic cells of the body, neutropenia is defined as a stage when the neutrophil count becomes less than 500 cells/mm^3 or neutrophil count is less than 1000 cells/mm^3 with a predicted decrease of less than 500 cells/mm^3. Neutropenia is one of the critical risk factor which leads to infection in cancer patients; seriousness of infection depends upon the grades and duration of chemotherapy and radiotherapy given to the patients (Gozalbo et al. 2004). In majority of the cases, chemotherapy accounts for a decrease in the neutrophil counts of the patient (Lionakis et al. 2005). In candidiasis there is decrease in the helper T cells producing IL-17 which plays an important role in switching from innate to adaptive immune response. In case of chronic mucocutaneous candidiasis there is infection in the skin, nails and mucous membrane and in the initial stages there is not much decrease in the cytokine level but as infection progresses, disease becomes severe and there is rapid decrease in IL-6, TNF-alpha, IL-12 (Ashley et al. 1997). Neutrophils producing IL-12 are released in response to low virulence *Candida* strain while IL-10 is released in response to high virulence strain (Cua and Tato 2010). IL-12 production is decreased due to overload of iron in the cells but can be overcome by priming IL-4 which helps in the memory of the Th-1 cells response to fungus (Mencacci et al. 2000). Hence candidiasis is known to modulate neutrophil function. Some antimicrobial agents like fosfomycin, clarithromycin have immunomodulatory role and can enhance lymphocyte production and synthesis of IL-10 and IL-12 and are given to the patients for a therapeutic role (Morikawa et al. 1996).

16.2.2 Damage of the Immune System

In most of the cases it is not cancer but cancer treatment that alters the immune system, the treatment can cause short term and long term damage (Verduyn et al. 2004). Radiation therapy causes short term damage while immune therapy and chemotherapy either alone or in combination causes long term damage. Splenectomy (surgical removal of spleen) is sometimes done to remove cancer as it is a lymphoid organ and plays an important role in development of the immune system. Bone marrow and stem cell transplantation therapy used to kill cancer cells, also kills bystander immune system cells which increases the risk of getting infection. Tumors may also reduce blood flow to the normal tissues by pressing blood capillaries and reducing blood supply. Tumors in the lungs may block normal mucous drainage which leads to infections. Cancer cells can also release chemicals that change normal immune cells such as lymphomas, leukaemia and multiple myeloma (Sauna et al. 2001).

16.3 Cancer Treatment That Increases Risk of Infection

16.3.1 Surgery

Surgery breaks the skin and mucous membrane which exposes internal tissues to microbial infections. This is a common site of infection of opportunistic pathogens. The extent of infection depends upon the extent of surgery, amount of bleeding during surgery, the person's nutritional status, medical problems such as diabetes or heart or lung anomalies etc. Infection of *Candida* is among one of the major causes of mortality in immuno-compromised patients, risk of fungal infections increases in intensive care units with widespread use of therapeutic and monitoring equipments (Vardakas et al. 2009). There may be formation of biofilms on the surface of medical device after cessation of antibiotic therapy; *Candida* cells within biofilms are more resistant to the fluconazole and amphotericin B treatment.

16.3.2 Chemotherapy

It is the most common cause of weakened immune system. Some drugs affect the bone marrow and hence immune system which develops from haematopoetic precursors is disrupted. Lymphocytes being highly dividing are greatly affected by chemotherapeutic drugs. Oral thrush is one of the major problems of the patients undergoing chemotherapy (White 1997). Patients suffer from a burning sensation for which Mycelex is given to reduce the pain. In some patients vaginal itching also occurs which can be prevented by giving probiotics like yoghurt, *Lactobacillus*, acidophilus etc. (Crawford et al. 2004).

16.3.3 Radiation Therapy

Effect of radiation therapy like that of chemotherapy also effects blood cell count and depends on many factors like the amount of radiation used, part of body being treated with radiation, surface area of the body exposed to radiation. Mostly it is given to just one part of the body so that whole immune system is not damaged by it. Radiotherapy is given to the patients suffering from oral cancer but cancer of neck and throat infections may be life threatening. Oesophageal candidiasis is most common in patients undergoing radiation therapy, patients with symptoms like chest pain, burning sensations; moreover fungus can also migrate to the gastrointestinal tract, liver, kidney etc. During radiation therapy, maintaining good hygienic conditions is very necessary like brushing 3–4 times daily, doing floss, use of medicated toothpaste, mouth wash is essential. Mucositis is a severe complication of radiotherapy of head and neck and in such conditions, regular checkups and taking antifungals like evoxac, salagen is recommended (Epstein et al. 2001).

16.3.4 Biotherapy or Immunotherapy

Immunotherapy is also known as biotherapy or biologic therapy. It is given to make immune system better to able to recognize and attack cancer cells. It is used along with many types of treatment but sometimes it changes the way in which the immune system works. People who get biologic therapies may be at risk of immune suppression. When lymphocyte levels are low, the chance of getting opportunistic fungal infection becomes very high, lymphocyte count may drop too. Most of the times lymphocytes return to the normal level after treatment stops, but it can stay low for months also (www.cancer.gov/publications/patient-education/infection, January 29, 2015).

16.3.5 Bone Marrow Transplant and Poor Nutrition

In hematopoietic stem cell transplant (HSCT), peripheral blood stem cell transplant (PBSCT) and umbilical cord blood cell transplant (UCBSCT), very high doses of chemotherapy and total body radiation is given to kill cancerous cells in the body, but in the process, the normal bystander stem cells are also killed and cause a severe damage to the system that lasts for a long time (National Cancer Institute, www.cancer.gov/publications/patient-education/infection, January 29, 2015). Immune system also gets weakened by lack of intake of vitamins, mineral, calories. Therefore patient develops opportunistic infection. So, food rich in proteins, minerals, vitamins, curd, yoghurt is given to the patients (Crawford et al. 2004).

16.4 Account of Virulence in Pathogenic Fungi

Major problem in health care is caused when commensal fungi become opportunistic and grow in uncontrolled manner (Vicente et al. 2006). Pathogenicity is defined as the ability of microbes to cause damage in the host (Casadevall et al. 1998). Fungi are opportunistic pathogens which prey on immuno-compromised individuals like HIV/AIDS, cancer patients or individuals receiving immunosuppressive drugs for organ transplantation (Pfaller and Diekema 2010). Some of the most deadly fungal pathogens are *Candida albicans*, *Cryptococcus neoformans*, *Aspergillus fumigates*. Mortality of immunocompromised individuals increase as fungi become resistant to drugs especially in patients suffering from cancer, tuberculosis, HIV (Gozlabo et al. 2004). Occurrence of *C. albicans* cells to build resistance to antifungal agents like azoles has grown appreciably, causing serious complications in chemotherapy. Both *C. albicans* as well as non-*albicans* species have developed a variety of mechanisms to fight against antifungal drugs (Prasad et al. 2005). Interplay between *C. albicans* cells and mammalian host tissues are highly complicated. *Candida* infections arise as a sequence of progressive steps (Ambudkar et al. 1999). In the first step adhesion to an epithelial surface is required to initiate the colonization on the actual surface. Penetration into the epithelial surfaces is the limit of the infectious process in most cases, leading to the establishment of a superficial candidiasis; normally the fungi are incapable of further invasion into the immunologically intact host (Prasad et al. 2005). In the case of chronic infection *Candida* sticks to the endothelial surface and invades tissue. For protection against *Candida* infection both antibody and cell mediated immune response is required (Bozza et al. 2005). One of the great interests in microbial pathogenesis is virulence factor because they are often the target of the immune response that neutralises the action of virulence factor (Casadevall et al. 2005). Pathogenicity in case of *Candida* also depends upon virulence factors such as adhensins, aspartyl proteases, phospholipases, phenotypic switching changes in tissue affinity and colony morphology (Calderone 2002). Strategy for evasion against chemotherapeutic agents like azoles is a serious problem during treatment (Prasad et al. 2005). To protect itself from host intracellular environment, *Candida* adapts strategies like dimorphic transition and phenotypic switching. Variants are formed with different morphological, physiological and pathogenic features which protects it from the altered environmental conditions (Ernst 2000).

16.5 Interaction Between Host and *Candida*

The relationship between *Candida* and its host can either be commensal or parasitic which completely depends upon the immune status of the host. If the person is healthy with strong immunity, then *Candida* dwells as commensal but if the host has impaired immunity, then *Candida* can cause great threat leading to candidiasis. Colonisation of *Candida* can be summarized as follows:

(i) **Adhesion**- The first step of infection is the adhesion of *Candida* to the epithelial surface and formation of colonies. Hydrolytic enzymes as well as adhesion molecules play a very important role (Naglik et al. 2003). Phospholipase, lysophospholipase, glycolytic proteins, lipase, chondritin sulfatase, HSP family like HSP70, HSP90 etc. are also associated with the penetration of *Candida* in the host and act as virulence factors (Tavares et al. 1993). Cell wall associated manno protein aid in the adhesion of the *Candida* to the host. There are proteinaceous adhesion molecules which interact with the glucosidic receptors like L-fucose, N- acetyl glucosamine on the epithelial surface of the host (Fukazawa and Kagaya 1997).

(ii) **Penetration**- When a strong adhesion between the epithelial surfaces takes place then the pathogen penetrates into deeper layers, it is the determining step of developing infection; in healthy individuals it remains only in the upper layer and is unable to penetrate further.

(iii) **Infection of the endothelium**- After penetrating into the epithelial surface it enters into the endothelial layer passing through blood borne cellular defence system and infects the tissue. *Candida* has receptors which recognize laminin, fibronectin and other ECM proteins and assist to penetrate the endothelial layer (Hostetter 1999).

(iv) **Types of infections caused by*Candida***-In immunocompromised individuals, *Candida* is a threat since it not only causes superficial infection but it also reaches the blood and causes deep seated infection (Mavor et al. 2005). There are three major types of infection caused by *Candida;* they are oropharyngeal candidiasis, vulvovaginal candidiasis and invasive candidiasis. In oropharyngeal candidiasis, infection occurs in mouth and throat area with formation of white patches called thrush. In vulvovaginal candidiasis, there is itching in the vaginal area with thick white discharge from vagina; men can acquire infection by sexual intercourse with infected women. In invasive candidiasis, fungus enters blood stream and spreads to organs like liver, kidney etc (Fig. 16.1).

16.6 Evasion Strategy Adapted by *Candida* Against Innate Immune System

Candida produces soluble factors which induce cytokines and also inactivates complement system (Vanadamme et al. 2009). Antimicrobial peptide (AMP) is produced to check multiplicity of *Candida*. These are lysine rich residues (Steinstraesser et al. 2010) which forms pore in the cell membrane leading to release of large amounts of ATP and ions that lead to cell death (Vanadamme et al. 2009). LL-37 is an AMP which causes rapid efflux of small molecules like ATP as well as larger molecules (Chang et al. 2011). Both innate and adaptive immune system weakens in case of immuno-compromised individuals; salivary amylase is considered as an important innate chemical barrier. It contains a chemical called histatin-5 and virulent strains have evolved mechanism to degrade it and cause oropharyngeal candidiasis (Timothy et al. 2009). Antimicrobial peptides are the major component of innate immune system and they recognise pathogens through pathogen associated molecular pattern (PAMP) and pathogen related receptors (PPR). Virulent strains of *Candida* have evolved mechanisms involving signalling pathways, AMP efflux pumps, peptide effectors to escape from the components of the innate immune system like cathelidicins, defencins etc. (Swidergall et al. 2015).

16.7 Escape Strategies from Antimicrobial Peptide(AMP)

Co evolution strategy is adapted by the fungus to protect itself from AMP. Three phase efflux pumps are developed to overcome the anti fungal activity of AMPs and regulation of the signalling pathways. *C. albican* uses the glycosylphosphatidylinositol (GPI) anchored proteases that leads to the inactivation of salivary peptidases (Meiller et al. 2009). It also degrades histatin 5 which results in the complete loss of the hydrolytic killing activity of AMP. Msb2 glycoprotein is secreted to provide a broad spectrum protection against peptide anti microbial inactivating wide range of the AMPs including LL-37, HST 5, hNP-1 and hBD1 (Cheng et al. 2012).

16.8 Drug Resistance Mechanisms Adapted by *Candida*

There are various mechanisms by which *Candida* acquires drug resistance which include modifying ergosterol pathway due to which they become resistant to azole drugs (White et al. 1998), by altering the sterol composition of the cells as in *ERG3* mutants which provides resistance to itraconazole, point mutation in the *ERG11* gene which results in fluconazole resistance (Joseph-Horne and Hollomon 1997), genetic alterations in the promoter, coding and terminator region of *THR1* gene which lies downstream of *ERG11* (White et al. 1997).

16.9 Evolution of Mutli Drug Resistance in Fungi an Emerging Strategy

In immune compromised individuals a great threat is caused due to various species of *Candida* as it causes opportunistic infections and the treatment is limited by the development of both narrow and broad drug resistance (Barelle et al. 2006). Development of the fungal infections are associated with the increase in the corticosteroid therapy, immunosupressent drugs, cancer treatment, cathereter etc. Clinical evidence of cross reactivity is shown by azole resistant strains of *Candida* which leads to the phenomenon of the multi drug resistance MDR (Franz et al. 1998).

16.10 Mechanism of the MDR

16.10.1 Over Expression of the ABC (ATP Binding Cassette) Efflux Pump Proteins and Major Facillator Super Family (MFS) Proteins

ABC pump proteins work by hydrolysis of ATP which acts as a driving force to efflux drugs. There is a complex interaction between the membrane sphingolipids and sterols which affect the drug efflux mechanism (Prasad et al. 2005). Multi drug resistance (MDR) is one of the major problems faced in the treatment of candidiasis. There are several mechanisms by which MDR resistance is explained such as decreased expression of metabolic enzymes, mutation of target enzymes, ROS detoxification, cellular repair mechanism, drug efflux pump mutations etc. (Prasad et al. 2005).

The cost of treatment is also increased due to MDR since pathogen becomes resistant to commercially available drugs triggering the use of more expensive therapies. The rate of success of present day medical applications like organ transplantation and cancer chemotherapy has been cntinuously influenced by development of MDR.

16.11 Classification of MDR

In past few years there is a marked increase in the multidrug resistance acquired by *Candida*. It is increasingly becoming resistant to antifungals which are administered to the patients. *Candida* is becoming a serious concern among the medical practioners now a days, approximately about 7 % of the species have become resistant to fluconazole which is commonly given to the patients. Sometimes antibacterial drugs given for long time reduces gut bacterial flora and increase the risk of infection. The type of resistance mechanism adapted by *Candida* can be classified as follows:

(i) **Primary resistance**- it occurs when organism has never encountered the drug of interest in a particular host.
(ii) **Secondary resistance**- also known as acquired resistance occurs when resistance arises in the organism after exposure to the drug. It is further classified as:

   (a) **Intrinsic resistance**- refers to the inherentin sensitivity of all microorganisms of a single species to certain common first line of drugs, which are used to treat disease based on the clinical symptoms of the patient. It is also known as MDR for eg, *Candida* to fluconazole.

(b) **Extrinsic resistance-** it defines the acquired ability of the organism to withstand the inhibitory effect of one or two most effective antimicrobial drugs.

(iii) **Clinical resistance**- it is defined by the situation in which the infected organism is inhibited by the concentration of an antimicrobial agent that is associated with high risk of the therapeutic failure or reappearance of infection within an organism due to impaired immune function.

16.12 Factors Influencing MDR and Pathogenicity

Microorganisms have evolved mechanisms to overcome the effectiveness of drugs. Cell wall in fungi plays a major role in their survival.

16.12.1 Reactive Oxygen Species

In eukaryotic cells mitochondria is a common organelle that represents an important source of reactive oxygen species (ROS), which are the key regulator of yeast apoptosis and mediates early and late steps of apoptosis. *Candida* possesses ROS scavenger, superoxide dismutase activity suggesting that they may acquire a cytoprotective mechanism against both exogenous and endogenous ROS (Kobayashi et al. 2002).

16.12.2 Iron

It is one of the important nutrients required by both host and the pathogen, its availability is tightly regulated since it is a transition metal and has high capacity to donate and accept pairs of electrons. It can participate in the formation of free radicals, therefore availability of iron in the host macrophage cells is tightly regulated (Radisky et al. 1999). Since iron in the macrophages is not freely available, therefore pathogen needs to exploit host iron reservoirs to make their own niches for their survival. But macrophages protect iron from intracellular pathogens. There is competition between pathogen and the host macrophages for iron which represents a critical virulence trait of many infectious fungi (Kontoyiannis et al. 2007). *C. albicans* has mechanism to gain iron from cellular proteins. It takes iron by binding to the extracellular glycosyl phosphotidylinositol-linked receptor Rbt5 and by degradation of the iron protoporphyrin IX by heme oxygenase CaHmx1 to alpha –biliverdin, free iron and CO. In case of excess iron, it uses siderophores, which have strong affinity for iron (Howard et al. 1999). It also relies on the extracellular ferric reductase (Fre) located in the plasma membrane that can reduce either free or complexes ferric (Fe^{3+}) ions to the soluble ferrous ion (Fe^{2+}) by reductive pathway utilizing high affinity permease (Ftr1p), the cell surface multicopper ferroxidase (Fet3p) along with intracellular copper transporter Ccc2p (Weissman et.al. 2002). Susceptibility to *C. albicans* infection is influenced by the iron content of the host. For e.g., pre-treatment of endothelial cells with iron chelator phenanthroline reduces the damage caused by *C. albicans*. Similar studies done on mouse model of systemic vulvovaginal candidosis (RVVC) have also shown the role of iron in candidiasis (Raman et al. 2000).

16.12.3 Hypoxia

Hypoxic areas are characterized by inadequate vascularisation and irregular blood flow controlled by complex hypoxia inducible factor-1 (HIF-1) controlling expression of most genes involved in adaptation to the hypoxic condition and causes resistance to the chemotherapy by inducing expression of human MDR1 in growing tumours. The transcriptome analysis of *C. albicans* reveals metabolic adaptation to the scarce oxygen conditions and is demonstrated by *Efg1* mutants. Under hypoxia, *Efg1* and all other major class of genes are required for biofilm production, thus adaptation to hypoxia forms an integral component of biofilm formation in *C. albicans* (Carvalho et al. 2010).

16.12.4 Steroids

They influence cell differentiation, morphogenesis and virulence. Steroids affect peliotropic drug resistance (PDR). *CDR1* promoter region contains steroid responsive elements (SRE) and drug response element. An inverted CCAAT box in a combination of conserved sequence is attributed towards *MDR1* responsiveness to cellular stresses (Banerjee et al. 2008).

16.12.5 Morphogenesis and pH

In response to environmental stress, *C. albicans* is able to switch from unicellular yeast into pseudo or true hyphae and this ability is an important virulence trait. It is established that Efg1p is a well known regulator of morphogenesis in *C. albicans* and is involved in resistance by regulating *ERG3* gene of ergosterol biosynthetic pathway. It is an important target affects sterol homeostasis of mutant cells resulting in disruption of the morphogenetic signalling (Banerjee et al. 2008).

Change in pH causes alteration in the electrochemical parameters, response of immunological agents and signal transduction pathways.

16.12.6 Development of MDR in ERG Mutants

Patients suffering from AIDS, cancer, burn victims and those who are undergoing transplantation therapy, are administered various immunosuppressive drugs along with antifungal drug, azole being one of them. This leads to cross resistance to many other drugs, a phenomenon called multi drug resistance (MDR) (White et al. 1998). MDR is reported to develop by the over expression of the efflux pump proteins belonging to the ABC Super family which include MDR1, FLU1, BENr (Ben-Yaacov et al. 1994). Alterations are reported in *ERG11* gene which produces enzyme lanosterol 14a –demythelase and is target for azole drugs (CYP51A1) (Mukhopadhyay et al. 2004). In azole resistant species, there is mutation from arginine to lysine residue at the position 467, which results in the over expression of the target enzyme lanosterol demethylase (White et al. 1997). *ERG11* belongs to the class of the cytochrome P450 enzymes which help in the metabolism of the organic compounds and also acts as a main check point for the formation of the sterols, vitamin and lipids. It helps in the removal of C-14α methyl group from the lanosterol (Lepesheva and Waterman 2008). Demethylation step is the critical step in the formation of the ergosterol which has the ability to alter the permeability and rigidity of the membrane, ion channel, enzyme activity (Lepesheva and Waterman 2008). Aneuploidy is also reported in the isochromosome by gene amplification of the *EGR11* and *TAC1* (Transcriptional regulator of drug efflux), which provides fluconazole resistance due to the presence of extra copy of chromosome 5 (Selmecki et al. 2008). Azoles target enzyme lanoterol demethylase due to which toxic sterols gets accumulated in the membrane and results in arrested growth, altered membrane function, change in the membrane fluidity and finally cell death (Kelly et al. 1997). To counter the effect of the azoles, *Candida* adapts a strategy to accumulate defective sterol biosynthetic pathway by mutations in desaturase encoded by *ERG3* gene, a similar phenomenon is seen in AIDS patients (Kelly et al. 1997). Drug resistance adapted by *Candida* result in change in the membrane permeability, alterations in the ergosterol and sphingolipid interactions, alterations in the morphogenic pathways which involve transcription factors *EFG1*, *SSK1* (Hameed et al. 2008, Mukhopadhyay et al. 2004, Prasad et al. 2005).

16.13 Working Mechanism of the Enzyme

Lanosterol demethylation occurs in the presence of molecular oxygen and one mole of the reducing agent, NADPH. In the first step, methyl group undergoes monoxygenation and then conversion of the sterol into the carboxy alcohol and subsequently to carboxy aldehyde occurs leading to the formation of the dymethylated product by evolution of the formic acid (Lepesheva and Waterman 2008).

16.14 Morphogenic regulators affect drug sensitivity

Candida has the ability for morphogenic transition between unicellular form to pseudohypae and true hypae, which depends on the environmental conditions. Efg1p is a common target required for this transition and responsible for morphogenesis (Hameed et al. 2008). Mutants of *EFG1* exhibit enhanced membrane fluidity which leads to higher drug diffusion and enhanced drug sensitivity (Hameed et al. 2008).

16.14.1 Candida *Drug Resistance Protein*

In fluconazole resistant isolates, there is over expression of the Cdr1p which helps in the drug efflux by the pathogen (Caledrone et al. 2002). P-gp (P glycoprotein) is overexpressed in cancer cells and is a homologue of Cdr1p. Overexpression of Pgp affects chemotherapy (Prasad et al. 2002). These efflux pump proteins utilize energy from the hydrolysis of the ATP to transport substance out from the membrane. NBD domains are the hub for the transportation activity and any disruption in the NBD results in the decrease in transport by the protein. There are certain conserved region in the NBD which have a consensus sequence GxxGxGKS/T, and called Walker A motif where x is any amino acid. Another conserved sequence is called Walker B motif hhhhD, where h is any aliphatic residue and have NBD signature sequence LSGGQQ/R/KQR. Lysine residue of the Walker motif binds to the beta and gamma phosphates of ribonucleotides and helps in the hydrolysis of the ATP; mutation in the lysine residue destroys the hydrolysis of the ATP (Schneider and Hunke 1998). Lysine residue is important for the hydrolysis of ATP and any mutation will lead to the impairment in functioning (Jha et al. 2004). In Cdr1p, Walker A (GRPGAGCST), Walker B (IQCWD) and ABC signature sequence (VSGGERKRVSIA) are present in the NBD region but lysine is replaced by cysteine residue in the Walker A sequence (Prasad et al. 1995). NBD domain is able to hydrolyse ATP along with UTP, CTP, GTP and thus shows general nucleosidase activity which also resembles the hydrolysis activity of the Cdr1p (Jha et al. 2004). Any mutation in the N terminal lysine residue leads to the loss of hydrolysis activity (Krishnamurthy et al. 1998).

16.14.2 Membrane Lipids Play Important Role in MDR

Lipids are the main structural components of the membrane which provide rigidity and play role in cell signalling, cell to cell communication and energy storage by accumulating cellular lipid droplets. Sterols are the major components of the membrane which are also the target of the azole drugs, therefore,any alteration in the sterol composition leads to the differential azole sensitivity (Prasad et al. 2004).

Rafts (membrane microdomians rich in ergosterol and sphingolipids) and proteins which are present in the membrane are upregulated during acquisition of MDR. Interactions between sphingolipid and sterol are very important for the proper functioning of the ABC drug efflux pumps, and such interactions play an important role in formation of hyphae and in morphogenesis (Prasad et al. 2005).

16.14.3 Treatment for Fungal Infection

Most of the infections are treated with antifungal agents and the formulation of the medication (pills, ointments, suppositories or powder) will largely depend on the clinical presentation of the infection. In case of mild oral candidiasis, clotrimazole lozenges or a nystatin suspension is used but may also require oral fluconazole for moderate to severe infection and recurrent cases (Kauffman et al. 2012). In case of skin infection, topical antifungal agents of the azole class (bifonazole or ketoconazole) is taken. Infection of the nails and toes are treated with both topical and systemic antifungals (Kauffman et al. 2012).

16.15 Nutritional Therapies for *Candida* Treatment

Rise in *Candida* infections and their increasing resistance against commonly used antifungal drugs, give rise to development of novel therapies for the prevention and management of these infections, these include:

16.15.1 Dietary Modifications

Limiting intake of refined carbohydrates (pasta, bread, sweets, soft drinks etc.) may be helpful for controlling *Candida* infections. Higher dietary sugar is associated with vulvovaginal *candidiasis* and abnormal glucose metabolism is associated with recurring vulvovaginal infections (Donders et al. 2010; Akpan et al. 2002).

16.15.2 Probiotics and Reversal

Probiotic such as *Lactobacillus* are beneficial against mucosal *Candida* infections (Mehra et al. 2012). They exert their beneficial actions by suppressing the growth of *Candida* and inhibiting *Candida's* ability to adhere to cell surfaces. Dietary products containing probiotic bacteria can help control *Candida* infection in the human body (Matsubara et al. 2016). Revestral, a compound found in the skin of grapes may contribute to the anti inflammatory property of red wine. In 2007, researchers investigated its fungicidal activity against *Candida albicans*; it has potent antifungal properties and appears to be safer than conventional drugs such as amphotericin B.

16.16 Detection of *Candida*

It is easy to diagnose candidiasis and severity of infection depends upon the immunity of the patient. In case of mild infections *Candida* has symptoms like itching and burning sensations, whitish or grey discharge from vagina, pain during sex which can be easily examined by the doctors by taking skin scrapings, nail clipping, plucked hair from affected area while in case of severe infections which is caused in patients suffering from cancer, aspartyl proteinases 2 (Sap2) serve as a potent biomarker for *Candida* infection which can be diagnosed by indirect competitive ELISA (Wang et al. 2013). Some cytokines like IL-8, TNF-alpha, enzymes like COX-2 and My88 signalling pathway are also activated and they can also be used to detect candidiasis (Mantovani et al. 2008). In patients with hematologic malignancies, candidiasis becomes severe and leads to various complications and their survival depends on early diagnosis; for this DNA based detection method like PCR-ELISA has been developed to offer more sensitive diagnosis of *Candida* (Badiee et al. 2009). Clinical detection of *Candida* is done microscopically in both unstained and gram-stained preparations (Salar Zai et al. 1997). In case of invasive *Candida infections,* meningitis circulating mannan in CSF is used for diagnosis (Frans et al. 2004).

16.17 Conclusion

Candida is a commensal organism residing in human host and do not cause much harm to healthy individuals but when the immune system of the individual is weakened which may be due to some medical treatment like surgery or regular intake of high doses of antibiotics for a long time or chemotherapy like in the case of cancer patients, immune system gets weakened due to which *Candida* causes opportunistic infections. In such patients, the type of infection may range from mild to severe depending upon how the patient is kept. Also *Candida* adapts several evasion strategies to escape from the immune system for which several antifungal drugs like azoles are given to prevent candidiasis. Therefore *Candida,* a commensal organismin most healthy individuals may however be responsible for severe complications in cancer patients and immuno-compromised individuals and thus it is absolutely essential for regular monitoring of cancer patients undergoing chemotherapy and radiation therapy for opportunistic fungal infections. In such patients *Candida* infection may lead to mild to severe problems and may even cause death.

References

Akpan A, Morgan R. Oral candidiasis-Review. Postgrad Med J. 2002;78:455–9.

Ambudkar SV, Dey S, Hrycyna CA, Ramachandra M, Pastan I, Gottesman MM. Biochemical, cellular, and pharmacological aspects of the multidrug transporter. Annu Rev Pharmacol Toxicol. 1999;39:361–98.

Ashley C, Morhart M, Rennie R, Ziola B. Release of *Candida albicans* yeast antigen upon interaction with human neutrophils *in vitro*. J Med Microbiol. 1997;46:747–55.

Badiee P, Alborzi A. Invasive fungal infections in renal transplant recipients. Exp Clin Transpl. 2011;9(6):355–62.

Banerjee D, Lelandais G, Shukla S, Mukhopadhyay G, Jacq C, Devaux F, Prasad R. Responses of pathogenic and nonpathogenic yeast species to steroids reveal the functioning and evolution of multidrug resistance transcriptional networks. Eukaryot Cell. 2008;7(1):68–77.

Barelle CJ, Priest CL, MacCallum DM, Gow NA, Odds FC, Brown AJ. Niche-specific regulation of central metabolic pathways in a fungal pathogen. Cell Microbiol. 2006;8:961–71.

Ben-Yaacov R, Knoller S, Caldwell GA, Becker JM, Koltin Y. *Candida albicans* gene encoding resistance to benomyl and methotrexate is a multidrug resistance gene. Antimicrob Agents Chemother. 1994;38:648–52.

Bozza S, Fallarino F, Pitzurra L, Zelante T, Montagnoli C, Bellocchio S, Mosci P, Vacca C, Puccetti P, Romani L. A crucial role for tryptophan catabolism at the host/Candida albicans interface. J Immunol. 2005;174(5):2910–8.

Brown AJ, Gow N. Regulatory networks controlling *Candida albicans* morphogenesis. Trends Microbiol. 1997;7:333–8.

Carvalho AP, Gursky LC, Rosa RT, Rymovicz AU, Campelo PM, Grégio AM. Non-steroidal anti-inflammatory drugs may modulate the protease activity of *Candida albicans*. Microb Pathog. 2010;49(6):315–22.

Casadevall A, Pirofski LA. Polysaccharide-containing conjugate vaccines for fungal diseases. Trends Mol Med. 2005;12(1):6–9.

Casadevall A, Cassone A, Bistoni F, Cutler JE, Magliani W, Murphy J. Antibody and/or cell-mediated immunity, protective mechanisms in fungal disease: an ongoing dilemma or an unnecessary dispute. Med Mycol. 1998;36:95–105.

Calderone RA. *Candida* and Candidiasis. 1st ed. Washington, DC: ASM Press; 2002.

Cheng SC, Joosten LA, Kullberg BJ, Netea MG. Interplay between *Candida albicans* and the mammalian innate host defense. Infect Immun. 2012;80:1304–13.

Cua DJ, Tato CM. Innate IL-17-producing cells: the sentinels of the immune system. Nat Rev Immunol. 2010;10(7):479–89.

Crawford J, Dale DC, Lyman GH. Chemotherapy-induced neutropenia. Cancer. 2004;100: 228–37.

Dean M, Annilo T. Evolution of the ATP-binding cassette (ABC) transporter super family invertebrates. Annu Rev Genomics Hum Genet. 2005;6:123–42.

Donders G, Bellen G, Byttebier G, Verguts L, Hinoul P, Walckiers R, Stalpaert M, Vereecken A, Van Eldere J. Individualized decreasing dose maintenance fluconazole regimen for recurrent vulvovaginal candidiasis (ReCiDiF trial). Am J Obstet Gynecol. 2010;199(6):613.e1–9.

Ernst J. Transcription factors in *Candida albicans* environmental control of morphogenesis. Microbiol. 2000;146:1763–74.

Epstein JB, Silverman Jr S, Paggiarino DA, Crocket S, Schubert MM, Senzer NN. Benzydamine HCl for prophylaxis of radiation induced oral mucositis: results from a multicenter, randomized, double-blind, placebo-controlled clinical trial. Cancer. 2001;92(4):875–85.

Fukazawa Y, Kagaya K. Molecular basis of adhesion of *Candida albicans*. J Med Vet Mycol. 1997;35:87–99.

Franz R, Kelly SL, Lamb DC, Kelly DE, Ruhnke M, Morschhauser J. Multiple molecular mechanisms contribute to a stepwise development of fluconazole resistance in clinical *Candida albicans* strains. Antimicrob Agents Chemother. 1998;42:3065–72.

Gozalbo D, Roig P, Villamón E, Gil ML. Candida and candidiasis: the cell wall as a potential molecular target for antifungal therapy. Curr Drug Targets Infect Disord. 2004;4(2):117–35.

Hameed S, Prasad T, Banerjee D, Chandra A, Mukhopadhyay CK, Goswami SK, Lattif AA, Chandra J, Mukherjee PK, Ghannoum MA, Prasad R. Iron deprivation induces *EFG1*-mediated hyphal development in *Candida albicans* without affecting biofilm formation. FEMS Yeast Res. 2008;8:744–55.

Hostetter MK. Integrin-like proteins in *Candida* spp. and other microorganisms. Fungal Genet Biol. 1999;28:135–45.

Jha S, Dabas N, Karnani N, Saini P, Prasad R. ABC multidrug transporter Cdr1p of Candida albicans has divergent nucleotide-binding domains which display functional asymmetry. FEMS Yeast Res. 2004;5(1):63–72.

Joseph-Horne T, Hollomon DW. Molecular mechanisms of azole resistance in fungi. FEMS Microbiol Lett. 1997;149:141–14.

Kelly SL, Lamb DC, Kelly DE, Manning NJ, Loeffler J, Hebart H. Resistance to fluconazole and cross resistance to amphotericin B in *Candida albicans* from AIDS patients caused by defective sterol delta 5,6-desaturation. FEBS Lett. 1997;400:80–2.

Kobayashi D, Kondo K, Uehara N. Endogenous reactive oxygen species is an important mediator of miconazole antifungal effect. Antimicrob Agents Chemother. 2002;46(10):3113–7.

Krishnamurthy S, Chatterjee U, Gupta V, Prasad R, Das P, Snehlata P. Deletion of transmembrane domain 12 of *CDR1*, a multidrug transporter from *Candida albicans*, leads to altered drug specificity: expression of a yeast multidrug transporter in baculovirus expression system.Yeast. 1998;14:535–50.

Lattif AA, Banerjee U, Prasad R, Biswas A, Wig N, Sharma N, Haque A, Gupta N, Baquer NZ, Mukhopadhyay G. Susceptibility pattern and molecular type of species-specific Candida in oropharyngeal lesions of Indian human immunodeficiency virus positive patients lesions of Indian human immunodeficiency virus positive patients. J Clin Microbiol. 2004;42:1260–2.

Lepesheva GI, Waterman MR. Sterol 14α-demethylase cytochrome P450 (*CYP51*), a P450 in all biological kingdoms. Biochim Biophys Acta. 2008;1770(3):467–77.

Lionakis MS, Lewis RE, Torres HA, Albert ND, Raad II, Kontoyiannis DP. Increased frequency of non-fumigatus Aspergillus species in amphotericin B or triazole-pre-exposed cancer patients with positive cultures for aspergilli. Diagn Microbiol Infect Dis. 2005;52:15–20.

Manzano-Alonso ML, Castellano-Tortajada G. Reactivation of hepatitis B virus infection after cytotoxic chemotherapy or immunosuppressive therapy. World J Gastroenterol. 2011;17(12): 1531–7.

Mantovani A, Allavena P, Sica A, Balkwill F. Cancer-related inflammation. Nature. 2008;454:436–44.

Matsubara VH, Bandara HM, Mayer MP, Samaranayake LP. Probiotics as Antifungals in Mucosal Candidiasis. Clin Infect Dis. 2016;62(9):1143–53.

Mavor AL, Thewes S, Hube B. Systemic fungal infections caused by *Candida* species: epidemiology, infection process and virulence attributes. Curr Drug Target. 2005;6(8):863–74.

Mehra T, Köberle M, Braunsdorf C, Mailänder-Sanchez D, Borelli C, Schaller M. Alternative approaches to antifungal therapies. Exp Dermatol. 2012;21(10):778–82.

Meiller TF, Hube B, Schild L, Shirtliff ME, Scheper MA, Winkler R, Ton A, Jabra-Rizk MA. A novel immune evasion strategy of *Candida albicans*: proteolytic cleavage of a salivary antimicrobial peptide. PLoS One. 2009;4:e5039.

Mencacci A, Bacci A, Cenci E, Montagnoli C, Fiorucci S, Casagrande A, Flavell RA, Bistoni F, Romani L. Interleukin 18 restores defective Th1 immunity to Candida albicans in caspase 1-deficient mice. Infect Immun. 2000;68(9):5126–31.

Morikawa N, Mori T, Abe T, Ghoda M, Takeyama M, Hori S. Pharmacokinetics of methotrexate in plasma and cerebrospinal fluid. Ann Pharmacother. 1997;31:1153–6.

Moriyama B, Ditullio B, Wilson E, Penzak SR, Danner RL, Pennick G, Rinaldi MG, Zelazny AM, Gea-Banacloche J. Pharmacokinetics of anidulafungin in pleural fluid during the treatment of a patient with Candida empyema. Antimicrob Agents Chemother. 2011;55(5):2478–80.

Mukhopadhyay K, Prasad T, Saini P, Pucadyil TJ, Chattopadhyay A, Prasad R. Membrane sphingolipid-ergosterol interactions are important determinants of multidrug resistance in *Candida albicans*. Antimicrob Agents Chemother. 2004;48(5):1778–87.

National Comprehensive Cancer Network. Prevention and treatment of cancer related infections. V. 2. 208/11/14. www.nccn.org/professionals/physician_gls/pdf/infections.pdf on 2 Feb 2015.

Naglik JR, ChallacombeS J, Hube B. *Candida albicans* secreted aspartyl proteinases in virulence and pathogenesis. Microbiol Mol Biol Rev. 2003;67:400–28.

Nikaido K, Liu PQ, Ames GF. Purification and characterization of HisP, the ATP-binding subunit of a traffic ATPase (ABC transporter), the histidine permease of *Salmonella typhimurium*. Solubility, dimerization, and ATPase activity. J. Biol. Chem. 1997;272:27745–52.

O'Connor R. A review of mechanisms of circumvention and modulation of chemotherapeutic drug resistance. Curr Cancer Drug Targets. 2009;9:273–80.

Pfaller MA, Diekema DJ. Epidemiology of invasive mycoses in North America. Crit Rev Microbiol. 2010;36(1):1–53.

Prasad T, Chandra A, Mukhopadhyay CK, Prasad R. Unexpected link between iron and drug resistance of Candida spp.: iron depletion enhances membrane fluidity and drug diffusion, leading to drug-susceptible cells. Antimicrob Agents Chemother. 2006;50:3597–606.

Prasad T, Saini P, Gaur NA, Vishwakarma RA, Khan LA, Haq QM. Functional analysis of CaIPT1, a sphingolipid biosynthetic gene involved in multidrug resistance and morphogenesis of *Candida albicans*. Antimicrob Agents Chemother. 2005;49:3442–52.

Prasad R, Kapoor K. Multidrug resistance in yeast *Candida*. Int Rev Cytol. 2005;242:215–48.

Prasad R, Panwar SL, Smriti. Drug resistance in yeasts – an emerging scenario. Adv. Microb Physiol 2002;46:155–201.

Prasad R, Worgifosse PD, Goffeau A, Balzi E. Molecular cloning and characterization of a novel gene of *Candida albicans*, *CDR1*, conferring multiple resistance to drugs and antifungals. Curr Genet. 1995;27(4):320–29.

Badiee P, Kordbacheh P, Alborzi A, Zakernia M, Haddadi P. Early detection of systemic candidiasis in the whole blood of patients with hematologic malignancies. Jpn J Infect Dis. 2009;62:1–5.

Sauna ZE, Smith MM, Muller M, Kerr KM, Ambudkar SV. The mechanism of action of multidrug resistance-linked P-glycoprotein. J Bioenerg Biomembr. 2001;33:481–91.

Schneider E, Hunke S. ATP-binding-cassette (ABC) transport systems: functional and structural aspects of the ATP-hydrolyzing subunits/domains. FEMS Microbiol Rev. 1998;22(1):1–20.

Selmecki A, Gerami-Nejad M, Paulson C, Forche A, Berman J. An isochromosome confers drug resistance *in vivo* by amplification of two genes, *ERG11* and *TAC1*. Mol Microbiol. 2008;68(3):624–41.

Shukla S, Wu CP, Ambudkar SV. Development of inhibitors of ATP-binding cassette drug transporters: present status and challenges. Expert Opin Drug Metab Toxicol. 2008;4:205–23.

Tavares D, Salvador A, Ferreira P, Arala-Chaves MP. Immunological activities of a *Candida albicans* protein which plays an important role in the survival of the microorganism in the host. Infect Immun. 1993;61:1881.

Vardakas KZ, Michalopoulos A, Kiriakidou KG, Siampli EP, Samonis G, Falagas ME. Candidaemia: incidence, risk factors, characteristics and outcomes in immunocompetent critically ill patients. Clin Microbiol Infect. 2009;15(3):289–92.

Verduyn Lunel FM, Voss A, Kuijper EJ, Gelinck LBS, Hoogerbrugge PM, Liem KL, Kullberg BJ, Verweij PE. Detection of the *Candida* antigen mannan in cerebrospinal fluid specimens from patients suspected of having *Candida* meningitis. J Clin Microbiol. 2004;42(2):867–70.

Vicente M, Hodgson J, Massidda O, Tonjum T, Henriques-Normark B, Ron EZ. The *fallacies of hope*: will we discover new antibiotics to combat pathogenic bacteria in time. FEMS Microbiol. 2006;30:841–52; PMID: 17064283; http://dx.doi.org/10.1111/j.1574–6976.2006.00038.x.

White TC, Kieren AM, Bowden RA Clinical, cellular, and molecular factors that contribute to antifungal drug resistance. Clin Microbiol Rev. 1998;11(2):382–402.

White TC. The presence of an R467K amino acid substitution and loss of allelic variation correlate with an azole-resistant lanosterol 14a-demethylase in *Candida albicans*. Antimicrob Agents Chemother. 1997a;41:1488–94.
White TC. Increased mRNA levels of *ERG16*, *CDR*, and *MDR1* correlate with increased azole resistance in *Candida albicans* isolates from a patient infected with human immunodeficiency virus. Antimicrob Agents Chemother. 1997b;41:14821487.
White TC, Pfaller MA, Rinaldi MG, Smith J, Redding SW. Stable azole drug resistance associated with a substrain of *Candida albicans* from an HIV-infected patient. Oral Dis. 1997;3:102–9.
Wang Y, Gao X, Zhi Gang JU, Liu. Detection of *Candida albicans* Sap2 in cancer patient serum samples by an indirect competitive enzyme-linked immunosorbent assay for the diagnosis of candidiasis. IndianJ Pathol Microbiol. 2013;56(3):243–7. doi:10.4103/0377-4929.120377.
Zai S, Khan SM, Gillani J. Candidiasis in cancer patients. JPMA. 1997;47:191.

Erratum to: DNA Barcoding for Diagnosis and Monitoring of Fungal Plant Pathogens

Prem Lal Kashyap, Pallavi Rai, Sudheer Kumar, Hillol Chakdar, and Alok K. Srivastava

Erratum to:
B.P. Singh, V.K. Gupta (eds.), *Molecular Markers in Mycology*, Fungal Biology, DOI 10.1007/978-3-319-34106-4_5

In the original version of Chapter 5, few of the illustrations had permission issues. In the current version, Chapter 5 has been replaced with a new chapter with original figure.

The updated online version of the original chapters can be found under DOI 10.1007/978-3-319-34106-4_5

B.P. Singh, V.K. Gupta (eds.), *Molecular Markers in Mycology*, Fungal Biology, DOI 10.1007/978-3-319-34106-4_17

Correction to: Genic Molecular Markers in Fungi: Availability and Utility for Bioprospection

Anirudha Chattopadhyay, Kapil Kumar Tiwari, Kul Bhushan, and Dharmendra Pratap

Correction to:
B.P. Singh, V.K. Gupta (eds.), *Molecular Markers in Mycology*, Fungal Biology, https://doi.org/10.1007/978-3-319-34106-4_7

In the original version of chapter 7, the author's complete name currently appears as Kulbhushan Chaudhary. The complete name has been corrected as Kul Bhushan and the email address has also been updated as lsr_bhushan@rediffmail.com✉

The updated online version of the original chapter can be found under https://doi.org/10.1007/978-3-319-34106-4_7

B. P. Singh, V. K. Gupta (eds.), *Molecular Markers in Mycology*, Fungal Biology, https://doi.org/10.1007/978-3-319-34106-4_18

Index

B.P. Singh, V.K. Gupta (eds.), *Molecular Markers in Mycology*, Fungal Biology,
DOI 10.1007/978-3-319-34106-4

MIX
Papier aus verantwortungsvollen Quellen
Paper from responsible sources
FSC® C105338

If you have any concerns about our products,
you can contact us on
ProductSafety@springernature.com

In case Publisher is established outside the EU,
the EU authorized representative is:
Springer Nature Customer Service Center GmbH
Europaplatz 3, 69115 Heidelberg, Germany

Printed by Libri Plureos GmbH
in Hamburg, Germany